T5-ANA-909

The Language of Mathematics

NELDA W. ROUECHE
BARBARA WASHBURN MINK

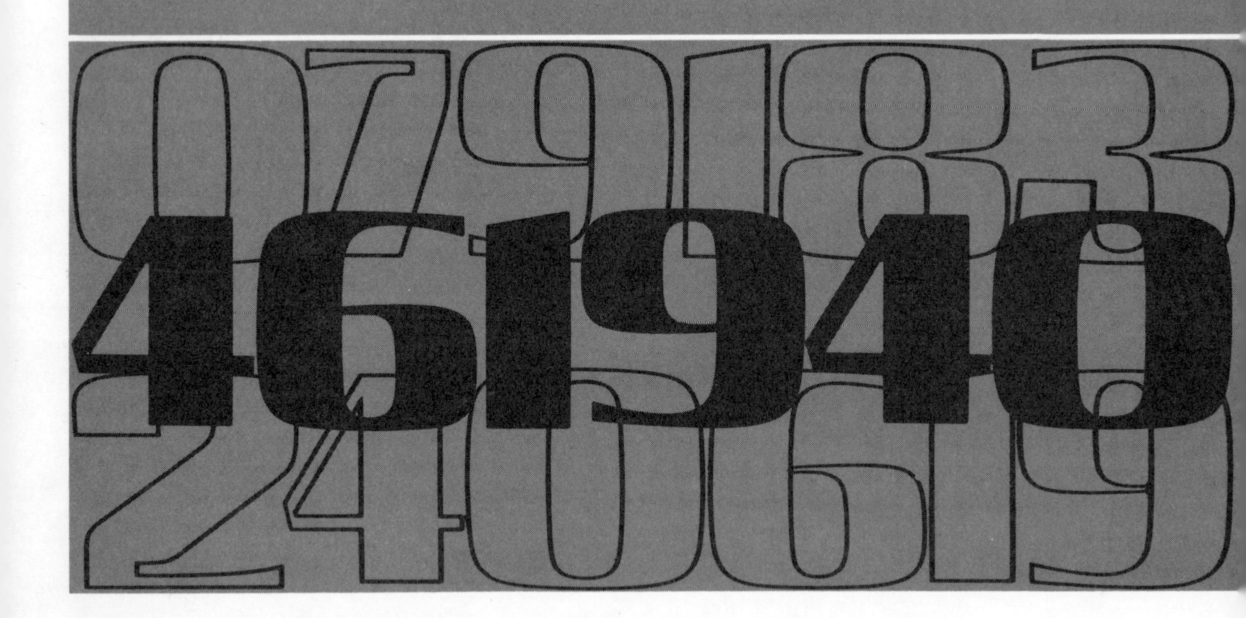

The Language of Mathematics

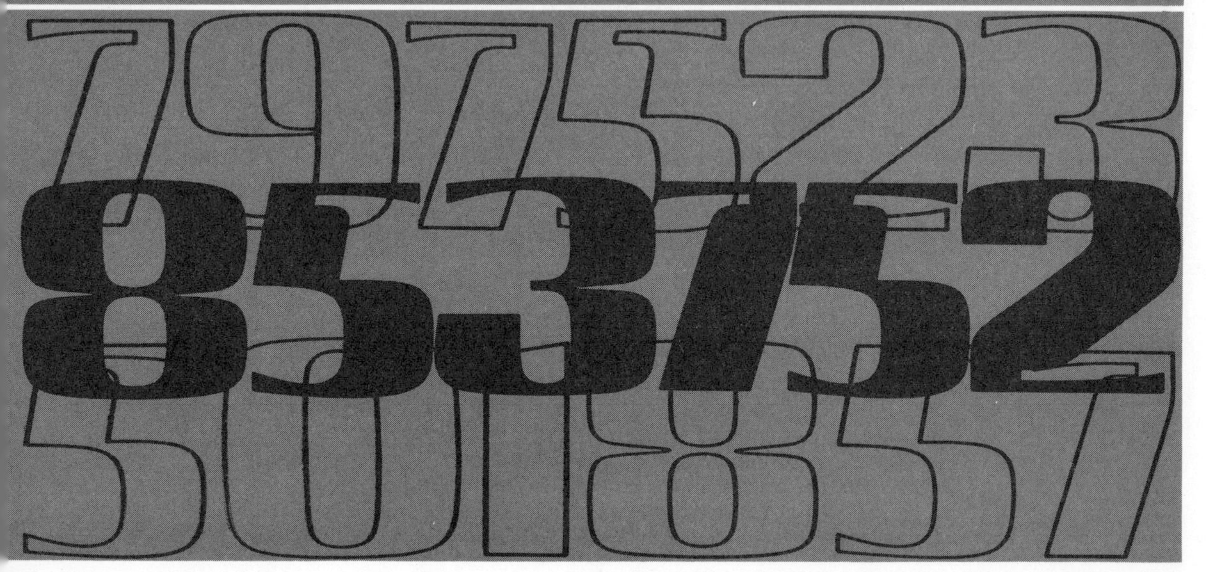

Prentice-Hall Inc., Englewood Cliffs, New Jersey 07632

Library of Congress Cataloging in Publication Data

ROUECHE, NELDA W
The language of mathematics.

Includes index.
1. Mathematics—1961- I. Mink, Barbara Washburn, (date) joint author. II. Title.
QA39.2.R67 510 78-13397
ISBN 0-13-522920-0

The Language of Mathematics

Nelda W. Roueche and Barbara Washburn Mink

Printed in the United States of America

10 9 8 7 6 5 4 3 2 1

Interior and cover design by Linda Conway
Page layout by Charles H. Pelletreau
Manufacturing buyer: Phil Galea

PRENTICE-HALL INTERNATIONAL, INC., *London*
PRENTICE-HALL OF AUSTRALIA PTY. LIMITED, *Sydney*
PRENTICE-HALL OF CANADA, LTD., *Toronto*
PRENTICE-HALL OF INDIA PRIVATE LIMITED, *New Delhi*
PRENTICE-HALL OF JAPAN, INC., *Tokyo*
PRENTICE-HALL OF SOUTHEAST ASIA PTE. LTD., *Singapore*
WHITEHALL BOOKS LIMITED, *Wellington, New Zealand*

Contents

Preface

The Language of Mathematics combines sound mathematical content with good instructional design in a course that provides both theory and practical applications. Directed to students in the general college, liberal arts, and elementary education fields, the text may be used for a survey course of either one term or an entire year. Standard survey topics are included, as well as several topics that are innovations for an introductory course, thus providing a more comprehensive background for the mathematical applications related to a student's chosen field.

The Language of Mathematics is designed to give a solid understanding of the basic language and structure of the real number system and other numerical systems and to provide the student with an opportunity to explore various branches of mathematics: logic, algebra, geometry, statistics, probability, the metric system, and business/consumer applications. The appendix contains a college-level presentation of percent.

The core course consists of the chapters on sets, logic, the real number system, and algebra. Additional chapters may be selected by the instructor or by individual students, according to their educational/career goals.

The text can be implemented in a variety of instructional settings: It can be used as the text in a course conducted in the standard lecture/discussion format, or as the basis for an individualized, competency-based mathematical program because of its several instructional design considerations. Some of the unique aspects of *The Language of Mathematics* are:

1. Each chapter contains a written list of explicit instructional objectives, including an index to the topics in the text where each objective is taught.
2. There are numerous examples and corresponding practice problems that are directly related to the skills stated in the objectives. Practice problems immediately follow the examples in the text. A large number of additional exercises are given at the end of each chapter and, again, are related directly to the objectives of the chapter.
3. Each new concept or skill is numbered for easy reference. The end-of-chapter exercises are coded to the text material so that the student can review any sections that the exercises reveal have not been mastered.
4. All concepts and examples are explained thoroughly in an informal, readable style. The use of mathematical symbols is minimized, since unfamiliar symbols often detract from the student's comprehension of otherwise understandable concepts.
5. Key phrases and objective numbers are set in the margins beside their corresponding explanations, to help the student use and understand the material.

The materials contained in *The Language of Mathematics* have been used (and revised) with over 700 freshman and sophomore college students. An Instructor's Manual provides helpful suggestions on strategies for implementing an individualized math program based on that experience, as well as chapter teaching suggestions, sample quizzes, and answers to even-numbered exercises.

Numerous instructors have taught the preliminary materials and made helpful suggestions for revisions. Special thanks go to Nancy G. Spann for development of Chapter 8 and to Diane A. McGowan for Chapter 14. The completed manuscript was later reviewed by the following professors: Caroline Sastello, Mount Olive College; Floyd S. Elkins, Northern Virginia Community College; Jack Gill, Miami Dade Community College; Lucille Groenke, Mesa Community College; Albert W. Liberi, Westchester Community College; David Russell, Prince George's Community College; and Ara Sullenberger, Tarrant County Junior College—South, whom we thank for their many additional recommendations for its refinement.

Our assistant Judy F. Frieling typed the entire manuscript and solutions materials and kept us organized through numerous revisions, for which we are grateful. Our thanks also to Prentice-Hall mathematics editor Harry H. Gaines for his support and especially to Zita de Schauensee for her competent production management.

This text is lovingly dedicated to our families, who light up our lives and who patiently endured our distractions during the long months of this project.

Austin, Texas

NELDA W. ROUECHE
BARBARA WASHBURN MINK

Part One

Working with Symbolic Notation

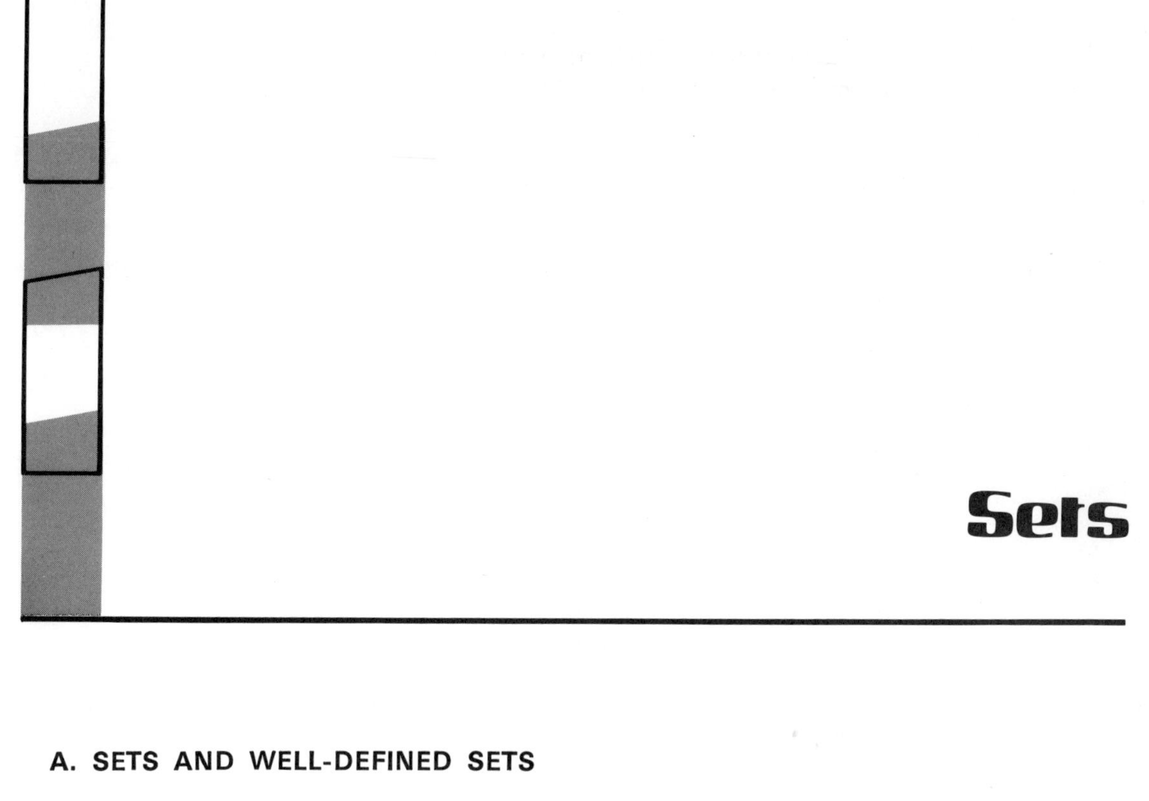

Sets

A. SETS AND WELL-DEFINED SETS

The concept of "set" is central in our development of the number system. The symbols, vocabulary, and processes studied in this chapter are important to master to ensure your success in subsequent chapters. "Set" is not a new or complex concept. A *set* is merely a collection of things (objects, concepts, people). An example of a set could be **(1)**

$$B = \{\text{cup, house, Tuesday}\}$$

Note that a set is represented by a *capital letter* (in this case B) and the *elements* or members (cup, house, Tuesday) are enclosed in *braces*, { }.

Objective 1: Well-defined sets

Most work with sets concerns *well-defined sets*. A set is well defined if, given any element, one can determine whether that element is or is not a member of the set. Set B above is well defined since, by inspection, we can easily decide whether "cup" or "Friday" is a member. The *set membership symbols* $\in$ (meaning "is a member of") and $\notin$ ("is not a member of" a given set) can be used to indicate **(2)**

$$\text{cup} \in B \qquad \text{and} \qquad \text{Friday} \notin B$$

Elements of most useful sets have some relationship to each other. A verbal description of a well-defined set must state that relationship so clearly that anyone could decide whether any potential element belongs in the set. If different people are asked to list the elements of a well-defined set, everyone should list the same elements.

A well-defined set might be described as "the set of all months of the year beginning with the letter J." If we call this set D, then

$$D = \{\text{January, June, July}\}$$

In contrast, the elements listed when describing the "set of the months of the year with hot weather" would vary depending upon who was listing the months and possibly with what area of the world the person was thinking about when listing the months. Different people have different ideas about what is "hot weather" and would thus list different months. In other words, this set would *not* be well defined.

Natural numbers

Before we further consider well-defined sets, let us define certain terms. The set of "natural numbers" (or "counting numbers") is the set of numbers one associates with objects when they are counted. The set of natural numbers would be denoted as

$$N = \{1, 2, 3, 4, 5, \ldots\}$$

Infinite and finite sets

The ellipsis marks, ". . .", are used to indicate that the set continues indefinitely, including all succeeding numbers beyond the ones listed. Any such set, which contains no last element or no greatest element, is called an *infinite* set. A set which has a limited number of elements (or no elements) is a *finite* set.

◆ **Example 1** Tell whether each of the following is or is not well defined. If well defined, write the elements of the set. (3)

(a) "Set of all natural numbers less than 7"
The natural numbers "less than 7" would be those numbers used in counting that are smaller than 7 and do *not* include 7. ("Less than" may also be indicated by the symbol "$<$.") Thus, the set of natural numbers less than 7 would be represented by

$$\{1, 2, 3, 4, 5, 6\}$$

Our set described above would be well defined. If we compared the numbers listed by several persons, the numbers listed would be the same. That is, everyone should name the set $\{1, 2, 3, 4, 5, 6\}$.

Whole numbers

(b) "Set of all whole numbers less than 2"
The set of whole numbers is composed of the natural numbers together with zero; that is, the set denoted $W = \{0, 1, 2, 3, 4, \ldots\}$. The "set of all whole numbers less than 2" is well defined. This set is represented by

$$\{0, 1\}$$

(c) "Set of all pleasant people living in Cleveland"
This is not a well-defined set. "Pleasant" means different things to different people. Not all persons asked would list the same people.

(d) "Set of all whole numbers between 3 and 11"

"Between" means those whole numbers that are both greater than 3 (not including 3) *and* less than 11 (not including 11). The given set is well defined and is denoted

$$\{4, 5, 6, 7, 8, 9, 10\}$$

● **Practice 1** Tell which of the following sets are well defined. Write the elements of the set in braces if the set is well defined. (4)

(a) Set of all letters in the alphabet that come before m.
(b) Set of all days of the week beginning with the letter T (if written in English).
(c) Set of all happy people taking math.
(d) Set of interesting books.
(e) Set of all whole numbers between 5 and 11.

See also Exercise 1.

B. SET NOTATION AND MEMBERSHIP

Suppose we had set A in which (5)

$$A = \{1, 2, 3, 4\}$$

Roster notation

Listing all the elements in a given set, as above, is called *roster notation.* But what are some of the ways in which we could describe Set A? The possibilities might include:

"Set A is composed of the elements 1, 2, 3, 4"

or

"Set A is composed of all counting numbers less than 5"

or

"Set A is composed of all whole numbers between 0 and 5"

Remember that *counting numbers* are denoted as $\{1, 2, 3, 4, \ldots\}$, whereas *whole numbers* include 0 and are indicated by $\{0, 1, 2, 3, 4, \ldots\}$.

Objective 2: Verbal descriptions to set notation

Set-builder notation

But suppose we had a description such as "Set B is the set of all whole numbers less than 997?" You can see it would take quite a while to list the elements of this set. Another method, called *set-builder notation,* expresses the same set, but without listing each element. For example, our set B could be expressed in set-builder notation as

$$B = \{x \mid x < 997, \text{ where } x \text{ is a whole number}\}$$

This is read: "Set B is equal to the set of all x, such that x is less than 997, where x is a whole number."

Note: It is necessary to define the set of numbers (in this case, whole numbers) with which we are working. As we expand the set of numbers we discuss, a description of "the set of all x such that x is less than 997" could mean including negative numbers, fractions, and other classes of numbers besides the whole numbers.

In summary, then, we will be describing sets in these three ways: with words, by listing all of the members (roster notation), and with a combination of symbols and words (set-builder notation).

◆ **Example 2** From the following verbal description of a set, indicate the set in both roster notation and set-builder notation. (6)

Verbal description: Set A is the set of all x such that x is less than or equal to 5 and x is a whole number.

Roster notation: $A = \{0, 1, 2, 3, 4, 5\}$

Set-builder notation:

A	$=$	$\{$	x	$\vert$	$x \leq 5,$	where x is a whole number$\}$
The set A	is (or is equal to)	the set of	all x	such that	x is less than or equal to 5	defines the set of numbers to which x must belong

Suppose you were given this verbal description: Set A is the set of all x such that x is less than zero and x is a whole number. What members would you list? Are there any whole numbers less than zero? No, the whole numbers are denoted by: (7)

$$\{0, 1, 2, 3, \ldots\}$$

So how would we represent the preceding situation where a set contains no members? The symbol used to denote such a set is $\varnothing$ or $\{\ \}$. It is called the *null set* or *empty set*.

Null or empty set

Note: Basically, the null or empty set is a set that contains no elements (or members). Be careful when you are using the null-set symbol that you write it correctly, as indicated by the following symbols:

$\varnothing$ or $\{\ \}$ This correctly denotes the null (or empty) set.

$\{0\}$ This denotes a set containing the number zero. Zero is an element; thus, $\{0\}$ does *not* denote a set containing no members.

$\{\varnothing\}$ This denotes the set of the symbol for null set. It is *not* the same as $\varnothing$. A verbal description of the set, $\{\varnothing\}$, would be "the set of the symbol used to represent the null set."

● **Practice 2** Given the following verbal descriptions, rewrite each set in both roster notation and set-builder notation; or else indicate that the set is empty: (8)

(a) Set D is the set of all x such that x is greater than ($>$) 7 and x is a whole number.

(b) Set E is the set of all x such that x is less than 7 and x is a natural number.

(c) Set A is the set of all x such that x is less than 1 and x is a natural number.

(d) Set B is the set of all x such that x is less than 3 and x is a natural number.

(e) Set B is the set of all x such that x is less than 10 and greater than 2 and x is a whole number.

(f) Set F is the set of all x such that x is less than 2 and x is a natural number.

See also Exercise 2.

C. SETS AND SUBSETS

(9)

Universal set

It is important to define the larger set which includes the set you are describing. In our preceding examples, whether a set was defined over the set of whole numbers or the set of natural numbers made a difference in the members (elements) of the set. The *universal set* is the basic set that includes all the elements from which the members of the indicated set may be chosen. The universal set is symbolized with the capital letter U. To illustrate, consider the following:

"Set A is the set of all x such that x is less than 4 and x is a whole number" or $A = \{x \mid x < 4$, where x is a whole number$\}$.

Subset

In this example, the set of whole numbers is the basic set, which includes all the elements that we might consider for set A. Thus, for this example, the set of whole numbers is our universal set, U. Any portion of a given universal set is called a *subset* of the given set. Set $A = \{0, 1, 2, 3\}$, which is a portion of the set of whole numbers, is thus a subset of the set of whole numbers.

Objective 3: Proper subset and subset

(10)

There are two variations of the term "subset." *Proper subset* describes a relationship between two sets in which one set is totally included within another. For example, set A above is a proper subset of the set of whole numbers. In general, we could write

$$P \subset R \qquad (P \text{ is a proper subset of } R)$$

This would mean that if we list all the elements of set P and all the elements of set R, then all the elements of set P are also elements of set R *and* there is at least one element of set R which is *not* an element of set P. Set A above is a proper subset of the set U of whole numbers because 0, 1, 2, and 3 are each whole numbers, but the set of whole numbers also includes other elements not in set A.

A set may also be considered a subset of another set when both sets contain the same elements. For example, if $C = \{3, 4, 5\}$ and $D = \{3, 4, 5\}$, then (11)

Set C is a subset of set D symbolized $C \subseteq D$

and also

Set D is a subset of set C symbolized $D \subseteq C$

However, C is *not* a proper subset of D, because D contains no additional elements not contained in set C. Similarly, D is not a proper subset of C. The symbol "$\subseteq$" is an interesting symbol and similar to some other mathematical symbols in that it can be read two ways. The symbol is a combination of the symbol $\subset$ (for proper subset) and the symbol $=$ (equals). Thus, $C \subseteq D$ indicates that *either* C is a proper subset of D *or else* C is equal to D. In general, set P is *equal* to set R if every element in set P is an element of set R *and* if every element in set R is an element of set P.

Thus, given $A = \{x \mid x < 4, \text{ where } x \text{ is a whole number}\}$, we could say that $A \subseteq U$ because set A is either a proper subset of the universal set *or* set A is equal to the universal set. (In this case, set A is a proper subset.) In summary, then,

	$A \subseteq U$	Set A is a subset of the universal set.
and	$A \subset U$	Set A is a proper subset of the universal set.
Also,	$C \subseteq D$	Set C is a subset of set D (because C is equal to D).
but	$C \not\subset D$	Set C is *not* a proper subset of set D.
Likewise,	$D \subseteq C$	Set D is a subset of set C (because the two sets are equal).
but	$D \not\subset C$	Set D is *not* a proper subset of set C.

(Observe that a slash through a mathematical symbol means that the relationship does *not* hold.)

When "subset" is used in reference to sets that contain the same elements, we could also say that every set is a subset of itself. That is,

	$C \subseteq C$	Set C is a subset of itself.
but	$C \not\subset C$	Set C is *not* a proper subset of itself.
Similarly,	$D \subseteq D$	Set D is a subset of itself.
but	$D \not\subset D$	Set D is *not* a proper subset of itself.

◆ **Example 3** Given sets U, A, B, and D below, list (a) all proper subset relationships and (b) all subset relationships. (12)

$$U = \{1, 2, 3, 4, 5\} \qquad B = \{2, 4, 5\}$$
$$A = \{1, 2, 4\} \qquad D = \{1, 2, 4, 5\}$$

(a) Proper subset relationships:

$A \subset U \quad A \subset D$
$B \subset U \quad B \subset D$
$D \subset U$

Consider the example $A \subset D$. Listing the elements of A we have $\{1, 2, 4\}$. The elements of D include $\{1, 2, 4, 5\}$. From the description of a proper subset, we see that all the elements of set A must also be elements of set D. Also, there must be an element of set D (in this case the element 5) which is *not* an element of set A.

(b) Subset relationships:

$A \subseteq U \quad U \subseteq U$
$B \subseteq U \quad A \subseteq A$
$D \subseteq U \quad B \subseteq B$
$A \subseteq D \quad D \subseteq D$
$B \subseteq D$

Remember that the symbol "$\subseteq$" can mean "is a proper subset" *or* "is equal to." All we did in column two is read "$\subseteq$" as "is equal to." (This "double" symbol allows us to state that each set is a subset of itself!)

The universal set is *not* a given set of numbers that never changes. That is, $\{1, 2, 3, 4, 5\}$ above is not the only set that can be named the universal set. The universal set can be any set that makes sense in a given situation. **(13)**

● **Practice 3** Given the following sets, list all *proper* subset and all subset relationships: **(14)**

(a) $U = \{a, b, c, d, e\}$
$A = \{a, c, d\}$
$B = \{a, b, c, d\}$
$C = \{c, d, e\}$
$D = \{d, e\}$

(b) $U = \{1, 10, 12\}$
$A = \{1, 10, 12\}$
$B = \{1\}$
$D = \{10, 12\}$

See also Exercise 3.

◆ **Example 4** Given the following set, write a universal set that will include the given set. **(15)**

Objective 4: Describing universal sets

"Set A is the set of all students enrolled in mathematics 101 at XYZ College who have class at 10:00 a.m."

Some of the possibilities for the universal set might be:

All students enrolled in XYZ College *or* U could be
All students enrolled in Mathematics 101 at XYZ College
or, All students enrolled in XYZ College who have class at 10:00 a.m.
or, All students enrolled in any college in the world!

When writing the universal set for a given set, you may name any set that contains the given set.

(16) Defining a universe and a subset of that universe will be used again in our introductory discussion of statistics. In statistics the concept will be similar, but different terms are used. In the study of statistics, the inclusive set of elements is termed a "population" and the subset in question is termed a "sample" of that given population.

(17) ● **Practice 4** Given the following sets, list three possible universal sets for each given set.

(a) The set of all people in Santa Fe, New Mexico, with a first name of James.
(b) The set of the numbers 3 and 4.
(c) The set of people in the room where you are now.

See also Exercise 4.

Objective 5: Subset relationship between sets

(18) Subset relationships may also exist between ordinary sets where neither is a universal set. A set B is said to be a subset of another set D if every element in set B is also an element of set D. If we want to show that set B is *not* a subset of set D, then there must be some element (or elements) of set B that is *not* an element of set D.

◆ **Example 5** Given sets E and F below, determine whether each is a subset of the other.

$$E = \{7, 10, 15, 32\} \qquad \text{and} \qquad F = \{4, 6, 7, 10, 14, 15, 32\}$$

We could write $E \subseteq F$. But F is *not* a subset of set E because there are some elements in F (namely 4, 6, 14) that are *not* elements in E. This could be symbolized

F	$\nsubseteq$	E
Set F	Is *not* a subset of	Set E

(Remember that a slash through a mathematical symbol negates the symbol.)

(19) ● **Practice 5** Given the sets below, determine whether each set is a subset of the other set:

(a) $A = \{1, 2, 3\}$ $\qquad B = \{1, 3\}$
(b) $C = \{2, 4, 8, 10, 15\}$ $\qquad D = \{1, 4, 8, 10, 16\}$
(c) $E = \{a, b\}$ $\qquad F = \{a, b\}$
(d) $G = \{1, 3, 5\}$ $\qquad H = \{1, 2, 3, 4, 5, \ldots\}$

See also Exercise 5.

The null set, $\varnothing$, has a place in the discussion of subsets. The null set is considered a subset of *every* set. This results from combining the meaning of the null set and the definition of subset. (20)

Null set: a set that contains no members.

Subset: a set B is said to be a subset of another set D if there are no elements in set B that are not in set D.

Putting the preceding two statements together, we can see that, since the null set has no elements, there cannot be any elements in the null set that are not elements of any other set. Thus, the null set is said to be a subset of every set.

◆ **Example 6**

Objective 6: All subsets of a set

(a) List all subsets of the set {2, 4, 6} (21)

Answer:

{2}	{2, 4}	{2, 4, 6}
{4}	{2, 6}	$\varnothing$
{6}	{4, 6}	

Note again that the null set symbol, $\varnothing$, is *not* enclosed in brackets. If it were written $\{\varnothing\}$, the literal interpretation would be "the set consisting of the symbol $\varnothing$."

It is often helpful in listing subsets of a given set to list the sets with one element first, then combinations of two elements, and so on. Note that after we listed all single elements and combinations, we listed the set itself, {2, 4, 6}, and the null set, $\varnothing$. The null set was listed because the null set is a subset of every set. The set itself was listed because of the special meaning of the subset symbol, $\subseteq$ (proper subset *or* equal to). The relationship $\{2, 4, 6\} \subseteq \{2, 4, 6\}$ is correct because in this case the subset symbol, $\subseteq$, would be interpreted using its "equal to" meaning.

(b) List all subsets of the set {1, 3, 5, 7}

(1) First, list all the subsets consisting of only one element:

{1} {3} {5} {7}

(2) Then, list subsets that are combinations of two elements. Make sure you have grouped every element with every other element.

{1, 3}	{3, 5}	{5, 7}
{1, 5}	{3, 7}	
{1, 7}		
These sets group 1 with all other elements.	These sets group 3 with all other elements with which it has *not* previously been grouped.	This set groups 5 with all other elements with which it has *not* previously been grouped.

Note that {3, 1} is the same as {1, 3}, so these are *not* listed as two separate subsets. The order of elements in a set has no significance. Two sets are equal whenever they contain exactly the same elements, regardless of the order of these elements. (22)

(3) List subsets that are combinations of 3 elements:

{1, 3, 5} {1, 5, 7} {3, 5, 7} {1, 3, 7}

(4) Finally, list the set itself and the null set:

{1, 3, 5, 7} ∅

(5) Therefore, the total number of subsets we have is 16:

{1}	{1, 3}	{1, 3, 5}	{1, 3, 5, 7}
{3}	{1, 5}	{1, 3, 7}	∅
{5}	{1, 7}	{1, 5, 7}	
{7}	{3, 5}	{3, 5, 7}	
	{3, 7}		
	{5, 7}		

Evaluating 2^n

One way to determine whether you have listed all subsets is to use the following information: The *number* of subsets of a given set with "n" elements can be found by evaluating the expression 2^n (2 to the "n" power). In Example 6b, the set {1, 3, 5, 7} has 4 elements and, therefore, we would expect it to have 16 subsets: (23)

$$2^n = 2^4 = 2 \cdot 2 \cdot 2 \cdot 2 = 16 \text{ subsets}$$

● **Practice 6** List all the subsets of the following sets. Also, evaluate 2^n to determine the correct number of subsets: (24)

(a) {1, 2} (b) {○, □, △} (c) {a}

(d) {1, 2, 3, 4} (e) {apple, pear, peach} (f) {Jane, Sarah}

See also Exercise 6.

D. SET OPERATIONS

There are three main operations (logical processes) that are performed with sets. They are complement, union, and intersection. The operation complement is an operation involving one set and would thus be termed a *unary* operation. In contrast, union and intersection are operations which involve associating two sets in certain ways to produce a unique third set. Since union and intersection involve two sets, they are termed *binary* operations. (25)

Unary and binary operations

Objective 7: Complement of a set

Let us examine complement first. If we define a set A, we say that the *complement* of A (symbolized A'), consists of all those elements of the given universe (U) that are *not* elements of set A. If you were given a set B (26)

such that $B = \{1, 3, 7\}$ and were asked to find B', you would have to respond that you do not have enough information to determine the answer. "Complement" means all the elements in the universe that are not included in the given set. You *could* solve the preceding problem if you were given the universal set that included set B.

◆ **Example 7** Given a universe and a set in the universe, find the complement of the given set: (27)

If $U = \{1, 3, 5, 6, 7, 10\}$ and $B = \{1, 3, 7\}$, find B'.

The elements in B' are those numbers that are in set U, but are not members of set B:

(given set)	$B = \{1, 3, \quad 7\}$
(universal set)	$U = \{1, 3, 5, 6, 7, 10\}$
(complement of given set)	$B' = \{ \quad 5, 6, \quad 10\}$

Thus, $B' = \{5, 6, 10\}$.

● **Practice 7** Given a universe and a set in the universe, find the complement of the given set. (28)

(a) If $U = \{1, 3, 5, 7, 9\}$ and
$A = \{1, 5, 9\}$, find A'.

(b) If $U = \{a, b, c, d, e, f, g\}$ and
$A = \{a, b, c, e, f, g\}$, find A'.

(c) If $U = \{1, 2, 3\}$ and
$A = \{1, 2, 3\}$, find A'.

See also Exercise 7.

Objective 8: Union of sets

The operation *union* ($\cup$) combines all of the elements of two (or more) sets into a single set. The union of two sets E and F is indicated by $E \cup F$. This union consists of all those elements that are in set E *or* in set F *or* in both sets (with *no* duplicate listing of elements from both sets). (29)

◆ **Example 8** Given two or more sets, list the elements in the union of these sets: (30)

(a) If $E = \{1, 3, 7, 9\}$ and $F = \{3, 6, 8, 10, 11\}$ then $E \cup F$ is determined as follows:

$$E \cup F = \{1, \quad 3, \quad 6, \quad 7, \quad 8, \quad 9, \quad 10, \quad 11\}$$

1,	3,	6,	7,	8,	9,	10,	11
↑	↑	↑	↑	↑	↑	↑	↑
$1 \in E$	$3 \in E$ and $3 \in F$ (but 3 is only listed once)	$6 \in F$	$7 \in E$	$8 \in F$	$9 \in E$	$10 \in F$	$11 \in F$

(b) If $A = \{1, 3, 8\}$, $B = \{1, 10, 14\}$, and $C = \{10, 14, 18\}$, find $(A \cup B) \cup C$.

Since union is a binary operation, only two sets can be combined at one time. The sets within parentheses are the first two sets to be combined. Thus, we will first determine $A \cup B$. Then, considering $A \cup B$ as a single set, we will perform the union operation with that set and set C. Thus,

$$\begin{aligned} A \cup B &= \{1, 3, 8, 10, 14\}, \quad \text{and} \\ (A \cup B) \cup C &= \{1, 3, 8, 10, 14\} \cup \{10, 14, 18\} \\ &= \{1, 3, 8, 10, 14, 18\} \end{aligned}$$

The concept of union, if thought of in an everyday sense, would be similar to two persons going on a picnic together. A conversation between the two persons planning the picnic might go like this: (31)

A: "I have cups, napkins, hot dogs, ketchup."
B: "I have cups, rolls, ketchup, relish, punch."
A: "OK, since we are going together I'll bring the cups, napkins, and hot dogs, and you bring the rolls, ketchup, relish, and punch."

or written in set notation:

$$A = \{\text{cups, napkins, hot dogs, ketchup}\}$$

$$B = \{\text{cups, rolls, ketchup, relish, punch}\}$$

$$A \cup B = \{\text{cups, napkins, rolls, hot dogs, ketchup, relish, punch}\}$$

● **Practice 8** Given two or more sets, list the elements in the union of these sets: (32)

(a) If $A = \{x, y\}$ and $B = \{a, x\}$, find $A \cup B$.

(b) If $D = \{1, 3, 7\}$ and $E = \{2, 4, 6\}$, find $D \cup E$.

(c) If $F = \{3, 6, 9, 12\}$ and $G = \varnothing$, find $F \cup G$.

(d) If $A = \{3, 8, 9\}$, and $B = \{1, 2, 3\}$, and $C = \{1, 2, 3, 5\}$, find $(A \cup B) \cup C$.

(e) If $A = \{3, 8, 9\}$, $B = \{1, 2, 3\}$, and $C = \{1, 2, 3, 5\}$, find $A \cup (B \cup C)$.

See also Exercise 8.

Objective 9: Intersection of sets

The operation *intersection* ($\cap$) identifies those elements that are common to two (or more) given sets. The intersection of two sets A and B is indicated by $A \cap B$. This intersection set is composed of those elements that belong to *both* set A *and* set B. (33)

◆ **Example 9** Given two or more sets, list the elements in the intersection: (34)

(a) If $A = \{1, 2\}$ and $B = \{2, 3, 6\}$, find $A \cap B$.
Looking at both sets A and B, we see that there is only one element, 2, that is common to both sets:

$$A = \{1, 2\}$$
$$/$$
$$B = \{2, 3, 6\}$$

Therefore, $A \cap B = \{2\}$.

(b) If $E = \{7, 8, 9\}$, $F = \{8, 9, 10, 11\}$ and $G = \{8, 9, 15\}$, find $(E \cap F) \cap G$.

Since intersection is a binary operation, only two sets can be associated in each step. Thus, we will first determine $E \cap F$; then, considering $E \cap F$ as a single set, we will repeat the intersection operation with that set and set G. Thus,

$$E \cap F = \{8, 9\}$$
$$(E \cap F) \cap G = \{8, 9\} \cap \{8, 9, 15\}$$
$$= \{8, 9\}$$

The concept of intersection in everyday conversation might involve two friends deciding what to do one Saturday afternoon: (35)

A: "I would like to go fishing, walking, or swimming."
B: "I would like to play tennis or go swimming."

Since swimming is an area that overlaps both of their interests, they decide to go swimming. In set notation, the situation above might look like:

$$A = \{\text{fishing, walking, swimming}\}$$
$$B = \{\text{tennis, swimming}\}$$
$$A \cap B = \{\text{swimming}\}$$

● **Practice 9** Given two or more sets, list the elements in the intersection: (36)

(a) If $A = \{2, 12, 27, 18, 3\}$ and $B = \{1, 18, 12\}$, find $A \cap B$.

(b) If $K = \{▭, ▯\}$ and $L = \{△, ▯\}$, find $K \cap L$.

(c) If $H = \{a, b, f\}$ and $I = \{c, d, e\}$, find $H \cap I$.

(d) If $M = \{\text{Ted, Mary, Janet}\}$ and $N = \{\text{Tom, Ted, Alice}\}$, find $M \cap N$.

(e) If $P = \{2, 4, 6, 8, 10, 12\}$ and $Q = \{4, 5, 6, 7, 8, 9\}$, find $P \cap Q$.

(f) If $A = \{6, 8, 10, 12\}$, $B = \{1, 2, 6, 12\}$ and $C = \{1, 10, 12\}$, find $(A \cap B) \cap C$.

(g) If $A = \{6, 8, 10, 12\}$,
$B = \{1, 2, 6, 12\}$
and $C = \{1, 10, 12\}$,
find $A \cap (B \cap C)$.

See also Exercise 9.

E. CARTESIAN PRODUCT OF SETS

Another operation between sets is called the *Cartesian product* (so named in honor of the French mathematician René Descartes, 1596–1650). It is an operation in which order is very important. To begin understanding what the Cartesian product is, we must understand what an ordered pair is, as well as the differences between a set and an ordered pair. An *ordered pair* is any pair of two elements (in parentheses), where one element is considered the first component and the other element is considered the second component. (37)

Ordered pairs

An example of an ordered pair is:

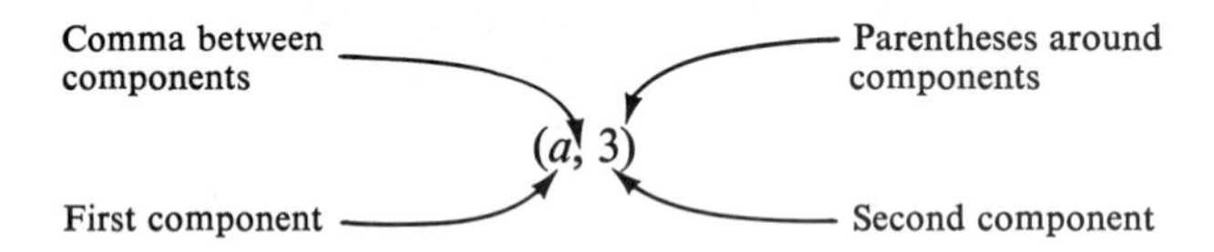

In contrast, the *set* with the elements a and 3 is written:

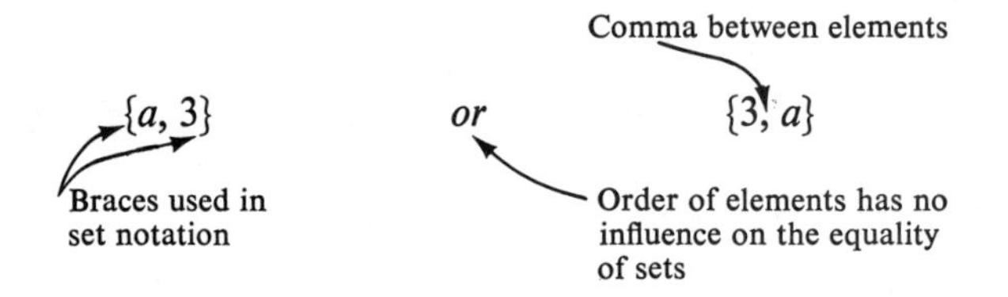

Also, a set is not restricted to only two elements as an ordered pair is. A set can have no members (null set), or a finite number, or an infinite number of elements.

In working with the Cartesian product of sets, we develop *a set of ordered pairs*. For example: (38)

$$\{(a, 1), (a, 2), (b, 1), (b, 2)\}$$

Knowing what we do so far about ordered pairs and sets, we can make some observations about the preceding set of the four ordered pairs:

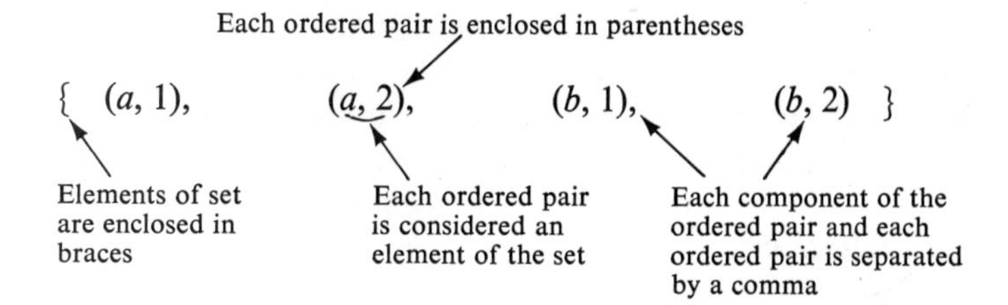

Elements in a set can be in any order without changing the meaning of the set. Thus,

$$\{(a, 1), (a, 2), (b, 1), (b, 2)\} = \{(a, 2), (b, 1), (b, 2), (a, 1)\}$$

But components of an ordered pair *cannot* be reversed:

$$\{(a, 1), (a, 2), (b, 1), (b, 2)\} \neq \{(1, a), (2, a), (1, b), (2, b)\}$$

Two ordered pairs, (a, b) and (c, d) are equal if and only if $a = c$ and $b = d$.

Objective 10: Cartesian product

Now let us consider how these sets of ordered pairs are obtained. The *Cartesian product* of set P and set Q (written "$P \times Q$") is the set of all ordered pairs (p, q) where $p \in P$ and $q \in Q$. In Cartesian products, the symbol "$\times$" is read "cross" (that is, "$P \times Q$" is read "P cross Q"). **(39)**

◆ **Example 10** Given two sets, perform the indicated Cartesian product (or cross product). **(40)**

(a) If $A = \{a, b, c\}$ and $B = \{1, 3\}$, find $A \times B$.

$$\{a, b, c\} \times \{1, 3\} = \{(a, 1), (a, 3), (b, 1), (b, 3), (c, 1), (c, 3)\}$$

Notice that each element in set A is paired with each element in set B, with the element from set A being listed *first* in the ordered pairs.

(b) If $A = \{a, b, c\}$ and $B = \{1, 3\}$, find $B \times A$.
In this instance, we are working with the same sets A and B, but the problem calls for us to take the first component of each ordered pair from set B.

$$\begin{aligned} B \times A &= \{1, 3\} \times \{a, b, c\} \\ &= \{(1, a), (1, b), (1, c), (3, a), (3, b), (3, c)\} \end{aligned}$$

● **Practice 10** List the set of ordered pairs that compose the following Cartesian products (sometimes called cross products): **(41)**

(a) If $A = \{a, b\}$ and $B = \{3, 7\}$, find $A \times B$.

(b) If $C = \{3, 7, 10\}$ and $D = \{1, 6\}$, find $D \times C$.

(c) If $A = \{7\}$ and $B = \{2\}$, find $A \times B$ and $B \times A$.

(d) If $E = \{a\}$ and $F = \{1, 2, 3, 4\}$, find $E \times F$ and $F \times E$.

See also Exercise 10.

Objective 11: Tree diagrams

One way to determine whether you have listed all the elements in a Cartesian product and whether the components of the ordered pairs are in the correct order is by using a *tree diagram*. A tree diagram is a diagram that uses connecting lines to associate the elements in an operation. It is used here to associate the components of ordered pairs in a Cartesian product. (42)

◆ **Example 11** Given two sets find the elements in the Cartesian product by using a tree diagram. (43)

(a) If $A = \{3, 7\}$ and $B = \{a, b, c\}$, find $A \times B$ by using a tree diagram.

Elements in set A	*Elements in set B*	*Ordered pairs in* $A \times B$
3	a	$(3, a)$
	b	$(3, b)$
	c	$(3, c)$
7	a	$(7, a)$
	b	$(7, b)$
	c	$(7, c)$

Therefore, $A \times B = \{(3, a), (3, b), (3, c), (7, a), (7, b), (7, c)\}$

(b) If $A = \{3, 7\}$ and $B = \{a, b, c\}$, find $B \times A$ by using a tree diagram.

Elements in set B	*Elements in set A*	*Ordered pairs in* $B \times A$
a	3	$(a, 3)$
	7	$(a, 7)$
b	3	$(b, 3)$
	7	$(b, 7)$
c	3	$(c, 3)$
	7	$(c, 7)$

Therefore, $B \times A = \{(a, 3), (a, 7), (b, 3), (b, 7), (c, 3), (c, 7)\}$

(c) If $P = \{1, 3\}$, find $P \times P$ by using a tree diagram.

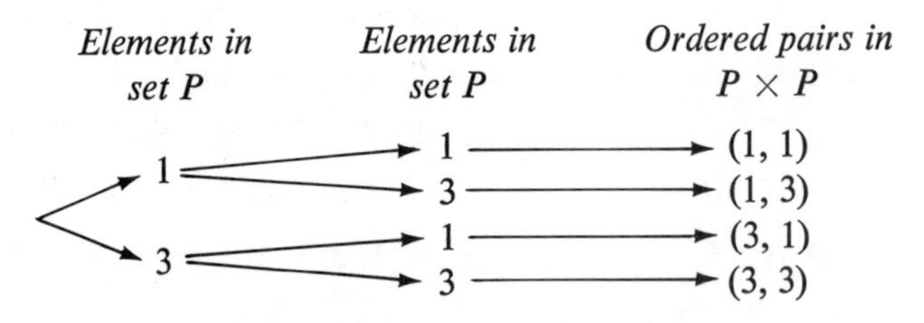

Therefore, $P \times P = \{(1, 1), (1, 3), (3, 1), (3, 3)\}$

● **Practice 11** Find the set of ordered pairs that compose the following Cartesian products by using tree diagrams: (44)

(a) If $A = \{7, 10\}$ and $B = \{a\}$, find $A \times B$ and $B \times A$.

(b) If $C = \{x\}$ and $D = \{4, 8, 9\}$, find $D \times C$.

(c) If $U = \{1, 2, 3\}$, find $U \times U$.

(d) If $E = \{1, 3, 7\}$ and $F = \{2, 4\}$, find $E \times F$.

See also Exercise 11.

Objective 12: Array

Cartesian products are often represented in an *array*. An array is a number of mathematical elements arranged in rows and columns. It is used here to represent the set of ordered pairs in a Cartesian product of two sets. In such an array the *first* components are listed horizontally and the *second* components are listed vertically, beginning at the bottom left-hand corner. (45)

◆ **Example 12** Given two sets, represent the Cartesian product in an array. (46)

(a) If $A = \{1, 3\}$ and $B = \{a, b, c\}$, represent the cross product $A \times B$ in an array.

Elements of set B →			
	c	$(1, c)$	$(3, c)$
	b	$(1, b)$	$(3, b)$
	a	$(1, a)$	$(3, a)$
		1	3 ← Elements of set A

$$A \times B = \{(1, a), (1, b), (1, c), (3, a), (3, b), (3, c)\}$$

(b) If $A = \{1, 3\}$ and $B = \{a, b, c\}$, represent the cross product $B \times A$ in an array:

Elements of set A →				
	3	$(a, 3)$	$(b, 3)$	$(c, 3)$
	1	$(a, 1)$	$(b, 1)$	$(c, 1)$
		a	b	c ← Elements of set B

$$B \times A = \{(a, 1), (a, 3), (b, 1), (b, 3), (c, 1), (c, 3)\}$$

Representing Cartesian products in an array and being able to read such a chart of ordered pairs are important skills in working with graphs, where position is important. Such charts are used extensively in statistics for presenting a summary of data collected about a problem. (47)

● **Practice 12** Given two sets, represent the indicated Cartesian products in an array: (48)

(a) If $A = \{1, 3, 5\}$ and $B = \{a, b\}$, represent $A \times B$ and $B \times A$.

(b) If $A = \{a\}$ and $B = \{1, 8\}$, determine $A \times B$.

See also Exercise 12.

F. REPRESENTING SETS WITH VENN DIAGRAMS

(49) Thus far, we have performed the set operations of Cartesian product, complement, intersection and union. It is also possible to show that properties of the number system (commutative, associative, and distributive) hold true for sets. We will now employ diagrams to illustrate set operations and properties. These diagrams can be used in illustrating overlapping categories and in determining the validity of an argument. These diagrams are called *Venn diagrams* (after John Venn, an English logician who lived from 1824 to 1923) or *Euler circles* (after Leonard Euler, a Swiss mathematician who lived from 1707 to 1783).

Objective 13: Venn diagrams

(50) In constructing Venn diagrams, the universal set is represented by a *rectangle* and the *sets* within that universe are represented by *circles* within the rectangle, The portion of the diagram that we are working with is usually highlighted in some way. Typical types of highlights are parallel lines (≡ or ||| or \\\ or ///), or dots (░), or colors.

We will illustrate the operations with two sets. It is also possible to illustrate operations with more than two sets using Venn diagrams. Venn diagrams can be used to illustrate an operation or situation either in general terms or with specific elements. We will first examine the operation *complement* with both general and specific Venn diagrams.

◆ **Example 13**

(51) *General:* Given a universe, U, and a set, A, within that universe, illustrate with a Venn diagram the complement of A (written, A').

The universe is represented by the rectangle and the letter $\mathcal{U}$.

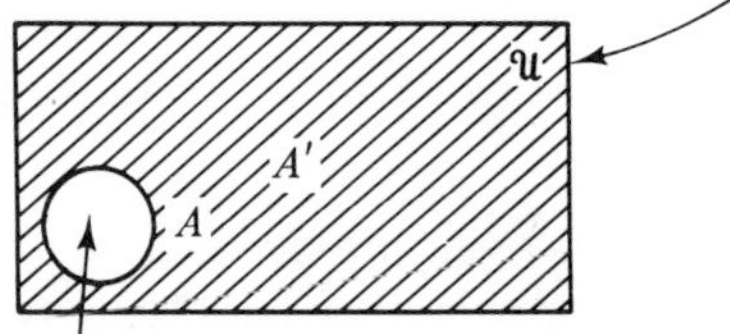

The *set A* is represented by a *circle* designated with the letter of the set (the letter is usually written on the edge of the circle).

The complement of A is those elements of the given universe that are *not* elements of set A. In the previous Venn diagram this is illustrated by the shaded (\\\\) portion of the diagram.

Specific: Given a universe $U = \{1, 2, 3, 4\}$ and a set $A = \{2, 3\}$ within that universe, illustrate with a Venn diagram the complement of A. (52)

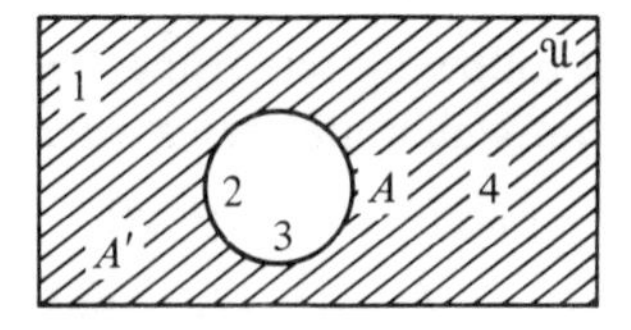

From the diagram we see that $A' = \{1, 4\}$. That is, A' is composed of those elements in $\{1, 2, 3, 4\}$ that are not elements of $\{2, 3\}$.

● **Practice 13** Illustrate the following sets with Venn diagrams. Shade in the portion of your diagram which illustrates complement: (53)

(a) $U = \{1, 2, 3, 4, 5, 6\}$, $A = \{1, 2\}$, and $A' = \{3, 4, 5, 6\}$.

(b) $U = \{1, 10, 12\}$, $B = \{10, 12\}$, and $B' = \{1\}$.

(c) $U = \{a, b, x, y\}$, $C = \{a\}$, and $C' = \{b, x, y\}$.

See also Exercise 13.

Next we will consider Venn diagrams illustrating the set operation of intersection, first in general terms and then with specific sets. (54)

◆ **Example 14**

General: Given a universal set, U, and two subsets A and B which have common elements, show $A \cap B$ with a Venn diagram. (55)

Venn diagrams and intersection

Set A is shown with horizontal lines.

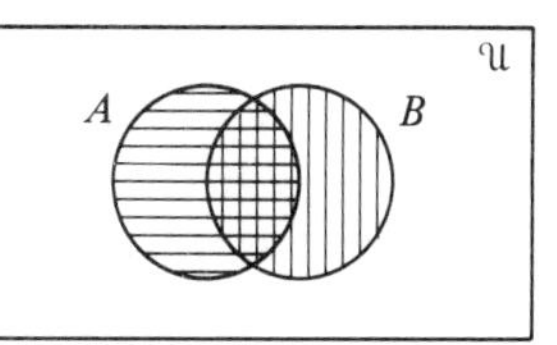

Set B is shown with vertical lines.

(|||||)

$A \cap B$ consists of those elements which are common to both set A (≡) and set B (|||). In our diagram, $A \cap B$ is that area where sets A and B overlap (#).

Specific: Given a universe $U = \{1, 2, 3, 4, 5, 6, 7\}$ and sets $A = \{1, 2\}$ and $B = \{2, 4, 6\}$ within that universe, illustrate $A \cap B$ with Venn diagrams. (56)

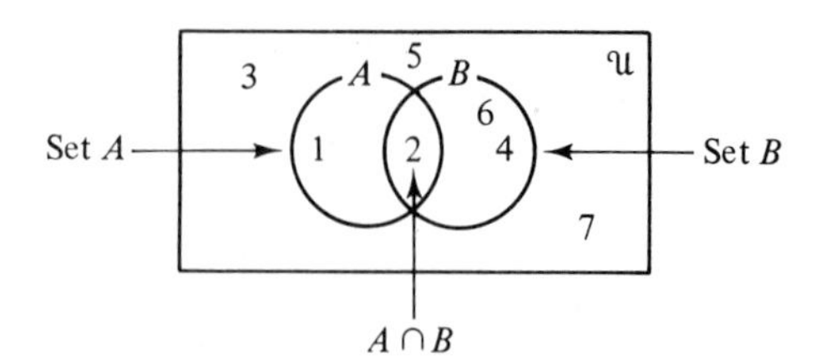

● **Practice 14** Illustrate the following sets using Venn diagrams. Shade in the portion of your diagram which illustrates intersection: (57)

(a) $U = \{a, b, c, d, e\}$, $A = \{a, b\}$, $B = \{b, d\}$, and $A \cap B = \{b\}$.

(b) $U = \{3, 10, 12\}$, $E = \{3\}$, $F = \{3, 12\}$, and $E \cap F = \{3\}$.

(c) $U = \{3, 4, 5, 6, 7, 8, 9\}$, $H = \{5, 6, 7, 8\}$, $I = \{3, 4, 5, 6, 7\}$, and $H \cap I = \{5, 6, 7\}$.

See also Exercise 13.

Union is the final set operation that we will illustrate with Venn diagrams, again both in general terms and with specific sets. (58)

◆ **Example 15**

(a) *General:* Given a universe, U, and two sets, A and B, within that universe, show $A \cup B$ with a Venn diagram. (59)

Venn diagrams and union

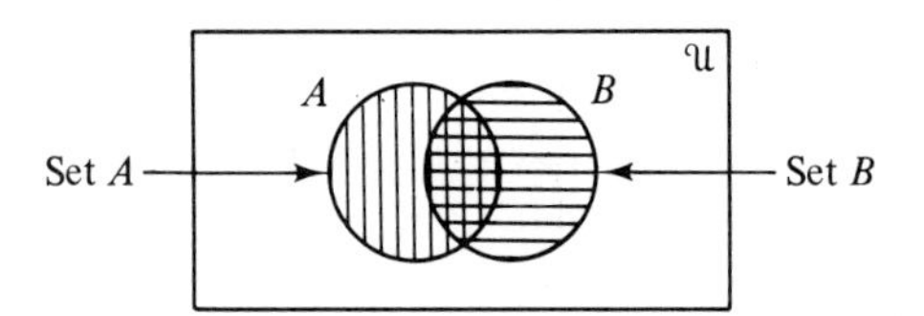

$A \cup B$ is all those elements that are in set A *or* in set B *or* in both sets A and B. Thus, the *entire* shaded portion of the above diagram is $A \cup B$. If we were representing $A \cup B$ without distinct lines for each set, $A \cup B$ would be represented as the shaded portion below.

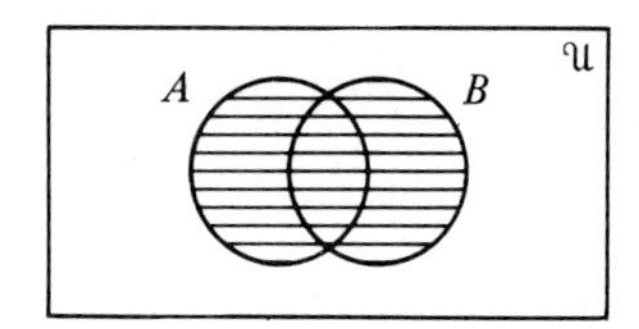

(b) *Specific:* Given a universe $U = \{1, 2, 3, 4, 5, 6, 7, 8\}$, $A = \{2, 4\}$, and $B = \{2, 6, 8\}$, show $A \cup B$ using a Venn diagram. (60)

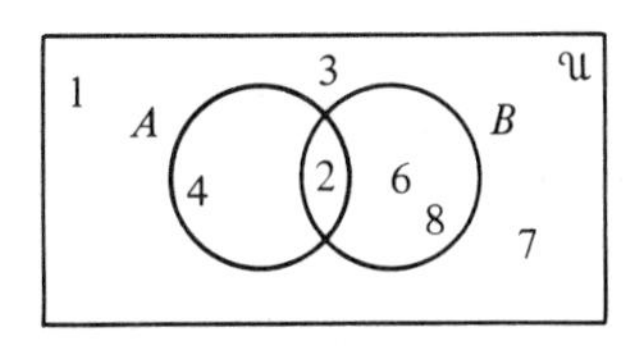

$$A \cup B = \{2, 4, 6, 8\}$$

● **Practice 15** Use a Venn diagram to illustrate each of the following set operations. Label each set, shade in the sets, and properly place the individual elements in the diagram: (61)

(a) $U = \{2, 4, 6, 8, 10, 12\}$, $A = \{4\}$, $B = \{4, 8, 12\}$. List the elements and diagram each operation (using a separate diagram for each exercise).

(1) $A \cup B$ (2) $A \cap B$ (3) A' (4) B'

(b) $U = \{a, b, c, d, e, f\}$, $E = \{e, f\}$, $F = \{a, b, e, f\}$. List the elements and diagram each operation (using a separate diagram for each exercise).

(1) E' (2) F'

See also Exercise 13.

We can now combine our knowledge of union, intersection, and complement to illustrate operations like $(A \cap B)'$ or $A' \cup B'$. (62)

◆ **Example 16** Given a universe $U = \{2, 4, 6, 8, 10, 12, 14\}$, $A = \{2, 6, 10, 12\}$, and $B = \{2, 8, 10, 14\}$, show the following using Venn diagrams: $A \cup B$, $(A \cup B)'$, and $A' \cap B'$. (63)

$A \cup B$

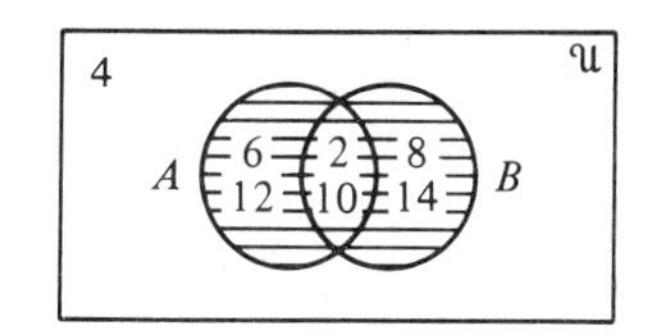

$$A \cup B = \{2, 6, 8, 10, 12, 14\}$$

$(A \cup B)'$

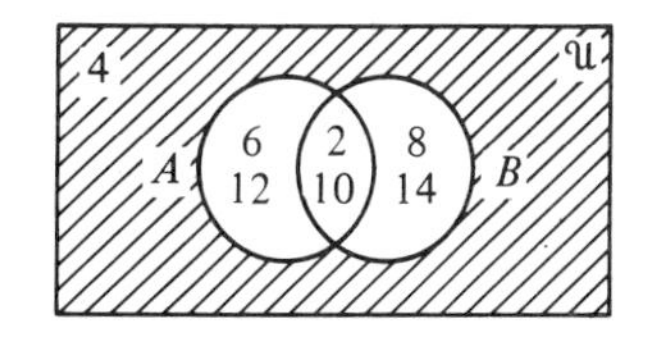

$$U = \{2, 4, 6, 8, 10, 12, 14\}$$
$$A \cup B = \{2, 6, 8, 10, 12, 14\}$$
$$(A \cup B)' = \{4\}$$

$A' \cap B'$

$$A' = \{4, 8, 14\} \quad \text{and}$$
$$B' = \{4, 6, 12\}, \quad \text{then}$$
$$A' \cap B' = \{4\}$$

● **Practice 16** Use a Venn diagram to illustrate each of the following set operations. Label each set, shade in the sets, and properly place the individual elements in the diagram. **(64)**

$U = \{a, b, c, d, e, f\}$, $E = \{e, f\}$, $F = \{a, b, e, f\}$

(a) $E \cup F$ (b) $(E \cup F)'$

(c) $E' \cap F'$ (d) $(E \cap F)$,

See also Exercise 13.

G. CARDINALITY AND EQUIVALENT SETS

Objective 14: One-to-one correspondence

To understand the number concept of a set, we need first to examine what is meant by one-to-one correspondence between sets. Two sets, E and F, are said to be in *one-to-one correspondence* if each element of set E can be paired with only *one* element of set F *and* each element of set F can be paired with only *one* element of set E. **(65)**

◆ **Example 17** If $A = \{3, 8, 10\}$ and $B = \{a, b, c, d\}$, are sets A and B in one-to-one correspondence? **(66)**

$A = \{3, 8, 10\}$
$\updownarrow \; \updownarrow \; \updownarrow$
$B = \{a, b, c, d\}$

3 in A is paired with a in B
8 in A is paired with b in B
10 in A is paired with c in B

(Note that the double arrow ($\updownarrow$) is used to indicate that an element of one set is paired with an element of the other set.)

What about the element d in set B? Is there an element in set A that can be paired with element d? No, not according to our definition of one-to-one correspondence. Review the two conditions of our definition and compare them with the example.

Two sets A and B are said to be in one-to-one correspondence if:

Condition 1: Each element of set A can be paired with *only one element* in set B:

$A = \{3, 8, 10\}$
$\updownarrow \; \updownarrow \; \updownarrow$
$B = \{a, b, c, d\}$

The first part of the definition holds, since each element in set A is paired with only one element in set B.

Condition 2: *Each* element in set B can be paired with *only one element* in set A.

$A = \{3, 8, 10\}$
$\updownarrow \; \updownarrow \; \updownarrow$
$B = \{a, b, c, d\}$

We might attempt to pair d with element 10 (or any other element in set A); but then the *first* condition would be violated, because we would have one element in set A paired with *two* elements in set B.

Therefore, we say that our two sets, A and B, are *not* in one-to-one correspondence.

● **Practice 17** Use the definition of one-to-one correspondence to determine whether the following sets are in one-to-one correspondence: (67)

(a) $\{6, 8, 10, 12\}$ and $\{6, 8, 10, 14\}$.

(b) $\{2, 3, 8\}$ and $\{a, x, \square\}$.

(c) $\{3, 10, 15\}$ and $\{e, f\}$.

(d) $\{x, y\}$ and $\{a, b, c\}$.

See also Exercise 14.

Objective 15: Ways of establishing a one-to-one correspondence

The following sets G and H can be shown to be in one-to-one correspondence: (68)

$G = \{7, \quad 10, \quad 27\}$
$H = \{\text{chair, table, desk}\}$

(7 ↔ chair, 10 ↔ table, 27 ↔ desk)

But, is that the only way one-to-one correspondence could be shown? No, our definition of one-to-one correspondence did not stipulate that the first element in set G must be matched with the first element in set H. There is no one way to show a one-to-one correspondence. As long as each element in the set is only matched with one element in the other set, it is a correctly diagramed one-to-one correspondence. Let us examine several ways to illustrate one-to-one correspondence.

◆ **Example 18** Given the two sets, $G = \{7, 10, 27\}$ and $H = \{\text{chair, table, desk}\}$, show all ways a one-to-one correspondence could be established between the two sets: (69)

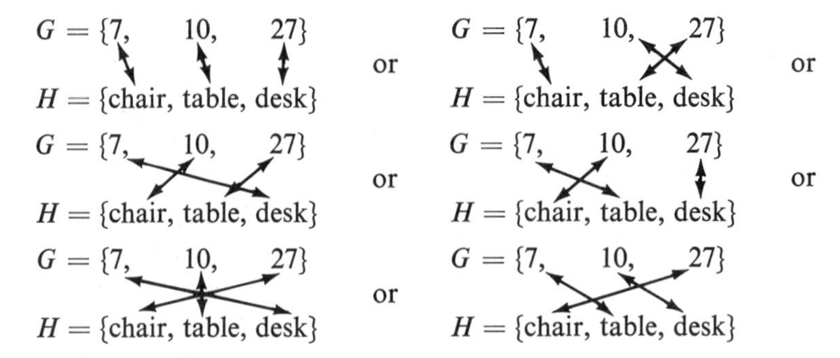

● **Practice 18** Show all possible ways of putting the following sets in one-to-one correspondence: (70)

(a) $\{1, 2, 8\}$ and $\{x, y, z\}$.
(b) {apple, pear} and $\{b, c\}$.
(c) $\{x\}$ and $\{a\}$.
(d) $\{1, 7\}$ and $\{7, 10\}$.

See also Exercise 15.

One-to-one correspondence might seem a rather complicated way to express relationships between sets, because you have already mastered the concept of "number." You could look at set G and see that there are three elements, so certainly each element in one set could be paired with only one element in the other set. However, before numbers were invented and became a means of communicating information, the idea of one-to-one correspondence was probably used for record keeping and reporting—a mark in the sand for each goat traded, a knot in a piece of fiber for each animal in a herd. Small children, too, before "number" is understood, can also match items. For example, small children might match candy bars with friends, socks with shoes. (71)

The number symbols need to be recognized and the concept of order understood before the natural numbers, $\{1, 2, 3, 4, 5, \ldots\}$, can be useful. And what a useful tool the set of natural numbers is! The pattern is so widely understood that virtually everyone could write the number that comes after 72 or after 1,753. Recall that the set of natural numbers, $\{1, 2, 3, 4, 5, \ldots\}$, is an infinite set, since it contains no last element. Also, a set with a limited number of elements is a finite set. (72)

Objective 16: Cardinal number of a finite set

We use a subset of natural numbers to establish a one-to-one correspondence with any finite set (except the null set). The *cardinal number* of a given finite set is the last-named natural number after a one-to-one correspondence is established between the given set and a subset of natural numbers. (73)

◆ **Example 19** Find the cardinal number of set B, where $B = \{a, b, d, f, g\}$, by establishing a one-to-one correspondence between the elements of set B and an *initial subset* (a set that occurs at the beginning) of the natural numbers: (74)

$$\begin{array}{l} B = \{a, b, d, f, g\} \\ \quad\;\; \updownarrow\;\updownarrow\;\updownarrow\;\updownarrow\;\updownarrow \\ N = \{1, 2, 3, 4, 5, 6, 7, 8, 9, \ldots\} \end{array}$$

We established a one-to-one correspondence of set B with the initial subset $\{1, 2, 3, 4, 5\}$. Since 5 is the last-named element of $\{1, 2, 3, 4, 5\}$, then the number of elements in set B is equal to 5. This is shown in symbolic notation as $n(B) = 5$. Stated in another way, 5 is the *cardinal number* of set B.

● **Practice 19** Find the cardinal number of the following sets by placing them in one-to-one correspondence with an initial subset of the set of natural numbers: (75)

(a) $A = \{a, b, c, d, e, f\}$
(b) $B = \{3, 10, 14, 16\}$
(c) $C = \{\text{glass, tire, aluminum}\}$
(d) $D = \{\text{David}\}$
(e) $E = \{3, \square, x, 7\}$
(f) $F = \{1, 2, 3, 4, 5\}$

See also Exercise 16.

Objective 17: Definition of equivalent sets

Cardinality and one-to-one correspondence can be used in identifying what is meant by *equivalent sets:* If the elements in two sets B and C can be placed in a one-to-one correspondence, the sets are said to be equivalent. Any two sets that are equivalent have the same *cardinal number.* (76)

◆ **Example 20** Are the following two sets equivalent? (77)

$$B = \{b, c, d\} \qquad C = \{\text{dog, cat}\}$$

The first part of the description of equivalent sets states: "If the elements in two sets B and C can be placed in a one-to-one correspondence, the sets are said to be equivalent."

$$\begin{array}{l} B = \{b, c, d\} \\ C = \{\text{dog, cat}\} \end{array}$$

Element d cannot be matched with an element in set C.
Therefore, sets B and C cannot be placed in one-to-one correspondence.

The second part of the description states: "Any two sets that are equivalent have the same cardinal number."

$$\begin{array}{l} B = \{b, c, d\} \\ \quad\;\; \updownarrow\;\updownarrow\;\updownarrow \\ \quad\;\; \{1, 2, 3, 4, 5, \ldots\} \qquad n(B) = 3 \end{array}$$

$$\begin{array}{l} C = \{\text{dog, cat}\} \\ \quad\;\; \{1, 2, 3, 4, 5, \ldots\} \qquad n(C) = 2 \end{array}$$

Is $n(B) = n(C)$?
Is $3 = 2$?
No, $3 \neq 2$.

Therefore, by both parts of our definition of equivalent sets, our sets fail to qualify as equivalent sets.

● **Practice 20** Use the two parts of the description of equivalent sets to determine whether the following sets are equivalent: (78)

(a) $A = \{5, 7, 10\}$ and $B = \{7, 10, 14\}$.

(b) $C = \{a, b, c\}$ and $D = \{\text{book, pen}\}$.

(c) $E = \{x, 7, b\}$ and $F = \{b, y, x\}$.

(d) $H = \{3\}$ and $I = \{7\}$.

(e) $X = \{1, 3, 7, 9\}$ and $Y = \{1, 3, 9, 10, 16\}$.

See also Exercise 17.

Objective 18: Equivalent versus equal sets

Note that we said nothing in our definition of equivalent sets about our sets having the *same* elements. We defined equivalent sets as having the same *number* of elements. If sets have exactly the same elements, they are called *equal*. In summary, then, (79)

Equivalent sets are sets that have the same number *of elements. Equal sets are sets that have exactly the* same *elements* (*in number and kind*).

◆ **Example 21** Indicate whether the following sets are equivalent and/or equal: (80)

	Equivalent?	*Equal?*
(a) $A = \{b, d\}$ and $B = \{1, 2\}$	Yes, because both set A and set B have the same number of elements.	No, because set A does not have the same elements as set B.
(b) $C = \{b, d\}$ and $D = \{d, b\}$	Yes, because both set C and set D have the same number of elements.	Yes, because set C is composed of the same elements as set D.
	(Remember that the order in which the elements are listed does *not* change the meaning of the set.)	
(c) $E = \{1, 8, 10\}$ and $F = \{a, b, c, d\}$	No, because set E and set F have a different number of elements.	No, because set E and set F are composed of different elements.

● **Practice 21** Indicate whether the following sets are equivalent and/or equal: (81)

(a) $A = \{7, 8, 9\}$ and
$B = \{8, 9, 7\}$

(b) $X = \{4, 10\}$ and
$Y = \{x, y\}$

(c) $D = \{6, 12, 18\}$ and
$E = \{6, 16, 18\}$

(d) $F = \{3, 7, 11\}$ and
$G = \{4, 9\}$

(e) $I = \{a, x, y\}$ and
$J = \{1, 2, 3\}$

(f) $K = \{c, d, e\}$ and
$L = \{d, e, c\}$

See also Exercise 18.

One area of special interest related to the cardinality of sets is the *cardinality of the null set*. Remember earlier we said that an initial subset of natural numbers can be used to establish a one-to-one correspondence with any finite set except the null set. The problem in defining the cardinal number of the empty (or null) set comes from the fact that the cardinal number of a set is determined by establishing a one-to-one correspondence between the elements of the given set and the elements of an initial subset of the set of *natural* numbers: (82)

$$A = \{x, y, z\}$$
$$\updownarrow \; \updownarrow \; \updownarrow$$
$$N = \{1, 2, 3, 4, 5, \ldots\}$$

Therefore, $n(A) = 3$

Now, the null set contains no elements. The number that denotes "nothing" is zero; but zero is not an element of the natural numbers. However, the cardinal number of the empty set is defined as being zero.

$$n(\varnothing) = 0$$

Furthermore, if we take the set of natural numbers, $\{1, 2, 3, 4, 5, \ldots\}$, and include zero, the resulting set is the set of *whole numbers*, $\{0, 1, 2, 3, 4, 5, \ldots\}$. Thus, the numbers in the set of whole numbers can be used to find the cardinal number of any finite set, including the null set.

EXERCISES

Objective 1; Example 1

1. Tell whether the following sets are well defined. If the set is well defined, list the elements in correct set notation:

(a) The set of natural numbers between 8 and 12.
(b) The set of states in the U. S. whose names begin with the letter "W."
(c) The set of tall people in Atlanta.
(d) The set of whole numbers less than 6.
(e) The set of letters in the word "friend."
(f) The set of natural numbers greater than 2.
(g) The set of the elements that are in both set A, where $A = \{8, 12, 13\}$, and set B, where $B = \{1, 13, 47\}$.

Objective 2; Example 2

2. Tell whether the following sets are empty or contain elements. If the set contains elements, list the elements in both roster notation and set builder notation:
   (a) The set of natural numbers less than 1.
   (b) The set of the whole numbers less than 14.
   (c) The set of the letters in the word "today."
   (d) The set of the whole numbers between 3 and 4.
   (e) The set of people living in Seattle whose age is greater than 175 years.
   (f) The set of the natural numbers greater than or equal to 10.
   (g) The set of whole numbers less than or equal to zero.
   (h) The set of the whole numbers greater than 18.
   (i) The set of natural numbers greater than 4.
   (j) The set of states in the United States that begin with the letter "X."

Objective 3; Example 3

3. Given the following sets, list all proper subset and subset relationships:
   (a) $U = \{1, 2, 3, 4, 5\}$
   $A = \{2, 4\}$
   $B = \{2, 3, 4\}$
   $C = \{1, 2, 4\}$
   $D = \{1, 2, 3, 4\}$
   (b) $U = \{x, y, z\}$
   $A = \{x, z\}$
   $B = \{y, x\}$
   $C = \{y, x, z\}$
   (c) $U = \{18, 12, 47\}$
   $A = \{18\}$
   $B = \{47\}$
   $C = \{47, 18\}$
   (d) $U = \{a, x, b, y\}$
   $A = \{x, a\}$
   $B = \{x, a, y\}$
   $C = \{a, x, b, y\}$

Objective 4; Example 4

4. Describe three possible universal sets for which each of the following would be a proper subset:
   (a) The set of people born in Dallas since 1900.
   (b) The set of the number 6.
   (c) The set of the letters a, b, and d.
   (d) The set of restaurants in Boston.
   (e) The set of the month of January.

Objective 5; Example 5

5. Given the sets below, determine whether each set is a subset of the other set.
   (a) $A = \{3, 6, 9\}$ $B = \{1, 3, 6, 9\}$
   (b) $C = \{a, b\}$ $D = \{b, a\}$
   (c) $E = \{1, 2, 3\}$ $F = \{3, 2\}$
   (d) $G = \{1, 2, 3, \ldots, 10\}$ $H = \{6\}$
   (e) $I = \{2, 0\}$ $J = \{0\}$
   (f) $K = \{x, y, a, b, c\}$ $L = \{y, b, a, x, c\}$

Objective 6; Example 6

6. List all the subsets of each of the following sets. Evaluate the expression "2^n" to determine whether the correct number of subsets is listed:

(a) $\{1, 2, 3\}$ (b) $\{7\}$
(c) $\{a, b\}$ (d) $\{\square, \bigcirc, \triangle\}$
(e) $\{4, 5, 6, 7\}$ (f) {Mary, Nancy}
(g) {Wednesday, Friday} (h) $\{x\}$

Objective 7; Example 7

7. List the elements in the complement of each of the following sets (with reference to the given universal sets):
(a) Find A', if $A = \{2, 3\}$ and $U = \{2, 3, 8, 10\}$.
(b) Find C', if $C = \{3, 8, 10\}$ and $U = \{3, 8, 10\}$.
(c) Find B', if $B = \varnothing$ and $U = \{1, 2, 3\}$.
(d) Find D', if $D = \{a, b, c, d, f\}$ and $U = \{a, b, c, d, e, f, g, h\}$.
(e) Find E', if $E = \{5\}$ and $U = \{1, 2, 3, 4, 5\}$.

Objective 8; Example 8

8. List the elements in the union of the given sets:
(a) If $A = \{1, 2\}$ and $B = \{2, 3, 4\}$, find $A \cup B$.
(b) If $E = \{a\}$ and $F = \{b, c\}$, find $E \cup F$.
(c) If $A = \{0\}$, $B = \{1, 2, 3\}$, and $C = \{4, 5, 6\}$, find $(A \cup B) \cup C$.
(d) If $C = \varnothing$ and $D = \{a, b, c, d\}$, find $C \cup D$.
(e) If $A = \{3, 8\}$ and $B = \{3, 8\}$, find $A \cup B$.
(f) If $E = \{8, 10, 12\}$, $F = \{8, 12\}$, $G = \{1, 2, 12\}$, find $E \cup (F \cup G)$.
(g) If $B = \{\square, \triangle, ▯\}$ and $C = \{\triangle\}$, find $B \cup C$.

Objective 9; Example 9

9. List the elements in the intersection of the given sets:
(a) If $A = \{1, 7, 10\}$ and $B = \{7, 12, 16\}$, find $A \cap B$.
(b) If $C = \{3, 10\}$, $D = \{8, 10\}$ and $E = \{9, 10\}$, find $(C \cap D) \cap E$.
(c) If $A = \{4, 16\}$ and $B = \{6, 8, 10, 12, 14\}$, find $A \cap B$.
(d) If $A = \{1, 6\}$, $B = \{2\}$ and $C = \{1, 2, 6\}$ find $A \cap (B \cap C)$.
(e) If $B = \{1, 3, 5, 7\}$ and $C = \{0, 2, 4, 6\}$, find $B \cap C$.
(f) If $A = \{1, 10, 15\}$ and $B = \{1, 10, 15\}$, find $A \cap B$.
(g) If $E = \{1, 2, 3, 4, 5\}$, $F = \{1, 3, 5\}$, and $G = \{2, 4\}$ find $(E \cap F) \cap G$.

Objective 10; Example 10

10. List the elements in the indicated Cartesian product:
(a) If $A = \{1, 2, 3\}$ and $B = \{6, 10\}$, find $A \times B$ and $B \times A$.
(b) If $C = \{c, d\}$ and $D = \{c, d\}$, find $C \times D$ and $D \times C$.
(c) If $B = \{x, y\}$ and $C = \{1, 7\}$, find $B \times C$ and $C \times B$.
(d) If $A = \{3, 6\}$ and $B = \{1, 2, 3, 4, 5\}$, find $A \times B$ and $B \times A$.

Objective 11; Example 11

11. Construct a tree diagram and use it to list the ordered pairs in the indicated Cartesian products:
(a) If $A = \{a, b\}$ and $B = \{c, d, e\}$, find $A \times B$ and $B \times A$.

(b) If $E = \{1, 2\}$ and $F = \{1, 2\}$, find $E \times F$ and $F \times E$.
(c) If $B = \{3\}$ and $C = \{5\}$, find $B \times C$ and $C \times B$.
(d) If $A = \{1, a\}$ and $B = \{b, 7, 10\}$, find $A \times B$ and $B \times A$.

Objective 12; Example 12

12. Use an array to determine the ordered pairs of each indicated Cartesian product:
(a) If $A = \{0, 1\}$ and $B = \{1, 2, 3\}$, find $A \times B$ and $B \times A$.
(b) If $C = \{e\}$ and $D = \{f, g, h\}$, find $C \times D$ and $D \times C$.
(c) If $A = \{3, 4\}$ and $B = \{3, 4\}$, find $A \times B$ and $B \times A$.

Objective 13; Examples 13, 14, 15, 16

13. Use a separate Venn diagram to illustrate each indicated operation:
(a) If $U = \{1, 2, 3, 10, 12, 16, 18\}$, $A = \{3, 10, 12\}$, and $B = \{12, 16, 18\}$, find and diagram $A \cup B$, $A \cap B$, A', and B'.
(b) If $U = \{1, 2, 3, 4, 5\}$, $C = \{3\}$, and $D = \{2, 4, 5\}$, find and diagram $C \cap D$, $(C \cup D)'$, and $C' \cup D'$.
(c) If $U = \{a, b, c\}$, $A = \{a\}$, $B = \{b\}$, and $C = \{c\}$, find and diagram $A \cap B$, $(A \cap C)'$, and $(A \cup B) \cup C$.
(d) If $U = \{1, 3, 5, 7, 9, 11, 14\}$, $C = \{1, 5\}$, $D = \{3, 5, 7, 9\}$ and $E = \{5, 7, 9, 11\}$, find and diagram $C \cap (D \cap E)$, $(C \cap D \cap E)'$, $C \cap (D \cup E)$, $(C \cap D) \cup (C \cap E)$, $C \cup (D \cap E)$, and $(C \cup D) \cap (C \cup E)$.

Objective 14; Example 17

14. Use the definition of one-to-one correspondence to determine whether the following sets are in one-to-one correspondence.
(a) $\{1, 8, 10\}$ and $\{10, 1, 8\}$.
(b) $\{a, b, c\}$ and $\{4, 6, 8\}$.
(c) $\{3, 7\}$ and $\{3, 7, 8\}$.
(d) $\{x, y, z, b\}$ and $\{x, b\}$.
(e) $\{c, d\}$ and $\{f, g\}$.

Objective 15; Example 18

15. Diagram all possible ways of establishing a one-to-one correspondence between each of the two given sets:
(a) $\{3, 10\}$ and $\{a, b\}$.
(b) $\{8\}$ and $\{10\}$.
(c) $\{a, b, c\}$ and $\{x, y, z\}$.
(d) $\{1, 2\}$ and $\{1, 2\}$.

Objective 16; Example 19

16. Write the cardinal number of each of the given finite sets by placing the elements of each set in a one-to-one correspondence with the initial elements of the set of natural numbers?
(a) $A = \{a, b, c\}$.
(b) $B = \{6\}$.
(c) $C = \{1, 2, 3, 4, 5, 6, 7\}$.
(d) $D = \{1, 6, 10, a, b\}$.
(e) $E = \{0, 1, 2, 3\}$.
(f) $F = \{x, y\}$.

Objective 17; Example 20

17. Use the two parts of the description of equivalent sets to determine whether the following sets are equivalent.
(a) $C = \{s, t, u, v\}$ and $D = \{1, 6, 7, 10\}$.
(b) $A = \{6\}$ and $B = \{6, 10\}$.
(c) $F = \{p, q, r\}$ and $G = \{q, r, p\}$.
(d) $X = \{3, 6, 9, 12\}$ and $Y = \{3, 9, 12\}$.
(e) $I = \{\square\}$ and $J = \{j\}$.
(f) $M = \{1, 3\}$ and $N = \{3\}$.

Objective 18; Example 21

18. Classify the following sets as equivalent and/or equal:
(a) $\{a, b, c\}$ and $\{x, y, z\}$.
(b) $\{1, 2\}$ and $\{1, 2\}$.
(c) $\{a\}$ and $\{b\}$.
(d) $\{6, 7, 8\}$ and $\{6, 7, 9\}$.
(e) $\{1, 2, 3\}$ and $\{4, 5, 6, 7\}$.
(f) $\{1, a, 7\}$ and $\{7, 1, a\}$.

OBJECTIVES

After completing this chapter, you should be able to

A. Sets and well-defined sets

1 Given a series of verbal descriptions of sets, indicate whether the indicated sets are "well defined." If a set is well defined, list the elements in correct set notation. **(1–4)**

B. Set notation and membership

2 Given a verbal description of a set, indicate whether the set is empty or contains elements. If it contains elements, list the elements in roster and set-builder notation. **(5–8)**

C. Sets and subsets

3 Given a universal set and sets within the universe, list all proper subset and all subset relationships. **(9–14)**

4 Given a verbal description of a set (or symbolic representation of a set), describe three possible universal sets that contain the given set. **(15–17)**

5 Given two sets, determine whether each set is a subset of the other set. **(18–19)**

6 Given a set with five members or less, list all subsets of the given set. Determine whether the number of subsets listed is correct by evaluating the expression 2^n. (20–24)

D. Set operations

7 Given a universe and a set in the universe, list the elements in the complement of the set. (25–28)

8 Given two (or more) sets, list the elements in the union of the sets. (29–32)

9 Given two (or more) sets, list the elements in the intersection of the sets. (33–36)

E. Cartesian product of sets

10 Given two sets, list the elements in the indicated Cartesian product. (37–41)

11 Given two sets, use a tree diagram to list the ordered pairs in the Cartesian product of the two sets. (42–44)

12 Represent a Cartesian product of two sets in an array. (45–48)

F. Representing sets with Venn diagrams

13 Given two or more sets, use Venn diagrams to illustrate the set operations of complement, intersection, and union. (49–64)

G. Cardinality and equivalent sets

14 Given two sets, use the definition of one-to-one correspondence to determine if the sets are in one-to-one correspondence. (65–67)

15 Diagram all possible ways of establishing a one-to-one correspondence between two given sets. (68–70)

16 Given a finite set, write its cardinal number by placing the elements of the set in a one-to-one correspondence with the initial elements of the set of natural numbers. (71–75)

17 Use the two parts of the description of equivalent sets to determine whether given sets are equivalent. (76–78)

18 Given two sets, classify them as equivalent and/or equal. (79–82)

Logic and Computer Programming

Decisions—both conscious and unconscious—are a basic part of everyday life. You consider facts, make judgments as to the truth of these statements, and draw conclusions. By this process you are using *logic*, the science of reasoning. (1)

In mathematics, we try to prove certain statements about numbers, points, and lines. Often the results are applied in other fields such as engineering, physics, chemistry, architecture, economics, and business. Even if your main interest is not mathematics, the logic used to obtain these results can be valuable. (2)

A. STATEMENTS

Definition of statement

In our discussion, we will examine statements and consider the possible truth value of these statements. In order to make the discussion easier, we will often use letters such as p and q in place of statements, just as x and y are used in algebra in place of numbers. The word "statement" has special meaning in logic. A *statement* that is either true or false is called a proposition. The truth or falseness of a statement (or proposition) is its *truth value*. The following sentences are examples of statements: (3)

1. Houston is the capital of Texas.
2. All squares have four equal sides.
3. Jack is older than Bill.
4. $3 + 2 = 4$.

Commands such as "shut the door" and "go to the store" are *meaningless sentences*—not statements—because they are neither true nor false.

(4) Another type of meaningless mathematical sentence is exemplified by $x + 3 = 8$. However, this sentence would have a truth value if numbers were substituted for x. If x is replaced by 2, the sentence becomes $2 + 3 = 8$, and the truth value is false. What number or numbers could be substituted for x in order to make the statement true? If x is replaced with 5, the sentence becomes $5 + 3 = 8$ and it is a statement with truth value of true. The equation $x + 3 = 8$ is an *open sentence*, one which would have a truth value if certain numbers were substituted for the unknown. The set of possible substitutions is the *domain*; in this case the domain is the set of real numbers (whole numbers, negatives, fractions, and decimals), since substitution of any real number makes $x + 3 = 8$ a statement.

Definition of open sentence

◆ Objective 1: Classifying sentences

(5) **Example 1** For each sentence given below, answer the following questions:

1. Is it a meaningless sentence?
2. Is it an open sentence? If yes, state the possible domain.
3. Is it a statement? If yes, tell whether the statement is true or false.

   (a) San Francisco is north of Los Angeles.
   (b) $8 + 2 = 4$.
   (c) $y - 4 < 6$.
   (d) A triangle has three sides.

(a) Since the sentence has a truth value the answer to questions (1) and (2) is no. The answer to (3) is yes; it is a true statement.

(b) $8 + 2 = 4$ is not a meaningless sentence because it has a truth value; thus, the answers to questions (1) and (2) are both no. Question (3) is answered by yes, it is a statement which is false because $8 + 2$ is not equal to 4.

(c) $y - 4 < 6$ is a meaningless sentence because it has no truth value. The answer to (1) is yes; therefore (3) is no; it is not a statement. However, (2) is yes; it is an open sentence because substitution of any real number for y would make the sentence be either true or false.

(d) The sentence has a truth value; thus, no is the answer to both (1) and (2). The answer to (3) is yes, it is a true statement.

● **Practice 1** Answer questions 1–3 of Example 1 for the following sentences: (6)

(a) $2x + 3 = 5$.
(b) The opposite sides of a rectangle are equal.
(c) 5 is an even number.
(d) Rhode Island is the smallest state in the United States.

See also Exercises 1 and 2.

B. NEGATION

Objective 2: Negating statements

The *negation* of a statement is formed by making the statement false if it is a true statement, or by making the statement true if it is a false statement. The letters p and q are often used to represent logical statements. Symbolically, the negation of a statement p is written (7)

$$\sim p \text{ which is read "not } p\text{."}$$

Negation of a verbal statement may be accomplished either by prefixing the statement with the phrase "it is not true that" or by negating the verb itself.

◆ **Example 2** Negate the statements below by (1) prefixing the statement with "it is not true that" and by (2) negating the verb: (8)

(a) Houston is the capital of Texas.
(b) All triangles do not have four equal sides.
(c) Jack is older than Bill.
(d) $3 + 2 = 4$.

(a) It is not true that Houston is the capital of Texas.
Houston is not the capital of Texas.

(b) It is not true that all triangles do not have four equal sides.
All triangles have four equal sides.

(c) It is not true that Jack is older than Bill.
Jack is not older than Bill.

(d) It is not true that $3 + 2 = 4$.
$3 + 2 \neq 4$.

● **Practice 2** Negate the following statements (1) by prefixing the statement with "it is not true that" and (2) by negating the verb: (9)

(a) I will go to the ballgame Friday.
(b) Benjamin Franklin was President of the United States.
(c) James is not sixteen years old.

See also Exercise 7.

Statement (d) in Example 2, $(3 + 2 = 4)$, is a mathematical statement—one written using numbers and symbols. Symbols are often used to represent words: $+$ means "plus" or "the sum of"; $-$ means "minus" or "the difference of"; $=$ means "equals"; $<$ means "less than"; $>$ means "greater than." The symbols for equality ($=$) and inequality ($<$ and $>$) are negated as follows: $\neq$ indicates "not equal"; $\not<$ means "not less than"; and $\not>$ means "not greater than." Negation of symbols was also used in Chapter 1: $\subseteq$ for "subset" became $\nsubseteq$ for "not a subset"; $\subset$ for "proper subset" became $\not\subset$ for "not a proper subset"; $\in$ for "is an element of" became $\notin$ for "is not an element of." (10)

◆ **Example 3** Write the following statements using symbols and then write the negation of the statement using symbols: (11)

(a) The sum of three and five is less than two.
(b) Six subtracted from three is not equal to four.
(c) Set A union set B is a subset of set C.
(d) C is not a subset of B.

(a) $3 + 5 < 2$ (statement)
$3 + 5 \not< 2$ (negation)

(b) $3 - 6 \neq 4$ (statement)
$3 - 6 = 4$ (negation)

(c) $A \cup B \subseteq C$ (statement)
$A \cup B \nsubseteq C$ (negation)

(d) $C \nsubseteq B$ (statement)
$C \subseteq B$ (negation)

● **Practice 3** Write the following statements using symbols and then negate the statement using symbols: (12)

(a) The intersection of sets A and B is a subset of C.

(b) Four subtracted from the product of two and five is greater than eight.

(c) Three is an element of set X.

See also Exercise 11.

C. COMPOUND STATEMENTS

Compound statement

The previous examples of statements have all been *simple statements*; that is, each statement has had a single subject and predicate. A *compound statement* is a statement formed by combining two or more simple statements. The following are examples of compound statements: (13)

(Statement 1) It snowed yesterday *and* I went skiing.
(Statement 2) Bill will go to college *or* he will go to work in Atlanta.
(Statement 3) *If* Karl gets a scholarship, *then* he will go to college.

Each of these statements can be represented symbolically, using the letters p and q to replace the statements.

Statement 1

Statement: It snowed yesterday and I went skiing.

Statement Components:

p and q
(It snowed yesterday) and (I went skiing).

Statement Symbolized: $p \wedge q$ (read "p and q")
Type of Statement: conjunction
Connective of Statement: and

Statement 2

Statement: Bill will go to college or he will go to work in Atlanta.
Statement Components:

p or q
(Bill will go to college) or (he will go to work in Atlanta).

Statement Symbolized: $p \vee q$ (read "p or q")
Type of Statement: disjunction
Connective of Statement: or

Statement 3

Statement: If Karl gets a scholarship, then he will go to college.
Statement Components:

if p , then q
If (Karl gets a scholarship), then (he will go to college).

Statement Symbolized: $p \rightarrow q$ (read "if p, then q")
Type of Statement: Conditional
Connective of Statement: if, then

In summary, the relation of these compound statements can be seen in the table below:

Statement	*Connective*	*Statement symbolized*	*Type of statement*
p and q	and	$p \wedge q$	conjunction
p or q	or	$p \vee q$	disjunction
if p, then q	if, then	$p \longrightarrow q$	conditional

◆ **Example 4**

Objective 5: Writing statements of conjunction, disjunction, and conditional

(a) Let the statement for p be "the dice roll is seven" and the statement for q be "I win the game." Write the statements using the given p and q which form a conjunction, a disjunction, and a conditional: (14)

Answer

Conjunction: The dice roll is seven and I win the game.

$(p \wedge q)$

Disjunction: The dice roll is seven or I win the game.

$$(p \lor q)$$

Conditional: If the dice roll is seven, then I win the game.

$$(p \rightarrow q)$$

(b) Write a verbal description of the following logical statements, when p is "I am at least 18 years old" and q is "I am allowed to vote."

(1) $p \rightarrow q$ (2) $\sim q$ (3) $\sim q \rightarrow \sim p$ (4) $\sim p \lor q$

Answer

(1) If I am at least 18 years old, then I am allowed to vote.
(2) I am not allowed to vote.
(3) If I am not allowed to vote, then I am not at least 18 years old.
(4) I am not at least 18 years old or I am allowed to vote.

(c) Let the statement for p be "I did not pass the final exam" and the statement for q be "I passed the course." Write the following statements using p and q and the logical connectives:

(1) If I did not pass the course, then I did not pass the final exam.
(2) I passed the course and I passed the final exam.
(3) I did not pass the course or I did pass the final exam.

Answer

(1) $\sim q \rightarrow p$ (2) $q \land \sim p$ (3) $\sim q \lor \sim p$

Practice 4

(a) Given the following statements for "p" and "q," write the statements which result from applying: conjunction, disjunction, and conditional: (15)

p—Bill is older than Jack.
q—Bill will inherit the estate.

(b) Let the statement for p be "a rectangle has 4 right angles" and the statement for q be "a square is a rectangle." Write a verbal description of the following:

(1) $p \rightarrow q$ (2) $\sim p$ (3) $\sim p \lor \sim q$ (4) $\sim p \land q$

(c) Let the statement of p be $8 + 3 = 11$ and q be $8 < 11$. Write the following statements using p and q and the logical connectives:

(1) If $8 + 3 = 11$, then $8 < 11$.
(2) $8 \not< 11$.
(3) $8 + 3 \neq 11$ and $8 < 11$.

See also Exercises 4, 5, and 6.

It is not easy to justify why the truth values of statements are defined as they are. However, logic is like a game: the definitions are the rules of the game, formed in order to make the game reasonable, workable, and nonconflicting. (16)

Truth table

In order to summarize the possible truth values of a statement, a truth table is used. A *truth table* is a chart that shows (a) the possible truth values of the simple statements and (b) the truth values resulting from negation or from combining statements into conjunctions, disjunctions, or conditionals. (17)

From the meaning of negation we could expect that: if our statement p is true, then the negation of p ($\sim p$) is false; conversely, if our statement p is false, then the negation of p is true. There are only two options for statement p: it is either true or false. The truth table for negation would therefore be (18)

Truth Table for Negation

p	$\sim p$
T	F
F	T

Given two statements, p and q, there are four possible combinations of truth value illustrated below: (19)

p	q	*Meaning*
T	T	p is true, q is true
T	F	p is true, q is false
F	T	p is false, q is true
F	F	p is false, q is false

Truth values for conjunctions

The connective "and" is illustrated by the example "It snowed yesterday and I went skiing." This is a compound statement formed by *conjunction*. This connective ($\wedge$) can be compared to the set operation of "intersection." The intersection of sets A and B is the set of elements common to *both* sets. If an element does not belong to both of the sets, it does not belong to the intersection. Similarly, the conjunction of two statements is true if and only if *both* statements are true. If either one or both statements are false, the conjunction is false. The truth table to illustrate conjunction is given below: (20)

Truth Table for Conjunction

p	q	$p \wedge q$
T	T	T
T	F	F
F	T	F
F	F	F

Truth values for disjunctions

The *disjunction* connective "or" is analogous to the set operation of "union." The union of two sets A and B is the set of all elements belonging to *either* set A or set B or to both. If an element does not belong to either set A or set B, it does not belong to the union. Similarly, the disjunction of two statements is true if *either* (or both) statement is true. Only if both of the statements are false is the disjunction itself false. (21)

Truth Table for Disjunction

p	q	$p \vee q$
T	T	T
T	F	T
F	T	T
F	F	F

Truth values for conditionals

The conditional statement is represented symbolically by $p \rightarrow q$. The statement p is called the *premise* (or hypothesis) and the statement q is called the *conclusion*. Let us consider a verbal example to show the reasoning behind the designation of truth values for conditionals. Suppose I tell you "If I get the job today, I will buy you dinner tonight." Under what circumstances will you consider that I have made a true statement? There are four possibilities: (22)

1. I get the job today and I buy you dinner tonight.
   (Promise kept. Statement is true.)
2. I get the job today and I do not buy you dinner tonight.
   (Promise broken. Statement is false.)

The other two situations are more complex. I did not say *what* I would do if I *do not* get the job. Promises were made contingent only upon my getting the job. Thus, if I *do not* get the job, no promise can be broken, hence the statement must be true. Therefore, when the premise of a conditional is false, the conditional itself is considered to be true.

3. I do not get the job today and I buy you dinner.
   (No promise broken. Statement is true.)
4. I do not get the job today and I do not buy you dinner.
   (No promise broken. Statement is true.)

The following truth table summarizes these four cases: (23)

Truth Table for Conditionals

p	q	$p \rightarrow q$
T	T	T
T	F	F
F	T	T
F	F	T

When a conditional statement is always true, the conditional may be called an *implication*. The logical connective "$\rightarrow$" is often called the "symbol of implication." Given the implication $p \rightarrow q$, we say "p implies q."

Other compound statements can be formed using various combinations of negation, conjunction, disjunction, and conditional. Truth tables can then be constructed to illustrate the possible truth values. When truth tables are constructed, a column is made for each negation or other combination before the whole statement is considered. (24)

Objective 6: Truth table construction

◆ **Example 5** Construct truth tables for each of the following situations: (25)
(a) $\sim p \wedge q$ (b) $(p \vee q) \rightarrow \sim p$

(a) First make a column for all components p, q, $\sim p$, and the whole statement $\sim p \wedge q$. Then fill in the table values for p, q, and $\sim p$.

p	q	$\sim p$	$\sim p \wedge q$
T	T	F	
T	F	F	
F	T	T	
F	F	T	

The statements q and $\sim p$ may then be treated as an ordinary conjunction. Consider the $\sim p$ and q columns to determine the truth possibilities of $\sim p \wedge q$. Remember a conjunction is true only when *both* parts of the conjunction (in our case, $\sim p$ and q) are true.

p	q	$\sim p$	$\sim p \wedge q$
T	T	F	F
T	F	F	F
F	T	T	T
F	F	T	F

(b) To construct a truth table for $(p \vee q) \rightarrow \sim p$, first make columns for p, q, $(p \vee q)$, $\sim p$, and $[(p \vee q) \rightarrow \sim p]$. Fill in the truth values for p, q, $(p \vee q)$, and $\sim p$:

p	q	$(p \vee q)$	$\sim p$	$(p \vee q) \rightarrow \sim p$
T	T	T	F	
T	F	T	F	
F	T	T	T	
F	F	F	T	

Consider the $(p \vee q)$ and $\sim p$ columns to determine the truth value for each conditional. Remember that a conditional is true *except* when the first statement is true and the second is false.

p	q	$(p \vee q)$	$\sim p$	$(p \vee q) \rightarrow \sim p$
T	T	T	F	F
T	F	T	F	F
F	T	T	T	T
F	F	F	T	T

● **Practice 5** Construct truth tables for the following situations: (26)

(a) $p \wedge \sim q$ (b) $(\sim p \vee q) \rightarrow q$

(c) $(\sim p \wedge q) \rightarrow p$ (d) $(p \rightarrow q) \vee (\sim p \wedge q)$

See also Exercise 8.

D. EQUIVALENT STATEMENTS

Another logical statement closely related to the conditional is the *biconditional* statement. This compound statement is written $p \leftrightarrow q$ and is read p if and only if q (often abbreviated p iff q). The biconditional statement $p \leftrightarrow q$ actually consists of the conjunction of the two conditional statements, $(p \rightarrow q)$ and $(q \rightarrow p)$; that is, $p \leftrightarrow q$ is the same as $(p \rightarrow q) \wedge (q \rightarrow p)$. Consider the following truth table: (27)

p	q	$p \rightarrow q$	$q \rightarrow p$	$(p \rightarrow q) \wedge (q \rightarrow p)$	$p \leftrightarrow q$
T	T	T	T	T	T
T	F	F	T	F	F
F	T	T	F	F	F
F	F	T	T	T	T

We will be concerned with biconditional statements that are true. Comparing the left- and right-hand columns of the truth table, we see that $(p \leftrightarrow q)$ is true only when p and q have the same values. By definition, a biconditional statement $p \leftrightarrow q$ is an *equivalence* (p is equivalent to q) when the two basic statements p and q are either both (28)

Definitions of equivalence

true or else both false. (Subsequent tests for equivalence will thus omit the three middle columns above.)

By a similar definition, two (compound) statements are logically equivalent whenever the final truth tables for each are identical (on every row). This test for equivalence is particularly useful, as illustrated by the following example.

Objective 7: Truth tables and logically equivalent statements

◆ **Example 6** Determine whether or not the following statements are logically equivalent by using a truth table: (29)

(a) Construct a truth table for $\sim(p \wedge q)$ and $(\sim p \vee \sim q)$.

p	q	$p \wedge q$	$\sim(p \wedge q)$	$\sim p$	$\sim q$	$(\sim p \vee \sim q)$	$\sim(p \wedge q) \leftrightarrow (\sim p \vee \sim q)$
T	T	T	F	F	F	F	T
T	F	F	T	F	T	T	T
F	T	F	T	T	F	T	T
F	F	F	T	T	T	T	T

Notice the separate columns under $\sim(p \wedge q)$ and $(\sim p \vee \sim q)$ contain the same values on each row (both true or both false). The last column thus indicates the biconditional is true in each case; hence, the two statements are logically equivalent (for all possible combinations of p and q).

(b) Construct a truth table for $(p \vee q)$ and $(p \rightarrow \sim q)$.

p	q	$p \vee q$	$\sim q$	$p \rightarrow \sim q$	$(p \vee q) \leftrightarrow (p \rightarrow \sim q)$
T	T	T	F	F	F
T	F	T	T	T	T
F	T	T	F	T	T
F	F	F	T	T	F

Since the individual truth tables for $p \vee q$ and $p \rightarrow \sim q$ are not identical, the biconditional statement (last column) is not true in each case. Hence, the biconditional is not an equivalence.

● **Practice 6** Determine whether or not the following statements are equivalent by using a truth table: (30)

(a) $(p \rightarrow q)$ and $(p \vee \sim p)$.

(b) $\sim(p \vee q)$ and $(\sim p \wedge \sim q)$.

(c) $\sim(p \wedge q)$ and $(\sim p \wedge \sim q)$.

(d) $(p \rightarrow q)$ and $(\sim q \rightarrow \sim p)$.

See also Exercise 9.

E. DeMORGAN'S LAWS

The equivalences below have special significance and illustrate *DeMorgan's laws.* (31)

$\sim(p \wedge q) \leftrightarrow (\sim p \vee \sim q)$ shown equivalent in Example 6(a)
$\sim(p \vee q) \leftrightarrow (\sim p \wedge \sim q)$ shown equivalent in Practice 6(b)

The two biconditional statements show the relationship between conjunctions and disjunctions as follows:

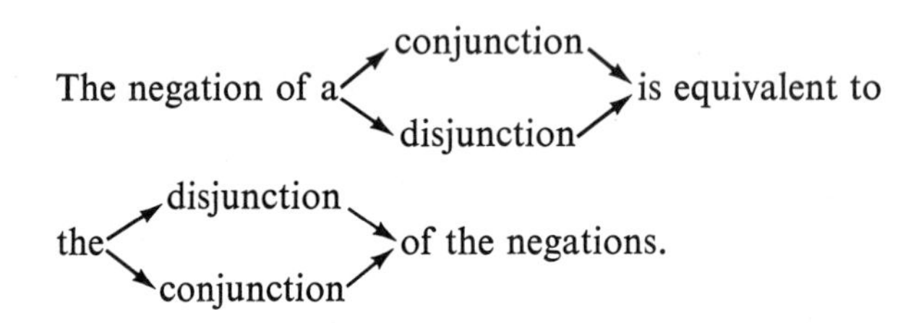

That is, the negation of the conjunction of two statements $[\sim(p \wedge q)]$ may be rewritten as the disjunction of the negation of each of the two statements $[\sim p \vee \sim q]$. Similarly the negation of the disjunction of two statements $[\sim(p \vee q)]$ may be rewritten as the conjunction of the negation of the two statements $[\sim p \wedge \sim q]$.

◆ **Example 7** Use DeMorgan's laws to write the logical equivalent of each given statement: (32)

(a) $2x - 4 \not> 0$ and $x \not> 0$.
(b) It is not true that $x \leq 5$ and $x \geq 0$.

(a) It is not true that $2x - 4 > 0$ or that $x > 0$.
(b) $x \not\leq 5$ or $x \not\geq 0$.

● **Practice 7** Use DeMorgan's laws to write the logical equivalent of each given statement: (33)

(a) $x \neq 6$ and $y \neq 4$.

(b) It is not true that $x > 6$ and that $x < -6$.

(c) $y < 6$ or $y \neq 2$.

(d) It is not true that Jack is older than Frank or that Jack is 25.

See also Exercise 12.

F. TAUTOLOGY

A statement which is always true regardless of the truth value of any of its parts is a *tautology*. To determine whether or not a given statement is a (34) tautology, construct a truth table. *If the statement is always true, it is a tautology.*

Definition of tautology

◆ **Example 8** Determine whether or not the following statements are tautologies: (35)

(a) $p \vee \sim p$ (b) $[(p \rightarrow q) \vee q] \rightarrow p$

(c) $[(p \vee q)] \leftrightarrow [(p \wedge \sim q) \vee q]$

(a) Construct a truth table.

Objective 9: Tautology and truth tables

p	$\sim p$	$p \vee \sim p$
T	F	T
F	T	T

Since the given statement is true in all cases, it is a tautology.

(b) Construct a truth table.

p	q	$p \rightarrow q$	$(p \rightarrow q) \vee q$	$[(p \rightarrow q) \vee q] \rightarrow p$
T	T	T	T	T
T	F	F	F	T
F	T	T	T	F ←
F	F	T	T	F ←

The statement is not a tautology because two parts of the statement are false.

(c) Construct a truth table for $[(p \vee q)] \leftrightarrow [(p \wedge \sim q) \vee q]$.

p	q	$(p \vee q)$	$\sim q$	$(p \wedge \sim q)$	$(p \wedge \sim q) \vee q$	$[(p \vee q)] \leftrightarrow [(p \wedge \sim q) \vee q]$
T	T	T	F	F	T	T
T	F	T	T	T	T	T
F	T	T	F	F	T	T
F	F	F	T	F	F	T

Since the biconditional statement is true in all cases, it is a tautology. Observe that the table also shows that the two statements, $(p \vee q)$ and $[(p \wedge \sim q) \vee q]$ are logically equivalent, since their individual truth tables contain identical values on each row. In general, any biconditional statement $p \leftrightarrow q$ is an equivalence if and only if $p \leftrightarrow q$ is also a tautology.* (Not every tautology is a biconditional, however, as illustrated in part (a) above.)

*Similarly, any conditional $p \rightarrow q$ is an implication if and only if $p \rightarrow q$ is a tautology.

● **Practice 8** Construct a truth table to determine whether or not the following statements are tautologies: (36)

(a) $\sim(p \wedge \sim p)$ (b) $\sim(\sim p)$

(c) $[(p \rightarrow q) \vee \sim p] \rightarrow \sim q$ (d) $[(p \rightarrow q) \wedge p] \rightarrow q$

See also Exercise 10.

G. CONVERSE, INVERSE, AND CONTRAPOSITIVE OF A CONDITIONAL

Objective 10: Writing the converse, inverse and contrapositive of a conditional

The conditional if p, then q symbolized by $p \rightarrow q$, has p as its premise (or hypothesis) and q as the conclusion. If the premise and the conclusion are interchanged, the resulting conditional is symbolized as $q \rightarrow p$ and is called the *converse*. If the premise and the conclusion of $p \rightarrow q$ are both negated, the conditional then becomes $\sim p \rightarrow \sim q$ and is called the *inverse*. A third related conditional may be obtained by both negating *and* interchanging the premise and the conclusion. The resulting conditional is symbolized by $\sim q \rightarrow \sim p$ and is called the *contrapositive*. A summary of these related conditionals is shown below: (37)

Statement	*Symbol*	
Conditional	$p \rightarrow q$	(Read "If p, then q.")
Converse	$q \rightarrow p$	(Read "If q, then p.")
Inverse	$\sim p \rightarrow \sim q$	(Read "If not p, then not q.")
Contrapositive	$\sim q \rightarrow \sim p$	(Read "If not q, then not p.")

◆ **Example 9** Write the converse, inverse, and contrapositive of the conditional, "If a quadrilateral is a rectangle, then the quadrilateral is a parallelogram." (38)

Converse: If a quadrilateral is a parallelogram, then the quadrilateral is a rectangle.

Inverse: If a quadrilateral is not a rectangle, then the quadrilateral is not a parallelogram.

Contrapositive: If a quadrilateral is not a parallelogram, then the quadrilateral is not a rectangle.

● **Practice 9** Write the converse, inverse, and contrapositive of the following conditionals: (39)

(a) If a number is positive, then a number is greater than 0.
(b) If it is raining, then we do not go to the ballgame.

See also Exercise 13.

From Example 9 and your practice, you may have observed that even when a conditional statement is true, its inverse and converse may not be (40) true. However, its contrapositive will be true. In fact, a conditional and its contrapositive are logically equivalent, since they have identical truth tables. These relationships are demonstrated by the following truth table:

		Conditional	*Converse*	*Inverse*	*Contra-positive*	
p	q	$p \to q$	$q \to p$	$\sim p \to \sim q$	$\sim q \to \sim p$	$[p \to q] \leftrightarrow [\sim q \to \sim p]$
T	T	T	T	T	T	T
T	F	F	T	T	F	T
F	T	T	F	F	T	T
F	F	T	T	T	T	T

You will observe that a conditional statement and its contrapositive are logically equivalent because they have the same truth values in all cases. However, this correspondence of truth values does not hold true for the conditional and either the inverse or the converse.

H. FLOWCHARTS

Objective 11: Flowcharts

A Venn diagram is a useful tool in depicting sets and logical statements. A flowchart is another tool which mathematicians find useful for depicting information with symbols. A *flowchart* is a diagram which gives a series of instructions. Below are some examples of simple flowcharts. Notice that each consists of three things: input, instructions, and output. In flowcharts, the circular areas denote input and output. Rectangles are used for instructions. (41)

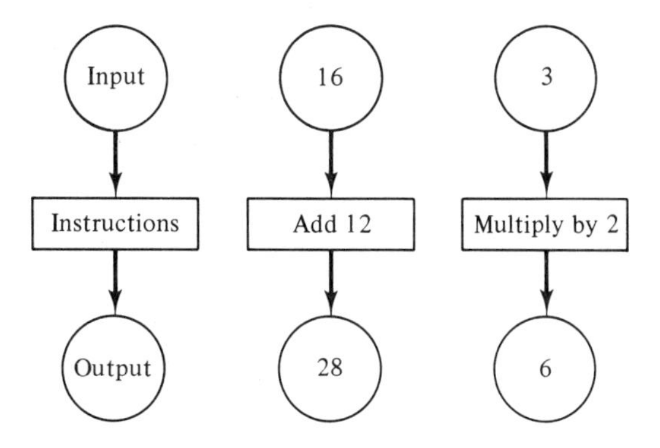

Some flowcharts have a series of instructions to follow as illustrated by the following: (42)

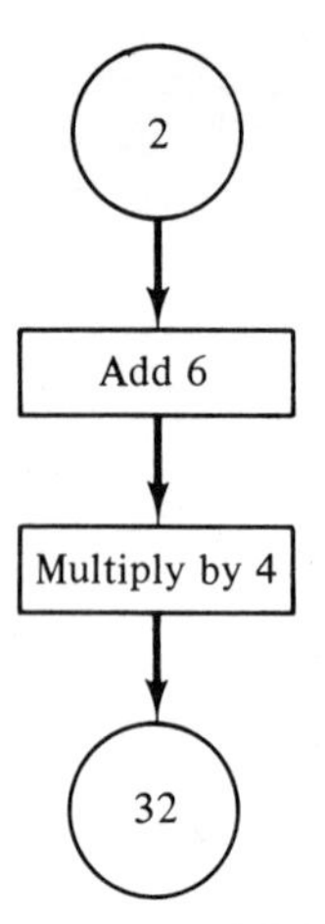

To give general instructions, a variable (letter used to represent a number) is used. The flowchart at left gives the general procedure. The flowchart on the right illustrates an application with a specific number substituted for the variable *a*. **(43)**

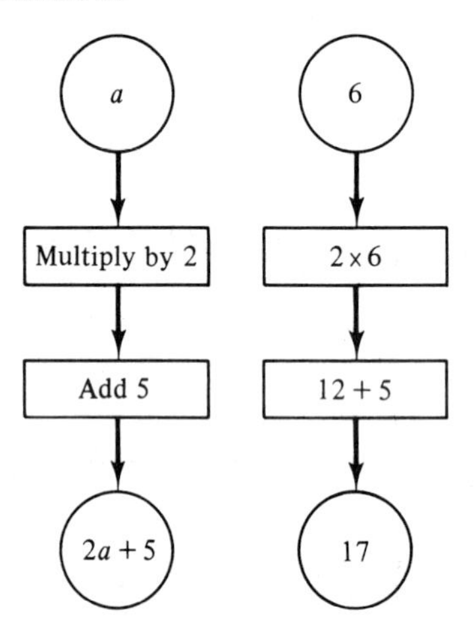

There are occasions when a decision must be made. In these cases, you must answer "yes" or "no" to a question and follow the arrows indicated by that answer. The "decision" box is indicated by a diamond shape. **(44)**

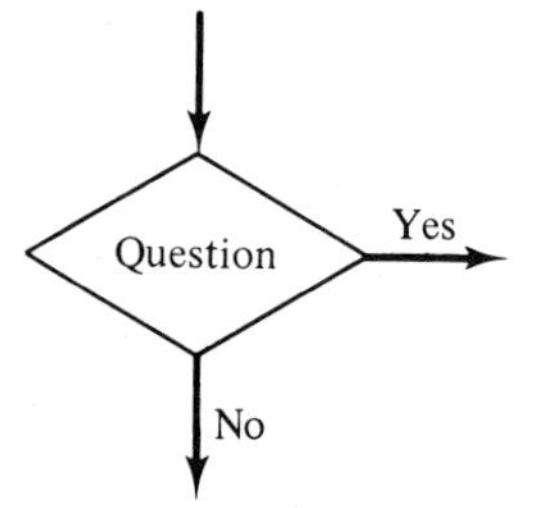

A flowchart with a decision box is shown on the left in the following diagram. In the flowchart on the right, these instructions are applied when the input is 5. **(45)**

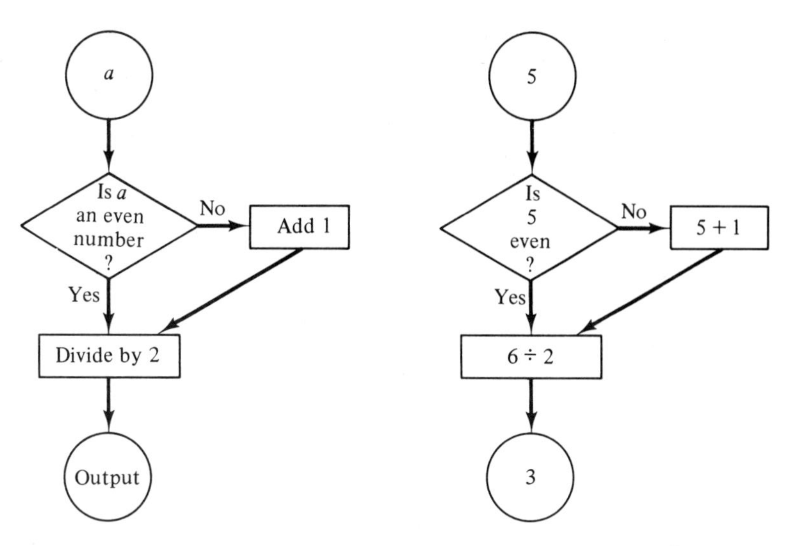

Flowcharts may even be used to give instructions for nonmathematical situations as, for example, procedure for operation of a copying machine: **(46)**

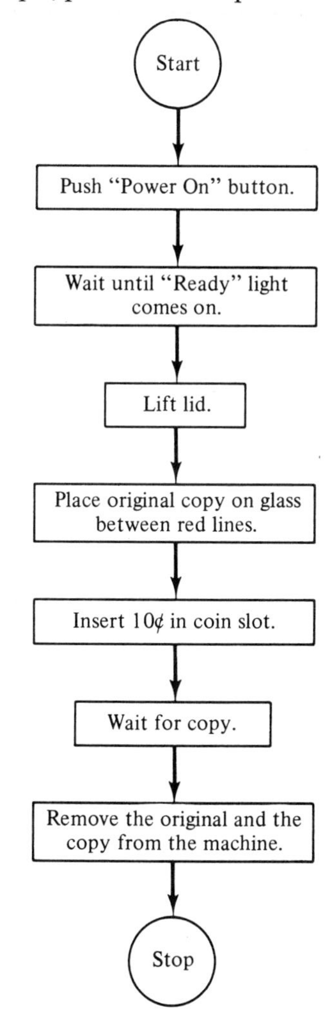

The following example further illustrates the use of flowcharts to organize a mathematical process. (47)

◆ **Example 10** Construct a flowchart to represent the following processes: (48)

(a) Given a number a, find $3a + 4$.
(b) Given a number a, find $3(a + 4)$.

(a) Notice that the number must first be multiplied by 3 and that this product will then have the number 4 added to it.

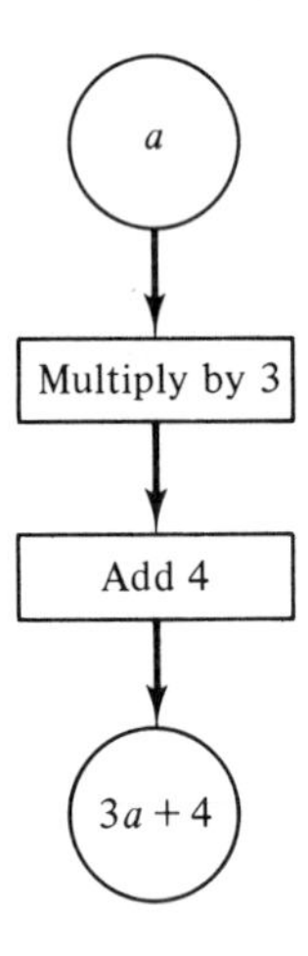

(b) The parentheses indicate that the number is first added to 4, and then this sum is multiplied by 3.

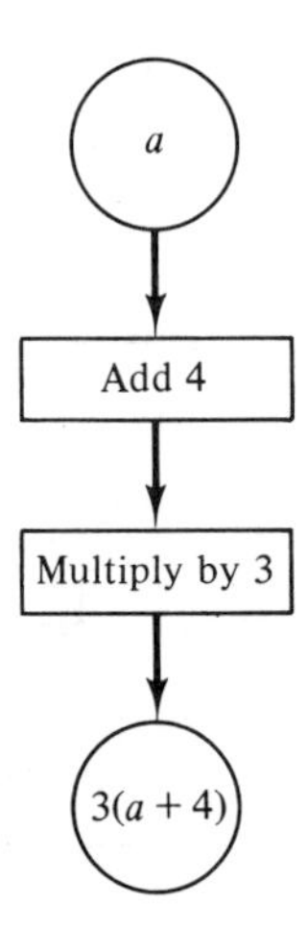

● **Practice 10** Construct a flowchart demonstrating the following procedures: (49)

(a) For a number n, find $2n - 8$.

(b) For a number n, find $n^2 + 5$.

(c) For a natural number n: if it is odd, add one and divide by 2; if it is even, divide by 2.

(d) Construct a flowchart to illustrate the process of operating a coin-operated washing machine.

See also Exercises 17 and 18.

I. COMPUTERS AND LOGICAL REASONING

Throughout history, man's ability to use tools has distinguished him from other animals. One kind of tool he has devised to aid him in problem solving is the computer. (50)

Objective 12: Analog computer and digital computer

There are two kinds of computers. The *analog computer* is designed to use the principle of measurement. The *digital computer* uses the concept of counting. To illustrate the difference, consider how the two types of computers could be used to find the sum $2 + 4$. A simple analog computer would be two rulers. Place the second ruler at the end of the "2" mark on the first ruler. (51)

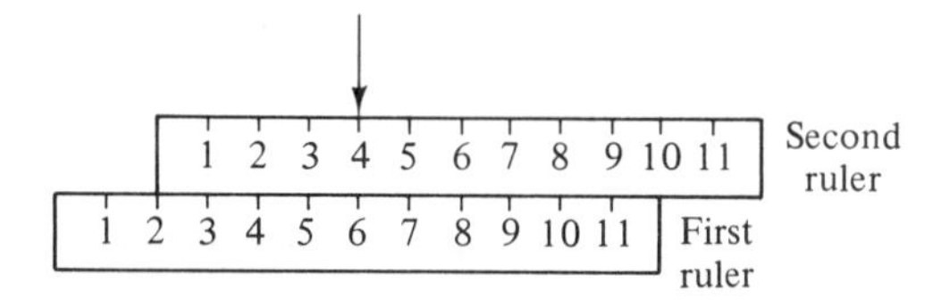

Find which numeral on the first ruler will correspond to the numeral 4 on the second ruler. The sum is 6. A digital approach to this problem $(2 + 4)$ is the use of poker chips. Place 2 chips on the table, then place 4 chips on the table. By counting, we obtain a sum of 6 as before.

Some common analog computers include the following: a car speedometer in which the distance that the needle moves indicates speed; a thermometer in which the distance the liquid rises indicates temperature; a watch in which the distance the hands move indicates the time; and a slide rule which works on a principle similar to the two rulers. Common digital computers are a cash register, an abacus, poker chips, and tally marks. Modern electronic computers are also digital computers. (52)

Objective 13: Computer operations

No matter which kind of computer is used, its operation must involve three steps: (53)

1. *Input:* information which includes data and instructions specifying what to do with the data.
2. *Processing:* performing the indicated operations on the given data.
3. *Output:* the final results of processing.

You may recognize these three steps as the parts of a flowchart. Flowcharts are used to organize the information given to the computer. The set of instructions "fed" to the computer is called the *program.* The person who prepares these instructions in a form that the computer will "understand" is called the *programmer.* A computer operates on the basis of logic, so the programmer must be able to reason logically. He/she first makes a flowchart to illustrate the step-by-step process and then writes the instructions in the "language" the computer understands. The input may be in the form of punched cards or magnetic tape or may be typed directly into the computer.

Objective 14: Computer control processing

The computer's central processing system consists of three connected divisions which operate together to process the information: the control unit, the memory unit, and the arithmetic unit. The *control unit* acts as a supervisor to see that the instructions are carried out as required and transfers data from one unit to another. The *memory unit* stores the data and the results. The *arithmetic unit* performs the required operations. The output of the computer may be in the form of punch cards, magnetic tape, printed information or even an audio response. (54)

◆ **Example 11**

(a) State whether the following are examples of an analog computer or a digital computer: (55)
(1) bathroom scale (2) abacus (3) slide rule

(b) Match each of the following to the word which best describes the unit:

(1) control unit	i. calculator
(2) arithmetic unit	ii. supervisor
(3) memory unit	iii. storer

(a) (1) The needle on a *bathroom scale* measures the distance from 0 to the weight; thus a bathroom scale is an example of an analog computer.
(2) The beads on an *abacus* are counted; hence, an abacus is a digital computer.
(3) A *slide rule* is an analog computer because its computations are based upon measurement.

(b) (1) ii (2) i (3) iii

● **Practice 11**

(a) Identify each of the following as an analog computer or a digital computer: (56)

(1) yardstick (2) watch (3) football scoreboard

(b) Match the following phrases with the appropriate word:

(1) The unit which transfers information from one unit to another unit.	i. input
(2) Instructions on operations to be performed.	ii. output
(3) The unit which performs calculations.	iii. control unit
(4) The unit which stores data.	iv. processing
(5) The final results of computer operations.	v. memory unit
(6) Performance by the computer of required operations.	vi. arithmetic unit

See also Exercises 19, 20 and 21.

J. BASIC LANGUAGE

Definition of machine language

Information given to a computer must be in a form the computer will understand. The *machine language* or specific code used by the machine must be learned by the programmer. To illustrate the logical reasoning involved in the use of a computer language, we will briefly consider the BASIC language. BASIC consists of a list of statements identified by number and written in specific code. The computer follows these instructions in numerical order. Several types of BASIC instructions including arithmetic statements, control statements, and input/output statements shall now be considered. (57)

Arithmetic statements instruct the computer to perform certain computations. Capital letters are used to represent variables. For example, the following symbols are used in BASIC: (58)

+	addition	=	equals
—	subtraction	>	greater than
*	multiplication	<	less than
/	division	>=	greater than or equal to
↑	raise base number to a power	<=	less than or equal to
( )	grouping		

The following expression illustrates a typical BASIC statement:

```
60  LET X = (2 * A + B)
```

In this example, the statement number is 60 and the computer is instructed to find $X = 2A$ plus B. The word "LET" indicates to the computer that the value of a letter is to be defined. In order for the computer to perform this calculation the values for A and B must have been input data or have been calculated by the computer in an earlier step in the program.

Another typical BASIC statement might be as follows:

```
30 LET Y = 5 * A - B/3
```

Remember that multiplication and division must be performed first unless grouped otherwise; then addition and subtraction are performed. Thus, the foregoing statement means "let $Y = 5A - \frac{B}{3}$." By contrast, the statement

```
40 LET Y = (5 * A - B)/3
```

means "let $Y = \frac{5A - B}{3}$."

◆ **Example 12**

Objective 17: BASIC and algebraic notation

(a) Translate the following expressions into normal algebraic notation: (59)

(1) $A/2 + 8 * B$ (2) $X = (A \uparrow 3 + B)/2$

(3) $W = (A + 7 * C) \uparrow 3$

(b) Translate the following algebraic expressions into BASIC language:

(1) $C = \frac{(A + B)}{2}$ (2) $Y = X^2 - Z^2$ (3) $M = \left(\frac{A}{2} - \frac{B}{3}\right)^3$

(a) (1) $\frac{A}{2} + 8B$ (2) $X = \frac{A^3 + B}{2}$ (3) $W = (A + 7C)^3$

(b) (1) $C = (A + B)/2$ (2) $Y = X \uparrow 2 - Z \uparrow 2$

(3) $M = (A/2 - B/3) \uparrow 3$

● **Practice 12**

(a) Rewrite the following BASIC expressions in normal algebraic notation: (60)

(1) $X = 3 * (5 * A - B \uparrow 2)$ (2) $Y = A/2 + B/2 + C/2$

(3) $C = 4 * (A \uparrow 2 - B \uparrow 2)$ (4) $A = (4 * X + 2 * Y)/6$

(5) $D = A \uparrow 2/(B \uparrow 2 - C \uparrow 2)$

(b) Translate the following algebraic expressions into legal BASIC language:

(1) $X = (2B)^3$ (2) $A = 3B + \frac{C}{2}$ (3) $X = (Y - 4)^2$

See also Exercises 22, 23, and 24

There are several kinds of BASIC statements which a computer will recognize. Each is preceded by a number which is the number of the (61) statement. *Arithmetic statements* instruct the computer to carry out some computation. LET directs the computer to evaluate a variable:

```
60  LET X = (A - B)/2
```

Control statements determine or change the sequence in which instructions are followed. IF . . . THEN and GO TO statements redirect steps of a program, or END would terminate it. For example,

```
40  IF A > 21 THEN 60
```

instructs the computer in statement 40 to go to statement 60 only if A is greater than 21. Otherwise, the computer proceeds directly to statement 41.

The control statements MAX and MIN instruct the computer to choose between two variables enclosed in parentheses. For instance,

```
70  LET C = (A MAX B)
```

instructs the computer to let C equal the larger of A and B. The operator MAX makes this a control statement despite the use of LET, which usually indicates an arithmetic statement.

Input statements feed in data, which will be processed by the computer. An input statement will be one such as **(62)**

```
70  INPUT A1, B1
```

When the computer comes to this statement, it turns on the terminal and prints a question mark. The programmer types in the numerical values assigned to the first variable (A1). The computer records this data and prints a second question mark. The programmer then types in the numerical value of the second variable (B1). The process is continued until each variable in the list has been assigned a numerical value.

The *output statement* instructs the computer to print data. Words which are to be printed are enclosed in quotation marks. For instance, the statement **(63)**

```
50  PRINT A, C
```

tells the computer to print the numerical value of A and then the numerical value of C. Similarly, the statement

```
80  PRINT A; "SQUARED IS"; A↑2
```

instructs the computer to print the numerical value of A, then the words "squared is" and then the numerical value of A^2. Also, the statement

```
95  PRINT X = (C MAX D)
```

would cause the computer to print out the larger value of the two variables C or D.

◆ **Example 13**

Objective 15: BASIC instructions

(a) Classify each of the following BASIC statements as an arithmetic instruction, a control instruction or an input/output statement: (64)

Statement	*Classification*
(1) 20 PRINT A↑2	(1) Input/output
(2) 34 LET B = 3 * C − 2	(2) Arithmetic
(3) 45 IF X <= 2 THEN 70	(3) Control

Objective 16: Legal BASIC statements

(b) State whether or not the following are legal BASIC statements:

Statement	*Analysis*
(1) 32 LET X = 2A + B	(1) Not legal, because 2 times A should be written 2 * A.
(2) 40 LET C = A↑3	(2) Legal
(3) 56 D = 3 * A/4	(3) Not legal, as the instruction statement LET has been omitted.
(4) 90 LET Y = (C MAX 4)	(4) Legal.

Objective 18: Compute execution

(c) Explain how the computer would execute each statement:

(1) 80 IF A >= 6 THEN 82
(2) 30 PRINT X = (A↑2 MAX B)

Explanation

(1) The computer would compare A to 6. If A is greater than or equal to 6, the computer will go to statement 82 and follow the instructions given there. If A is less than 6, the computer will go to the next statement (81).
(2) The computer will find A^2, compare this number with B, and print the larger numerical value.

Objective 19: Writing BASIC statements

(d) Write BASIC statement 40 which tells the computer to go to statement 42 if A squared is less than 16:

```
40 IF A↑2 < 16 THEN 42
```

● **Practice 13**

(a) Match the following: (65)

(1) 32 LET X = (2 MIN Y) i. control statement
(2) 17 LET Y = (2 * C) + 3 ii. input/output
(3) 72 PRINT Y iii. arithmetic statement

(b) State whether or not the following is a legal BASIC statement:
  (1) `111  LET C = B/3`
  (2) `80  IF X < 2 THEN C = 3`
  (3) `70  LET N = 3M`

(c) Explain how the computer would execute each statement:
  (1) `70  LET X = (2 * A MAX B)`
  (2) `40  PRINT A; "CUBED IS"; A↑3`

(d) Write BASIC statement 45 which would instruct the computer to find twice B, compare this result with D, and print the smaller of the two numbers.

See also Exercises 25, 26, 27, 28, and 29.

EXERCISES

Objective 1; Example 1

1. Indicate whether or not the following sentences are statements:
   (a) Norway has a greater population than Sweden.
   (b) Read this book.
   (c) $3x + 7 > 14$.
   (d) Alvin is 5′ 10″ tall.
   (e) $4 + 7 = 11$.

Objective 1; Example 1

2. For each sentence given below, answer the following questions:
   Is it a meaningless sentence?
   Is it an open sentence? If yes, state the possible domain.
   Is it a statement? If yes, tell whether the statement is true or false.
   (a) Atlanta is the capital of Georgia.
   (b) $7 + 3 = 9$.
   (c) $x + 6 = 18$.
   (d) Division by zero is permitted in mathematics.
   (e) $3 < 5$

Objective 4: Logical symbols and words

3. Write a verbal description of each of the following logical symbols:
   (a) $\sim p \vee q$ (b) $p \rightarrow \sim q$ (c) $p \wedge q$
   (d) $\sim q \rightarrow \sim p$ (e) $p \leftrightarrow q$ (d) $(p \rightarrow q) \wedge \sim q$

Objective 4; Example 4

4. Write each of the following symbolically:
   (a) p or q (b) not q implies p
   (c) p and not q (d) If p and q, then q.
   (e) If not q, then not p or q. (f) not p if and only if q

Objective 5; Example 4

5. Let p be the statement "It is raining" and q be the statement "I will go shopping." Write symbolically:
   (a) If I go shopping, then it is not raining.
   (b) It is not raining and I will not go shoppoing.
   (c) I will go shopping or it is raining.

Objective 5; Example 4

6. Write a verbal description of the following logical statements, where p is "I like dogs" and q is "I like cats."
(a) $p \vee q$ (b) $p \rightarrow \sim q$ (c) $\sim(\sim p)$
(d) $\sim p \wedge \sim q$

Objective 2; Example 2

7. Negate the following statements (1) by prefixing the statement with "it is not true that" and (2) by negating the verb:
(a) I am older than my sister.
(b) George is not taking history this semester.
(c) The number 3 is a whole number.
(d) Seven times six is equal to fifty.

Objective 6; Example 5

8. Construct a truth table for each of the following:
(a) $\sim p \rightarrow q$ (b) $(p \vee q) \rightarrow q$
(c) $\sim p \rightarrow (p \wedge \sim q)$ (d) $(p \wedge q) \rightarrow (p \vee q)$

Objective 7; Example 6

9. Determine whether or not the given statements are equivalent by using a truth table:
(a) $\sim p \rightarrow q$ and $\sim q \rightarrow p$ (b) $\sim(p \rightarrow q)$ and $(p \wedge \sim q)$

Objective 9; Example 8

10. Use a truth table to show whether or not the given statement is a tautology:
(a) $\sim p \vee (q \rightarrow p)$ (b) $p \rightarrow (\sim p \rightarrow q)$
(c) $[(p \rightarrow q) \wedge \sim q] \rightarrow \sim p$

Objective 3; Example 3

11. Write the following statements using symbols and then write the negation of each using symbols:
(a) The union of set A and set D is the set of elements 4, 7, 10, 12, and 14.
(b) Six times three is less than 22.
(c) Set D is a subset of set E.
(d) Eleven is not an element of set F.

Objective 8; Example 7

12. Use DeMorgan's laws to write each logical equivalent:
(a) It is not snowing and I will not go skiing.
(b) It is not true that $x \geq 2$ or that $x \leq 6$.
(c) $x < 6$ or $x \neq 5$.
(d) It is not true that $2y < 5$ and that $y = 4$.

Objective 10; Example 9

13. Write the converse, inverse and contrapositive of the statement:
(a) If two lines are parallel, they are not perpendicular.
(b) If I join a car pool, my travel expenses will be lower.

Item (40)

14. Construct a truth table for a conditional statement, its converse, inverse, and contrapositive. From your truth table, which of the three is equivalent to the original conditional?

Objective 11: Flowcharts

15. Find the input in the following flowchart.

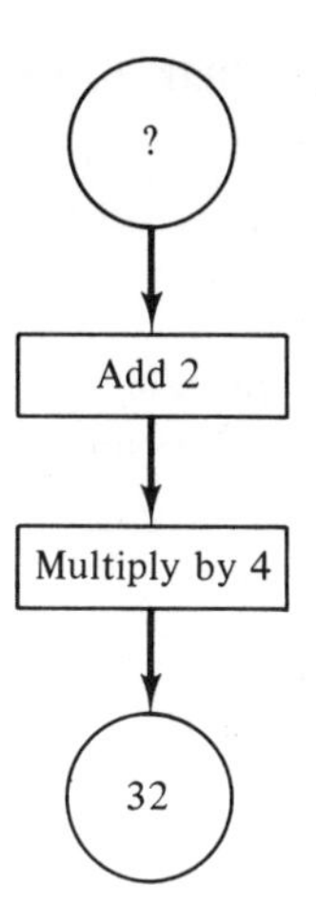

Objective 11: Flowcharts

16. Use the flowchart below for (a) $n = 5$ and (b) $n = 6$:

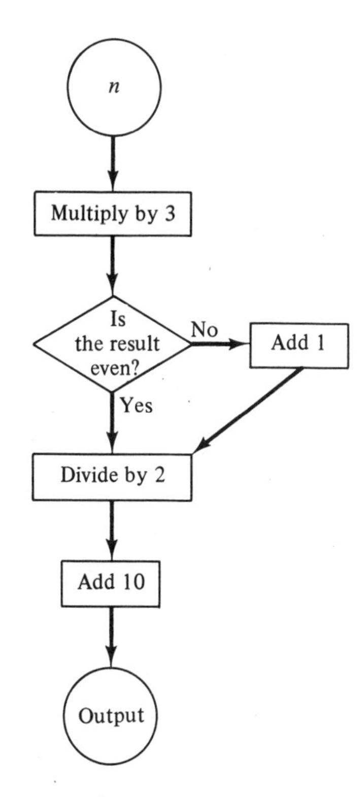

Objective 11; Example 10

17. Construct a flowchart to show the process for (a) $3(2n^2 - 1)$ and (b) $3(2n - 1)^2$.

Objective 11; Example 10

18. Construct a flowchart that shows the process to determine whether or not a number is an element of the intersection of two sets.

Objective 12; Example 11

19. What type of computer is each of the following? Give a reason for your answer and name two other examples of the same type of computer:

(a) an hour glass (b) automobile mileage gauge

Objective 13; Example 11

20. Describe each of the three steps in computer operation.

Objective 14; Example 11

21. Name the three divisions of the central processing unit of a computer and describe the function of each division.

Objective 16; Example 12

22. Name the arithmetic operation or relation indicated by each of the following BASIC symbols:

(a) $*$ (b) $<=$ (c) $\neq$ (d) $-$ (e) $+$
(f) /

Objective 17; Example 12

23. Translate the following expressions to algebraic notation:

(a) $Y = 2 * A + B \uparrow 3$ (b) $X = (4 * A - C)/3$
(c) $X <= 2 * B \uparrow 3$ (d) $X = 3 * B + 7 * D/4$
(e) $X = 3 * (A \uparrow 3 + B \uparrow 2)$ (f) $Y = B/2 + 3 * C$

Objective 17; Example 12

24. Rewrite the following algebraic notations in BASIC language:

(a) $X = 2Y + Z$ (b) $Y = \frac{(A^2 + B^2)}{4}$
(c) $X = 3(X + Y)^2$ (d) $Y = \frac{3}{A} + (7B)^3$
(e) $X \leq 4C^2 - (2B)^3$

Objective 15; Example 13

25. Classify the following BASIC statements as arithmetic statements, control statements or input/output statements:

(a) 56 PRINT A = (B MIN C)
(b) 70 IF X < 5 THEN 50
(c) 50 LET X = C ↑ 2 + B/2

Objective 16; Example 13

26. State whether or not each of the following statements is a legal BASIC notation:

(a) 20 LET Y = 3A − 4 (b) 42 IF X ≤ Y THEN C = Y
(c) 32 IF Y < −2 (d) 37 PRINT Y = 3 * X

Objective 18; Example 13

27. Explain how each statement would be executed by the computer:

(a) 60 LET X = (A ↑ 3 + B ↑ 3) (b) 70 PRINT X = (20 MAX C)
(c) 62 IF B >= 7 THEN 65 (d) 75 LET Y = 3 * X ↑ 2

Objective 19; Example 13

28. Write a BASIC statement 17 which would tell the computer to jump to statement 24 provided the numerical value of X is equal to 0.

Objective 19; Example 13

29. Write a BASIC statement 80 which would instruct the computer to print the words "the area is" and then to print the numerical value of the variable A^2.

OBJECTIVES

After completing this chapter, you should be able to

A. Statements

1 Given a sentence, respond to each of the following questions about the sentence: **(1–6)**
(a) Is it a meaningless sentence?
(b) Is it an open sentence? If yes, state the possible domain.
(c) Is it a statement? If yes, tell whether the statement is true or false.

B. Negation

2 Negate verbal statements in two ways: **(7–9)**
(a) by prefixing the statement with "it is not true that."
(b) by negating the verb.

3 Given a verbal statement describing mathematical symbols, write the mathematical symbols and negate the symbol statement. **(10–12)**

C. Compound statements

4 Translate between logical symbols and words. **(13)**

5 Given a statement for "p" and a statement for "q," write the statement that results from applying conjunction, disjunction, and conditional. **(13–15)**

6 Construct truth tables for two or more statements which involve negation, conditional, disjunction, and conjunction. **(17–26)**

D. Equivalent statements

7 Show by constructing truth tables whether given statements are logically equivalent. **(27–30)**

E. DeMorgan's laws

8 Use DeMorgan's laws to write the logical equivalent of a given statement. **(31–33)**

F. Tautology

9 Determine whether or not a given statement is a tautology by constructing a truth table. **(34–36)**

G. Converse, inverse, and contrapositive of a conditional

10 Given a conditional, write its converse, inverse, and contrapositive. **(37–40)**

H. Flowcharts

11 Construct flowchart diagrams using the correct symbols to illustrate mathematical and everyday processes. **(41–49)**

I. Computers and logical reasoning

12 Describe the difference between analog computers and digital computers. Give three examples of each group of computers. **(50–52)**

13 Name and describe the three basic steps in the operation of a computer. **(53)**

14 Name and describe the function of the three divisions of the central processing unit of a computer. **(54)**

J. BASIC *language*

15 Given a list of BASIC instructions, classify them as: arithmetic statements, control statements, or input/output statements. **(58, 61–62)**

16 Given a list of computer symbols or instructions, identify those that are legal BASIC statements. **(64)**

17 Translate expressions written using the BASIC arithmetic symbols into the normal algebraic notation; or translate algebraic expressions into legal BASIC language. **(58–59)**

18 Explain how a given BASIC statement would be executed by the computer if it appeared as part of a program. **(58–65)**

19 Given a series of statements, write a BASIC statement that will produce the indicated result. **(58–65)**

Part Two

Structure of the Number System and Other Systems

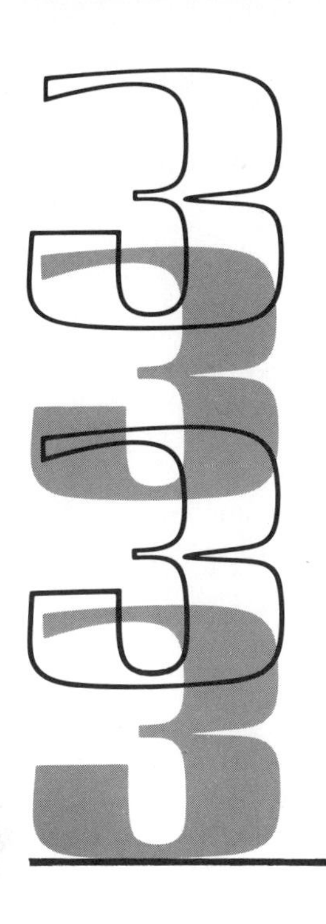

The Natural Numbers and the Whole Numbers

Mathematics, like language, is said to be the invention of man, rather than having been created by God. Unlike language, however, mathematics is quite consistent and predictable. It operates on a well-organized structure wherein each more intricate part is based on less complicated aspects already developed. (1)

The topics of this chapter and the next five will trace the development of our number system from the most basic aspects through the more intricate ones. Hopefully, this process will enable you to gain both an understanding of this development and an appreciation of the soundness of the structure.

Objective 1
Mathematical system

Any *mathematical system* (or number system) must meet four basic requirements: it must consist of (1) a set of *elements* (such as the numbers you use to count with); (2) one or more *operations* (such as addition, subtraction, multiplication, and division); and (3) some *properties* (axioms or rules) that define how the elements and operations will behave. These facts are sometimes stated as follows: A mathematical system is a set of elements together with one or more operations defined on the set. Also, (4) one or more *relations* (principally the relation of *equality*, such as $2 + 3 = 5$) are considered basic to any number system. (2)

We will start with simple number systems and add on more complete systems until we derive the real number system, which is often informally called "the" number system. This process utilizes the set concepts introduced in Chapter 1. At each step, we will study the elements, operations, properties, and relations that compose the system. We will also learn to solve problems that involve the numbers in each system.

THE NATURAL NUMBERS

The *natural* or *counting numbers* were previously defined as the set of numbers one associates with objects when they are enumerated or counted: (3)

$$N = \{1, 2, 3, 4, 5, 6, \ldots\}$$

Notice that the set of natural numbers is an infinite set, having a smallest element but not a greatest element. We may illustrate the set of natural numbers as follows:

Natural
numbers
1, 2, 3, 4, . . .

Before defining the operations of addition and multiplication, we need to consider certain characteristics of an operation. The operations in number systems combine only two elements (numbers) at one time and hence are called *binary* operations. (Similarly, recall that the set operations of union and intersection were binary operations because they combined two sets at one time.) Furthermore, when combining any two elements of a set produces a result that is also a member of the set, then the set is said to be *closed* under the operation. Stated differently, we say that the set has *closure* under the operation. (4)

Objective 2: Closure

A closely related concept is that an operation must produce *one and only one result*. That is, we say one result must *exist* (must be an element of the set) and there must be a *unique* (only one) result. Both existence and uniqueness are also generally considered to be essential aspects of a closure property. That is, a *closure property* holds when, for any two elements in a set, there *exists* in the set a *unique* element that is the result of the operation.

Indeed, this concept of closure is of such significance that it is considered a major property of a number system. Thus, the first two properties of the system of natural numbers are closure under (or "for") addition and closure for multiplication, which may be stated as follows: (5)

Objectives 5 and 6: Closure for addition and for multiplication

If a and b are each a natural number, then $(a + b)$ is a unique natural number. Similarly, $(a \times b)$ is a unique natural number.

To illustrate, consider the examples $4 + 3 = 7$ and $4 \times 3 = 12$:

4	4 is a natural number	4
$+3$	3 is a natural number	$\times$ 3
7	← is a natural number →	12

Thus, we have verified (*not* proved) that the system of natural numbers is closed under the operations of addition and multiplication. In each case, the operation produces one and only one result. That is, each result exists in the set and is unique.*

A. ADDITION IN THE NATURAL NUMBERS

We are now ready to define addition and multiplication. This is done in terms of sets, as introduced in Chapter I. The definition of addition utilizes the concepts of union and the cardinal number of two disjoint $(A \cap B = \varnothing)$ sets. (6)

Definition of addition

In general, if $n(A) = a$ and $n(B) = b$ (where $A \cap B = \varnothing$), then the binary operation of addition, $a + b$, produces one and only one element, c, which is the $n(A \cup B)$. That is, (7)

$$n(A) + n(B) = n(A \cup B)$$

or

$$a + b = c$$

The definition of addition can be described in words: if the number of elements in one set is added to the number of elements in a second, disjoint set, the result is the same as the number of elements in the union of the two sets. (8)

◆ **Example 1** Let $A = \{g, h\}$ and $B = \{w, x, y, z\}$. (9)

Objective 3: Addition

Then $A \cup B = \{g, h, w, x, y, z\}$.

Also, $n(A) = 2$, the $n(B) = 4$, and $n(A \cup B) = 6$.

Thus
$$n(A) + n(B) = n(A \cup B)$$

indicates
$$2 + 4 = 6$$

If $a + b = c$, recall that a and b are each called an *addend* and c is called the *sum*.

● **Practice 1** To be sure you understand Example 1 and the definition of addition, (10)

(a) Work through Example 1 using these sets: $D = \{c, r, z\}$ and $E = \{b, k, q, t, w\}$.

(b) Let $n(D \cup E) = f$ and give the general definition of addition in terms of these sets.

See also Exercises 2 and 3.

*By contrast, the operation 5–8 would *not* produce a result that is an element of the original set of natural numbers.

B. MULTIPLICATION IN THE NATURAL NUMBERS

Although multiplication is often introduced to young students as repeated addition, we will define multiplication as a binary operation (involving two elements), using the set concepts of the Cartesian product, $(A \times B)$, and cardinal number, as introduced in Chapter I. (11)

Definition of multiplication

In general, if $n(A) = a$ and $n(B) = b$, then the binary operation of multiplication, $a \times b$, produces one and only one element, c, which is the $n(A \times B)$. That is, (12)

$$n(A) \times n(B) = n(A \times B)$$

or

$$a \times b = c$$

The definition of multiplication can be stated in words: if the number of elements in one set is multiplied times the number of elements in a second set, the result is the same as the number of elements (the number of ordered pairs) in the Cartesian product of the two sets. (13)

◆ **Example 2** Again, let $A = \{g, h\}$ and $B = \{w, x, y, z\}$. (14)

Objective 4: Multiplication

Then $A \times B = \{(g, w), (g, x), (g, y), (g, z), (h, w)\ (h, x), (h, y), (h, z)\}$

Also, $n(A) = 2$, the $n(B) = 4$, and $n(A \times B) = 8$.

Thus,

$$n(A) \times n(B) = n(A \times B)$$

indicates

$$2 \times 4 = 8$$

If $a \times b = c$, recall that a and b are each called a *factor* and c is called the *product*.

● **Practice 2** To be certain you understand Example 2 and the definition of multiplication, (15)

(a) Work through Example 2 using these sets: $T = \{g, h, k\}$ and $V = \{r, s\}$.

(b) Let $n(T \times V) = u$ and write a general definition of multiplication using sets T and V.

See also Exercises 4 and 5.

Our development of the system of natural numbers is progressing. We now have a set of elements, represented by $N = \{1, 2, 3, 4, 5, 6, \ldots\}$. We have defined two binary operations, addition and multiplication. We have also included the equals relation and the properties of closure under addition and multiplication. A final step is to describe other properties the elements obey when the operations are performed. (16)

C. COMMUTATIVE AND ASSOCIATIVE PROPERTIES OF ADDITION

Objective 7: Commutative property of addition

(17) Addition was defined as a binary operation involving two elements. The *commutative property of addition* states that the *order* in which these *two* addends are taken does not affect the sum. In other words, for any two natural numbers a and b,

$$a + b = b + a$$

For example,

$$5 + 4 = 4 + 5$$
$$9 = 9$$

(18) If addition is binary (two elements), can three (or more) elements be added? Yes. Consider the addition procedure

$$a + b + c = (a + b) + c$$

By the closure property, $(a + b)$ is a unique number; thus, combining that unique number with c is simply another binary operation. (The addition of more than three elements would be explained similarly.)

Objective 8: Associative property of addition

(19) We now know that three elements may be added. The *associative property of addition* applies to the *grouping* of these *three* elements. In general terms, for any natural numbers a, b, and c,

$$(a + b) + c = a + (b + c)$$

For example,

$$(8 + 4) + 6 = 8 + (4 + 6)$$
$$12 \quad + 6 = 8 + \quad 10$$
$$18 = 18$$

(20) Long before students learn the name of the associative property, they are taught to look for groups that total 10 (or 100 or 1,000) and to use these groupings to speed addition. Think about the difference between $(37 + 75) + 25$ and $37 + (75 + 25)$.

(21) To review: the *commutative property* illustrates that, for any *two* elements, the *order* (left to right) changes as the elements are commuted. Mentally identify "commutative" with "commute"; like people commuting to work, the elements in the commutative property are moved back and forth. The *associative property*, on the other hand, describes how *three* elements may be *grouped* (or associated) together. The left-to-right order of the three elements does not change—only the grouping within parentheses changes.

(22) Recall that, by the closure property, $(3 + 7)$ would denote a unique number (element). Thus, given $a + b = b + a$, the a or the b could represent the single element $(3 + 7)$, for example. Thus, at first glance, it may not be obvious whether an equality illustrates the com-

Objective 13: Applying addition properties

mutative or the associative property. Therefore, the following example identifies the parts that make each illustration commutative or associative, starting with the basic form of each property:

◆ **Example 3**

(23)

Commutative, +	*Associative,* +
(a) $5 + 7 = 7 + 5$	$(5 + 7) + 2 = 5 + (7 + 2)$
$a + b = b + a$	$(a + b) + c = a + (b + c)$
(b) $2 + (5 + 6) = 2 + (6 + 5)$	$2 + (5 + 6) = (2 + 5) + 6$
$(a + b) = (b + a)$	$a + (b + c) = (a + b) + c$
Order changes.	Grouping changes; order (left to right) does not change.
(c) $(4 + 8) + 3 = 3 + (4 + 8)$	$(4 + 8) + 3 = 4 + (8 + 3)$
$a + b = b + a$	$(a + b) + c = a + (b + c)$
Order changes.	Grouping changes, not order.

(24)

(d) $(4 + 3) + (5 + 2) = (5 + 2) + (4 + 3)$

$a + b = b + a$

The expressions in parentheses above are each treated as a single element and thus the commutative property is illustrated. There are two ways this example could be treated to illustrate the associative property. If we let $P = (4 + 3)$ and $Q = (5 + 2)$, we could say either

$$(4 + 3) + Q = 4 + (3 + Q)$$
$$(4 + 3) + (5 + 2) = 4 + [3 + (5 + 2)]$$

or

$$P + (5 + 2) = (P + 5) + 2$$
$$(4 + 3) + (5 + 2) = [(4 + 3) + 5] + 2$$

● **Practice 3** Indicate which addition property of the natural numbers is illustrated by each of the following equalities: (25)

(a) $8 + 3 = 3 + 8$

(b) $(5 + 4) + 1 = 5 + (4 + 1)$

(c) $(9 + 5) + 2 = (5 + 9) + 2$

(d) $4 + (7 + 6) = (7 + 6) + 4$

(e) $(1 + 2) + (3 + 4) = (3 + 4) + (1 + 2)$

(f) $(2 + 4) + (3 + 5) = [(2 + 4) + 3] + 5$

(g) $2 + (6 + 3) = (2 + 6) + 3$

(h) $2 + (6 + 3) = (6 + 3) + 2$

(i) $2 + (6 + 3) = 2 + (3 + 6)$

(j) $(2 + 5) + (6 + 3) = (6 + 3) + (2 + 5)$

(k) $(2 + 5) + (6 + 3) = 2 + [5 + (6 + 3)]$

(l) $9 + 7 = 7 + 9$

(m) $(4 + 1) + (2 + 6) = [(4 + 1) + 2] + 6$

(n) $(8 + 3) + (1 + 5) = 8 + [3 + (1 + 5)]$

See also Exercise 14.

In order to add several numbers efficiently, we often by intuition make use of both the commutative and associative properties. The following example analyzes the use of these properties. Here we apply properties in succession to the left-hand member (the left side) of each equality until we obtain the right-hand member (the right side), and name the property which permits each step: (26)

◆ **Example 4** (27)

(a) Show that $(8 + 5) + 2 = 5 + (8 + 2)$.

$$(8 + 5) + 2 = (5 + 8) + 2 \quad \text{Commutative, } +$$
$$= 5 + (8 + 2) \quad \text{Associative, } +$$

(b) Show that $(2 + 5) + 4 = (2 + 4) + 5$: (28)

First solution		*or*	*Second solution*	
$(2 + 5) + 4 = 2 + (5 + 4)$	A, +	*or*	$(2 + 5) + 4 = 4 + (2 + 5)$	C, +
$= 2 + (4 + 5)$	C, +		$= (4 + 2) + 5$	A, +
$= (2 + 4) + 5$	A, +		$= (2 + 4) + 5$	C, +

Either solution here can be obtained in three steps and either is perfectly acceptable. The first solution may be easier to follow, however, since the first step keeps the number 2 in the left-hand position, where it will also be in the third step.

(c) Show that $(4 + 3) + (6 + 7) = (4 + 6) + (3 + 7)$. (29)

To do this, first think of $(4 + 3)$ as a single element; that is, as $P + (6 + 7)$:

$$(4 + 3) + (6 + 7) = [(4 + 3) + 6] + 7 \quad \text{Associative, } +$$
$$= [4 + (3 + 6)] + 7 \quad \text{Associative, } +$$
$$= [4 + (6 + 3)] + 7 \quad \text{Commutative, } +$$
$$= [(4 + 6) + 3] + 7 \quad \text{Associative, } +$$
$$= (4 + 6) + (3 + 7) \quad \text{Associative, } +$$

Practice 4

(1) Name the property which permits each step: **(30)**

(a) $(4 + 8) + 6 = (8 + 4) + 6$
$= 8 + (4 + 6)$

(b) $(4 + 8) + 6 = 6 + (4 + 8)$
$= (6 + 4) + 8$

(c) $(3 + 9) + 7 = 3 + (9 + 7)$
$= 3 + (7 + 9)$
$= (3 + 7) + 9$

(d) $4 + (1 + 8) = (1 + 8) + 4$
(See note.)
$= 1 + (8 + 4)$
$= 1 + (4 + 8)$

(e) $(2 + 6) + (8 + 4) = [(2 + 6) + 8] + 4$
$= [2 + (6 + 8)] + 4$
$= [2 + (8 + 6)] + 4$
$= [(2 + 8) + 6] + 4$
$= (2 + 8) + (6 + 4)$

Note: This solution keeps the element 1 in the first position throughout.

(2) Apply successive number properties to each left-hand member of the equalities below in order to obtain each right-hand member. Show each step and name each property: **(31)**

(f) $(1 + 7) + 9 = 1 + (9 + 7)$

(g) $(3 + 5) + 2 = 3 + (2 + 5)$

(h) $(7 + 9) + 3 = (7 + 3) + 9$

(i) $4 + (2 + 6) = 2 + (4 + 6)$

(j) $(8 + 1) + (2 + 9) = (8 + 2) + (1 + 9)$

See also Exercise 15.

D. COMMUTATIVE AND ASSOCIATIVE PROPERTIES OF MULTIPLICATION

Objective 9: Commutative property of multiplication

(32) There are also corresponding commutative and associative properties which apply to the operation of multiplication with natural numbers. The *commutative property of multiplication* states that the *order* in which *two* factors is taken does not affect the product. Thus, for any two natural numbers a and b,

$$a \times b = b \times a^*$$

For example,

$$4 \times 5 = 5 \times 4$$
$$20 = 20$$

*Recall that $a \times b$ may also be indicated by ab, by $a \cdot b$, by $a(b)$, or by $(a)b$.

Objective 10: Associative property of multiplication

As in the case of addition, the associative property of multiplication pertains to the *grouping* of *three* elements. The *associative property of multiplication* implies that, when three elements are to be multiplied, the product will be the same regardless of which two are grouped together to be multiplied first. Thus, for any natural numbers a, b, and c, we can say that (33)

$$(a \times b) \times c = a \times (b \times c)$$

or $$(a \cdot b)c = a(b \cdot c)$$

illustrated by $$(2 \cdot 3)4 = 2(3 \cdot 4)$$

$$(6)\ 4 = 2\,(12)$$

$$24 = 24$$

Objective 13: Applying multiplication properties

The following example identifies the parts that make each illustration commutative or associative, starting with the basic form of each property. (34)

◆ **Example 5**

Commutative, $\times$	*Associative,* $\times$
(a) $5 \cdot 7 = 7 \cdot 5$	$(5 \cdot 7)2 = 5(7 \cdot 2)$
$a \cdot b = b \cdot a$	$(a \cdot b)c = a(b \cdot c)$
(b) $2(5 \cdot 6) = 2(6 \cdot 5)$	$2(5 \cdot 6) = (2 \cdot 5)6$
$a \cdot b = b \cdot a$	$a(b \cdot c) = (a \cdot b)c$
Order changes.	Grouping changes; order (left to right) does not change.
(c) $(4 \cdot 8)3 = 3(4 \cdot 8)$	$(4 \cdot 8)3 = 4(8 \cdot 3)$
$(\ a\ )b = b(\ a\ )$	$(a \cdot b)c = a(b \cdot c)$
Order changes.	Grouping changes.

(35)

(d) $(4 \cdot 3) \times (5 \cdot 2) = (5 \cdot 2) \times (4 \cdot 3)$ (36)

$a \times b = b \times a$

If we let $P = (4 \cdot 3)$ and $Q = (5 \cdot 2)$, the associative property could be illustrated in two ways:

$$(4 \cdot 3) \times Q = 4 \times (3 \cdot Q)$$

$$(4 \cdot 3) \times (5 \cdot 2) = 4 \times [3 \times (5 \cdot 2)]$$

or

$$P \times (5 \cdot 2) = (P \cdot 5) \times 2$$

$$(4 \cdot 3) \times (5 \cdot 2) = [(4 \cdot 3) \times 5] \times 2$$

● **Practice 5** Indicate which multiplication property of the natural numbers is illustrated by each of the following equalities: (37)

(a) $8 \cdot 3 = 3 \cdot 8$ (b) $(5 \cdot 4)1 = 5(4 \cdot 1)$

(c) $(9 \cdot 5)2 = (5 \cdot 9)2$ (d) $4(7 \cdot 2) = (7 \cdot 2)4$

(e) $(3 \cdot 5)2 = (5 \cdot 3)2$ (f) $(3 \cdot 5)2 = 2(3 \cdot 5)$

(g) $(3 \cdot 5)2 = 3(5 \cdot 2)$ (h) $4 \times 9 = 9 \times 4$

(i) $4(25 \cdot 5) = (4 \cdot 25)5$

See also Exercise 14.

As in the case of addition, we may often combine the commutative and associative properties in order to multiply more efficiently. In the following example, we apply properties of the natural numbers in individual steps until we obtain each given right-hand member. (38)

◆ **Example 6** (39)

(a) Show that $(5 \cdot 8)2 = 8(5 \cdot 2)$.

$$(5 \cdot 8)2 = (8 \cdot 5)2 \quad \text{Commutative, } \times$$
$$= 8(5 \cdot 2) \quad \text{Associative, } \times$$

(b) Show that $(4 \cdot 7)25 = (4 \cdot 25)7$

$$(4 \cdot 7)25 = 4(7 \cdot 25) \quad \text{A, } \times \qquad \textit{or} \qquad (4 \cdot 7)25 = 25(4 \cdot 7) \quad C, \times$$
$$= 4(25 \cdot 7) \quad \text{C, } \times \qquad\qquad = (25 \cdot 4)7 \quad A, \times$$
$$= (4 \cdot 25)7 \quad \text{A, } \times \qquad\qquad = (4 \cdot 25)7 \quad C, \times$$

● **Practice 6** Apply properties of the natural numbers to each left-hand member of the given equalities to obtain each right-hand member. Show each step and name each property. (40)

(a) $(5 \cdot 7)2 = 7(5 \cdot 2)$ (b) $3(4 \cdot 8) = (3 \cdot 8)4$

(c) $(6 \cdot 9)5 = 9(6 \cdot 5)$ (d) $(8 \cdot 7)5 = (8 \cdot 5)7$

(e) $6(2 \cdot 7) = 2(6 \cdot 7)$ (f) $25(3 \cdot 4) = 3(25 \cdot 4)$

See also Exercise 15.

E. DISTRIBUTIVE PROPERTY OF MULTIPLICATION OVER ADDITION

In developing properties for our number system, we have found that the operations of addition and multiplication each have closure, commutative, and associative properties. The *distributive property* (of multiplication over addition) next describes how the operations of addition and multipli- (41)

Objective 11: Distributive property

cation are related to each other. In general terms, for any natural numbers a, b, and c,

$$a(b+c) = a \cdot b + a \cdot c$$

illustrated by

$$2(3+5) = 2 \cdot 3 + 2 \cdot 5$$
$$2\ (8) = 6 + 10$$
$$16 = 16$$

Objective 13: Applying distributive property

The following example identifies the commutative or distributive parts of some equalities that appear to be similar. (42)

◆ **Example 7**

Commutative, + or ×	*Distributive, × over +*
(a) $5(8+7) = 5(7+8)$	$5(8+7) = 5 \cdot 8 + 5 \cdot 7$
$a+b = b+a$	$a(b+c) = ab + ac$
Order changes.	Basic distributive form.
(b) $3(4+9) = (4+9)3$	$3(4+9) = 3 \cdot 4 + 3 \cdot 9$
$a(\ b\ ) = (\ b\ )a$	$a(b+c) = ab + ac$

● **Practice 7** Name the property of natural numbers that each equality illustrates: (43)

(a) $6(9+2) = 6 \cdot 9 + 6 \cdot 2$ (b) $4(5+3) = 4(3+5)$

(c) $5(7+4) = (7+4)5$ (d) $8(7+3) = (7+3)8$

(e) $8(7+3) = 8 \cdot 7 + 8 \cdot 3$ (f) $8(7+3) = 8(3+7)$

See also Exercises 14 and 15.

Suppose $8 + 4 \times 2$ were given without any parentheses or other symbols to indicate which operation should be performed first. If it were performed $(8+4) \times 2 = (12) \times 2 = 24$, we have a different solution from $8 + (4 \times 2) = 8 + 8 = 16$. In order to avoid confusion, mathematicians have agreed on the following *order of operations: multiplication* (and then *division*) should be performed first from left to right, and then any *addition* (and finally *subtraction*) should be performed from left to right. This is not a basic property of numbers that could be proved; it is simply an agreed order accepted for convenience. Thus, $8 + 4 \times 2 = 8 + (4 \times 2) = 8 + 8 = 16$. (44)

Order of operations

F. MULTIPLICATIVE IDENTITY

The natural number 1 has a particular characteristic for multiplication that applies *only* for the number 1. That is, for any natural number a, (45)

Objective 12: Identity element for multiplication

$$a \times 1 = 1 \times a = a$$

For example, $4 \times 1 = 1 \times 4 = 4$

This property of 1 is called the *multiplicative identity property*; the number 1 itself is called the *identity element for multiplication*, or the *multiplicative identity*.

◆ **Example 8** Show that $2(25 + 1) = 2 \cdot 25 + 2$. (Show each step and name each property.) (46)

$2(25 + 1) = 2 \cdot 25 + 2 \cdot 1$	Distributive, $\times$ over $+$
$= 2 \cdot 25 + 2$	Identity, $\times$

● **Practice 8** Apply number properties to the left-hand member of each equality to obtain each right-hand member. Show each step and name each property. (47)

(a) $5(8 + 1) = 5 \cdot 8 + 5$ (b) $6(5 + 1) = 6 \cdot 5 + 6$

See also Exercises 14 and 15.

Our development of the system of natural numbers is now complete. Although the system permits a number of very helpful computations, there is no natural number we can use to tell us what $5 - 8$ or $5 \div 3$ would equal.* We thus say the natural numbers are *not* closed under the operations of subtraction or division. (Observe that one contradictory example—called a counterexample—is sufficient to prove that a property does *not* hold.) Furthermore, although addition and multiplication generally have corresponding properties, there is no identity element for addition in the system of natural numbers. (48)

Page 434 contains a summary chart listing the four basic requirements for *any* mathematical system, along with characteristics of the *set* of natural numbers which fulfill those requirements, thereby comprising the *system* of natural numbers. Similar information is also given for the subsequent systems we will study, along with references to the paragraphs in each chapter where each topic is discussed. (49)

EXERCISES

Objective 1; Item (2)

1. (a) Name the four requirements that must be met in order to have a mathematical (number) system.

*Limited definitions for subtraction and division, which are suitable for natural numbers, are given in the unit on whole numbers.

Item (49)

(b) Identify the characteristics of the set of natural numbers that fulfill each requirement of part (a).

Objective 2; Item (4)

(c) Tell what is meant by a "binary" operation.

Objective 3; Example 1

2. (a) Let $K = \{g, e, t\}$ and $P = \{b, u, s, y\}$. Use the definition of addition and the given sets to find the sum of $3 + 4$.
   (b) If $n(K \cup P) = r$, use sets K and P to express the definition of addition in general terms for any natural numbers.

Example 1

3. (a) Let $M = \{s, t, e, a, d, y\}$ and $R = \{o, n\}$. Use the definition of addition and the given sets to find the sum $6 + 2$.
   (b) If $n(M \cup R) = z$, use sets M and R to express the definition of addition in general terms for any natural numbers.

Objective 4; Example 2

4. (a) Let $L = \{g, o\}$ and $P = \{t, r, 4\}$. Use the definition of multiplication and the given sets to find the product 2×3.
   (b) If $n(L \times P) = q$, use sets L and P to express the definition of multiplication in general terms for any natural numbers.

Example 2

5. (a) Let $U = \{b, 2, y\}$ and $T = \{1, r, 3, z\}$. Use the definition of multiplication and the given sets to find the product 3×4.
   (b) If $n(U \times T) = v$, use sets U and T to express the definition of multiplication in general terms for any natural numbers.

Number properties: Objectives 5–12; Item (5)

6. Use natural numbers x, y, and z (if needed) to express the following properties of the natural numbers:
   (a) Closure property of addition. Verify the property using 2 and 5.
   (b) Closure property of multiplication. Verify this property using 7 and 3.

Item (17)

   (c) Commutative property of addition. Verify this property using 8 and 5.

Item (19)

   (d) Associative property of addition. Verify this property using 7, 6, and 4.

Item (32)

   (e) Commutative property of multiplication. Verify the property using 7 and 9.

Item (33)

   (f) Associative property of multiplication. Verify the property using 3, 5, and 2.

Item (41)

   (g) Distributive property of multiplication over addition. Verify this property using 4, 3, and 7.

Item (45)

   (h) Multiplicative property of 1. Verify the property using 6.

Objective 4

7. Example 2 (item 14) gave sets $A = \{g, h\}$ and $B = \{w, x, y, z\}$. The Cartesian product $A \times B$ was found to show that $2 \times 4 = 8$.
   (a) Show that multiplication is commutative by finding the Cartesian product $B \times A$ and using the definition of multiplication to find the product 4×2.
   (b) Is $A \times B$ equal to the set $B \times A$?

Objective 2; Item (4)

8. (a) Tell what is meant in saying that the set of natural numbers are "closed" under the binary operations of addition and multiplication.

Inductive reasoning

(b) A general conclusion that is reached on the basis of testing several examples illustrates the use of *inductive reasoning*. Most mathematical conclusions were originally developed through the use of inductive reasoning, but this does not constitute a proof. After a reasonable conclusion (called a *hypothesis*) is developed, then the mathematician must prove (or disprove) the hypothesis. For this he/she uses *deductive reasoning*, a formal proof that step-by-step makes statements and gives reasons supporting them until the conclusion to be proved is finally reached.

Use inductive reasoning yourself to examine the following sets and decide whether you think they are closed under the operation of addition. (If they are not closed, give an example to illustrate that.)

(1) $\{2, 4, 6, 8, 10, \ldots\}$ (2) $\{1, 3, 5, 7, 9, \ldots\}$
(3) $\{5, 10, 15, 20, \ldots\}$ (4) $\{1, 2, 4, 8, 16, 32, 64, 128, \ldots\}$

(c) Use inductive reasoning again to test the same sets and form a hypothesis as to whether they are closed under the operation of multiplication. (Give an example for any that are not closed.)

Proof by counter example

9. Recall that a single counterexample is sufficient proof that a statement does *not* hold. Use this approach for the following problems.

(a) Prove that subtraction is not commutative; $(a - b \neq b - a)$.
(b) Prove that division is not commutative; $(a \div b \neq b \div a)$.

10. (a) Show that subtraction is not associative; $(a - b) - c \neq a - (b - c)$.
(b) Show that division is not associative; $(a \div b) \div c \neq a \div (b \div c)$.

11. The statement $2 + (5 \times 8) = (2 + 5) \times (2 + 8)$ would be a way of writing a distributive property of addition over multiplication. Test the given example to determine whether you think such a property exists.

Item (45)

12. (a) What is the identity element for multiplication?
(b) What do we mean when we say the set of natural numbers has a multiplicative identity element?

Multiplicative identity

(c) If we know that $a \times b = 5$ illustrates the multiplicative identity property, then what can we conclude about a and b?

Subset number systems?

13. Perhaps a subset of the natural numbers could also possess all the properties that hold for the set of natural numbers. Test the following sets against the list of properties of the system of natural numbers (see page 434), and decide whether you think each set is a system of numbers that includes the operations of addition and multiplication. (If not, state which properties it lacks.)

(a) $\{2, 4, 6, 8, 10, \ldots\}$
(b) $\{1, 3, 5, 7, 9, \ldots\}$
(c) $\{10, 20, 30, 40, 50, \ldots\}$
(d) Can you think of a subset of natural numbers for which all 8 properties hold?

Objective 13; Examples 3, 5, 7, and 8

14. The following equalities are an assortment of those types presented in the text. Identify the property of natural numbers that each example illustrates:

(a) $5 \times 9 = 9 \times 5$
(b) $4 + 3 = 3 + 4$
(c) $(2 + 6) + 4 = 4 + (2 + 6)$
(d) $(3 \cdot 2)5 = 3(2 \cdot 5)$
(e) $(3 + 8) \times 1 = 1 \times (3 + 8)$
(f) $35 \times 1 = 35$
(g) $6(9 \cdot 2) = 6(2 \cdot 9)$
(h) $5(4 + 8) = 5 \cdot 4 + 5 \cdot 8$
(i) $4(2 + 7) = 4(7 + 2)$
(j) $9 + (3 + 6) = (9 + 3) + 6$
(k) $(4 + 7) \quad \times \quad (3 + 5) \quad = \quad (3 + 5) \quad \times \quad (4 + 7)$
(l) $2(8 + 5) = (8 + 5)2$
(m) $(3 + 4) \quad + \quad (1 + 8) = [(3 + 4) + 1] \quad + \quad 8$
(n) $7 + (5 \cdot 4) = (5 \cdot 4) + 7$

Objective 13; Examples 4, 6, 7, and 8

15. Apply properties of the natural numbers to each left member of the following equalities to obtain each right member. Show each step and name each property.

(a) $4 + (8 + 6) = (4 + 6) + 8$
(b) $5(7 \cdot 8) = (5 \cdot 8)7$
(c) $(9 + 6) + 1 = (9 + 1) + 6$
(d) $(6 \cdot 3)5 = (6 \cdot 5)3$
(e) $4(25 + 1) = 4 \cdot 25 + 4$
(f) $(4 + 5) \quad + \quad (6 + 5) = (4 + 6) \quad + \quad (5 + 5)$
(g) $3 + (5 + 7) = 5 + (3 + 7)$
(h) $5(20 + 1) = 5 \cdot 20 + 5$
(i) $(4 \cdot 9)5 = 9(4 \cdot 5)$
(j) $(1 + 2) \quad + \quad (9 + 8) = (1 + 9) \quad + \quad (2 + 8)$
(k) $(3 + 9) + 7 = 9 + (3 + 7)$
(l) $(8 \cdot 7)5 = (8 \cdot 5)7$

THE WHOLE NUMBERS

Objective 14: Difference between naturals and wholes

Having found that the system of natural numbers has some distinct limitations, we are ready to proceed to the next set in our development of the real number system—the whole numbers. Actually, the one significant difference between elements of the two sets is that the set of whole numbers includes the number zero, as illustrated by the following graph: (50)

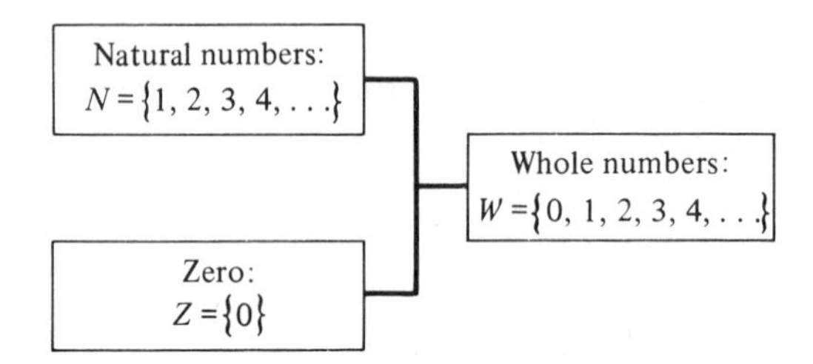

We see by this representation that the set W of whole numbers is the union of the set N of natural numbers and zero: (51)

$$W = N \cup \{\text{ZERO}\}$$

Since every natural number is also a whole number, we may further say that the set of natural numbers is a subset of the set of whole numbers:

$$N \subset W$$

Like the set of natural numbers, the set of whole numbers* is also an infinite set, as it is impossible to name the last member of the set. The subset relationship of set N to set W is depicted below.

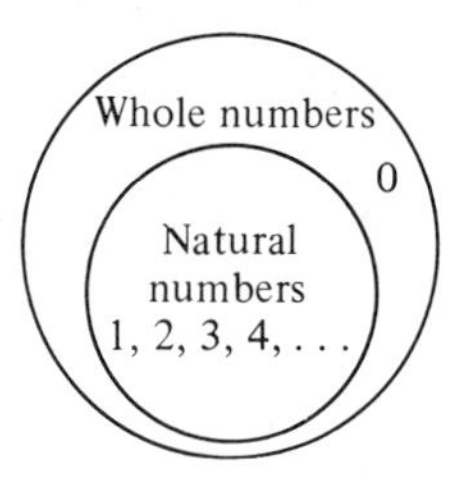

G. ADDITION AND MULTIPLICATION IN THE WHOLE NUMBERS

Our previous definitions for the operations of addition and multiplication employed the cardinal number (the number of elements) of sets. This was appropriate since the natural numbers are used to indicate the cardinal number of finite sets. In naming the cardinal number of the empty set, we (52)

*Because of the similarity between the sets of natural and whole numbers, development of the real number system is sometimes begun with the set of whole numbers.

Objective 15: Addition with zero

use the number zero. The following example shows that our definition of addition also applies for empty sets.

◆ **Example 1**

(a) Suppose $A = \{j, k\}$ and $Z = \varnothing$. (53)

Then, $A \cup Z = \{j, k\} = A$.

As before,

$$n(A) + n(Z) = n(A \cup Z)$$
$$2 + 0 = 2$$

(b) Thus, to express the definition in general terms that apply for any whole numbers a and 0, we say, (54)

Definition of addition for zero

If $n(A) = a$ and $n(Z) = 0$, then the binary operation of addition, $a + 0$, produces one and only one element, a, which is the $n(A \cup Z)$. That is,

$$n(A) + n(Z) = n(A \cup Z)$$
$$a + 0 = a$$

● **Practice 1**

(a) Suppose $P = \{a, r, t\}$ and $Z = \varnothing$. Use the given sets and the definition of addition to find the sum $3 + 0$. (55)

(b) Let P be any nonempty set. Use P and the given set Z to express the definition of addition for any whole number and zero.

See also Exercises 1 and 2.

Objective 16: Multiplication with zero

By a similar example, we show that multiplication by zero also conforms to our earlier definition that used the Cartesian product. (56)

◆ **Example 2** Again, let $A = \{j, k\}$ and $Z = \varnothing$. Since there are no elements in set Z, it is impossible to complete any ordered pair of a Cartesian product; thus, $A \times Z$ is also an empty set: $A \times Z = \varnothing$.

Then, $n(A) = 2$, $n(Z) = 0$, and $n(A \times Z) = 0$.

So,

$$n(A) \times n(Z) = n(A \times Z)$$
$$2 \times 0 = 0$$

Thus, multiplication may be defined for any whole numbers a and 0: (57)

Definition of multiplication by zero

If $n(A) = a$ and $n(Z) = 0$, then the binary operation of multiplication, $a \times 0$, produces one and only one element, 0, which is the $n(A \times Z)$. That is,

$$n(A) \times n(Z) = n(A \times Z)$$
$$a \times 0 = 0$$

● **Practice 2**

(a) Consider again the sets $P = \{a, r, t\}$ and $Z = \varnothing$. Use the given sets and the definition of multiplication to find the product 3×0. (58)

(b) Let P be any nonempty set. Use P and given set Z to express the general definition of multiplication for any whole number and zero.

See also Exercises 3 and 4.

H. ADDITIVE IDENTITY

Our building-block development of the real number system has shown that our earlier definitions for addition and multiplication still apply. Likewise, the properties of the set of natural numbers also still apply for the set of whole numbers. (59)

Objective 17: Identity element for addition

You will recall, however, that the system of natural numbers contained an identity element for multiplication but not for addition. The presence of 0 in the set of whole numbers now makes possible an *additive identity property*. That is, for any whole number, a, (60)

$$a + 0 = 0 + a = a$$

For example, $$7 + 0 = 0 + 7 = 7$$

The number zero itself is called the *additive identity* or the *identity element for addition.*

I. SUBTRACTION

Objective 18: Definition of subtraction for whole numbers

When $a - b = k$, recall that a is called the *minuend*; b is called the *subtrahend*, and k is called the *difference*. (61)

A limited amount of subtraction may be defined as follows: for any whole numbers* a and b,

$a - b = k$ if and only if k is a whole number such that $a = b + k$.

Or, substituting the abbreviation "iff" for the terminology "if and only if,"

$$a - b = k \quad \text{iff} \quad a = b + k$$

For example, $$9 - 6 = 3 \quad \text{iff} \quad 9 = 6 + 3$$

The phrase "if and only if" denotes an "if—then" relationship that works in both directions. Or, we could say that each part "implies" (62)

*This definition also applies for the set of natural numbers, if the term "natural number" is substituted for "whole number."

the other. For instance *if* $9 - 6 = 3$, *then* $9 = 6 + 3$. Similarly, $9 = 6 + 3$ *implies* $9 - 6 = 3$.

Subtraction is basically a variation of addition—the "undoing" of addition. Since subtraction is defined only in terms of addition, it is therefore not considered a fundamental operation of the system of whole numbers. The operation of subtraction is *not closed* for the set of whole numbers (for example, given $4 - 7 = k$, *no* k *exists* such that $4 = 7 + k$). Most other properties similarly fail to hold, as previous exercises have shown. **(63)**

Properties of subtraction?

Distributive property

It may come as something of a surprise to learn that a *distributive property* (of multiplication over subtraction) does hold—and is the *only* basic property that holds. For any whole numbers a, b, and c (where $b - c$ exists), **(64)**

$$a(b - c) = a \cdot b - a \cdot c$$

For example,

$$3(8 - 6) = 3 \cdot 8 - 3 \cdot 6$$
$$3 \quad (2) = 24 - 18$$
$$6 = 6$$

J. DIVISION

Objective 19: Definition of division for whole numbers

Like subtraction, the operation of division has a limited definition: For any whole numbers* a and b (where $b \neq 0$), **(65)**

$$\frac{a}{b} = q \text{ if and only if } q \text{ is a whole number* such that } a = bq.$$

In abbreviated form,

$$\frac{a}{b} = q \quad \text{iff} \quad a = bq$$

For example,

$$\frac{8}{2} = 4 \quad \text{iff} \quad 8 = 2 \cdot 4$$

Recall that, in any division problem represented by $\frac{a}{b} = q$ (or by $a \div b = q$), the element a is called the *dividend*, b is called the *divisor*, and q is called the *quotient*. **(66)**

Objective 20: No division by zero

Notice the requirement that $b \neq 0$ in the definition of division; this is essential to the definition. Let us consider some examples to see why this must be so. Assume $b = 0$; we will first consider $a \neq 0$ (let $a = 3$), and secondly consider $a = 0$: **(67)**

(1) $$\frac{3}{0} = q \quad \text{iff} \quad 3 = 0 \cdot q$$

*The definition also applies when "natural number" is substituted for "whole number."

There is no whole number q such that $3 = 0 \cdot q$. Thus, if $a \neq 0$ and $b = 0$, the operation of division is not closed because the solution does *not exist*.

(2) $$\frac{0}{0} = q \quad \text{iff} \quad 0 = 0 \cdot q \qquad \textbf{(68)}$$

Given $\frac{0}{0}$, we would not know which q to choose for our result, because $0 = 0 \cdot q$ for *every* whole number q. In this case, the operation of division is not closed because the solution is *not unique*.

(69) So we see that whether $a = 0$ or $a \neq 0$, in either case when $b = 0$ the operation is not closed. Only when $b \neq 0$ can we possibly obtain a solution that *exists* and that is *unique*. We therefore do not define the operation of division by zero under any circumstances.

(70) Since division is defined in terms of multiplication, it (like subtraction) is not considered a fundamental operation of the system of whole numbers. Division also fails to meet the basic properties of the system of whole numbers: For instance, division is *not closed* for the entire set of whole numbers (for example, given $\frac{15}{4} = q$, no q exists such that $15 = 4 \cdot q$). Also, previous exercises have already determined that division is neither commutative nor associative. Other basic properties similarly fail to hold.

Properties of division?

(71) There is, however, an interesting variation of the distributive property which *may* apply for the set of whole numbers (provided the indicated division quotients exist). This is called the *right distributive property of division over addition*, which may be expressed

Objective 21: Right distributive property (÷ over +)

$$(a + b) \div c = a \div c + b \div c$$

Or, similarly,
$$\frac{a + b}{c} = \frac{a}{c} + \frac{b}{c}$$

For example,
$$\frac{4 + 8}{2} = \frac{4}{2} + \frac{8}{2}$$
$$\frac{12}{2} = 2 + 4$$
$$6 = 6$$

(72) The system of whole numbers has thus extended the capabilities of the system of natural numbers. However, this new system continues to have many limitations. There is still no whole number we can use to name $5 - 8$ or $5 \div 3$; hence, we have very restricted definitions of subtraction and division since these operations are *not closed* for the system of whole numbers. It will thus be necessary to further extend our system of numbers in order to develop a more useful system.

(73) In reviewing this chapter, keep in mind that the system of natural numbers is a subset of the system of whole numbers. This means that every characteristic and property of the system of natural numbers also applies in the system of whole numbers. You may refer to the summary chart on page 434 for paragraph references pertaining to the system of whole numbers.

EXERCISES

Objective 15; Example 1

1. (a) Let $D = \{w, o, r, k\}$ and $Z = \varnothing$. Use the definition of addition and the given sets to find the sum $4 + 0$.
(b) Use any nonempty set D and the given set Z to express the definition of addition as it applies for the addition of any whole number and zero.

Example 1

2. Suppose we have two sets $E = \varnothing$ and $Z = \varnothing$. Show that the definition of addition applies for two empty sets. That is, using the given sets and the definition of addition, work through Example 1 to find the sum $0 + 0$.

Objective 16; Example 2

3. (a) Consider the same sets given in Exercise 1: set $D = \{w, o, r, k\}$ and $Z = \varnothing$. Use these sets and the definition of multiplication to find the product 4×0.
(b) Use any nonempty set D and the given set Z to express the definition of multiplication for any whole number by zero.*

Example 2

4. Consider again sets $E = \varnothing$ and $Z = \varnothing$. Show that the definition of multiplication applies for two empty sets. That is, using the given sets and the definition of multiplication, work through Example 2 to find the product 0×0.

5. Tell the distinction between elements of the set of natural numbers and the set of whole numbers.

Objective 14a: Additive identity

6. (a) What is the identity element for addition?
(b) Tell what is meant when we say the set of whole numbers has an additive identity element.
(c) If we know that $a + b = 5$ illustrates the additive identity property, then what can we conclude about a and b?

Objective 17: Number properties

7. Use whole numbers r, s, and t (if needed) to express the following properties of the whole numbers:
(a) Closure property of addition. Verify this property using 6 and 0.
(b) Closure property of multiplication. Verify the property using 0 and 4.
(c) Commutative property of addition. Verify this property using 6 and 7.

*This characteristic of multiplication by zero, although not a fundamental property of the mathematical system for whole numbers, is often called the *multiplication property of zero*: $a \times 0 = 0 \times a = 0$.

(d) Associative property of addition. Verify this property using 5, 0, and 9.
(e) Commutative property of multiplication. Verify the property using 3 and 7.
(f) Associative property of multiplication. Verify the property using 8, 4, and 2.
(g) Distributive property of multiplication over addition. Verify the property using 4, 5, and 7.

Item (60)

(h) Additive identity property. Verify the property using 9.
(i) Multiplicative identity property. Verify the property using 4.

Objective 14b: Characteristics of natural and whole numbers

8. (a) Why is it impossible to verify all instances (examples) of the additive identity property (or any other property) for the set of whole numbers?
(b) Is there a least element in the set of whole numbers?
(c) Is there a greatest element in the set of whole numbers?
(d) Show that there is one-to-one correspondence between the first three natural numbers and the first three whole numbers.
(e) Show that there is one-to-one correspondence between the entire set of natural numbers and the set of whole numbers.

9. (a) Is every natural number a whole number?
(b) Is every whole number a natural number?

Subset number systems?

10. Let us repeat a previous experiment. Compare each of the following subsets of the whole numbers against the list of properties of the system of whole numbers (page 434). Use inductive reasoning to determine whether each is a system of numbers that includes the operations of addition and multiplication. (If not, state the property or properties that it lacks.)
(a) $\{1, 3, 5, 7, 9, \ldots\}$ (b) $\{0, 2, 4, 6, 8, \ldots\}$
(c) $\{0, 1\}$ (d) $\{0, 1, 2\}$

Objective 13: Applying number properties

11. In each of the following equalities, tell which property or properties is applied to each left-hand member to obtain each right-hand member. (Show each step, if more than one step is required.)
(a) $7(6 + 4) = 7 \cdot 6 + 7 \cdot 4$
(b) $(5 + 9) + 1 = 5 + (9 + 1)$
(c) $8 \cdot 5 = 5 \cdot 8$
(d) $(5 \cdot 9)4 = 9(5 \cdot 4)$
(e) $1(6 + 2) = (6 + 2)$
(f) $4 + (3 + 6) = 3 + (4 + 6)$
(g) $3(2 + 4) = 3(4 + 2)$
(h) $(7 \cdot 5)4 = 7(5 \cdot 4)$
(i) $1(0 + 1) = 1$
(j) $(8 \cdot 9)5 = (8 \cdot 5)9$
(k) $6(3 + 7) = (3 + 7)6$
(l) $2 + (7 + 8) = (2 + 8) + 7$

Objective 18: Subtraction

12. (a) Define the operation of subtraction for the set of whole numbers.
(b) If $7 - 2 = 5$, what addition fact is implied by the definition of subtraction?
(c) If $9 = 1 + 8$, what subtraction fact is implied?

13. (a) Why is subtraction not considered a fundamental operation of the system of whole numbers?
(b) Is the operation of subtraction closed for the set of natural numbers? Explain.

Objective 18

14. Which of the number properties below hold for the operation of subtraction for whole numbers? Give an example to support your answer:
(a) Closure?
(b) Commutative? Does $a - b = b - a$?
(c) Associative? Does $(a - b) - c = a - (b - c)$?
(d) Distributive? Does $a(b - c) = a \cdot b - a \cdot c$?
(e) Identity? Does $a - 0 = 0 - a$ or $0 - a = a$?

Distributive property of × over −

15. (a) Illustrate the distributive property of multiplication over subtraction using r, s, and t.
(b) Verify this property using 9, 7, and 3.

Objective 19: Division

16. (a) Define the operation of division for the set of whole numbers.
(b) If $\frac{12}{6} = 2$, what multiplication fact is implied by the definition of division?
(c) If $15 = 3 \cdot 5$, what division fact is implied?

17. (a) Why is division not considered a fundamental operation of the system of whole numbers?
(b) Does the definition of division apply for the set of natural numbers? Why?
(c) Which part of the definition of division for whole numbers could be omitted for the set of natural numbers?
(d) Why is the operation of division not closed for the system of whole numbers?

Objective 20: Division by zero

18. Any operation in mathematics must produce a *result* that *exists* and that is *unique*. Use this information and the definition of division to explain below why division by zero is not permitted:

Item (67)

(a) What kind of result is indicated by $\frac{a}{0}$, if $a \neq 0$?

Item (68)

(b) What kind of result is indicated by $\frac{0}{0}$?

Objective 19: Division

19. Which of the following basic number properties of the system of whole numbers apply for the operation of division? Give an example to support your answer:
(a) Closure?
(b) Commutative? Does $a \div b = b \div a$?
(c) Associative? Does $(a \div b) \div c = a \div (b \div c)$?

(d) Distributive? Does $a \div (b + c) = a \div b + a \div c$?
(e) Identity? Does $a \div 1 = 1 \div a$ or $1 \div a = a$?

20. (a) Does $24 \div (6 + 2) = 24 \div 6 + 24 \div 2$?
(b) Does $(24 + 6) \div 2 = 24 \div 2 + 6 \div 2$?
(c) What property of division by whole numbers is represented by $(a + b) \div c = a \div c + b \div c$, or by $\frac{a + b}{c} = \frac{a}{c} + \frac{b}{c}$?

Objective 21: Right distributive property of division over addition

(d) Why may this property often fail to hold?
(e) Illustrate the right distributive property of division over addition using r, s, and t. Verify this property using 9, 6, and 3.

OBJECTIVES

Unit I

The Natural Numbers

After completing this unit, you should be able to

A. Addition in the natural numbers

1 Identify the four basic characteristics of a number system. Tell how the set of natural numbers fulfills each requirement. (2, 49)

2 Explain the meaning of a "binary" operation and of the "closure" property of addition and multiplication. (4–5)

3 Use the definition of addition to illustrate the binary operation for a specific case (such as, $4 + 2$) or for a general case (such as, $a + b$). (6–10)

B. Multiplication in the natural numbers

4 Use the definition of multiplication to illustrate the binary operation for a specific case (such as, 4×2) or for a general case (such as, $a \times b$). (11–15)

D, E, F. Properties of the natural numbers

For objectives 5–12, express with variables (that is, letters) and verify using numbers each of the following properties:

5 Closure property of addition. (5)

6 Closure property of multiplication. (5)

7 Commutative property of addition. (17)

8 Associative property of addition. (19)

9 Commutative property of multiplication. (32)

10 Associative property of multiplication. (33)

11 Distributive property (of multiplication over addition). (41)

12 Multiplicative identity property. (45)

13 Given a numerical equality, tell which property (or properties) was applied to each left-hand member to obtain each right-hand member. (Show all steps.) (22–31, 34–40, 42–47)

Unit II The Whole Numbers

After completing this unit, you should be able to

H. Addition and multiplication in the whole numbers

14 (a) Tell the distinction between elements of the set of whole numbers and the set of natural numbers.
(b) Describe characteristics of the sets of whole numbers and natural numbers (subset relationships, correspondence of elements, infiniteness, least or greatest elements, etc.). (50–51)

15 Apply the definition of addition for addition with zero. (52–55)

16 Apply the definition of multiplication for multiplication by zero. (56–58)

I. Properties of the whole numbers

17 (a) Repeat objectives 5–12 using whole numbers.
(b) Express with variables and verify using numbers the additive identity property. (59–60)

J. Subtraction

18 Define subtraction for the set of whole numbers. Tell which number properties hold for subtraction of whole numbers and give an example to support your answer. (61–67)

K. Divison

19 Define division for the set of whole numbers. Tell which number properties hold for division of whole numbers and give an example to support your answer. (68–75)

20 Explain why division by zero $(a \div 0)$ is not permitted, both when $a \neq 0$ and when $a = 0$. (70–72)

21 Identify an illustration of the right distributive property of division over addition. (76)

The Integers

Until this point we have studied two systems of numbers—the natural numbers and the whole numbers. We found that the system of whole numbers incorporated everything from the system of natural numbers and also established some new characteristics. Specifically, the number zero was included, which allowed an additive identity property. However, we found that this system still had many limitations. For instance, the operations of subtraction and division were not closed; there were no whole numbers we could use to name $5 - 8$ or $5 \div 3$. Let us see, then, how the next step in our system of numbers, the integers, can be developed. In order to do this, we must first understand the number line and the concept of order for whole numbers. (1)

A. THE NUMBER LINE

Objective 1: The number line

Our development of the systems of natural numbers and whole numbers was based on the concept of sets. For further development of our number system, it will be helpful to adapt the set idea to a number line. In associating the set of whole numbers with a number line, we first mark a *point of origin*, which we designate as 0. To the right of point 0, at any convenient distance, we mark another point which we designate as point 1. The distance between point 0 and point 1 is called the *unit distance*, which (2) is then used to mark off successive points to the right. These points are identified as 2, 3, 4, . . .

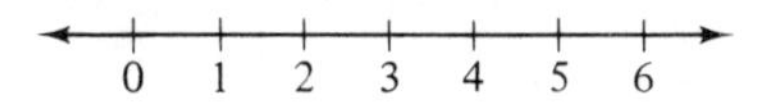

In this manner we establish a one-to-one correspondence between the elements of the set of whole numbers and a set of points on the number line.

A line in geometry is a set of infinitely many points aligned in a plane (a flat surface). The concept of a line implies that the points representing the whole numbers are only some of the points of the entire line. It is extremely important to realize that there are many more points—an infinite number more—that are also part of the line. Later, we will assign matching numbers to these other points as well. (3)

◆ **Example 1**

(a) Graph on a number line the set of whole numbers *between* 1 and 6: (4)

(b) Graph on a number line the set of whole numbers between 1 and 6 *inclusive* (that is, "including 1 and 6."):

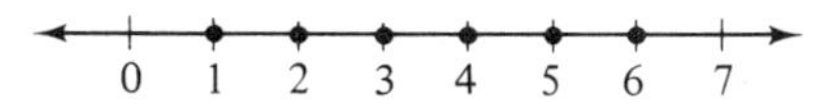

(c) Graph the set $\{0, 2, 4\}$ on a number line:

● **Practice 1**

(a) Graph on a number line the set of whole numbers between 0 and 4. (5)
(b) On a number line, graph the set of whole numbers between 0 and 4, inclusive.
(c) Graph the set $\{0, 1, 2, 5\}$ on a number line.
(d) Graph on a number line the set of whole numbers between 1 and 3.

See also Exercise 6.

B. ORDER OF WHOLE NUMBERS

Until now, we have simply used intuitively the important concept of order which the number line helps us visualize. Consider the alphabet as a comparison: There is no apparent reason for the order of letters in the alphabet and no significance attached when one letter follows another. (6)

Objective 2: Order on the number line

By contrast, each number on the number line is larger than those numbers to the left of it and smaller than those to the right of it. This is significant because, given two whole numbers, we can immediately tell which is larger and which is smaller.

Inequality symbols

To indicate that two numbers are not equal, we use the *symbols of inequality*, "<" or ">", as follows: (7)

$3 < 5$ "3 is less than 5"

or $7 > 0$ "7 is greater than 0."

The concept of order for the set of whole numbers is defined in the following manner: (8)

Definition of order (a)

For any whole numbers a, b, and k (where $k \neq 0$),

$$\textit{if} \quad a + k = b, \quad \textit{then} \quad a < b$$

◆ **Example 2** Given the numbers 9 and 4, use the definition of order to correctly place the symbol "<" between the two numbers. (9)

We know that, for some $k \neq 0$,

$$\text{either} \quad 9 + k = 4 \quad \text{or} \quad 4 + k = 9.$$

There is no whole number k for which $9 + k = 4$. When $k = 5$, however, $4 + k = 9$. Thus, by definition,

$$\text{if} \quad 4 + 5 = 9, \quad \text{then} \quad 4 < 9$$

● **Practice 2** Given the numbers 8 and 3, use the definition of order to correctly place the symbol "<" between the two numbers. (10)

See also Exercise 8.

Notice that the symbols of inequality are related. Thus, if $7 < 12$ then also $12 > 7$. That is, for any whole numbers a and b, (11)

Definition of order (b)

$$\text{if} \quad a < b \quad \text{then} \quad b > a.$$

Recall that a mathematical system requires one or more relations, and the equals relation was cited as an example. We now have defined another such relation—the *order* (or *inequality*) *relation*.

Now, in an equals relation, when $a = b$ then also $b = a$. Notice, however, that $a < b$ and $b < a$ cannot both exist. Consider then the following possibilities concerning $(3 + 4)$ and (2×4): (12)

1. Is $(3 + 4) = (2 \times 4)$?
2. Is $(3 + 4) < (2 \times 4)$?
3. Is $(3 + 4) > (2 \times 4)$?

Objective 3: Trichotomy law

We find that only one of the three relations is correct. This characteristic of numbers is expressed by the *trichotomy law*:

For any whole numbers a and b, one and only one of the following will hold:

$$a = b \quad \text{or} \quad a < b \quad \text{or} \quad a > b$$

(13) Knowing by the trichotomy law that only one of the relations will hold for two numbers, we may combine the symbols in the following manner. If we write

$6 \leq (4 + 3)$ "6 is less than or equal to $4 + 3$"

this implies that

either $6 < (4 + 3)$ *or* $6 = (4 + 3)$

In this case, the correct interpretation is $6 < (4 + 3)$, since $6 < 7$.

(14) ◆ **Example 3** Insert the correct symbol $<$, $>$, or $=$ between each of the following pairs of numbers:

(a) 1 and 5:

$1 < 5$

(b) 9 and $(5 + 2)$:

$9 > (5 + 2)$, since $9 > 7$

Evaluate the following inequalities in order to determine whether each is true or false:

(c) $(4 + 1) \leq (7 - 2)$

$5 \leq 5$

This means $5 < 5$ *or* $5 = 5$. The second case is correct; hence, the expression is true.

(d) $(9 - 3) \geq 7$

$6 \geq 7$

This says $6 > 7$ *or* $6 = 7$. Neither of these is correct, because actually $6 < 7$. Thus, the expression is false.

(15) ● **Practice 3** Place the correct symbol $<$, $>$, or $=$ between each of the following pairs:

(a) 3 and 9

(b) 6 and $(4 + 2)$

(c) 8 and 1

(d) $(3 + 5)$ and $(6 + 1)$

(e) $(2 + 6)$ and $(3 + 5)$

(f) $(5 + 1)$ and $(4 + 3)$

Evaluate each of the following expressions and tell whether each is true or false:

(g) $(5 + 2) \geq 6$

(h) $(2 + 3) \leq (1 + 4)$

(i) $9 \leq (3 + 5)$

(j) $(8 - 2) \geq (2 + 4)$

(k) $(3 + 1) \geq (9 - 4)$

(l) $(7 - 2) \leq (6 + 2)$

See also Exercises 9 and 10.

We have previously established a one-to-one correspondence between the set of whole numbers and a set of points on a number line. We have also noted that these points represent only some of the infinite number of points that compose the line. But geometry also describes a line as extending indefinitely in both directions. Therefore, if a line is an accurate representation of the concept of numbers, should not numbers be graphed to the left of zero as well as to the right? This seems reasonable to do, using the same unit distance as before, and using the symbol "−" to identify these new *negative* numbers.* (16)

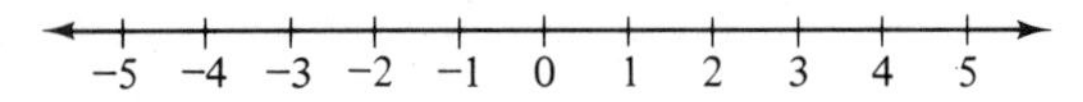

C. THE SET OF INTEGERS

Objective 4: Our extended number line now more closely corresponds to the geometric concept of a line. It includes the elements from the set of whole numbers, $W = \{0, 1, 2, 3, 4, \ldots\}$, as well as the set of newly invented negative numbers, $M = \{-1, -2, -3, -4, \ldots\}$. The new set formed by the union of sets W and M is known as the *set of integers* (I) (17)

Characteristics of set of integers

$$\boldsymbol{I = \{\ldots, -4, -3, -2, -1, 0, 1, 2, 3, 4, \ldots\}}$$

The set of whole numbers is thus a proper subset of the set of integers:

$$\boldsymbol{W \subset I}$$

This subset relationship may be diagramed in the following manner:

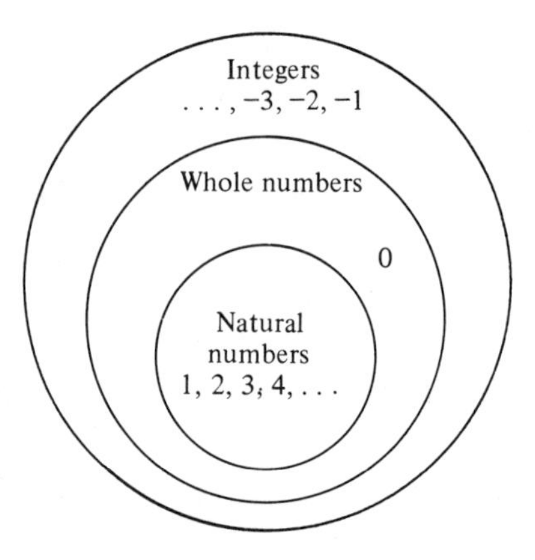

*It is interesting to note that negative numbers were not generally accepted as "real" until the latter part of the seventeenth century, which makes them a rather recent development compared to much of mathematics.

(18) Like the sets of natural and whole numbers, the set of integers is also an infinite set. There is a distinction, however. The sets of natural and whole numbers contained no largest elements, but they did each have a smallest element. The set of integers, however, contains neither a largest element nor a smallest element.

(19) The negative elements of the set of integers are commonly called the *negative integers*, while the elements from the set of whole numbers are known as the *nonnegative integers*. The nonnegative integers greater than zero are frequently referred to as the *positive integers*. The set of integers may then be classified in either of these two ways:

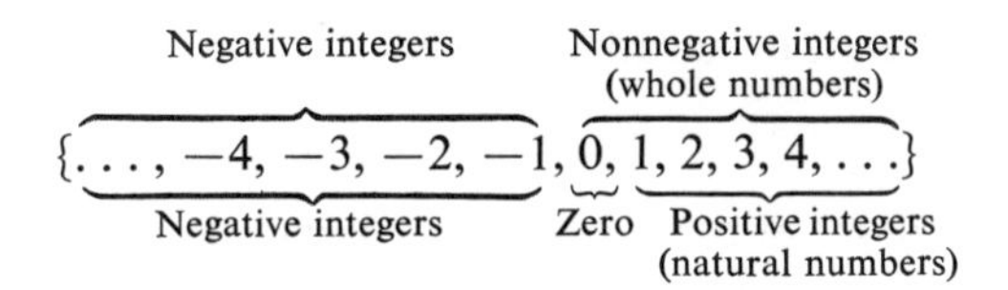

(20) Observe that, when negative integers are included on the number line, the order of numbers still follows the same pattern. That is, any given integer is larger than those to the left of it and is smaller than the integers to the right of it. The question remaining is whether the negative integers can be used in the operations we have defined.

(21) ◆ **Example 4** Place the correct symbol $<$ or $>$ between each of the following pairs of integers:

(a) 4 and -2:

$4 > -2$

(b) -5 and -1:

$-5 < -1$

Number lines are drawn below to graph each set given on the right:

(c)

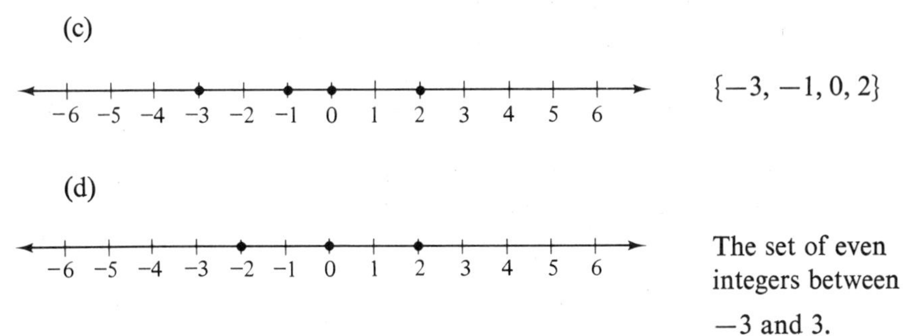

$\{-3, -1, 0, 2\}$

(d)

The set of even integers between -3 and 3.

(e)

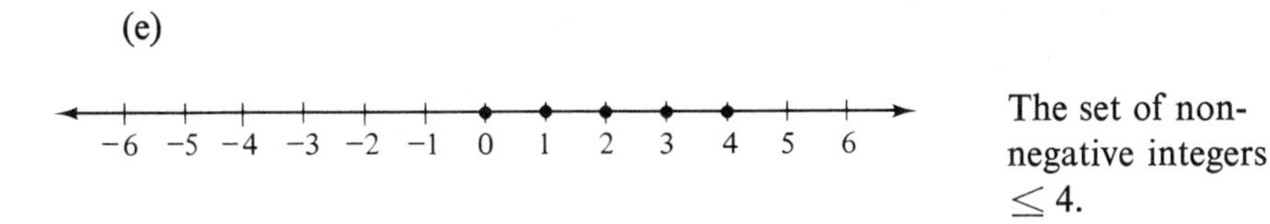

The set of nonnegative integers ≤ 4.

● **Practice 4** Place the correct symbol $<$ or $>$ between each of the following pairs of integers: (22)

(a) 0 and 5 (b) 0 and -3

(c) -2 and 2 (d) 3 and -4

(e) -7 and -1 (f) -3 and -6

Draw a number line and graph each of the following sets:

(g) $\{-4, -2, 0, 1, 3\}$

(h) the set of odd integers between -4 and 4.

(i) the set of negative integers ≥ -5.

(j) the set of nonnegative integers ≤ 1.

(k) the set of even negative integers ≥ -6.

See also Examples 6 and 9.

D. ADDITIVE INVERSE

We saw earlier that the operation of subtraction was not closed for the system of whole numbers. This was probably the key factor that led to the invention of negative numbers. For instance, it was possible centuries ago for a merchant who only had 10 coins to order goods worth 15 coins. It was apparent to the merchant, after he paid the 10 coins he had, that he still owed 5 more coins. It seemed there should be some operation that would show the amount of his obligation. In effect, it seemed reasonable that $10 - 15 = -5$. But how could this operation be defined? (23)

Since our present number system does not permit its operations to have negative results, the closest we could come to a negative result would be zero. That is, zero is the smallest difference it is possible to obtain: $a - a = 0$. If we wish eventually to be able to obtain negative results, then zero seems the place to start in order to define the fundamental operations of addition and multiplication for the set of integers. In other words, it seems there should be some integer with the property that (24)

$$a + (\text{some integer}) = 0$$

Of course, there is no whole number which, for example, when added to 4 gives zero.

On the expanded number line, each whole number a is the same distance to the right of zero as the negative integer $-a$ is located to the left of zero. (25)

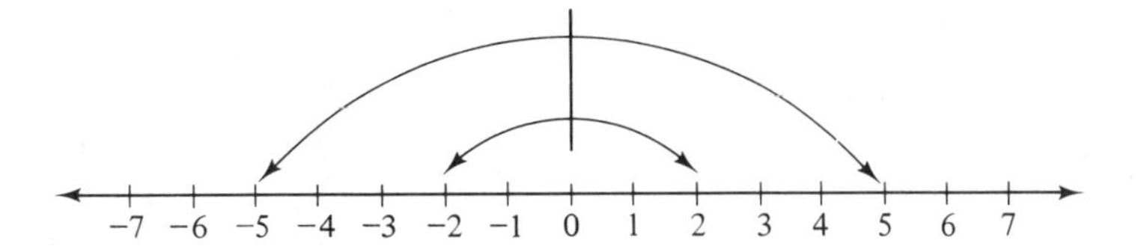

Each integer $-a$ is therefore called the *opposite* of a. For example,

-4 is the opposite of 4

and similarly, 4 is the opposite of -4

In order to make possible the property $a + (\text{some integer}) = 0$, this concept of opposites is used. Since each a and $-a$ are opposites (and each the same distance from zero), it makes sense that, when added, they would offset each other and total zero: **(26)**

For example, $a + (-a) = 0$

implies $4 + (-4) = 0$ and $(-4) + 4 = 0$

Additive inverse

We say that each opposite is the *additive inverse* of the other. Thus,

-4 is the additive inverse of 4; and

4 is the additive inverse of -4

Notice that every nonnegative integer a has an additive inverse $-a$ except the number zero; there is no point on the number line designated -0. What would be the additive inverse of zero? Now, we know that zero is its own additive *identity*; that is, that $0 + 0 = 0$. Therefore, we conclude and agree that zero is also its own additive *inverse*. Zero is thus unique in that it is the only integer which is its own additive inverse. The additive inverse property for the set of integers is now defined as follows: **(27)**

Objective 5b: Additive inverse property

For every element, a, of the set of integers there exists a unique element, $-a$, such that **(28)**

$$a + (-a) = (-a) + a = 0$$

For example, $7 + (-7) = (-7) + 7 = 0$

or, $-5 + (5) = 5 + (-5) = 0$

Having defined additive inverses, we can now say that the set of integers is the union of the set of whole numbers and the set of additive inverses of the natural numbers. **(29)**

Note: it is important to understand "$-a$" as either "the additive inverse of a" or as "the opposite of a" (rather than as "negative a"), since the variable a may itself represent a negative integer.

◆ **Example 5**

(30)

Objective 6: Additive inverses

(a) Evaluate "$-a$" when $a = 5$.
$-a = -(5) = -5$ The additive inverse of a equals the additive inverse of 5 which equals negative 5.

(b) Evaluate "$-a$" when $a = -7$.
$-a = -(-7) = 7$ The additive inverse of a equals the additive inverse of negative 7, which equals 7.

● **Practice 5** What integer is represented by $-x$ when: (31)

(a) $x = 2$ (b) $x = -3$

(c) $x = 0$ (d) $x = 4 + 3$

(e) $x = -(2 + 2)$ (f) $x = 6 + (-6)$

See also Exercise 11.

◆ **Example 6** Evaluate the following pairs and tell whether each pair represents additive inverses: (32)

(a) -4 and 4 (b) 6 and $-(-6)$

(c) $[8 + (-8)]$ and 0 (d) -2 and $-(-2)$

(a) Yes; -4 and 4 are additive inverses.

(b) No; $-(-6) = 6$; hence, 6 and 6 are not additive inverses.

(c) Yes; $8 + (-8) = 0$; zero is its own additive inverse.

(d) Yes; $-(-2) = 2$; thus, -2 and 2 are additive inverses.

● **Practice 6** Evaluate the following pairs and tell whether each pair represents additive inverses: (33)

(a) -3 and 3 (b) 0 and $5 + (-5)$

(c) $-(3 + 1)$ and -4 (d) $-(-7)$ and -7

(e) 1 and $-(-1)$ (f) $-(-6)$ and $-(4 + 2)$

See also Exercise 12.

E. ADDITION IN THE INTEGERS

We have now developed several of the essentials for a *system* of integers. The elements defined for this system are the set of integers. We have applied the equals relation from our earlier systems and have also used the (34) operation of addition in defining our first property—the property of additive inverses: $a + (-a) = (-a) + a = 0$. At this point, however, *we have not shown whether addition for the set of integers applies for any elements except additive inverses.* That is, we do not know yet whether addition for the integers is closed.

(35) In expanding our properties for the system of integers, we will include all the properties from our earlier systems. Although these properties can be proved, we will use them here without proof. Exercise problems will be included, however, to verify that the number properties hold in the set of integers.

(36) Assuming the commutative, associative, and identity properties of addition, and applying our new additive inverse property, we will now consider an example to verify that the binary operation of addition may be performed using a positive and a negative integer: What is the sum of $7 + (-4)$?

Addition verified

(37) To perform the addition $7 + (-4)$, we first rename the positive element 7 so that it is a sum which includes the additive inverse of -4. Succeeding steps are verified using familiar addition properties.

$7 + (-4)$	$= (3 + 4) + (-4)$	Renaming the number 7
	$= 3 + [4 + (-4)]$	Associative property of addition
	$= 3 + 0$	Additive inverse
	$= 3$	Additive identity

Thus, $7 + (-4) = 3$.

Closure for addition

(38) Other possible combinations of positive and negative integers* can similarly be used to verify that addition may be performed using *any two elements* of the set of integers. We thus conclude that addition is closed for the system of integers.

Objective 7: Addition with integers

(39) We can now define the binary operation of addition for the integers as follows: To add two integers with

1. *Same signs: Add* their numerical components and affix the same sign.
2. *Opposite signs:* Find the *difference* between their numerical components (disregarding signs) and affix the sign of the larger component.

(40) ◆ **Example 7** Use the definition of addition for integers to find the following sums:

(a) $-3 + (-8)$

*In our example, (1) the positive integer has the larger numerical component (disregarding the positive or negative signs). Other combinations include (2) the larger numerical component is negative and the smaller is positive; and (3) both integers are negative. The case when (4) both integers are positive (that is, whole numbers) has already been considered. Combinations (2) and (3) are given as Exercises 23 and 24.

Since both addends have the same sign, we *add* their numerical components $(3 + 8 = 11)$; the sum also has the *same sign* (negative):

$$-3 + (-8) = -11$$

(b) $-3 + 8$

The addends have different signs, so we find the *difference* between the numerical components $(8 - 3 = 5)$ and affix the sign of the larger. The sum will thus be *positive*:

$$-3 + 8 = 5$$

(c) $3 + (-8)$

Again, we find the *difference* between the components $(8 - 3 = 5)$. The sum will be *negative*, since the larger numerical component is negative.

$$3 + (-8) = -5$$

● **Practice 7** Use the definition of addition for integers to find the sums: (41)

(a) $4 + 5$ (b) $-3 + (-5)$

(c) $5 + (-2)$ (d) $4 + (-7)$

(e) $-3 + 9$ (f) $-8 + 6$

See also Exercise 13.

Addition properties

(42) We have now accepted that addition in the system of integers includes the same properties found in the system of whole numbers. Specifically, the integers are closed for the operation of addition; also, the operation is commutative, associative, and includes the additive identity. Furthermore, the integers include a new additive property not found in previous systems —the property of additive inverses.

F. MULTIPLICATION IN THE INTEGERS

Closure for multiplication

(43) Having covered the additive properties, we are now ready to consider the operation of multiplication and its properties in the system of integers. That the binary operation of multiplication is closed for the set of integers can be verified by considering all combinations of positive and negative integers.* As an example, we will consider the combination that has always been most perplexing to students: a negative times a negative.

*The possible combinations are (1) both factors may be positive (the whole numbers already considered); (2) one factor may be positive and the other negative; or (3) both may be negative (as shown). In the second case, either factor may be the negative, since multiplication is commutative. The second combination would ordinarily be verified first and then applied in the third case. Case (2) is given as Exercise 25.

(44) In order to visualize the operation, consider the following series of multiplications. (We will assume that a positive times a negative factor has already been verified.)

$$5 \times -4 = -20$$
$$4 \times -4 = -16$$
$$3 \times -4 = -12$$
$$2 \times -4 = -8$$
$$1 \times -4 = -4$$
$$0 \times -4 = 0$$

(45) Notice that as each first factor decreases (from 5 to 4 to 3, etc.), the product *increases* by 4 each time (from -20 to -16 to -12, etc.) Continuing this same pattern, suppose we decrease our first factors below zero. We would assume that the products would continue to increase by 4 each time, as follows:

$$2 \times -4 = -8$$
$$1 \times -4 = -4$$
$$0 \times -4 = 0$$
$$(-1) \times -4 = 4?$$
$$(-2) \times -4 = 8?$$
$$(-3) \times -4 = 12?$$
$$(-4) \times -4 = 16?$$
$$(-5) \times -4 = 20?$$

(46) It seems extremely likely that these products are correct. To verify them, we will consider a specific case: what is the product $-4 \times (-2)$?

(47) To find the product $-4 \times (-2)$, we substitute in the distributive property, $a(b + c) = ab + ac$, letting the $(b + c)$ portion be $(2 + {}^{-}2)$,* and then apply the property of additive inverses:

$-4(2 + {}^{-}2)$	$=$	(-4×2)	$+$	$(-4 \times {}^{-}2)$	Distributive property, $\times$ over $+$
$-4(0)$	$=$	(-4×2)	$+$	$(-4 \times {}^{-}2)$	Additive inverse
0	$=$	(-4×2)	$+$	$(-4 \times {}^{-}2)$	Renaming (-4×0)
0	$=$	-8	$+$	$(-4 \times {}^{-}2)$	Renaming (-4×2)

Thus, the additive inverse of -8 is $(-4 \times {}^{-}2)$, since their sum equals zero. But we know the additive inverse of -8 is 8. Therefore, we have also verified that

$$-4 \times (-2) = 8$$

Hence, we have verified that a negative times a negative is a positive.

*To avoid excessive parentheses, -2 is written as ${}^{-}2$.

Having concluded that multiplication is closed for the set of integers, (since multiplying any combination of integers yields an integer), we are now ready to express a general definition of multiplication: Given two integers a and b, (48)

Definition of multiplication for integers

1. If both integers have the *same sign* (both positive or else both negative), then the product $(a \times b)$ will be positive.
2. If the integers have *different signs* (one positive and the other negative), then the product will be negative.
3. If either (or both) of the integers are equal to zero, then the product will be zero.

◆ **Example 8** Tell what integer is equal to each of the following expressions: (49)

Objective 8: Multiplication

(a) -6×3 (b) $-8 \times (-7)$

(c) $-(5 \times {}^{-}2)$ (d) $-4(3 + {}^{-}3)$

(a) When one factor is negative and the other is positive, the product is negative. Thus $-6 \times 3 = -18$.

(b) Here the two factors have the same sign (both negative), therefore the product $(a \times b)$ is positive: $-8 \times (-7) = 56$.

(c) The expression within parentheses must first be evaluated, and then we find the additive inverse of that integer. Thus, $-(5 \times {}^{-}2) = -(-10) = 10$.

(d) Since the parentheses contain additive inverses, we now have a multiplication by zero, which we know will always equal zero. Therefore, $-4(3 + {}^{-}3) = -4(0) = 0$.

● **Practice 8** Evaluate the following expressions and name the integer each represents: (50)

(a) $5 \times (-8)$ (b) -4×9

(c) $-7 \times (-6)$ (d) 4×0

(e) $(6 + {}^{-}6) \times 7$ (f) $-(8 \times {}^{-}3)$

(g) $-(-2 \times {}^{-}7)$ (h) $-3(4 + {}^{-}2)$

(i) $8(2 + {}^{-}6)$ (j) $-4(6 + {}^{-}9)$

(k) $(5 + {}^{-}2) \times (6 + {}^{-}3)$ (l) $(7 + {}^{-}3) \times (-5 + {}^{-}2)$

See also Exercise 14.

To this point, then, our new system of integers includes exactly the same operations and properties as did the earlier systems, as well as the new (51) property of additive inverses. It was noted, however, that a main reason for the invention of negative integers was the fact that subtraction for the system of whole numbers was not closed. With this in mind, let us now consider the operation of subtraction for the system of integers.

G. SUBTRACTION IN THE INTEGERS

(52) Subtraction for any whole numbers a, b, and k was defined as

$$a - b = k \quad \text{if and only if} \quad a = b + k$$

Subtraction with integers

This definition of subtraction for whole numbers already applies for the set of nonnegative integers. Therefore, it seems logical to determine whether this same definition could be altered in order to apply also to the operation of subtraction with negative integers. To accomplish this, we will use the property of additive inverses, which thus far applies uniquely to the set of integers.

(53) Using $a = b + k$ from above, suppose we add $(-b)$ to both sides of the equality. Then

$$a = b + k$$

becomes

$$a + (-b) = (b + k) + (-b)$$

We know that the sums are equal because the closure property says that the addition of two elements produces a unique (one and only one) result. These are the first two steps in the following proof:

(54)

$a - b = k$ implies $a = b + k$	Definition of subtraction for whole numbers
Then $a = b + k$ implies	
$a + (-b) = (b + k) + (-b)$	Closure for addition
$= b + [k + (-b)]$	Associative property of addition
$= b + [(-b) + k]$	Commutative property of addition
$= [b + (-b)] + k$	Associative property of addition
$= [0] + k$	Additive inverse
$= k$	Additive identity

Thus, $a + (-b) = k$ or, conversely, $k = a + (-b)$. Now we have

$$a - b = k \quad \text{and also} \quad k = a + (-b)$$

Therefore, $$a - b = a + (-b)$$

Objective 9: Definition of subtraction for integers

(55) The foregoing procedure thus gives us a definition for subtraction that we can apply to negative as well as nonnegative integers: For any integers a and b,

$$a - b = a + (-b)$$

Some examples are

$$7 - 3 = 7 + (-3) = 4$$
$$6 - 9 = 6 + (-9) = -3$$
$$-4 - 6 = -4 + (-6) = -10$$

Closure for subtraction

(56) Since the subtraction operation applies for all integers and is defined in terms of addition (which is closed for the system of integers), the operation of subtraction is also closed for the system of integers. This was the primary reason for the invention of the negative integers and we see now that they accomplished the purpose: subtraction is finally closed. It is also significant that the definition of subtraction for the integers utilizes the concept of the additive inverse: $a - b = a + (-b)$, regardless of whether a and b are positive or negative.

◆ **Example 9** Perform the indicated subtractions. In all cases, $a - b = a + (-b)$: (57)

(a) $5 - 9 = 5 + (-9) = -4$

(b) $-3 - 5 = -3 + (-5) = -8$

We can read "$a + (-b)$" as "a plus (the opposite of b)." Then,

(c) $2 - (-6) =$ 2 plus (the opposite of -6) $= 2 + 6 = 8$

(d) $-8 - (-4) = -8$ plus (the opposite of -4) $= -8 + 4 = -4$

(e) $-3 - (-7) = -3$ plus (the opposite of -7) $= -3 + 7 = 4$

(58) This concept of "adding the opposite" is the source of an informal students' definition: To subtract, you simply "change the second sign and add." And that is essentially what adding the inverse (or opposite) involves.

● **Practice 9** Perform the indicated subtractions: (59)

(a) $5 - 7$	(b) $9 - 3$	(c) $0 - 4$
(d) $-5 - 8$	(e) $-6 - 2$	(f) $-3 - 0$
(g) $8 - (-2)$	(h) $2 - (-4)$	(i) $-7 - (-5)$
(j) $-4 - 7$	(k) $6 - (-3)$	(l) $-1 - (-8)$

Subtract each of the following (by changing the sign of the subtrahend and adding):

(m)	(n)	(o)	(p)
4	4	-4	-4
9	-9	9	-9

See also Exercise 20.

(60) In conclusion, the only two basic number properties which hold for the operation of subtraction in the integers are our important new property of

Closure; distributive

closure and the one property observed previously—the distributive property (of multiplication over subtraction). The closure property in particular makes subtraction a very helpful and frequently used operation. However, the fact that subtraction is defined in terms of addition means that we still cannot list subtraction as a fundamental operation of the system of integers.

H. DIVISION IN THE INTEGERS

Division with integers

(61) Finally, how does the invention of negative integers affect our definition of division? We said that for any whole numbers a, b ($b \neq 0$), and q,

$$\frac{a}{b} = q \qquad \text{if and only if} \qquad a = bq$$

Objective 10:

(62) We now find that this definition applies quite well when negative integers are involved:

◆ **Example 10** Find the integer that represents each quotient:

(a) $\frac{-12}{3} = -4$ since $-12 = 3(-4)$

(b) $\frac{8}{-4} = -2$ since $8 = (-4)(-2)$

(c) $\frac{-6}{-2} = 3$ since $-6 = (-2)3$

Determining the sign of a quotient

Notice that the operation of division for the system of integers, since it is defined in terms of multiplication, follows similar rules for determining the sign of the result.

(63) ● **Practice 10** Find the divisor and dividend (if necessary) and compute each quotient.

(a) $\frac{16}{-4}$ (b) $\frac{-10}{5}$ (c) $\frac{-12}{-2}$

(d) $\frac{0}{-8}$ (e) $\frac{4 + (-2)}{-2}$ (f) $\frac{-6 + (-2)}{7 + (-3)}$

(g) $\frac{6 - 12}{-5 + 3}$ (h) $\frac{8 - (-4)}{-9 - (-3)}$

See also Exercise 22.

(64) With the invention of the negative integers, the operation of subtraction became closed. But does this extension of the number system also allow closure for division? Consider the following:

$$\frac{15}{-4} = q \qquad \text{iff} \qquad 15 = (-4)q$$

Limitations of integers

Even with the added flexibility that the integers permit, there is still no integer q—positive or negative—for which $15 = (-4)q$. That is, the operation of division still is not closed in the system of integers. Also, we have seen that each property of addition generally has a corresponding property of multiplication. The system of integers, however, includes an additive inverse property but without any corresponding multiplicative inverse.

(65) Thus, the development of the system of integers gave additional capabilities to the system of whole numbers—the most significant being closure for the operation of subtraction. However, two noticeable properties are still lacking—there is no closure for the operation of division and there exists no multiplicative inverse property. It will thus be desirable to continue the development of our number system in order to provide these properties.

(66) Since the system of whole numbers is a subset of the system of integers, and the system of natural numbers in turn is a subset of the whole numbers, then every characteristic of these systems is also included in the system of integers:

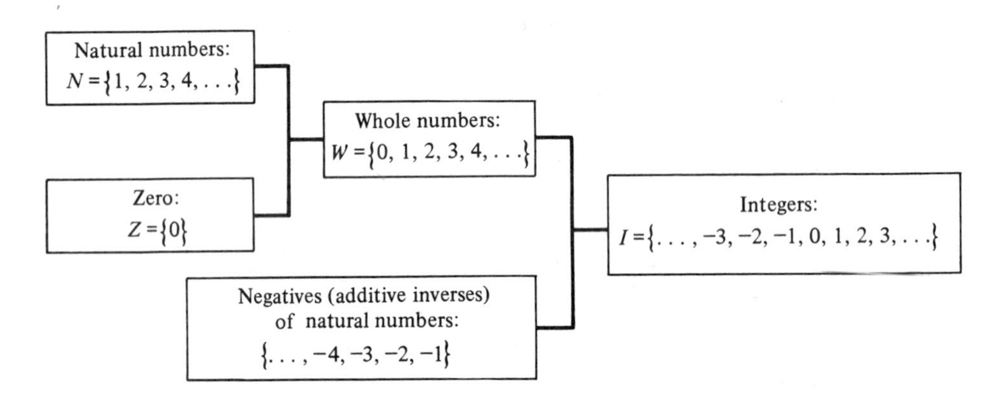

You may wish to refer to the summary chart on page 434 to review how the system of integers fulfills each of the basic requirements for a mathematical system.

EXERCISES

Objective 4

1. (a) Does the set of natural numbers have a smallest element? a largest element? If so, what is each?
   (b) Does the set of whole numbers have a smallest element? a largest element? If so, name each.
   (c) Does the set of integers possess a smallest element? a largest element? If so, what is each?

Objective 4

2. (a) The set of integers is formed by the union of which two sets?
   (b) What is the subset relation between the sets of integers and whole numbers? That is, which is a subset of the other?
   (c) Are the natural numbers a subset of the integers?
   (d) What is the distinction between the nonnegative integers and the positive integers?
   (e) Does the union of the set of positive integers and the set of negative integers yield the set of integers? Explain.
   (f) What other set that we have studied is the same as the set of nonnegative integers?
   (g) What other set is the same as the set of positive integers?
   (h) Show that there is one-to-one correspondence between the elements of the set of positive integers and the elements of the set of negative integers.
   (i) What is the intersection of the sets of positive and negative integers? of the sets of nonnegative and negative integers? of the sets of nonnegative and positive integers?

Objective 4

3. Answer the following. Explain each "no" answer:
   (a) Is every natural number a whole number?
   (b) Is every natural number an integer?
   (c) Is every whole number an integer?
   (d) Is every integer a whole number?
   (e) Is every integer either positive or negative?
   (f) Is every whole number a positive integer?
   (g) Is every positive integer a whole number?
   (h) Is every natural number a positive integer?

Additive inverse

4. Answer the following. Explain each "no" answer.
   (a) What fundamental property of the number system do the integers possess that was missing in the system of whole numbers?
   (b) Which element(s) of the set of whole numbers possess an additive inverse?
   (c) Which element(s) of the set of integers have an additive inverse?
   (d) Which element(s) of the set of integers are their own additive inverses?
   (e) For any given element, x, of the set of integers, how many additive inverses does x have?
   (f) The sum of an integer plus its additive inverse is what?
   (g) Is every additive inverse a negative integer?
   (h) Is every integer the opposite of some integer?
   (i) Is every negative integer the opposite of some nonnegative integer? the opposite of some whole number? the opposite of some natural number?
   (j) Is every natural number the additive inverse of some negative integer?

(k) Is every whole number the additive inverse of some negative integer?
(l) Does every negative integer have an opposite that is a whole number?
(m) Does every nonnegative integer have an opposite in the negative integers?
(n) Does every positive integer have an additive inverse in the negative integers?
(o) Is the set of integers equal to the set of additive inverses of the integers?

Number line

5. (a) The number line establishes a one-to-one correspondence between each integer and some ____ on the line.
(b) Does every point on the number line match with some integer? Explain.
(c) The number associated with each point on the number line is ____ than those numbers to the left of it and ____ than those to the right of it.

Objective 1; Examples 1 and 4

6. For each of the following, draw a number line and graph the indicated set:
(a) The set of integers between -2 and 3.
(b) The set of even integers between -4 and 4, inclusive.
(c) The set $\{-3, -1, 0, 2\}$.
(d) The set of whole numbers that have an additive inverse in the whole numbers.
(e) The set of additive inverses of $\{-5, -2, 1, 3\}$.
(f) The set of odd integers between -5 and 4, inclusive.
(g) The set of positive integers ≤ 5.
(h) The set of negative integers > -5.
(i) The set of nonnegative integers ≤ 3.
(j) The additive inverses of the set of positive integers ≤ 4.

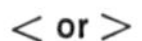

7. The following questions pertain to equality or inequality of numbers. Place the correct words and/or symbol in each blank.
(a) By the definition of order, for any integers a, b, and k (where $k > 0$), if $a + k = b$, then $a \_ b$.
(b) For any integers a and b, if $a < b$ then $b \_ a$.
(c) By the trichotomy law, we know that exactly ____ (how many?) of three relations will hold: $a \_ b$ or $a \_ b$ or $a \_ b$.
(d) If a and b are integers and $a \neq b$, then either ____ or ____.
(e) If a, b, and c are points on the number line as shown,

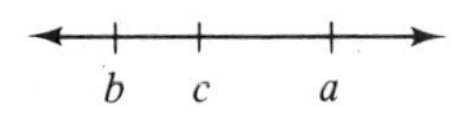

then we know $a \_ b$.
(f) Referring to the number line in part (e), we also know $b \_ c$.

Objective 2; Example 2

8. Use the definition of order for integers (Exercise 7a) to correctly place the symbol "$<$" between the following pairs of numbers.

(a) 2 and 6 (b) 7 and 2
(c) -5 and -2 (d) 3 and -1

Objective 3; Examples 3 and 4

9. Insert the correct symbol $<$, $>$, or $=$ between each of the following pairs of numbers:

(a) -1 and 4 (b) 0 and -2
(c) -8 and -3 (d) -2 and -6
(e) 4 and -8 (f) $(4 + 2)$ and $(7 - 2)$
(g) $(-6 + 3)$ and $(6 - 9)$ (h) $(-2 - 3)$ and $(-5 + 5)$
(i) $-(-7 + 2)$ and $(2 + {}^{-}5)$ (j) $-(4 + {}^{-}6)$ and $(-7 - {}^{-}9)$

Objective 3; Example 3

10. Evaluate each of the following expressions and tell whether it is true or false:

(a) $(3 + 2) \geq (8 - 5)$ (b) $4 + (-4) \leq -2 + 6$
(c) $5 - 4 \leq 4 - 5$ (d) $4 + (-7) \geq -2 + (-1)$
(e) $2 - (-4) \leq -3 + (-4)$ (f) $7 + (-9) \geq -5 + 3$
(g) $-4 - 5 \leq 6 + (-8)$ (h) $-8 + 2 \geq -4 - (-9)$

Objective 6; Example 5

11. Evaluate $-x$ when:

(a) $x = 4$ (b) $x = 0$
(c) $x = -1$ (d) $x = 3 + 2$
(e) $x = -(-2)$ (f) $x = (3 - 7)$
(g) $x = 3 + (-3)$ (h) $x = -2 + (-5)$
(i) $x = -7 + 2$ (j) $x = -(2 - 6)$
(k) $x = -3 - (-6)$ (l) $x = 8 + (-4)$

Objective 6; Example 6

12. Evaluate the following pairs and tell whether each represents additive inverses:

(a) 5 and -5 (b) $2 + (-2)$ and 0
(c) 4 and $-(8 - 4)$ (d) 3 and $-(-3)$
(e) $6 + (-4)$ and $-7 + 5$ (f) $-(4 + 3)$ and $2 + (-9)$
(g) $-(-8)$ and $-3 + (-5)$ (h) $-5 - (-4)$ and $-8 + 7$

Objective 7; Example 7

13. Use the definition of addition to find the following sums:

(a) $5 + 1$ (b) $-3 + (-7)$ (c) $8 + (-4)$
(d) $5 + (-9)$ (e) $-8 + 6$ (f) $-2 + 7$

Objective 8; Example 8

14. Evaluate each of the following expressions:

(a) $7 \times (-3)$ (b) $(-8) \times 9$
(c) $-4 \times (-5)$ (d) -6×0
(e) $5(2 + {}^{-}2)$ (f) $-(6 \times {}^{-}8)$
(g) $-(-9 \times {}^{-}2)$ (h) $-4(7 + {}^{-}3)$
(i) $6(4 + {}^{-}9)$ (j) $-3(-2 + {}^{-}6)$
(k) $2(7 - 9)$ (l) $-5(-4 - 3)$
(m) $-8(3 - {}^{-}5)$ (n) $-7(-2 \times {}^{-}3)$
(o) $(6 + {}^{-}3) \times (9 - 4)$ (p) $(4 + {}^{-}8) \times (-3 - 2)$
(q) $(2 - {}^{-}6) \times (-8 + 1)$ (r) $(-9 + 2) \times (-1 - {}^{-}5)$

Closure

15. Is the system of integers closed under the operation of
(a) addition? (b) multiplication?
(c) subtraction? (d) division?

Objective 5

16. Use integers g, h, and k (if needed) to express the following properties of the integers:
(a) Closure property of addition. Verify this property using 5 and -9.
(b) Closure property of multiplication. Verify this property using 6 and -7.
(c) Commutative property of addition. Verify this property using -8 and 2.
(d) Associative property of addition. Verify this property using 4, -6, and -7.
(e) Commutative property of multiplication. Verify the property using -3 and 8.
(f) Associative property of multiplication. Verify the property using -5, -2, and -3.
(g) Distributive property of multiplication over addition. Verify the property using -2, 4, and -7.
(h) Additive identity property. Verify the property using -4.
(i) Multiplicative identity property. Verify the property using -7.
(j) Additive inverse property. Verify the property using 8 and -8.

Closure

17. When we studied the system of whole numbers (or nonnegative integers), we found that those numbers were closed under the operations of addition and multiplication. Consider now just the set of *negative* integers: Use inductive reasoning to determine:
(a) Is the set of negative integers closed under the operation of addition? Explain.
(b) Is the set of negative integers closed under the operation of multiplication? Explain.

Subtraction

18. (a) If the operation of subtraction is closed for the system of integers, why is subtraction not listed as a fundamental operation of the system of integers?
(b) State the definition of subtraction for the system of integers.
(c) The definition of subtraction for integers involves concepts from two of the basic number properties. Which properties are used?

19. Which of the number properties below hold for the operation of subtraction for the integers? Give an example (that includes negative integers) to support your answer.
(a) Closure? (b) Commutative?
(c) Associative? (d) Distributive, $\times$ over $-$

Objective 9; Example 9

20. Convert each of the following subtraction problems to its related addition problem and compute the solution:

(a) $7 - 2$ (b) $5 - 9$ (c) $-3 - 5$

(d) $-8 - 2$ (e) $-4 - (-5)$ (f) $6 - (-2)$

(g) $\begin{array}{r} 2 \\ \underline{8} \end{array}$ (h) $\begin{array}{r} -3 \\ \underline{5} \end{array}$ (i) $\begin{array}{r} 7 \\ \underline{-8} \end{array}$ (j) $\begin{array}{r} -9 \\ \underline{-5} \end{array}$

Division

21. (a) Does the definition of division for whole numbers have to be changed in order to apply for the integers?

(b) Does the invention of negative integers help in any way to bring about closure for the operation of division?

(c) If $a \div b = b \div a$, then a is related to b in one of two ways. What are the two possible ways that a is related to b?

Objective 10; Example 10

22. Find the integer that represents each quotient:

(a) $\frac{-15}{3}$ (b) $\frac{0}{-2}$ (c) $\frac{12}{-4}$

(d) $\frac{-24}{-8}$ (e) $\frac{7 + (-3)}{-2}$ (f) $\frac{-4 + 9}{-3 + (-2)}$

(g) $\frac{7 + (-7)}{8 + (-2)}$ (h) $\frac{-2 + (-8)}{-9 + 4}$ (i) $\frac{3 - 9}{8 + (-5)}$

(j) $\frac{4 - (-8)}{-1 + 7}$ (k) $\frac{2 \times (-8)}{5 - (-3)}$ (l) $\frac{-(8 + 4)}{(-3) \times (-4)}$

(m) $\frac{(-3) \times 8}{2 \times (-2)}$ (n) $\frac{1 - 9}{-4 - (-2)}$ (o) $\frac{-2 + 8}{-1 - 2}$

(p) $\frac{(-8) \times 3}{-(4 - 8)}$

Objective 11; Items (36–38)

***23.** *Verify the sum of two negatives*: Aiming toward $11 + (-11) = 0$, start with $(7 + 4) + ({}^-7 + {}^-4)$ and use various number properties to prove the sum is actually 0. Which number property will then verify that $-7 + (-4) = -11$?

Items (36–38)

***24.** *Verify a negative plus a positive* (where the negative has the larger numerical component), based on Problem 23: Start with $-7 + 4$, rename -7 as the sum of two negatives, and use number properties to obtain the sum $-7 + 4 = -3$.

Items (43–47)

***25.** *Verify a positive times a negative*: Aiming toward $0 = 12 + (-12)$, start with $4(3 + {}^-3)$ and use various number properties to prove $0 = 12 + (4 \times {}^-3)$. Which number property will then verify that $4 \times (-3) = -12$?

*Exercises marked with an asterisk require more independent thinking by the student.

OBJECTIVES

After completing this unit, you should be able to

A. and G. The number line and addition

1 Draw a number line and use it to graph sets of points. **(2–5, 20–22)**

B. Order of whole numbers

2 Use the definition of order to correctly place the symbol "$<$" between two whole numbers or integers. **(6–10, 20–22)**

3 Use the trichotomy law to place the correct symbol $<$, $>$, or $=$ between two numbers; or tell whether given statements of inequality are true or false. **(11–15)**

C. The set of integers

4 Identify the characteristics of the set of integers and tell the distinctions between the integers and the sets of whole and natural numbers. **(17–22)**

Properties of the integers

5 (a) Express with variables and verify with integers the following properties from the natural and whole numbers: closure, commutative, and associative of addition and multiplication; distributive; and identity for addition and multiplication. **(Exercise 16)**

(b) Express with variables and verify with integers the additive inverse property. **(28)**

D. Additive inverse

6 Use the concept of additive inverses (a) to find the additive inverse of a given integer or (b) to determine whether the elements of a given pair are opposites. **(23–33)**

E. Addition

7 Use the definition of addition for integers in order to add any two integers. **(34–42)**

F. Multiplication

8 Find the product of any two integers. (When asked, show steps and name the property that permits each.) **(43–51)**

G. Subtraction

9 Use the definition of subtraction to convert any subtraction problem to the related addition problem, and compute the result. (52–60)

H. Division

10 Use the definition of division to find the quotient of integers. (61–66)

*11 Show steps and name the properties that verify the sum or product of two numbers. (Exercises *23–*25)

The Rational Numbers

The most basic system of numbers (the natural numbers) has now been extended two times—to become first the whole numbers and then the integers. In the system of natural numbers, the fundamental operations of addition and multiplication were closed, commutative, associative, distributive, and included the multiplicative identity. The whole numbers incorporated the element zero and thereby included the additive identity property. The invention of the negative integers permitted the additive inverse property and thereby allowed closure for the operation of subtraction. **(1)**

However, the system of integers still had two basic limitations: the principal limitation was that the operation of *division was not closed*, and also there was *no multiplicative inverse* property to correspond to the additive inverse property of the integers. It is thus desirable to further extend our number system.

In extending the system of integers, we will again employ the concept of the number line. The geometric concept of a line implies that, between the points already identified as integers, the number line also contains an infinite number of other points. This leads us to wonder if it would not be possible to invent another kind of number which could be matched with some of these points. **(2)**

As in an earlier example, practical experience indicated that some other type of number should exist. Suppose, for example, that an ancient **(3)** woodsman took a single stick and broke it into two equal pieces. Could he designate the whole stick as "one" and the equal pieces as "two" and say

$$1 = 2?$$

Or, could he consider each broken piece as "one" and the original whole piece as "one" and say

$$1 + 1 = 1?$$

Since the numbers he knew were not adequate to represent the situation, some type of new number was needed.

A. MULTIPLICATIVE INVERSE

(4) We have seen that each additive property seems to have a corresponding multiplicative property. Since it was possible to obtain closure for subtraction by inventing an additive inverse, perhaps closure for division might be achieved if we invented an inverse element for multiplication.

Now, the sum of the additive inverses equals the additive identity element, zero; that is, $a + (-a) = 0$. Using this same pattern as a guide, we would say "the product of the multiplicative inverses should equal the multiplicative identity element, one." That is, for some multiplicative inverse element, j,

$$a \times j = 1$$

There exists no integer j for which $-4 \times j = 1$, for example.

(5) Now, by the definition of division, for each integer $a, b\ (b \neq 0)$, and k,

$$\frac{a}{b} = k \quad \text{iff} \quad a = b \times k$$

(6) In searching for a multiplicative inverse, we would need two elements whose product is 1. This product results when $a = 1$; the second part of our definition then becomes

$$1 = b \times k$$

The entire definition could then be written (in reverse order) as

$$b \times k = 1 \quad \text{iff} \quad k = \frac{1}{b}$$

This manipulation would also imply, by substitution, that

$$b \times \frac{1}{b} = 1$$

(7) Thus, the multiplicative inverse of b could be some new number of the form $\frac{1}{b}$. The multiplicative inverse would then conform to the following general definition:

Multiplicative inverse property

For each element a $(a \neq 0)$ there exists a *multiplicative inverse*, $\frac{1}{a}$, such that,

$$a \times \frac{1}{a} = \frac{1}{a} \times a = 1$$

No multiplicative inverse for zero

(8) Observe the restriction that $a \neq 0$. This restriction traces directly back to our original definition of division, since division by zero is not defined. We thus conclude that there is no multiplicative inverse for zero. That is, we are developing a set of elements such that, for every integer $a \neq 0$, there is an element $\frac{1}{a}$:

$$\left\{\ldots, \frac{1}{-3}, -3, \frac{1}{-2}, -2, \frac{1}{-1}, -1, 0, 1, \frac{1}{1}, 2, \frac{1}{2}, 3, \frac{1}{3}, 4, \frac{1}{4}, \ldots\right\}$$

B. RATIONALS ON THE NUMBER LINE

(9) Let us return to the number line to determine how these elements might be matched with points on the line. Consider the new element, $\frac{1}{4}$, for example. Since this number resembles a division problem, it seems logical to locate the element on the number line by dividing one unit into four equal parts. The first point may be identified as $\frac{1}{4}$, but then two other points remain unidentified:

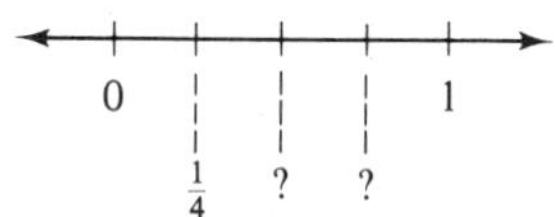

(10) It would seem logical that perhaps these unidentified points might be multiples of the first point. Could the second point, for example, be

$$2 \times \frac{1}{4} \qquad \text{and identified} \qquad \frac{2}{4}?$$

If this were the case, then the point following the fourth part should be

$$4 \times \frac{1}{4} \qquad \text{and identified as} \qquad \frac{4}{4}$$

But this point has already been identified as 1. If our scheme for identifying points were correct, then it would have to hold that $\frac{4}{4} = 1$.

(11) Now, the new elements that we have devised look very much like division problems. Furthermore, we can use the definition of division to state that

$$\frac{4}{4} = 1 \qquad \text{since} \qquad 4 = 4 \times 1$$

Objective 1:

(12) Thus, we conclude that our new method for identifying points on the number line does indeed work. That is, for any integer n (either positive or

negative), each unit distance may be divided into n equal parts and these parts then identified as $\frac{1}{n}, \frac{2}{n}, \frac{3}{n}, \ldots, \frac{|n|}{n}$. Points obtained by choosing a negative integer n would be marked off in the negative direction on the number line. Successive points beyond the unit distance are similarly identified as multiples of $\frac{1}{n}$:

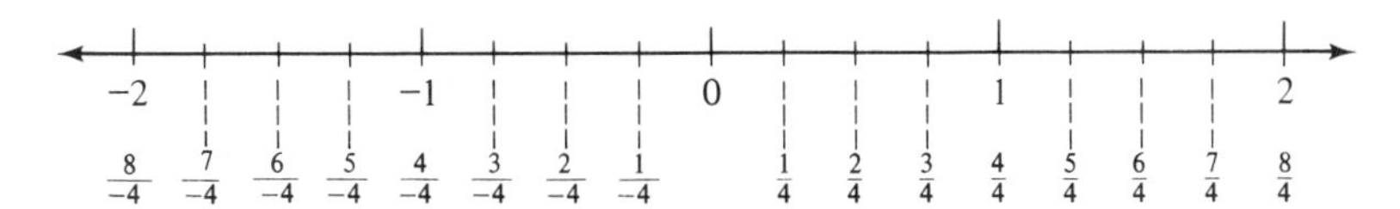

As before, we can verify for example that $\frac{8}{-4} = -2$, since $8 = (-4) \times (-2)$.

These new elements are defined as the set of *rational numbers*, as follows: (13)

Definition of rational numbers

The set, F, of rational numbers consists of all elements which can be expressed in the form $\frac{a}{b}$, where a and b are integers and $b \neq 0$.

As we have observed, a rational number expressed in the form $\frac{a}{b}$ bears a strong resemblance to the operation of division. Because of this similarity, a rational number is sometimes defined as "any number which can be expressed as the quotient of two integers, $\frac{a}{b}$, where $b \neq 0$." Thus, all *common fractions* may also be identified as rational numbers. Recall for every common fraction (or rational number) $\frac{a}{b}$, the element a is called the *numerator* and b is called the *denominator*. (14)

However, we must emphasize the word "can" in the definition of a rational number as "any number which *can* be expressed as the quotient of two integers." You have previously learned, for instance, that 0.5 can be expressed as $\frac{1}{2}$. By this definition, 0.5 is a rational number because it "can be expressed" as $\frac{1}{2}$. The symbols 0.5 and $\frac{1}{2}$ are simply two different names for the same rational number. This concept will be studied in more detail in the chapter on real numbers.

We have seen that, for every integer $n \neq 0$, each unit distance may be divided into a set of points designated as rational numbers. Since the set of integers contains an infinite number of elements, the set, F, of rational numbers is also an infinite set. It was noted earlier that the number line contains an infinite number of points besides those identified as integers. We now see that the set of rational numbers may be matched with many of these points. (15)

Integers as rational numbers

Every integer (including zero) may be expressed in the form $\frac{a}{b}$ where $b \neq 0$. For example, **(16)**

$$4 = \frac{4}{1} \quad \text{or} \quad -5 = \frac{-5}{1} \quad \text{or} \quad 0 = \frac{0}{1}$$

Each integer may therefore be identified as a rational number. Hence the set, I, of integers is a proper subset of the set, F, of rational numbers. That is,

$$\boldsymbol{I} \subset \boldsymbol{F}$$

Since the natural numbers and whole numbers have also been shown to be proper subsets of the integers, then it follows that they are also proper subsets of the rational numbers. Thus, we could say

$$\boldsymbol{N} \subset \boldsymbol{W} \subset \boldsymbol{I} \subset \boldsymbol{F}$$

These relationships are depicted in the following diagram.

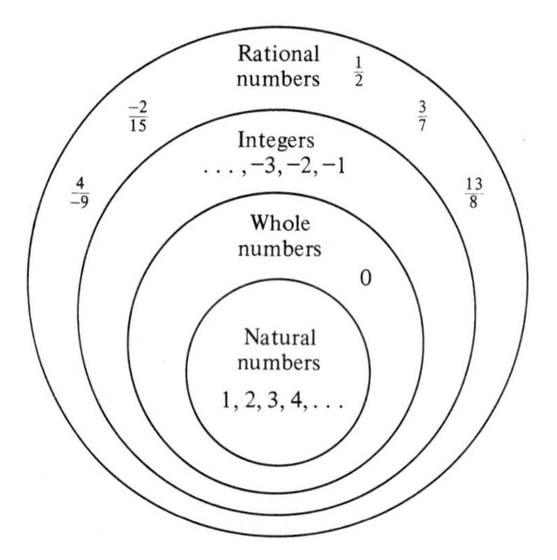

Exercises will be given to determine that all properties from the system of integers still hold. The investigation of new properties will also be of interest, and the multiplicative inverse property has already been identified: for any integer* $a \neq 0$, **(17)**

Objective 2b

$$a \times \frac{1}{a} = \frac{1}{a} \times a = 1$$

For example, $$5 \times \frac{1}{5} = \frac{1}{5} \times 5 = 1$$

or, $$-2 \times \frac{1}{-2} = \frac{1}{-2} \times -2 = 1$$

*The multiplicative inverse property will be considered further, beginning at item (34), in order to apply it for all rational numbers, rather than to just integers.

We will be particularly interested, however, to determine whether the set of rational numbers can be used to bring about closure for the operation of division.

C. EQUALS RELATION IN THE RATIONALS

(18) When the unit distance on the number line is divided into n equal parts, we find that many points representing rational numbers that are not integers appear to coincide with each other. For instance, from the diagram

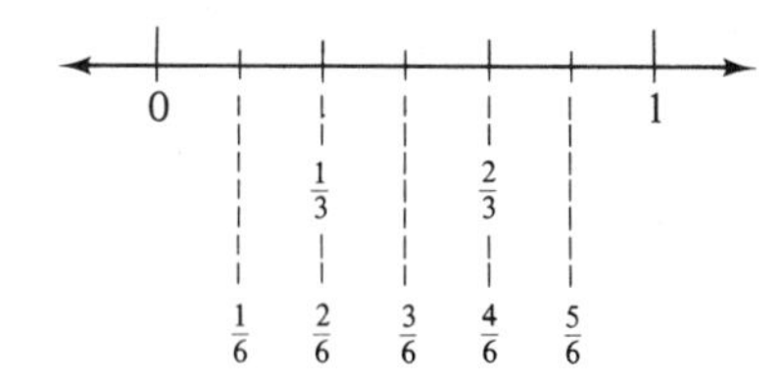

it appears that

$$\frac{1}{3} = \frac{2}{6} \quad \text{and} \quad \frac{2}{3} = \frac{4}{6}$$

(19) Now, the equals relation holds $(a = b)$ when the symbols a and b are two names for the same number. By drawing a number line we have seen that two fractions may seem to be names for the same rational number. The equality (or inequality) of two common fractions may be definitely determined using the following definition:

Objective 3: Definition of equality for common fractions

For any two common fractions $\frac{a}{b}$ and $\frac{c}{d}$, (where $b \neq 0$ and $d \neq 0$),

$$\frac{a}{b} = \frac{c}{d} \quad \textit{if and only if} \quad ad = bc$$

(20) ◆ **Example 1** Use the definition of equality to determine whether the following fractions are names for the same rational number:

(a) Does $\frac{1}{3} = \frac{2}{6}$? (b) Does $\frac{4}{8} = \frac{15}{32}$?

(a) $\frac{1}{3} = \frac{2}{6}$ since $1 \cdot 6 = 3 \cdot 2$

$6 = 6$

(b) $\frac{4}{8} \neq \frac{15}{32}$ since $4 \cdot 32 \neq 8 \cdot 15$

$128 \neq 120$

(21) ● **Practice 1** Use the definition of equality to determine whether each of the given pairs of common fractions is equal:

(a) Does $\frac{1}{4} = \frac{4}{16}$? (b) Does $\frac{3}{5} = \frac{16}{25}$?

(c) Does $\frac{8}{17} = \frac{48}{102}$? (d) Does $\frac{6}{11} = \frac{75}{132}$?

See also Exercise 20.

(22) In developing a *system* of rational numbers, we have thus far devised a set of elements (the rational numbers themselves) and have identified the equals relation. In considering operations for the system, we will first discuss multiplication and then addition (as well as their variations, division and subtraction). We will also discuss which number properties would hold for each operation.

D. MULTIPLICATION IN THE RATIONALS

(23) Our initial development of the set of rational numbers was oriented towards obtaining multiplicative inverses $\left(\frac{1}{n}\right)$ for the set of integers. Having derived on the number line elements of the type $\frac{1}{n}$, we then located other elements which we considered multiples of the first points. Thus, we designated the second point as

$$2 \times \frac{1}{n} \quad \text{identified as} \quad \frac{2}{n}$$

and so on.

Objective 4: Definition of multiplication (a)

(24) The concept that was used in naming these points thus provides the basis for the first of three definitions of multiplication for rational numbers:

For any integers a and b $(b \neq 0)$,

$$a \times \frac{1}{b} = \frac{a}{b}$$

Definition of multiplication (b)

(25) The second definition pertains to the multiplication of two factors which are both of the form $\frac{1}{n}$:

For each rational number $\frac{1}{a}$ and $\frac{1}{b}$ (where $a \neq 0$ and $b \neq 0$),

$$\frac{1}{a} \times \frac{1}{b} = \frac{1}{ab}$$

(26) A unit-square diagram will help visualize this concept of the multiplication of rational numbers. Suppose we divide one side of a unit

square into halves and one side into thirds. The shaded portion illustrates that $\frac{1}{2} \times \frac{1}{3}$ represents $\frac{1}{6}$ of the total unit.

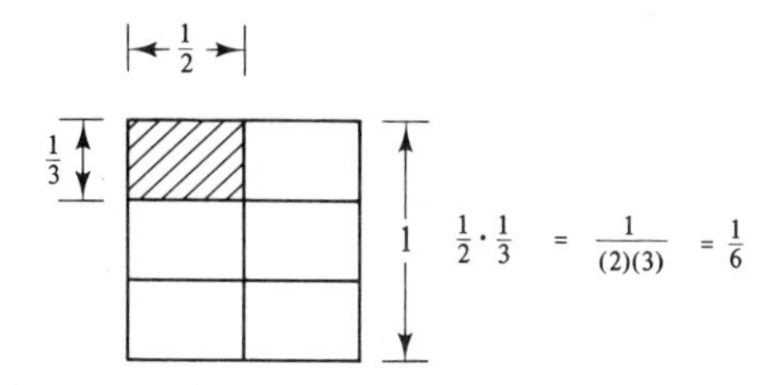

The third definition of multiplication covers the multiplication of any rational numbers, rather than just special cases such as the first two definitions. However, the third definition is derived from the first two, and thus is often given as a theorem (a statement that must be proved). (27)

Definition of multiplication (c)

For any rational numbers $\frac{a}{b}$ and $\frac{c}{d}$,

$$\frac{a}{b} \times \frac{c}{d} = \frac{ac}{bd}$$

We can prove that $\frac{a}{b} \times \frac{c}{d} = \frac{ac}{bd}$, using number properties and the definitions for $a \times \frac{1}{b}$ and for $\frac{1}{a} \times \frac{1}{b}$, as follows: (28)

Multiplication verified

$\frac{a}{b} \times \frac{c}{d}$	$= \left(a \cdot \frac{1}{b}\right) \times \left(c \cdot \frac{1}{d}\right)$	Definition of $a \times \frac{1}{b}$
	$= a\left[\frac{1}{b} \times \left(c \cdot \frac{1}{d}\right)\right]$	Associative, $\times$
	$= a\left[\left(\frac{1}{b} \times c\right) \cdot \frac{1}{d}\right]$	Associative, $\times$
	$= a\left[\left(c \times \frac{1}{b}\right) \cdot \frac{1}{d}\right]$	Commutative, $\times$
	$= a\left[c \times \left(\frac{1}{b} \cdot \frac{1}{d}\right)\right]$	Associative, $\times$
	$= (ac) \times \left(\frac{1}{b} \cdot \frac{1}{d}\right)$	Associative, $\times$
	$= (ac) \times \left(\frac{1}{bd}\right)$	Definition of $\frac{1}{a} \times \frac{1}{b}$
	$= \frac{ac}{bd}$	Definition of $a \times \frac{1}{b}$

The shaded portion of the following unit square illustrates that $\frac{2}{3} \times \frac{2}{5}$ is $\frac{4}{15}$ of the total unit: (29)

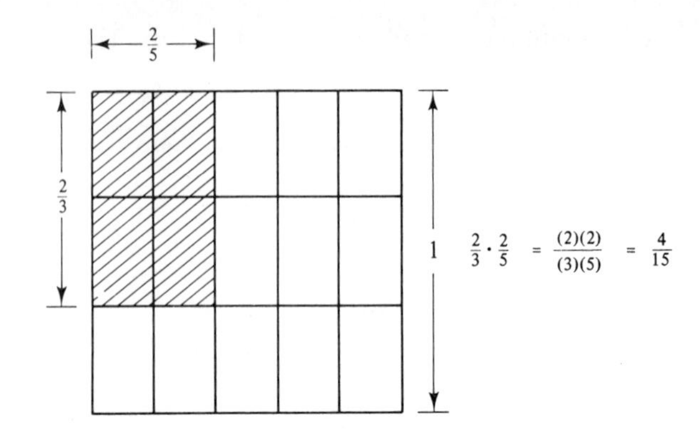

◆ **Example 2** The definition of multiplication is used to find the indicated products: **(30)**

(a) $3 \times \frac{1}{7} = \frac{3}{7}$

(b) $\frac{1}{-3} \times \frac{1}{-2} = \frac{1}{(-3)(-2)} = \frac{1}{6}$

(c) $\frac{2}{5} \times \frac{-3}{5} = \frac{(2)(-3)}{(5)(5)} = \frac{-6}{25}$

(d) $\frac{2}{-3} \times \frac{-4}{7} = \frac{(2)(-4)}{(-3)(7)} = \frac{-8}{-21} = \frac{(-1)8}{(-1)21} = \frac{8}{21}$

Multiplication for negative rational numbers thus follows the previous rules for multiplication of negative integers. (Notice that these problems did not have to be "reduced." This concept is included later in Chapter 6, The Theory of Numbers.)

● **Practice 2** Use the definitions of multiplication to find the following products: **(31)**

(a) $4 \times \frac{1}{5}$ (b) $-3 \times \frac{1}{5}$ (c) $-2 \times \frac{1}{-3}$

(d) $\frac{1}{2} \times \frac{1}{5}$ (e) $\frac{1}{4} \times \frac{1}{-8}$ (f) $\frac{1}{-6} \times \frac{1}{-2}$

(g) $\frac{-2}{9} \times \frac{4}{-5}$ (h) $\frac{-3}{4} \times \frac{3}{-5}$ (i) $\frac{2}{-5} \times \frac{-3}{-7}$

See also Exercise 21.

As would be expected, all the familiar multiplicative properties from the earlier number systems also apply in the system of rational numbers. It is customary to express each property in the same form as it was previously written; for example, a rational number would ordinarily be represented by a, rather than by $\frac{a}{b}$. **(32)**

Multiplicative inverse property (cont.)

At this point, then, we need to consider further the multiplicative inverse property: $a \times \frac{1}{a} = \frac{1}{a} \times a = 1$ for every $a \neq 0$. Students often think of the element a as representing only an integer and the element $\frac{1}{a}$ as being the rational part of the property. It is essential to realize that *every* rational number $\neq 0$ (not just the integers) has a *multiplicative inverse* or *reciprocal*. (33)

Objective 5:

◆ **Example 3** The reciprocal of 4 is $\frac{1}{4}$ (34)

of $\frac{1}{3}$ is $\frac{3}{1}$

of $\frac{-2}{3}$ is $\frac{3}{-2}$

In each case, any element ($\neq 0$) times its inverse equals one. That is, for any rational number $\frac{a}{b}$ $\left(\text{where } \frac{a}{b} \neq 0\right)$

$$\frac{a}{b} \times \frac{b}{a} = 1$$

For example, $\frac{-2}{3} \times \frac{3}{-2} = \frac{(-2)(3)}{(3)(-2)} = \frac{-6}{-6} = 1$

It should be emphasized again that any rational number which equals zero will *not* have a multiplicative inverse.

● **Practice 3** Illustrate the multiplicative inverse property by multiplying each of the following rational numbers times its multiplicative inverse: (35)

(a) 5 (b) -8 (c) $\frac{1}{2}$ (d) $\frac{1}{-4}$

(e) $\frac{3}{5}$ (f) $\frac{8}{5}$ (g) $\frac{-3}{7}$ (h) $\frac{4}{-3}$

See also Exercise 22.

E. DIVISION IN THE RATIONALS

We are now ready to consider the operation of division in the system of rational numbers, for which the new concept of the multiplicative inverse is a basic part. Division in the rationals is defined as follows: (36)

Objective 6: Definition of division for rational numbers

For any rational numbers $\frac{a}{b}$ and $\frac{c}{d}$ $\left(\text{where } b \neq 0 \text{ and } \frac{c}{d} \neq 0\right)$,

$$\frac{a}{b} \div \frac{c}{d} = \frac{a}{b} \cdot \frac{d}{c}$$

For example,

$$\frac{1}{4} \div \frac{2}{3} = \frac{1}{4} \cdot \frac{3}{2} = \frac{3}{8}$$

(37) As in the case of our definition of multiplication, this definition could have been stated as a theorem, since it can be proved, as we will show. Since $\frac{a}{b}$ means $a \div b$, the definition of division for integers,

$$\frac{a}{b} = k \quad \text{iff} \quad a = b \times k$$

is first restated by replacing integers a and b with rational numbers $\frac{a}{b}$ and $\frac{c}{d}$ $\left(\text{where } \frac{c}{d} \neq 0\right)$, as follows:

$$\frac{a}{b} \div \frac{c}{d} = q \quad \text{iff} \quad \frac{a}{b} = \frac{c}{d} \times q$$

This definition doesn't provide much help in finding the *value* of q, however! But, since $\frac{c}{d} \neq 0$, it has an inverse, $\frac{d}{c}$, which we use in the second step below:

Division verified (38)

$$\frac{a}{b} \div \frac{c}{d} = q \quad \text{implies}$$

$$\frac{a}{b} = \left(\frac{c}{d} \times q\right) \quad \text{Definition of division (for integers)}$$

Then

$$\left(\frac{c}{d} \times q\right) = \frac{a}{b} \quad \text{implies}$$

$$\frac{d}{c}\left(\frac{c}{d} \times q\right) = \frac{d}{c} \cdot \frac{a}{b} \quad \text{Closure for multiplication}$$

$$\left(\frac{d}{c} \cdot \frac{c}{d}\right)q = \frac{d}{c} \cdot \frac{a}{b} \quad \text{Associative property of multiplication}$$

$$1 \cdot q = \frac{d}{c} \cdot \frac{a}{b} \quad \text{Multiplicative inverse property}$$

$$q = \frac{d}{c} \cdot \frac{a}{b} \quad \text{Multiplicative identity}$$

$$q = \frac{a}{b} \cdot \frac{d}{c} \quad \text{Commutative property of multiplication}$$

Now we have

$$\frac{a}{b} \div \frac{c}{d} = q \quad \text{and also} \quad q = \frac{a}{b} \cdot \frac{d}{c}$$

Therefore, by substitution, we verify that

$$\frac{a}{b} \div \frac{c}{d} = \frac{a}{b} \cdot \frac{d}{c}$$

This definition is the basis of the rule students often use: "To divide one fraction by another, invert the divisor and multiply." However, the definition has more significance than simply being a method for performing division. Concepts from *two number properties* are employed in this definition: Notice that the operation of division is expressed in terms of multiplication, using the *multiplicative inverse** or reciprocal. Therefore, since we have *closure for multiplication* in the system of rational numbers, we can now say that the operation of *division is closed* (except by 0) for the system of rational numbers. (39)

Closure for division

This was a primary goal in developing the system of rational numbers and it is especially significant: now that division is closed, we have closure for all four functions. This succeeds in making the system of rational numbers a "finished" system in the sense that any operation can be begun and we know it will have a solution that is a rational number. (40)

◆ **Example 4** In each case below, $\frac{a}{b} \div \frac{c}{d} = \frac{ad}{bc}$ provided $\frac{c}{d} \neq 0$: (41)

(a) $\frac{-4}{1} \div \frac{5}{1} = \frac{-4}{1} \cdot \frac{1}{5} = \frac{-4}{5}$

(b) $\frac{-4}{5} \div \frac{-3}{2} = \frac{-4}{5} \cdot \frac{2}{-3} = \frac{-8}{-15} = \frac{8}{15}$

(c) $\frac{0}{5} \div \frac{1}{-2} = \frac{0}{5} \cdot \frac{-2}{1} = \frac{0}{5} = 0$

(d) $\frac{2}{5} \div \frac{0}{3}$ is impossible since $\frac{c}{d} = 0$

● **Practice 4** Apply the definition of division for rational numbers to solve the following: (42)

(a) $\frac{1}{4} \div \frac{1}{3}$ (b) $5 \div \frac{1}{-2}$ (c) $\frac{1}{-3} \div 2$

(d) $\frac{2}{3} \div \frac{3}{4}$ (e) $\frac{1}{-5} \div \frac{2}{3}$ (f) $-4 \div (-3)$

(g) $\frac{4}{7} \div \frac{0}{5}$ (h) $\frac{0}{2} \div \frac{2}{3}$ (i) $\frac{-2}{5} \div \frac{3}{-7}$

See also Exercise 23.

*We have said that the multiplicative inverse of any rational number a is represented by $\frac{1}{a}$. This implies that the reciprocal of $\frac{2}{3}$ is $\frac{1}{2/3}$. However, we then expressed the reciprocal of $\frac{2}{3}$ as $\frac{3}{2}$—an apparent contradiction. Using the definition of division for rationals, however, notice that

$$\frac{1}{2/3} = 1 \div \frac{2}{3} = 1 \times \frac{3}{2} = \frac{3}{2}$$

Thus, $\frac{1}{2/3}$ and $\frac{3}{2}$ are two different names for the same rational number; both represent the reciprocal of $\frac{2}{3}$.

Just as there are corresponding properties for the operations of addition and multiplication, so is there certain correspondence between subtraction and division. In particular, closure for the operation of subtraction was obtained by applying the additive inverse. Similarly, closure for the operation of division was obtained by applying the multiplicative inverse. (43)

Despite the achievement of closure, division (like subtraction) is still not listed as a fundamental operation of the system, since division is defined in terms of multiplication. As you would also probably expect, the achievement of closure for division still does not allow any of the other number properties to hold. (44)

F. ADDITION IN THE RATIONALS

Addition for rational numbers

Objective 7:

Addition of natural numbers was first introduced as the union of two sets. Later, the number line was used to develop the concept of addition for integers. The definition of addition for the system of rational numbers retains the same procedure used with integers in order to determine whether the sum is positive or negative. However, for the first time, the sum of two numbers now can be expressed by simple equations: (45)

Definition of addition (a)

For any rational numbers $\frac{a}{c}$ and $\frac{b}{c}$ (where $c \neq 0$), (46)

$$\frac{a}{c} + \frac{b}{c} = \frac{a+b}{c}$$

For example,

$$\frac{2}{7} + \frac{3}{7} = \frac{2+3}{7} = \frac{5}{7}$$

This definition of addition can be derived and verified as follows: (47)

Addition verified (a)

$\frac{a}{c} + \frac{b}{c}$	$= \left(a \cdot \frac{1}{c}\right) + \left(b \cdot \frac{1}{c}\right)$	Definition of multiplication: $a \times \frac{1}{b}$
	$= \left(\frac{1}{c} \cdot a\right) + \left(\frac{1}{c} \cdot b\right)$	Commutative property of multiplication
	$= \frac{1}{c} \cdot (a+b)$	Distributive property, $\times$ over $+$
	$= (a+b) \cdot \frac{1}{c}$	Commutative property of multiplication
	$= \frac{a+b}{c}$	Definition of multiplication: $a \times \frac{1}{b}$

The preceding definition for addition may be extended to rational numbers with different denominators as follows: (48)

Definition of addition (b)

For any rational numbers $\frac{a}{b}$ and $\frac{c}{d}$ (where $b \neq 0$ and $d \neq 0$),

$$\frac{a}{b} + \frac{c}{d} = \frac{ad + bc}{bd}$$

Exercise 10 contains the proof of this statement, which applies the concept of the multiplicative identity $(a \times 1 = a)$, two definitions of multiplication $\left(\frac{a}{b} \times \frac{c}{d} \text{ and } a \times \frac{1}{b}\right)$, and other familiar properties.

Perhaps, when you first learned to add fractions, you thought of it as a mysterious new procedure unrelated to anything else you had done in arithmetic. We see now that the operation of addition for the system of rational numbers is based on the same basic (even "obvious") properties of the earlier systems. (49)

◆ **Example 5** Use $\frac{a}{c} + \frac{b}{c} = \frac{a+b}{c}$ to find the following sums: (50)

(a) $\frac{1}{5} + \frac{3}{5} = \frac{1+3}{5} = \frac{4}{5}$

(b) $\frac{6}{7} + \frac{-3}{7} = \frac{6+(-3)}{7} = \frac{3}{7}$

Apply the definition $\frac{a}{b} + \frac{c}{d} = \frac{ad+bc}{bd}$ for fractions having different denominators:

(c) $\frac{2}{3} + \frac{2}{5} = \frac{(2)(5)+(3)(2)}{(3)(5)} = \frac{10+6}{15} = \frac{16}{15}$

(d) $\frac{-3}{8} + \frac{-2}{5} = \frac{(-3)(5)+(8)(-2)}{(8)(5)} = \frac{-15+(-16)}{40} = \frac{-31}{40}$

● **Practice 5** Apply the definition of addition to determine the sums of the following rational numbers: (51)

(a) $\frac{3}{7} + \frac{2}{7}$ (b) $\frac{-3}{5} + \frac{4}{5}$ (c) $\frac{2}{9} + \frac{-7}{9}$

(d) $\frac{1}{3} + \frac{2}{5}$ (e) $\frac{-5}{7} + \frac{2}{5}$ (f) $\frac{-2}{7} + \frac{-2}{3}$

(g) $\frac{-2}{3} + \frac{1}{-5}$ (h) $\frac{2}{-7} + \frac{-3}{5}$ (i) $\frac{2}{-5} + \frac{1}{-3}$

See also Exercise 24.

Development of the system of rational numbers includes all previous addition properties but did not bring about any new properties pertaining to the operation of addition. (52)

G. SUBTRACTION IN THE RATIONALS

Objective 8: Definition of subtraction (a)

Earlier, the operation of subtraction for integers was defined as: $a - b = a + (-b)$. This same definition may be restated as follows: (53)

For any rational numbers $\frac{a}{c}$ and $\frac{b}{c}$, (where $c \neq 0$),

$$\frac{a}{c} - \frac{b}{c} = \frac{a}{c} + \frac{(-b)}{c}$$

Or, applying the definition of addition for rational numbers,

Same denominator

$$\frac{a}{c} - \frac{b}{c} = \frac{a + (-b)}{c}$$

Definition of subtraction (b)

Similarly, for any rational numbers $\frac{a}{b}$ and $\frac{c}{d}$ (where $b \neq 0$ and $d \neq 0$), (54)

$$\frac{a}{b} - \frac{c}{d} = \frac{a}{b} + \frac{-c}{d}$$

Different denominators

Or,

$$\frac{a}{b} - \frac{c}{d} = \frac{ad + b(-c)}{bd}$$

Justification for these procedures is not included here because they are almost identical to the proofs for addition with rational numbers. (The subtraction proofs would require one extra step to show, for example, that $\frac{a}{c} - \frac{b}{c} = \frac{a}{c} + \frac{-b}{c}$, by the definition of subtraction.)

◆ **Example 6** Use the definition $\frac{a}{c} - \frac{b}{c} = \frac{a + (-b)}{c}$ to subtract: (55)

(a) $\frac{5}{7} - \frac{2}{7} = \frac{5 + (-2)}{7} = \frac{3}{7}$

By the definition $\frac{a}{b} - \frac{c}{d} = \frac{ad + b(-c)}{bd}$ for different denominators

(b) $\frac{2}{3} - \frac{3}{5} = \frac{(2)(5) + (3)(-3)}{(3)(5)} = \frac{10 + (-9)}{15} = \frac{1}{15}$

(c) $\frac{-1}{4} - \frac{-2}{3} = \frac{(-1)(3) + (4)(2)}{(4)(3)} = \frac{-3 + 8}{12} = \frac{5}{12}$

● **Practice 6** Apply the definition of subtraction for rational numbers to solve the following: (56)

(a) $\frac{4}{5} - \frac{2}{5}$ (b) $\frac{-1}{7} - \frac{2}{7}$ (c) $\frac{-7}{9} - \frac{-2}{9}$

(d) $\frac{4}{7} - \frac{1}{2}$ (e) $\frac{3}{5} - \frac{3}{8}$ (f) $\frac{2}{5} - \frac{-2}{7}$

(g) $\frac{-3}{7} - \frac{1}{4}$ (h) $\frac{-2}{9} - \frac{-2}{5}$ (i) $\frac{-3}{4} - \frac{2}{3}$

See also Exercise 25.

Subtraction properties?

As before, the only properties of the operation are that subtraction is closed and that multiplication is distributive over subtraction in the rational numbers. (57)

H. DENSITY OF RATIONAL NUMBERS

In developing the set of rational numbers, so many new points have been identified on the number line that it is now said to be *dense*; that is, between any two given points representing rational numbers, there is always another point associated with another rational number. This property can be stated as follows: (58)

Definition of density

For any two rational numbers a and b where $a < b$, there exists another rational number c, such that (59)

$$a < c < b$$

This definition utilizes the concept of order ($a < b$ iff $a + k = b$, where $k > 0$), as well as the trichotomy law ($a = b$ or $a < b$ or $a > b$), which apply in the rational numbers as in the integers.

Objective 9:

The density property may be illustrated by finding the *arithmetic mean* of two given numbers, which can be represented by the expression, $\frac{1}{2}(a + b)$. On the number line, the point representing the arithmetic mean is half way between the given points. (60)

◆ **Example 7**

(a) Use $\frac{1}{2}(a + b)$ to locate another rational number between $\frac{1}{11}$ and $\frac{2}{11}$: (61)

$$\frac{1}{2}\left(\frac{1}{11} + \frac{2}{11}\right) = \frac{1}{2}\left(\frac{3}{11}\right) = \frac{3}{22}$$

(b) Find another rational number between $\frac{1}{11}$ and $\frac{3}{22}$:

$$\frac{1}{2}\left(\frac{1}{11} + \frac{3}{22}\right) = \frac{1}{2}\left(\frac{5}{22}\right) = \frac{5}{44}$$

These points could be depicted on the number line as shown: (62)

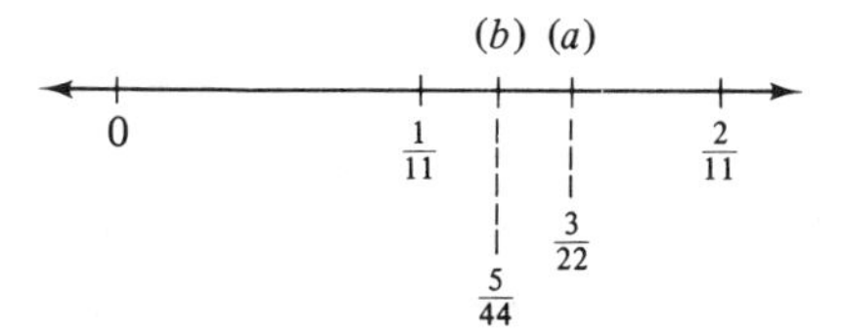

From (a): $\frac{1}{11} < \frac{3}{22} < \frac{2}{11}$

and in (b): $\frac{1}{11} < \frac{5}{44} < \frac{3}{22}$

By the density property, this process could be continued indefinitely. Each new number would be successively closer to $\frac{1}{11}$, but would never be quite that small in value.

(c) Find the arithmetic mean for $\frac{2}{3}$ and $\frac{3}{5}$: (63)

$$\frac{1}{2}(a+b) = \frac{1}{2}\left(\frac{2}{3}+\frac{3}{5}\right) = \frac{1}{2}\left(\frac{10+9}{15}\right) = \frac{1}{2}\left(\frac{19}{15}\right) = \frac{19}{30}$$

● **Practice 7** Illustrate the density property by locating another rational number between the following pairs: (64)

(a) $\frac{1}{13}$ and $\frac{2}{13}$ (b) $\frac{1}{13}$ and $\frac{3}{26}$ (c) $\frac{7}{17}$ and $\frac{8}{17}$

(d) $\frac{7}{17}$ and $\frac{15}{34}$ (e) $\frac{1}{2}$ and $\frac{2}{3}$ (f) $\frac{3}{8}$ and $\frac{4}{7}$

(g) $\frac{3}{4}$ and $\frac{4}{5}$

I. PROPERTIES OF THE RATIONAL NUMBERS

As noted previously, addition and multiplication for rational numbers include the fundamental number properties from earlier number systems. As an example, consider the closure properties, which are easily verified: for any sum $\frac{a}{b}+\frac{c}{d}=\frac{ad+bc}{bd}$ or any product $\frac{a}{b}\cdot\frac{c}{d}=\frac{ac}{bd}$, each a, b, c, and d are integers (where $b \neq 0$, $d \neq 0$, and $bd \neq 0$). Because the integers are closed under addition and multiplication, the numerator and denominator of each result is also an integer. Each result therefore represents a rational number (that is, the quotient of two integers). Thus, the system of rational numbers is closed under the operations of addition and multiplication. (65)

Closure for addition and multiplication

Our development of the *system* of rational numbers has now included all of the elements, operations, relations, and properties of the earlier systems: the closure, commutative, and associative properties of addition and multiplication, the distributive property, and the multiplicative identity, additive identity, and additive inverse properties. The rational numbers have also extended the system to include the multiplicative inverse property. These preceding eleven properties are called the *field properties*. Accordingly, any set of elements which has two operations for (66)

Field properties

which these properties apply is called a *field*. The rational numbers are thus the first system we have studied that constitutes a field.

(67) The other important achievement of the system of rational numbers is closure for division. The rational numbers are thus a "finished" system in that all of the operations of the system are finally closed. Also, the number line is so dense with rational numbers that between any two elements there is always another rational number.

Limitations of the rational numbers

(68) Knowing the foregoing facts, it might seem that every point on the number line has now been identified. In spite of the density of the rational numbers, however, there still remain an infinite number of unidentified points. In the next extension of our number system—the real numbers—we will finally establish a one-to-one correspondence between all the points on the number line and the new set of numbers.

(69) Before considering these new numbers, however, we will first study some applications of the numbers we have developed thus far. This topic on number theory will also include some additional exercises in operations with rational numbers, similar to examples presented in this section but utilizing the new theory.

(70) Properties of the system of rational numbers are summarized in the table on page 434. Reference may also be made to the diagram on p. 435 to illustrate development of the system of rational numbers.

EXERCISES

Subsets; greatest and least element; density

1. (a) What do we call a number that can be expressed as the quotient of two integers?
(b) Is every integer a rational number?
(c) Is every rational number an integer?
(d) Is the system of rational numbers a subset of the system of natural numbers?
(e) Is the system of whole numbers a subset of the system of rational numbers?
(f) Is there a greatest rational number? A least rational number? If so, name each.
(g) Is there a least positive rational number? If so, what is it? Which property of the rational numbers answers this question?
(h) Is there a least nonnegative rational number? If so, name it.
(i) How many rational numbers could be found between 0 and $\frac{1}{100}$?
(j) Part (i) is an example of which property of the rational numbers? Explain.
(k) Is there a one-to-one correspondence between the elements of the rational numbers and a set of points on the number line?

(l) Is there a one-to-one correspondence between all the points on the number line and the set of rational numbers? Explain.

2. Answer the following questions. Explain each "no" response.

Reciprocals

(a) What fundamental property of the system of rational numbers did not exist for the system of integers?
(b) Does every rational number have a multiplicative inverse that is a rational number?
(c) Does every negative rational number have a reciprocal that is a rational number?
(d) Is the reciprocal of a positive rational number always a negative rational number?

3. Answer the following, explaining any "no" answers.

Parts of a rational number

(a) Express the element "1" in rational form (the quotient of two integers).
(b) Express the element "0" in rational form.
(c) If $\frac{a}{b}$ is any rational number, can $a = 0$?
(d) Can $b = 0$?
(e) If rational number $\frac{a}{b} = 0$, does $a = 0$?
(f) Does $b = 0$? If not, what numbers are permitted as values for b?
(g) If rational number $\frac{a}{b}$ has a multiplicative inverse $\frac{b}{a}$, can $\frac{b}{a} = 0$?
(h) For any integer, a, does $\frac{a}{a}$ always equal 1?
(i) If any nonzero integer is represented in rational form as $\frac{a}{b}$, what numbers are permitted as values for b?
(j) For any rational numbers, $\frac{a}{b} \div \frac{c}{d}$, can $\frac{a}{b} = 0$?
(k) Can $\frac{c}{d} = 0$?
(l) Referring to part (j), can $c = 0$?
(m) Can $d = 0$?

4. (a) What is the identity element for addition in the integers?

Identity and inverse

(b) What is the identity element for addition in the rational numbers?
(c) If $a - 0 = a$ for any rational number a, why do we say there is no identity property for the operation of subtraction?
(d) What is the identity element for multiplication in the rational numbers?
(e) If $a \div 1 = a$ for any rational number a, why do we say there is no identity property for the operation of division?

(f) There is correspondence between the operations of subtraction and division in that closure for subtraction was obtained by using the additive ______ and closure for division was obtained by applying the ______.

(g) If we should invent an inverse property for subtraction so that $a - j = j - a = 0$, what would that inverse element j have to be?

(h) Similarly, if we invent an inverse property for division so that $a \div j = j \div a = 1$, what would that inverse element j have to be? (You can understand that (g) and (h) would not contribute significantly to the development of the number system; therefore, inverses for subtraction and division are not considered as properties.)

(i) The sum of additive inverses equals the ______.

(j) The product of multiplicative inverses equals the ______.

Objective 10: Proofs

5. The following proof uses the general definition of addition $\frac{a}{b} + \frac{c}{d} = \frac{ad + bc}{bd}$. Name the definition, property, or "renaming" which permits each step. (If necessary, refer to the proofs in this section to find similar steps.)

For rational numbers $\frac{x}{y}$ and $\frac{0}{z}$ where $y \neq 0$ and $z \neq 0$,

$$\frac{x}{y} + \frac{0}{z} = \frac{xz + y \cdot 0}{yz} \quad \text{(a)}$$

$$= \frac{xz + 0}{yz} \quad \text{(b)}$$

$$= \frac{xz}{yz} \quad \text{(c)}$$

$$= \frac{x}{y} \cdot \frac{z}{z} \quad \text{(d)}$$

$$= \frac{x}{y} \cdot 1 \quad \text{(e)}$$

$$= \frac{x}{y} \quad \text{(f)}$$

6. Now use the definition $\frac{a}{b} \cdot \frac{c}{d} = \frac{ac}{bd}$ and steps similar to those of Problem 5 in order to prove the following yourself:

For any rational numbers $\frac{x}{z}$ and $\frac{z}{y}$ where $y \neq 0$ and $z \neq 0$,

$$\frac{x}{z} \cdot \frac{z}{y} = \frac{x}{y}$$

7. Using a procedure similar to problems 5 and 6, prove that (where $x \neq 0$ and $y \neq 0$),

$$\frac{x}{y} \cdot \frac{y}{x} = 1$$

8. This problem shows another way to prove Problem 7, using the more fundamental definition, $a \cdot \frac{1}{b} = \frac{a}{b}$ (where $b \neq 0$). Justify each step:

For any rational number $\frac{x}{y}$ (where $x \neq 0$ and $y \neq 0$),

$$\frac{x}{y} \cdot \frac{y}{x} = \left(x \cdot \frac{1}{y}\right)\left(y \cdot \frac{1}{x}\right) \quad \text{(a)}$$

$$= x\left[\frac{1}{y} \cdot \left(y \cdot \frac{1}{x}\right)\right] \quad \text{(b)}$$

$$= x\left[\left(\frac{1}{y} \cdot y\right)\frac{1}{x}\right] \quad \text{(c)}$$

$$= x\left[\left(y \cdot \frac{1}{y}\right)\frac{1}{x}\right] \quad \text{(d)}$$

$$= x\left[1 \cdot \frac{1}{x}\right] \quad \text{(e)}$$

$$= x \cdot \frac{1}{x} \quad \text{(f)}$$

$$= 1 \quad \text{(g)}$$

9. Now use the definition $a \cdot \frac{1}{b} = \frac{a}{b}$ to prove the following yourself: For any integers a and k (where $k \neq 0$),

$$\frac{ak}{k} = a$$

10. Proved below is the general definition of addition for rational numbers: $\frac{a}{b} + \frac{c}{d} = \frac{ad + bc}{bd}$. Justify each step. (Refer to item 48.)

For any rational numbers $\frac{a}{b}$ and $\frac{c}{d}$ (where $b \neq 0$ and $d \neq 0$),

$$\frac{a}{b} + \frac{c}{d} = \left(\frac{a}{b} \cdot 1\right) + \left(\frac{c}{d} \cdot 1\right) \quad \text{(a)}$$

$$= \left(\frac{a}{b} \cdot \frac{d}{d}\right) + \left(\frac{c}{d} \cdot \frac{b}{b}\right) \quad \text{(b)}$$

$$= \left(\frac{a}{b} \cdot \frac{d}{d}\right) + \left(\frac{b}{b} \cdot \frac{c}{d}\right) \quad \text{(c)}$$

$$= \frac{ad}{bd} + \frac{bc}{bd} \qquad \text{(d)}$$

$$= \left(ad \cdot \frac{1}{bd}\right) + \left(bc \cdot \frac{1}{bd}\right) \qquad \text{(e)}$$

$$= \left(\frac{1}{bd} \cdot ad\right) + \left(\frac{1}{bd} \cdot bc\right) \qquad \text{(f)}$$

$$= \frac{1}{bd} \cdot (ad + bc) \qquad \text{(g)}$$

$$= (ad + bc) \cdot \frac{1}{bd} \qquad \text{(h)}$$

$$= \frac{ad + bc}{bd} \qquad \text{(i)}$$

11. The definition $\frac{a}{c} + \frac{b}{c} = \frac{a + b}{c}$ was proved in item 47. Here, we show a second proof, which applies the general definition $\frac{a}{b} + \frac{c}{d} = \frac{ad + bc}{bd}$. Name the reason for each step:

For any rational numbers $\frac{a}{c}$ and $\frac{b}{c}$ (where $c \neq 0$),

$$\frac{a}{c} + \frac{b}{c} = \frac{ac + bc}{cc} \qquad \text{(a)}$$

$$= \frac{ca + cb}{cc} \qquad \text{(b)}$$

$$= \frac{c(a + b)}{cc} \qquad \text{(c)}$$

$$= \frac{c}{c} \cdot \frac{a + b}{c} \qquad \text{(d)}$$

$$= 1 \cdot \frac{a + b}{c} \qquad \text{(e)}$$

$$= \frac{a + b}{c} \qquad \text{(f)}$$

12. Use the definition $\frac{a}{c} + \frac{b}{c} = \frac{a + b}{c}$ to prove that (where $c \neq 0$),

$$\frac{a}{c} + \frac{0}{c} = \frac{a}{c}$$

13. Name the steps that prove (where $c \neq 0$) $\frac{a+b}{c} - \frac{b}{c} = \frac{a}{c}$:

$$\frac{a+b}{c} - \frac{b}{c} = \frac{a+b}{c} + \frac{(-b)}{c} \quad \text{(a)}$$

$$= \frac{[a+b]+(-b)}{c} \quad \text{(b)}$$

$$= \frac{a+[b+(-b)]}{c} \quad \text{(c)}$$

$$= \frac{a+0}{c} \quad \text{(d)}$$

$$= \frac{a}{c} \quad \text{(e)}$$

14. Use Problem 13 as a guide to prove that (where $c \neq 0$),

$$\frac{a}{c} - \frac{a}{c} = \frac{0}{c}$$

15. We have studied four sets of numbers, and you should be able to identify the set (or sets) to which any given number belongs. To verify that you can, copy and complete the following chart. Beside each number, put a check under each set of which the number is an element.

Objective 1b

Number	*Natural numbers*	*Whole numbers*	*Integers*	*Rational numbers*
(a) -3	___	___	___	___
(b) $\frac{-6}{7}$	___	___	___	___
(c) 0	___	___	___	___
(d) -2	___	___	___	___
(e) 12	___	___	___	___
(f) $\frac{7}{10}$	___	___	___	___
(g) 6	___	___	___	___
(h) $\frac{3}{4}$	___	___	___	___
(i) $\frac{8}{5}$	___	___	___	___
(j) -1	___	___	___	___
(k) 2	___	___	___	___

16. Use rational numbers x, y, and z (if needed) to express the following properties of the rational numbers:

Objective 2a

(a) Closure property of addition. Verify this property using $\frac{1}{2}$ and $\frac{1}{3}$.

(b) Closure property of multiplication. Verify this property using $\frac{2}{3}$ and $\frac{-2}{5}$.

(c) Commutative property of addition. Verify this property using $\frac{-1}{4}$ and $\frac{2}{3}$.

(d) Commutative property of multiplication. Verify this property using $\frac{-3}{4}$ and $\frac{-5}{7}$.

(e) Associative property of addition. Verify the property using $\frac{1}{3}$, $\frac{1}{5}$, and $\frac{3}{4}$.

(f) Associative property of multiplication. Verify this property using $\frac{-3}{5}$, $\frac{1}{-4}$, and $\frac{-3}{2}$.

(g) Distributive property of multiplication over addition. Verify the property using $\frac{2}{3}$, $\frac{1}{7}$, and $\frac{4}{7}$.

(h) Additive identity property. Verify the property using $\frac{5}{6}$.

(i) Multiplicative identity property. Verify this property using $\frac{3}{-8}$.

(j) Additive inverse property. Verify the property using $\frac{-4}{7}$ and its additive inverse.

Objective 2b

(k) Multiplicative inverse property. Verify the property using $\frac{-2}{9}$ and its multiplicative inverse.

(l) The preceding group of properties are identified by what name?

Objective 2c

17. We have also studied four number systems, and you should have learned which of the number properties hold in each system. To verify that you have, copy and complete the following chart. Beside each property, put a check under each number system to which the property applies.

Property	*Natural numbers*	*Whole numbers*	*Integers*	*Rational numbers*
Closure, +	___	___	___	___
Closure, ×	___	___	___	___
Commutative, +	___	___	___	___
Commutative, ×	___	___	___	___
Associative, +	___	___	___	___
Associative, ×	___	___	___	___
Distributive, × over +	___	___	___	___
Identity, +	___	___	___	___
Identity, ×	___	___	___	___
Inverse, +	___	___	___	___
Inverse, ×	___	___	___	___

Subtraction and division

18. The operations of addition and multiplication were closed in each number system, but the operations of subtraction and division were not.

(a) For which number system(s) is the operation of subtraction closed?
(b) For which number system(s) is the operation of division closed?
(c) State the general definition of subtraction for the system of rational numbers.
(d) State the definition of division for the system of rational numbers.
(e) When we "invert the divisor," what is the name for this new element?
(f) Concepts from two number properties are applied in the definition of division. Which properties are these?
(g) Since both subtraction and division are closed for the system of rational numbers, why is neither considered a fundamental operation of the system?

19. We developed the set of rational numbers after our system already included negatives. Consider now just the set of *positive* rational numbers. Use inductive reasoning to determine the following and explain each response:

Is the set of positive rational numbers closed under the operation of
(a) addition? (b) subtraction?
(c) multiplication? (d) division?

Again, use inductive reasoning to answer the following and explain each response: Is the set of *negative* rational numbers closed under the operation of
(e) addition? (f) subtraction?
(g) multiplication? (h) division?

Objective 3; Example 1

20. Show by the definition of equality whether each pair of common fractions is equal:

(a) Does $\frac{3}{8} = \frac{9}{24}$? (b) Does $\frac{-2}{5} = \frac{-4}{10}$?
(c) Does $\frac{2}{-3} = \frac{-8}{12}$? (d) Does $\frac{-3}{-4} = \frac{6}{8}$?
(e) Does $\frac{-3}{5} = \frac{-9}{-15}$? (f) Does $\frac{-4}{7} = \frac{18}{-28}$?
(g) Does $\frac{-8}{-15} = \frac{-40}{75}$? (h) Does $\frac{-7}{-19} = \frac{-28}{-76}$?
(i) Suppose $\frac{p}{r} = \frac{s}{q}$. Then, by the definition of equality, what else may we say?
(j) Does $\frac{m}{-n} = \frac{-m}{n}$?

Objective 4; Example 2

21. Use the definition of multiplication to find the following products:

(a) $-4 \times \frac{1}{7}$ (b) $\frac{1}{4} \times \frac{1}{-2}$ (c) $\frac{1}{-3} \times \frac{1}{-4}$

(d) $\frac{2}{-7} \times \frac{4}{3}$ (e) $\frac{-3}{-5} \times \frac{-4}{-7}$ (f) $\frac{-6}{7} \times \frac{-3}{-7}$

(g) $\frac{-5}{7} \times \frac{0}{2}$ (h) $\frac{g}{h} \times \frac{j}{k}$

Objective 5; Example 3

22. Demonstrate the multiplicative inverse property by multiplying each of the following rational numbers times its multiplicative inverse:

(a) 7 (b) -3 (c) $\frac{1}{5}$ (d) $\frac{1}{-6}$

(e) $\frac{-2}{7}$ (f) $\frac{0}{8}$ (g) $\frac{-9}{-4}$ (h) $\frac{t}{s}$ $(\neq 0)$

Objective 6; Example 4

23. Apply the definition of division for rational numbers to perform the indicated divisions:

(a) $\frac{1}{9} \div \frac{1}{4}$ (b) $\frac{1}{-5} \div -2$ (c) $\frac{3}{8} \div \frac{-2}{5}$

(d) $\frac{0}{4} \div \frac{3}{7}$ (e) $\frac{5}{-7} \div \frac{3}{-4}$ (f) $\frac{-6}{11} \div \frac{5}{3}$

(g) $\frac{-7}{-9} \div \frac{-4}{5}$ (h) $\frac{p}{q} \div \frac{r}{s}$ $\left(q, \frac{r}{s} \neq 0\right)$

Objective 7; Example 5

24. Apply the definition of addition for rational numbers in order to find the following sums:

(a) $\frac{5}{-8} + \frac{2}{-8}$ (b) $\frac{-2}{7} + \frac{-4}{7}$ (c) $\frac{3}{4} + \frac{1}{5}$

(d) $\frac{5}{9} + \frac{-2}{5}$ (e) $\frac{-4}{5} + \frac{2}{-3}$ (f) $\frac{5}{-6} + \frac{2}{7}$

(g) $\frac{2}{-5} + \frac{3}{-7}$ (h) $\frac{r}{s} + \frac{t}{v}$ $(s, v \neq 0)$

(i) $\frac{r}{s} + \frac{r}{v}$ $(s, v \neq 0)$

Objective 8; Example 6

25. Use the definition of subtraction for rational numbers to find the indicated differences:

(a) $\frac{7}{9} - \frac{5}{9}$ (b) $\frac{2}{5} - \frac{-2}{5}$ (c) $\frac{-3}{5} - \frac{1}{4}$

(d) $\frac{2}{-3} - \frac{3}{-8}$ (e) $\frac{-5}{7} - \frac{2}{-3}$ (f) $\frac{-5}{-8} - \frac{2}{5}$

(g) $\frac{4}{-5} - \frac{-4}{7}$ (h) $\frac{w}{x} - \frac{y}{z}$ (i) $\frac{w}{x} - \frac{-w}{z}$

Objective 9; Example 7

26. Locate another rational number between the following pairs:

(a) $\frac{1}{19}$ and $\frac{2}{19}$ (b) $\frac{1}{19}$ and $\frac{3}{38}$

(c) $\frac{5}{11}$ and $\frac{6}{11}$ (d) $\frac{5}{11}$ and $\frac{11}{22}$

(e) Locate three rational numbers between 0 and $\frac{1}{100}$.

Objective 9; Example 7

27. Find the arithmetic mean for the following pairs of rational numbers:

(a) $\frac{1}{4}$ and $\frac{1}{3}$ (b) $\frac{2}{3}$ and $\frac{7}{8}$

(c) $\frac{2}{5}$ and $\frac{3}{4}$ (d) $\frac{4}{9}$ and $\frac{7}{10}$

Objective 10 Proofs

***28.** Using as a guide the proof of multiplication, $\frac{a}{b} \cdot \frac{c}{d} = \frac{ac}{bd}$ (item 28), prove the following:

For rational numbers $\frac{q}{r}$ and $\frac{s}{t}$ (where $r \neq 0$ and $t \neq 0$),

$$\frac{q}{r} \times \frac{s}{t} = \frac{qs}{rt}$$

***29.** The proof for division, $\frac{a}{b} \div \frac{c}{d} = \frac{a}{b} \cdot \frac{d}{c}$, is given in item 38. Using this example as a guide, prove the following:

For any rational numbers $\frac{w}{x}$ and $\frac{y}{z}$ (where $x \neq 0$ and $\frac{y}{z} \neq 0$),

$$\frac{w}{x} \div \frac{y}{z} = \frac{w}{x} \cdot \frac{z}{y}$$

***30.** Item 47 shows that $\frac{a}{c} + \frac{b}{c} = \frac{a+b}{c}$. Use a similar method to prove:

For any rational numbers $\frac{n}{d}$ and $\frac{m}{d}$ (where $d \neq 0$),

$$\frac{n}{d} + \frac{m}{d} = \frac{n+m}{d}$$

***31.** Refer to problem 30 and item 54 in order to prove the following definition for subtraction:

For any rational numbers $\frac{a}{c}$ and $\frac{b}{c}$ (where $c \neq 0$),

$$\frac{a}{c} - \frac{b}{c} = \frac{a+(-b)}{c}$$

*Exercises marked with an asterisk require more independent thinking by the student.

OBJECTIVES

After completing this unit, you should be able to

A. Multiplicative inverse
B. Rationals on the number line

1 (a) Identify the characteristics of the set of rational numbers and tell the distinctions between rational numbers and the sets of integers, whole numbers, and natural numbers. (Characteristics of the rationals include definition of the elements, representation on the number line, representation of integers as rationals, rationals as common fractions, etc.) **(12–17)**

(b) Given a list of numbers, indicate to which set (or sets) each belongs—set of natural numbers, whole numbers, integers, or rational numbers. **(Exercise 15)**

I. Properties of the rationals

2 (a) Express with variables and verify with rational numbers the following properties from the natural and whole numbers and the integers: closure, commutative, and associative of addition and multiplication; distributive; identity for addition and multiplication; and additive inverse. **(Exercise 16)**

(b) Express with variables and verify with rational numbers the multiplicative inverse property. **(7)**

(c) Given a list of number properties, indicate to which number system (or systems) each belongs—system of natural numbers, whole numbers, integers, or rational numbers. **(Exercise 17)**

C. Equals relation in the rationals

3 Use the definition of equality in order to determine whether two fractions represent the same rational number. **(18–21)**

D. Multiplication in the rationals

4 Use the definition of multiplication to find the product of any two rational numbers. **(23–31)**

5 Find the multiplicative inverse (reciprocal) of any rational number and use it to illustrate the multiplicative inverse property. **(33–35)**

E. Division in the rationals

6 Apply the definition of division to find the quotient of any two rational numbers (where the divisor does not equal zero). **(36–44)**

F. Addition in the rationals

7 Use the definition of addition to find the sum of any two rational numbers. **(45–52)**

G. Subtraction in the rationals

8 Apply the definition of subtraction to find the difference of any two rational numbers. **(53–57)**

H. Density of rationals

9 Illustrate the density property by finding a rational number between any two given rational numbers, using the arithmetic mean. **(60–64)**

Proofs of equalities involving rationals

10 Name the reason that permits each step of a given proof, or develop a complete proof that justifies basic operations (addition, subtraction, multiplication, division) with rational numbers. **(Exercises 5–14 and *28–*31.)**

The Theory of Numbers

Since the time of ancient China and Egypt, mankind has been fascinated by numbers. Both professional mathematicians and amateurs spent many hours puzzling over peculiar characteristics of numbers. As a result, superstitions arose which attributed special powers to certain numbers. For instance, the number 13 was believed to be unlucky. (Even in modern times, very few buildings have a thirteenth floor.) By contrast, the number 7 was deemed to represent good luck, and it is still closely associated with games of chance. (1)

The *theory of numbers* was originally the study of the properties and relationships of the natural numbers. Since the acceptance of negative numbers, however, the theorems of number theory have also been extended to include the set of integers. Some of the study of number theory has no direct applications other than the fascination it provides. Other portions, as we shall see, have practical applications in finding a least common denominator or for reducing rational numbers, which is the principal reason we have waited until now to consider a few topics from the theory of numbers. (2)

It is possible to partition (separate) the natural numbers into several different categories or subsets. We will first consider a partitioning of the set of natural numbers into prime and composite numbers, which requires a knowledge of "divisors" and "multiples." (3)

A. DIVISORS

In the operation of multiplication, if $r \times s = t$, then r and s are called *factors* of t. In the study of number theory, r and s are also called *divisors* of t, and t is said to be *divisible by* r and s. We say: The number b is a *divisor* of a if and only if (4)

Definition of divisor

$$\frac{a}{b} = c, \qquad \text{(where } a, b, \text{ and } c \text{ are natural numbers).}$$

Thus, 2 is a divisor of 8, since $\frac{8}{2} = 4$. Likewise, 4 is a divisor of 8. But 3 is not a divisor of 8, since $\frac{8}{3}$ is not a natural number. (Recall that 0 could never be a divisor in any number system. Zero may be a factor, however.) (5)

Sets of divisors

The number of divisors that a natural number may have varies greatly from one number to another. Any natural number has at least two divisors: itself and one. However, some numbers have many other divisors as well. Let us now consider the set of all the divisors of a natural number: (6)

Objective 1:

◆ **Example 1**

(a) Find D_{12} and D_{18}, where D_{12} denotes the set of all divisors of 12, and D_{18} represents the set of all divisors of 18. (7)

The divisors of any number may be found by listing all the pairs whose product equals that number. That is,

12	18
1×12	1×18
2×6	2×9
3×4	3×6
4×3	6×3
6×2	9×2
12×1	18×1

However, since 3×4 is not distinct from 4×3, for example, it is only necessary to list pairs until they begin to repeat. Thus, the sets of all divisors of 12 and 18 would be

$$D_{12} = \{1, 2, 3, 4, 6, 12\} \quad \text{and}$$
$$D_{18} = \{1, 2, 3, 6, 9, 18\}$$

(b) The intersection set of D_{12} and D_{18} would be

$$D_{12} \cap D_{18} = \{1, 2, 3, 6\}$$

The least element is 1. The greatest element common to D_{12} and D_{18} is 6.

(c) Now the set of all divisors of 6 would also be

$$D_6 = \{1, 2, 3, 6\}$$

Thus, $D_{12} \cap D_{18} = \{1, 2, 3, 6\} = D_6$

● **Practice 1**

(a) Let D_{20} denote the set of all divisors of 20 and D_{30} represent all factors of 30. Find D_{20} and D_{30}. (8)

(b) What is the smallest element common to both sets? What is the largest common element?

(c) Find $D_{20} \cap D_{30}$. Examine this set and rename it as the set of all divisors of some number.

See also Exercise 8.

Objective 2:

In the preceding example we listed the set of all divisors of 12 and 18. Their intersection set, $D_{12} \cap D_{18} = \{1, 2, 3, 6\} = D_6$, indicates that the greatest element common to both sets is 6. (9)

Greatest common divisor

The *greatest common divisor* of two natural numbers is the largest number that evenly divides each of the given numbers. (This element may also be called the *greatest common factor.*) Greatest common divisor may be indicated using the symbols: (10)

$$\text{GCD}(12, 18) = 6$$

This is read "the greatest common divisor of 12 and 18 is 6."

It may happen that two numbers have no common divisors larger than one. In that case, the two numbers are said to be *relatively prime.* (This does *not* imply that either number is a prime number, which will be discussed later.) (11)

◆ **Example 2** The set of all divisors of 8 and 21, denoted by D_8 and D_{21}, are (12)

$$D_8 = \{1, 2, 4, 8\} \quad \text{and}$$
$$D_{21} = \{1, 3, 7, 21\}$$

Thus, the greatest common divisor is

$$\text{GCD}(8, 21) = 1 \qquad \text{(Relatively prime)}$$

● **Practice 2** List the sets of all divisors for each pair and tell the greatest common divisor: (13)

(a) 8 and 12 (b) 16 and 24
(c) 18 and 27 (d) 26 and 35

See also Exercise 9.

B. MULTIPLES

Sets of multiples

When $a \times b = c$, we say that c is a *multiple* of a and a multiple of b. In general, the set of all multiples of a can be represented by $\{a \cdot n\}$, where n is replaced in succession by each of the elements of the set of natural numbers (or whole numbers or integers). It then follows that each set of multiples is an infinite set. (14)

Objective 3:

For example, the set of multiples of the natural number 2, represented by $\{2n\}$, can be denoted as (15)

$$M_2 = \{2, 4, 6, 8, 10, 12, \ldots\}$$

Multiples of the natural numbers 3 and 6 are represented as follows:

$$M_3 = \{3, 6, 9, 12, 15, 18, \ldots\}$$
$$M_6 = \{6, 12, 18, 24, 30, 36, \ldots\}$$

◆ **Example 3**

Intersection of sets of multiples

(a) $M_2 \cap M_3 = \{6, 12, 18, 24, 30, 36, \ldots\}$. But this set is equivalent to the set of multiples of 6; therefore, we could denote $M_2 \cap M_3$ as M_6. That is, (16)

$$M_2 \cap M_3 = \{6, 12, 18, 24, \ldots\} = M_6$$

(b) In practical terms, this means any number that is a multiple of both 2 and 3 is also a multiple of 6:

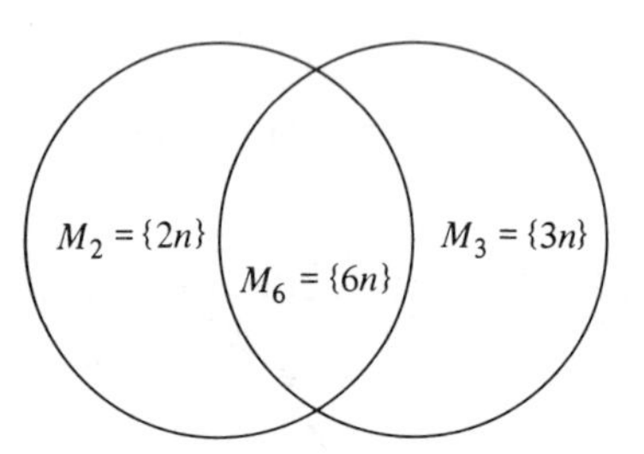

● **Practice 3**

(a) Indicate the set, M_5, of multiples of 5. What is the smallest element that M_2 and M_5 have in common? (17)

(b) Find $M_2 \cap M_5$. What other name would identify the set $M_2 \cap M_5$?

(c) What does this relationship mean in common terms?

See also Exercise 11.

Objective 4:

In the preceding example, the intersection of two infinite sets of multiples was also an infinite set containing no largest element. However, we did identify a smallest element of the intersection set, as follows: (18)

Least common multiple

The *least common multiple* of two natural numbers is the smallest number that is a multiple of both the given numbers. (19)

◆ **Example 4** The set of multiples of 3, M_3, and the multiples of 4, M_4, would be indicated (20)

$$M_3 = \{3, 6, 9, 12, 15, \ldots\} \quad \text{and}$$
$$M_4 = \{4, 8, 12, 16, 20, \ldots\}$$

Thus, "the least common multiple of 3 and 4 is 12," denoted

$$\text{LCM}(3, 4) = 12$$

● **Practice 4** Indicate the sets of multiples for each pair, and give the least common multiple: (21)

(a) 3 and 5 (b) 6 and 8

(c) 6 and 9 (d) 5 and 7

See also Exercise 12.

C. PRIME NUMBERS

Whether a natural number has only two divisors or has more than two divisors is the aspect that partitions the set of natural numbers as prime or composite: (22)

Definition of prime

A natural number, p, is said to be *prime* if and only if $p > 1$ and p is divisible only by itself and 1.

Definition of composite

A natural number, p, is said to be *composite* if and only if $p > 1$ and p is not prime.

For example, the natural numbers 2, 3, 5, and 7 are primes according to these definitions, while the numbers 4, 6, 8, 9, and 10 are composites. (23)

Prime excludes 1 (and 0)

Notice that these definitions of prime* and composite exclude both 0 and 1. As we have seen, both numbers possess certain properties that are unique. Consequently, many theorems in number theory hold for other numbers but not for 0 and 1. (24)

*A prime number may also be defined as "a natural number that has exactly two distinct (different) divisors." A composite number would then have "more than two distinct divisors." These definitions likewise eliminate the numbers 0 and 1. (A negative integer is prime if its additive inverse is prime, and composite if its additive inverse is composite.)

Throughout history, primes have provided a particular fascination for mathematicians. Many persons have attempted to devise a formula that will generate (produce) primes. Some of their formulas will produce a large number of primes, but no formula has ever been devised that will generate all the prime numbers. For example, the formula $n^2 - n + 41$ will yield a prime when n is any natural number between 1 and 40. However, this formula omits the first twelve primes and fails to produce a prime when $n = 41$. Other suggested formulas have similarly failed. (25)

Small numbers can easily be identified as prime or composite by testing to determine whether each has divisors other than itself and 1. As numbers become increasingly larger, however, testing for divisors can be an exhausting process. For instance, it would take considerable experimentation in order to determine that 4757 is not prime, because 67 is its smallest factor other than one $(4757 = 67 \cdot 71)$. Thus, the most efficient way to determine whether a given number is prime is to consult a table of primes. (26)

It can be proved that there exists an infinite number of primes, and the list of known primes continues to increase, aided greatly by the use of computers. The number $2^{11213} - 1$ was discovered to be prime in 1963 and remained the largest known prime until 1971, when the prime number $2^{19937} - 1$ was reported. (The expression $2^{19937} - 1$ represents a number which has 6,002 digits.) (27)

Objective 5(a): Sieve of Eratosthenes

Surprisingly, a table of primes can be constructed using a rather simple method (not a formula). This somewhat inefficient but extremely accurate method was devised by the Greek mathematician Eratosthenes some two thousand years ago and is known as the *Sieve of Eratosthenes*. The procedure is to list all the numbers $\leq$ any given number and then to delete all of those numbers that are not prime, as demonstrated: (28)

◆ **Example 5** Use the Sieve of Eratosthenes to find all the prime numbers ≤ 50. (29)

First, list all of the numbers from 1 through 50. Since 1 is not prime, it may be deleted immediately. The first prime is 2, so it is circled to remain in the sieve. Now, a prime can have only itself and 1 as factors. Thus, any multiple of 2 cannot be prime, because it would also have the number 2 as a factor. Therefore, we delete all of the remaining multiples of 2; that is, we delete 4, 6, 8, 10, etc. (Notice that the total of numbers that must be considered as primes is immediately cut in half as a result of this step.)

We next circle 3 and then delete all the remaining multiples of 3, since none of them would have only itself and 1 as divisors. (Some multiples of 3 are also multiples of 2; these will already have been deleted.) We then progress to the next undeleted number, which is 5, circle it, and delete all its remaining multiples. Each time that we proceed to the next undeleted (30) number, this number will always be a prime. Upon completion, the sieve will appear as follows:

~~1~~	(2)	(3)	~~4~~	(5)	~~6~~	(7)	~~8~~	~~9~~	~~10~~
(11)	~~12~~	(13)	~~14~~	~~15~~	~~16~~	(17)	~~18~~	(19)	~~20~~
~~21~~	~~22~~	(23)	~~24~~	~~25~~	~~26~~	~~27~~	~~28~~	(29)	~~30~~
(31)	~~32~~	~~33~~	~~34~~	~~35~~	~~36~~	(37)	~~38~~	~~39~~	~~40~~
(41)	~~42~~	(43)	~~44~~	~~45~~	~~46~~	(47)	~~48~~	~~49~~	~~50~~

● **Practice 5a** Rework example 5 yourself to verify that you obtain the same primes. (Practice 5 continues at item 37.) **(31)**

In using the sieve, we find as we proceed that fewer and fewer multiples remain, since many have been eliminated as multiples of some previous prime. In fact, we found that 7 had only one multiple remaining (49), and that none of the larger primes had any multiple remaining. This would make us wonder whether it is possible to know that, after a certain point, every remaining number is a prime. **(32)**

Objective 5(b): How to know when only primes remain

When computing a Sieve of Eratosthenes, it is actually only necessary to *find multiples of the primes up through the square root* of the largest number, n, in the table*. That is, any composite number less than n will be a multiple of some number less than $\sqrt{n}$. For example, in a Sieve of Eratosthenes with the largest number 36, it is only necessary to find multiples of the primes up through 6. Then, every undeleted number larger than 6 will be prime. **(33)**

To understand why every composite number less than n will have a divisor less than $\sqrt{n}$, recall our listing of pairs of factors and our observation that the pairs repeat. Study all the pairs of factors of 36: **(34)**

$$1 \times 36,\quad 2 \times 18,\quad 3 \times 12,\quad 4 \times 9,\quad \underset{\uparrow \atop \sqrt{36}}{6 \times 6},\quad 9 \times 4,\quad 12 \times 3,\quad 18 \times 2,\quad 36 \times 1$$

As larger numbers are used for the first factor of 36, smaller numbers are used for the second factor. This continues until we reach $\sqrt{36} = 6$, when both factors are equal. After that point, the same factors begin repeating in reverse order.

*For any natural numbers p and n, if $p \times p = n$ then p is said to be a *square root* of n. The symbol, $\sqrt{n} = p$, indicates "the principal (positive) square root of n is p." (Although $(-p) \times (-p)$ also $= n$, the $(-p)$ is not the principal square root and is not indicated by the $\sqrt{\ }$ symbol.) Also, in number theory we are only concerned with positive numbers.

However, suppose we are still concerned, for instance, whether 31 might be a multiple of one of the primes greater than 6. But we know (35)

$$31 < 36$$

therefore $\sqrt{31} < \sqrt{36}$

or $\sqrt{31} < 6$

Thus, when we reach 6 in the sieve, we have already passed $\sqrt{31}$. But we have seen that all the distinct factors of any number have been identified by the time we reach the square root of the number. Therefore, if we have passed $\sqrt{31}$ and found no divisors of 31, then 31 must be prime. We can thus accept that $\sqrt{n}$ is the largest number we must compute in any Sieve of Eratosthenes that contains n numbers.

Objective 5(c): Twin primes

Two of the curiosities related to primes are twin primes and reversible primes. *Twin primes* are two primes with a difference of 2, such as 5 and 7 or larger ones such as 2,129 and 2,131. Since it has been proved that there is no largest prime, there are thought to be no largest twin primes. But no mathematician has been able either to prove or disprove this. *Reversible primes* are pairs whose digits are the same but in reverse order. Examples of these primes would be 13 and 31, or 107 and 701. (36)

Reversible primes

● **Practice 5b**

(a) In a Sieve of Eratosthenes with 100 as the largest number, at what number can you stop computing multiples and know that all remaining undeleted numbers are prime? (37)
(b) Use a Sieve of Eratosthenes to find all primes up to 100.
(c) Find eight pairs of twin primes remaining in this sieve and list them.
(d) Find and list four pairs of two-digit reversible primes.

See also Exercises 14 and 15.

D. PRIME FACTORIZATION

We have seen previously that any prime number has only two divisors: itself and 1. Every composite number has more than two factors. In fact, some may be expressed as the product of several different sets of factors. For example, the pairs of divisors of 12 would be as follows: (38)

$$1 \times 12, \quad 2 \times 6, \quad 3 \times 4, \quad 4 \times 3, \quad 6 \times 2, \quad 12 \times 1$$

Factorization

Each of these indicated products is called a *factorization* of 12. As we noted earlier, however, 3×4 is not distinct from 4×3, etc. Thus, 12 has three distinct pairs of factors. We will now consider how any composite number can be distinctly expressed as a product of primes.

(39) Some composite numbers can easily be expressed as a product of primes. For example,

$$6 = 2 \times 3 \qquad \text{or} \qquad 15 = 3 \times 5$$

Now, the divisors listed for 12 are not limited to primes: 1×12, 2×6, and 3×4. Thus, the composite factors can themselves be factored. We will use a branching tree to continue factoring the composite factors. (We will not concern ourselves with the factorization 1×12, since 1 is neither prime nor composite.)

Branching tree

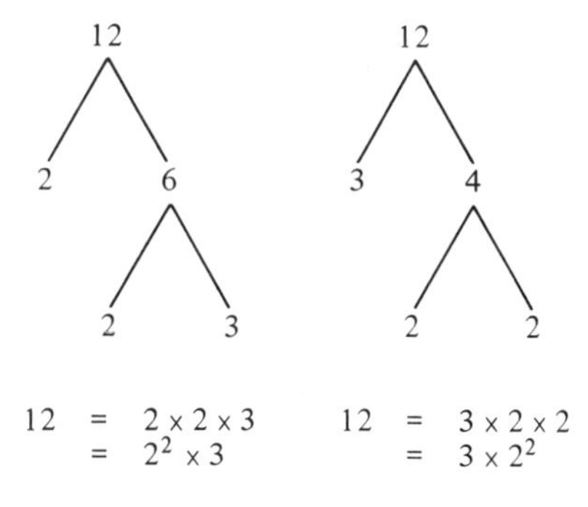

(40) The branching tree is continued until each branch has terminated in a prime number. The product of all these primes is the *prime factorization* (or *complete factorization*) of the given number.* Notice, however, that $2^2 \times 3$ is not distinct from 3×2^2. As this example illustrates, there is only *one* distinct way to factor any composite number into prime factors. This fact is of such great significance that it is designated the *Fundamental Theorem of Arithmetic*, as follows:

Prime factorization

Objective 6: Fundamental Theorem of Arithmetic

(41) Every composite natural number can be expressed as the product of primes in one and only one way, if the order of factors is disregarded. (This theorem is also called the *Unique Prime Factorization Theorem.*)

The fact that this statement is called a "theorem" rather than a "conjecture" or "hypothesis" tells us that it has been proved, although the proof will not be included here. Notice that the Fundamental Theorem of Arithmetic does *not* specify *how many* primes a prime factorization might contain. Thus, a prime factorization may contain two, three, or more different primes.

*It is customary to use an *exponent* to express a prime factor that is repeated. The exponent, n, in the expression 2^n means that the base, 2, is used as a factor n times. For example, $2^3 = 2 \cdot 2 \cdot 2$. The term *power* refers to the value of a number (or expression) written with an exponent. For example, 2^3 (or 8) represents the third power of 2.

◆ **Example 6** Show that 24 has a unique prime factorization. (42)

The number 24 can be factored as 2×12, 3×8, or 4×6. Prime factors are obtained from a branching tree as follows:

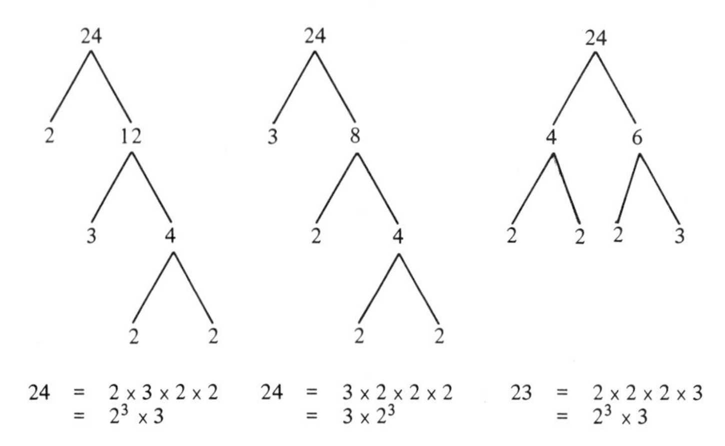

Thus, we see that each original factorization results in the same (except for order) prime factorization.

● **Practice 6** Find the prime factorization of the following numbers: (43)

(a) 8 (b) 10 (c) 21

Use branching trees to show that each of the following numbers can be expressed as a product of primes in only one distinct way:

(d) 18 (e) 30 (f) 36

See also Exercise 16.

E. GREATEST COMMON DIVISOR AND REDUCING

Objective 7: Greatest common divisor (by prime factorization)

Prime factorization may be used to find the greatest common divisor of two numbers. Examine the prime factors of two numbers to determine all the prime factors (including powers of primes) that occur in *both* of the numbers. The greatest common divisor (factor) will be the product of all the prime factors that are common to *both* numbers. For example, (44)

◆ **Example 7** Find the greatest common divisor for the following, using prime factorization: (45)

(a) When there are identical prime factors in two numbers, the product of these factors is the greatest common divisor. That is, for 30 and 42,

$30 = 2 \times 3 \times 5$ Thus, the greatest common factor is $2 \times 3 = 6$.

$42 = 2 \times 3 \times 7$

Or, $\text{GCD}(30, 42) = 6$

(b) $72 = 2 \cdot 2 \cdot 2 \cdot 3 \cdot 3$ $\text{GCD}(72, 180) = 2 \cdot 2 \cdot 3 \cdot 3$

$180 = 2 \cdot 2 \cdot 3 \cdot 3 \cdot 5$ $= 36$

It may not be necessary for you to write out each individual prime factor. The GCD may be determined by selecting the highest power of each prime that appears in *both* prime factorizations. Thus, (46)

$72 = 2^3 \times 3^2$ Then, $\text{GCD}(72, 180) = 2^2 \times 3^2$

$180 = 2^2 \times 3^2 \times 5$ $= 36$

Relatively prime

(c) If two numbers have no identical prime factors, then the numbers are *relatively prime*: (47)

$8 = 2 \times 2 \times 2$

$15 = 3 \times 5$ Relatively prime

Observe that neither 8 nor 15 is a "prime" number itself; however, the two numbers are "relatively prime" to each other.

● **Practice 7** Find the GCD of each pair using the prime factorization of each number: (48)

(a) 15 and 35 (b) 12 and 20
(c) 30 and 70 (d) 36 and 90
(e) 300 and 360 (f) 32 and 63

See also Exercise 17.

Reducing fractions

Students learn to "reduce" fractions by finding the "largest number that divides into both the numerator and denominator." Perhaps they think this process is accomplished by magic! Actually, this "largest number" is the greatest common divisor, and the process is possible because of the multiplicative identity property of one. For instance, (49)

$$\frac{12}{18} = \frac{2 \cdot 2 \cdot 3}{2 \cdot 3 \cdot 3} = \frac{2 \cdot (2 \cdot 3)}{(2 \cdot 3) \cdot 3} = \frac{2 \cdot (2 \cdot 3)}{3 \cdot (2 \cdot 3)} = \frac{2}{3} \cdot \frac{(2 \cdot 3)}{(2 \cdot 3)} = \frac{2}{3} \cdot 1 = \frac{2}{3}$$

Objective 8:

When the numerator and denominator of a fraction are relatively prime, the number is said to be in *lowest terms*. (50)

$$\frac{8}{15} = \frac{2 \cdot 2 \cdot 2}{3 \cdot 5} \quad \text{is in lowest terms}$$

because the numbers 8 and 15 are relatively prime; that is, 8 and 15 have no common factor. (Neither 8 nor 15 is a prime number, however.) To *reduce* or *simplify* a fraction means to change it to its lowest terms, by applying the greatest common factor and the multiplicative identity property.

◆ **Example 8** Use the greatest common divisor to simplify (or reduce) $\frac{45}{60}$ to lowest terms: (51)

$$\frac{45}{60} = \frac{3\cdot 3\cdot 5}{2\cdot 2\cdot 3\cdot 5} = \frac{3(3\cdot 5)}{2^2(3\cdot 5)} = \frac{3}{4}$$

● **Practice 8** Find the prime factorization of the following and apply the greatest common factor to reduce each to lowest terms: (52)

(a) $\frac{9}{15}$ (b) $\frac{12}{30}$ (c) $\frac{32}{48}$

(d) $\frac{54}{72}$ (e) $\frac{84}{108}$ (f) $\frac{126}{196}$

See also Exercise 18.

F. LEAST COMMON MULTIPLE (AND DENOMINATOR)

Objective 9: LCM by prime factorization

Prime factorization may also be used to find the least common multiple of two numbers. The least common multiple is the product of every prime factor that occurs in *either* of the prime factorizations. (53)

◆ **Example 9** Use prime factorization to determine the least common multiple for each pair below: (54)

(a) Select *every* prime that occurs in *either* factorization, being careful to choose the *highest power* of each prime. The product of these primes will then be the least common multiple.

$6 = 2\cdot 3$ Thus, $\text{LCM}(6, 10) = 2\cdot 3\cdot 5$

$10 = 2\cdot 5$ $= 30$

6
2·3·5
10

(b) $18 = 2\cdot 3^2$ $\text{LCM}(18, 30) = 2\cdot 3^2\cdot 5$

$30 = 2\cdot 3\cdot 5$ $= 90$

18
2·3·3·5
30

(c) $8 = 2^3$ $\text{LCM}(8, 15) = 2^3\cdot 3\cdot 5$

$15 = 3\cdot 5$ $= 120$

8
2·2·2·3·5
15

From the diagrams, we see that every factor of the LCM is essential, otherwise each given number could not be produced. Observe that the same factor may sometimes be used twice—as a factor of each given number. (55)

This "double duty" is the aspect that makes the LCM the *least* of the many possible common multiples. When the given numbers have no prime factors in common, notice that the LCM is the product of the two numbers.

(56) Since GCD and LCM both apply prime factorization, students sometimes forget which process is associated with which name. The key is to remember that, typically, divisors (GCD) are *smaller* than a given number while multiples (LCM) are *larger*. In summary, then,

GCD compared to LCM

$\text{GCD}(a, b) \leq a$ or b $\qquad$ $\text{LCM}(a, b) \geq a$ or b

GCD = only primes common to *both* factorizations

LCM = every prime found in *either* factorization, each to its highest power.

(57) ● **Practice 9** Find the least common multiple for each pair, using prime factorization:

(a) 10 and 14 $\qquad$ (b) 8 and 12

(c) 18 and 24 $\qquad$ (d) 30 and 45

(e) 56 and 60 $\qquad$ (f) 9 and 16

See also Exercise 19.

(58) A familiar process used in adding fractions is to "find the least common denominator." This least common denominator is actually the least common multiple of the two denominators. Consider $\frac{1}{4} + \frac{1}{6}$. Since $4 = 2^2$ and $6 = 2 \cdot 3$, then $\text{LCM}(4, 6) = 2^2 \cdot 3 = 12$. Thus, 12 is also the least common denominator, which is obtained by applying the multiplicative property of one. (Notice below that each fraction is multiplied by the primes of the LCM that were *not* factors of that given denominator. Thus, $\frac{1}{4}$ is multiplied by $\frac{3}{3}$, and $\frac{1}{6}$ by $\frac{2}{2}$.) Hence,

Objective 10: Least common denominator

$$\begin{aligned} \frac{1}{4} + \frac{1}{6} &= \frac{1}{4}(1) + \frac{1}{6}(1) = \frac{1}{4} \cdot \frac{3}{3} + \frac{1}{6} \cdot \frac{2}{2} \\ &= \frac{3}{12} + \frac{2}{12} = \frac{3+2}{12} \\ &= \frac{5}{12} \end{aligned}$$

(59) ◆ **Example 10** Use the concept of LCM to find $\frac{4}{15} + \frac{3}{20}$. Reduce the sum to lowest terms.

$15 = 3 \cdot 5$ and $20 = 2^2 \cdot 5$. Thus, $\text{LCM}(15, 20) = 2^2 \cdot 3 \cdot 5 = 60$.

$$\frac{4}{15} + \frac{3}{20} = \frac{16}{60} + \frac{9}{60} = \frac{25}{60}$$

$$\frac{25}{60} = \frac{5(5)}{2^2 \cdot 3(5)} = \frac{5}{12}$$

● **Practice 10** Use the concept of LCM to perform the following operations with rational numbers. If the results are not relatively prime, use the GCD to reduce the number: (60)

(a) $\frac{1}{3} + \frac{1}{4}$ (b) $\frac{2}{3} - \frac{2}{5}$ (c) $\frac{5}{8} - \frac{1}{12}$

(d) $\frac{19}{24} - \frac{1}{6}$ (e) $\frac{2}{5} + \frac{4}{15}$ (f) $\frac{7}{18} + \frac{11}{30}$

(g) $\frac{5}{12} - \frac{3}{20}$ (h) $\frac{15}{28} + \frac{7}{20}$ (i) $\frac{21}{50} + \frac{26}{75}$

See also Exercise 20.

Objective 11: Relation between GCD and LCM

The greatest common divisor is usually easier to determine mentally than is the least common multiple. For this reason, the following relationship may prove helpful: for two natural numbers a and b (61)

$$\frac{a \times b}{\text{GCD}(a, b)} = \text{LCM}(a, b)$$

◆ **Example 11** Find the least common denominator for $\frac{x}{6} + \frac{x}{9}$. We can easily determine that GCD(6, 9) = 3. Thus, (62)

$$\frac{6 \times 9}{\text{GCD}(6, 9)} = \frac{54}{3} = 18 = \text{LCM}(6, 9)$$

Therefore, the least common denominator is 18.

● **Practice 11** Use the greatest common divisor of the following numbers to determine the least common multiple: (63)

(a) 4 and 6 (b) 8 and 12
(c) 9 and 15 (d) 10 and 15

See also Exercise 21.

We have thus concluded our partitioning of the set of natural numbers into prime and composite numbers. We found the most significant aspect of primes to be the Fundamental Theorem of Arithmetic—the fact that every composite number can be expressed as a unique product of primes. We then used this prime factorization to find the greatest common divisor and the least common multiple. (64)

Further, although the theory of numbers strictly pertains to the set of natural numbers, we found that it also has application in the rational numbers—that the convenient procedures of "reducing" and "least common denominator" are possible only because of the Fundamental Theorem of Arithmetic. (65)

G. ODD AND EVEN NUMBERS

We have previously assumed another way of partitioning the set of natural numbers—as even or odd—which again applies the concept of multiples: (66)

Definition of even

Any natural number, a, is *even* if and only if a is an element of the set of multiples of 2.

Definition of odd

If any natural number, a, is not an element of the set of multiples of 2, then a is *odd*.

Thus, every even number is a member of the set, $E = \{2, 4, 6, 8, 10, \ldots\}$. Using our previous notation, any even number can be represented as $2k$, where k is a natural number. Similarly, every odd number is an element of the set $O = \{1, 3, 5, 7, 9, \ldots\}$. Any odd number may then be represented by the expression $2k + 1$. Every element of the set of natural numbers* may thus be designated as either even or odd. (67)

Objective 12:

The concept of even and odd is the basis for observing some interesting patterns in the operations of addition and multiplication, as the following example illustrates: (68)

◆ **Example 12** Prove that the product of any even number times any odd number equals an even number. (69)

Let $(2k)$ represent any even number and $(2p + 1)$ represent any odd number. Then,

$(2k)(2p + 1)$	$= (2k)(2p) + (2k)(1)$	Distributive property
	$= (2k)(2p) + 2k$	Multiplicative identity
	$= 2(k \cdot 2p) + 2k$	Associative property of multiplication
	$= 2[(k \cdot 2p) + k]$	Distributive property
But $[(k \cdot 2p) + k]$ represents a natural number, say K. Then,		Closure for addition and multiplication
	$= 2K$	Substitution

Thus, the product of an even number times an odd number equals an even number, since $2K$ is even.

*The concept of even and odd may also be applied to the set of integers.

● **Practice 12** Prove that the sum of an even and an odd number is an odd number. (70)

See also Exercises 23–26.

EXERCISES

Finite or infinite sets

1. Tell whether each of the following sets is finite or infinite:
   (a) The set of all divisors of a natural number
   (b) The set of multiples of a natural number
   (c) The set of prime numbers
   (d) The set of prime factors of a natural number
   (e) The set of composite numbers
   (f) The set of odd numbers
   (g) The set of even numbers

Divisors (not necessarily prime)

2. Consider the set of *all* (not necessarily just prime) divisors of a natural number. Explain each answer:
   (a) How many such divisors does a prime number have?
   (b) What is the least number of these divisors any composite number may have? Give an example.
   (c) Is there a largest number of divisors possible for any composite number? (That is, a maximum number of divisors such that no composite number would have more?)
   (d) If the set of *all* divisors of a number (greater than one) contains an odd number of elements, then the given number is a perfect square. Explain why. (Hint: choose some perfect squares such as 25 or 36, and study the list of their factors.

Prime and composite; prime factorization

3. Consider the prime factorization of a natural number. Explain each answer:
   (a) Is every natural number either prime or composite?
   (b) How many prime factors does a prime number have?
   (c) Why does the Fundamental Theorem of Arithmetic (Unique Factorization Theorem) pertain only to composite numbers?
   (d) Is there a largest number of prime factors that a composite number can have?
   (e) Why is the pair of factors $(1 \times n)$ not considered when a branching tree is used to find prime factors?
   (f) Can some natural numbers be expressed as the sum of primes in more than one way? If so, give an example.
   (g) Can a natural number be expressed as the product of primes in more than one way? If so, give an example.

GCD and LCM

4. Supply the correct inequality relation, $\leq$ or $\geq$ for (a) and (b); then answer the remaining questions:

(a) The greatest common divisor (factor) of two numbers is ____ either given number. Give an example of two numbers where the GCD equals one of the numbers.
(b) The least common multiple of two numbers is ____ either given number. Give an example of two numbers where the LCM equals one of the numbers.
(c) What two main concepts (or properties) are used to reduce a rational number?
(d) Which concept corresponds to the least common denominator of fractions?

Relatively prime; GCD; LCM

5. Answer as indicated:
(a) Two numbers are relatively prime if their greatest common divisor is ____.
(b) Two numbers are ____ if they have no prime factors in common.
(c) Can two composite numbers be relatively prime? Explain.
(d) When a numerator and denominator are relatively prime, the fraction is in ____.
(e) The greatest common divisor (factor) of two numbers is the product of all prime factors from ____ numbers.
(f) The least common multiple of two numbers is the product of all prime factors from ____ number, with each factor to its highest power.
(g) Is there a least common divisor of two numbers? Explain.
(h) Is there a greatest common multiple of two numbers? Explain.

Odd and even; prime and composite

6. Answer the following. Explain each "no" answer with an example.
(a) Is every natural number either odd or even?
(b) Is every odd number a prime?
(c) Is every prime an odd number?
(d) Is every composite number even?
(e) Is every even number (greater than two) a composite? (Explain either answer.)

Odd and even; prime and composite

7. Let $E =$ the set of all even natural numbers, $O =$ the set of all odd natural numbers, $P =$ the set of all prime natural numbers, and $C =$ the set of all composite natural numbers. List or describe in words the elements of the following sets:
(a) What is $E \cup O$? (b) What is $E \cap O$?
(c) What is $P \cup C$? (d) What is $P \cap C$?
(e) What is $E \cap P$? (f) What is $O \cap P$?

Example 1

8. Let D_{16} represent the set of all divisors (not necessarily prime factors) of 16, and let D_{28} denote the set of all divisors of 28.
(a) Find D_{16} and D_{28}.
(b) What is the smallest element common to both sets? What is the largest common element?

(c) Find $D_{16} \cap D_{28}$. Rename this set as the set of all divisors of some number.

Example 2

9. Indicate the sets of all divisors for each pair below, and name the greatest common divisor:
(a) 24 and 32 (b) 36 and 48
(c) 20 and 27 (d) 45 and 56

Divisors; GCD

10. Show the set of all divisors of the following numbers: 2, 3, 4, 5, 6, 9, 10, 15.
(a) Find $D_2 \cap D_4$; $D_2 \cap D_6$; and $D_2 \cap D_{10}$.
(b) Use inductive reasoning to determine the GCD(2, n) when n is even.
(c) Find $D_2 \cap D_5$; $D_2 \cap D_9$; and $D_2 \cap D_{15}$.
(d) By inductive reasoning, what is the GCD(2, n) when n is odd?
(e) Find $D_2 \cap D_4$; $D_3 \cap D_9$; and $D_5 \cap D_{10}$.
(f) Using inductive reasoning, how would you define GCD(a, b) when b is a multiple of a?

Objective 3; Example 3

11. Let M_3 represent the set of multiples of 3 and M_7 denote the set of multiples of 7.
(a) Show M_3 and M_7 in set notation.
(b) Find $M_3 \cap M_7$. Study this set and rename it as the set of multiples for some number.
(c) Explain the meaning of the relationship in part (b).

Multiples; LCM

12. (a) Indicate the set of multiples for each of the following numbers: 2, 3, 4, 5, 6, 9, 10, 12, 15.
(b) Find $M_2 \cap M_3$; $M_2 \cap M_5$; and $M_3 \cap M_5$.
(c) By inductive reasoning how would you define LCM(a, b) when a and b are both prime?
(d) Find $M_2 \cap M_6$; $M_3 \cap M_6$; and $M_4 \cap M_{12}$.
(e) Use inductive reasoning to define LCM(a, b) when b is a multiple of a.
(f) Find $M_4 \cap M_9$; $M_8 \cap M_9$; and $M_9 \cap M_{10}$.
(g) Using inductive reasoning, how would you define LCM(a, b) when a and b are relatively prime?
(h) Find $M_4 \cap M_6$; $M_6 \cap M_9$; and $M_{10} \cap M_{15}$.
(i) Is the least common multiple of two composite numbers always less than the product of the numbers? Explain.

Objective 4; Example 4

13. Use set notation to indicate the multiples of each number below, and find the least common multiple for each pair:
(a) 6 and 15 (b) 8 and 12
(c) 9 and 16 (d) 7 and 13

14. Suppose a Sieve of Eratosthenes is being computed, with a largest

Objective 5b; Example 5b

number as shown below. In each case, at what number would we know that all composite numbers had been deleted?

(a) 400 (b) 900 (c) 1600

Objective 5a; Example 5b

15. (a) Use a Sieve of Eratosthenes to determine all prime numbers less than 200. (At what number do you stop computing multiples of primes?)

(b) How many prime numbers do you find between 1 and 50? Between 50 and 100? Between 100 and 150? Between 150 and 200?

(c) List 7 pairs of twin primes between 100 and 200.

Objective 5c: Primes

(d) List 5 three-digit reversible primes between 100 and 200. (Primes such as these are also called *symmetrical* primes.)

(e) No reversible prime can begin or end with 2, 4, 6, 8, (or 0). Can you explain why?

(f) Three distinct primes with a difference of 2 between the first and second and between the second and third are *triplet primes*. Find a set of triplet primes early in the sieve. Can you explain why this is the only set of triplet primes? (Hint: the explanation has to do with multiples.)

Objective 6; Example 6

16. Use a branching tree to find the unique prime factorization of the following:

(a) 20 (b) 42 (c) 45

(d) Make a list of the pairs of factors that equal 48. Use branching trees to show that these factorizations all result in the same prime factorization.

Objective 7; Example 7

17. Use prime factorization to determine the greatest common divisor for each pair:

(a) 42 and 70 (b) 30 and 75
(c) 54 and 81 (d) 45 and 56
(e) 90 and 108 (f) 120 and 252
(g) 360 and 450 (h) 63 and 110

Objective 8; Example 8

18. Apply the greatest common divisor in order to reduce the following to lowest terms:

(a) $\frac{30}{42}$ (b) $\frac{45}{60}$ (c) $\frac{36}{54}$ (d) $\frac{126}{168}$

(e) $\frac{27}{81}$ (f) $\frac{108}{132}$ (g) $\frac{56}{84}$ (h) $\frac{72}{112}$

Objective 9; Example 9

19. Use prime factorization to find the least common multiple of each pair:

(a) 14 and 21 (b) 15 and 25
(c) 16 and 21 (d) 54 and 72
(e) 24 and 42 (f) 75 and 90
(g) 48 and 60 (h) 50 and 63

Objective 10; Example 10

20. Apply the LCM to perform the following operations. When the results are not in lowest terms, use the GCD to reduce the fraction:

(a) $\frac{3}{4} + \frac{1}{6}$ (b) $\frac{3}{5} - \frac{1}{4}$ (c) $\frac{1}{6} + \frac{4}{9}$

(d) $\frac{5}{6} - \frac{2}{15}$ (e) $\frac{8}{15} + \frac{3}{10}$ (f) $\frac{7}{24} + \frac{7}{40}$

(g) $\frac{9}{10} - \frac{3}{14}$ (h) $\frac{13}{28} - \frac{7}{20}$ (i) $\frac{2}{15} + \frac{12}{35}$

Objective 11; Example 11

21. Use the greatest common divisor of the following numbers in order to find their least common multiple (or, the least common denominator for two fractions with the given numbers as denominators):

(a) 6 and 8 (b) 9 and 12
(c) 16 and 20 (d) 25 and 30

Odd and even; prime and composite

22. Try some examples and use inductive reasoning to determine whether the following results are even or odd natural numbers:

(a) The sum of two even numbers
(b) The sum of two odd numbers
(c) The sum of an even and an odd number
(d) The product of two even numbers
(e) The product of two odd numbers
(f) The product of an even and an odd number

Answer the following based on your results above. Explain or give a counter example for each answer:

(g) Does $13{,}576{,}896^2$ produce an even or an odd number?
(h) Does $37{,}954{,}277^3$ produce an even or an odd number?
(i) Is the sum of two primes (both greater than two) always an even number?
(j) Is the sum of a prime and a composite number (both greater than two) always an odd number?
(k) Is the product of two primes (both greater than two) always an odd number?
(l) Is the product of two composites always an even number?
(m) We have said that any even number can be expressed in the form of $2n$, where n is a natural number. Show 476 in the form $2n$.
(n) Similarly, every odd number can be expressed in the form $2n + 1$. Demonstrate 579 as $2n + 1$.

Objective 12; Example 12

23. (a) Prove that the set of even numbers is closed under the operation of addition. (That is, prove that the sum of *any* two even numbers is an even number.)
(b) Is the set of odd numbers closed under the operation of addition? Explain.

Example 12

24. Prove that the sum of any even number plus itself is an even number.

Example 12

25. Prove that the square of an even number is an even number.

Example 12

26. Prove that the set of even numbers is closed under the operation of multiplication. (That is, prove that the product of any two even numbers is an even number.)

OBJECTIVES

After completing this chapter, you should be able to

A. Divisors

1 Find the set of all divisors of any natural number. **(4–8)**

2 Use the intersection of two sets of divisors to determine the greatest common divisor (factor), or to determine when two numbers are relatively prime. **(9–13)**

B. Multiples

3 Use set notation to represent the set of all multiples of any natural number. **(14–17)**

4 Use the intersection of two sets of multiples to determine the least common multiple of two numbers. **(18–21)**

C. Prime numbers

5 (a) Prepare a Sieve of Eratosthenes to find all the prime numbers up to any given number. **(28–31)**

(b) Know at which number in the sieve all of the primes will have been determined. **(32–35)**

(c) Identify twin primes and reversible primes from the sieve (as well as triplet primes and symmetrical primes in the exercises). **(36–37; Exercise 15)**

D. Prime factorization

6 Use a branching tree to demonstrate that the prime factors of any composite number are unique. (Fundamental Theorem of Arithmetic.) **(38–43)**

E. Greatest common factor; reducing

7 Find the greatest common divisor (factor) of two numbers using prime factors. **(44–48)**

8 Apply the greatest common divisor in order to reduce any rational number to lowest terms. **(49–52)**

F. Least common multiple (and denominator)

9 Find the least common multiple of two numbers using prime factors. **(53–57)**

10 Use the least common multiple in order to add (or subtract) any two rational numbers. (Also, apply the greatest common divisor to reduce the sum to lowest terms.) **(58–60)**

11 Use the greatest common divisor to find the least common multiple of two numbers. **(61–63)**

G. Odd and even numbers

12 Prove given statements about the sum or the product of even and odd numbers. **(66–70)**

General **13** Identify the characteristics and distinctions of prime and composite numbers, even and odd numbers, greatest common divisor, and least common multiple. **(22–24; 56; 64–67)**

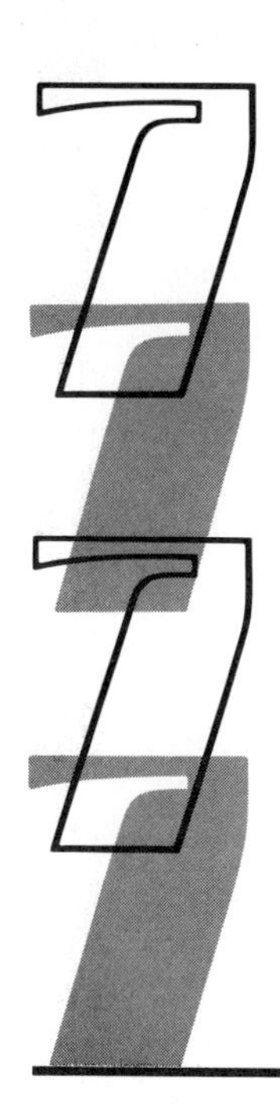

The Real Numbers and the Complex Numbers

Previous units have now introduced four systems of numbers: the natural numbers, whole numbers, integers, and rational numbers. Each new number system included all the elements, operations, relations, and properties of the earlier systems; it then added one or more new concepts, until closure was finally achieved for four operations. Indeed, the number line became so dense that, between any two given rational numbers, it was always possible to locate another rational. (1)

It would seem that surely an ultimate number system has been achieved. As we shall soon discover, however, mixed in among the dense rational numbers of the number line, there is yet another infinite set of numbers. (2)

In fact, these new numbers are impossible to represent accurately using ordinary written numerals.* When we represent them at all, we often use some symbol that does not even look like a number. In this sense, the new numbers are a completely different kind of set that mixes in among the rational numbers. But we can prove that these unusual numbers—the irrational numbers—do exist, and we can determine where they would be located on the number line. Before we do that, however, it will be helpful (3)

*Strictly speaking, "number" is only a mental concept. Any symbol written on paper is a "numeral." For practical purposes, however, the two are often used interchangeably.

to review how rational numbers may be expressed in the form of decimal fractions.

A. RATIONAL NUMBERS AS DECIMAL FRACTIONS

Decimal fractions are rational numbers

Any decimal number is actually an abbreviated way of writing a fraction, where only the numerator is written. The decimal point serves as a reference point that enables us to determine the value of the denominator, which in our base 10 system is always some power of 10. Because decimal numbers are indicated fractions, they are frequently called *decimal fractions*. Furthermore, since a decimal fraction can be expressed as the quotient of two integers, *any decimal fraction is therefore a rational number*. (4)

Objective 3: Converting fractions to decimals

When you learned to "convert a fraction to its decimal equivalent," you were really converting the given fraction to an equivalent fraction (decimal) that had some power of 10 as its denominator. For instance, (5)

$$\frac{2}{5} = \frac{2}{5}\cdot 1 = \frac{2}{5}\cdot\frac{2}{2} = \frac{4}{10} = .4$$

Recall, however, that a rational number may be defined as "the quotient of two integers." The word "quotient" implies division; that is, (6)

$$\frac{a}{b} = a \div b$$

Thus, to convert a fraction to its decimal equivalent, we divide the numerator by the denominator.

◆ **Example 1** Observe the pattern that develops in each example below: (7)

(a)
```
    .875
 8)7.000
   6 4
    60
    56
     40
     40
```
$\frac{7}{8} = 0.875$

(b)
```
    .8333 . . .
 6)5.0000
   4 8
    20
    18
     20
     18
      20
      18
       2
```
$\frac{5}{6} = 0.83\overline{3}$

(c)
```
     .6363 . . .
 11)7.0000
    6 6
     40
     33
      70
      66
       40
       33
        7
```
$\frac{7}{11} = 0.6\overline{363}$

Objective 1a: Terminating decimal

These examples illustrate the two types of decimals that denote rational numbers. As shown in the first case, a decimal value that ends after one or more decimal places is called a *terminating decimal*. (8)

Nonterminating, repeating decimal

In examples (b) and (c), on the other hand, observe that a pattern develops where the same digit or group of digits repeats over and over.* A decimal such as these that never ends but which repeats the same cycle of digits infinitely is called a *nonterminating, repeating decimal.* **(9)**

As illustrated by the examples above, *every rational number can be expressed either as* (1) *a terminating decimal or as* (2) *a nonterminating, repeating decimal.* Stated conversely, any terminating decimal or any nonterminating, repeating decimal represents some rational number. **(10)**

● **Practice 1** Find the decimal representation of the following rational numbers (using division): **(11)**

(a) $\frac{4}{5}$ (b) $\frac{3}{8}$ (c) $\frac{13}{20}$ (d) $\frac{5}{16}$

(e) $\frac{7}{9}$ (f) $\frac{5}{12}$ (g) $\frac{6}{11}$ (h) $\frac{2}{7}$

See also Exercise 7.

When $\frac{a}{b}$ terminates

Let us consider the two decimal representations of rational numbers. *The decimal equivalent of a rational number $\frac{a}{b}$ will terminate provided the denominator has some multiple that is a power of 10.* (This happens when the only prime factors of the denominator are some powers of 2 and 5.) For example, **(12)**

$$\frac{1}{2} = \frac{1}{2}\cdot 1 = \frac{1}{2}\cdot\frac{5}{5} = \frac{5}{10} = 0.5$$

or

$$\frac{7}{8} = \frac{7}{8}\cdot 1 = \frac{7}{8}\cdot\frac{125}{125} = \frac{875}{1{,}000} = 0.875$$

When $\frac{a}{b}$ is nonterminating, repeating

Otherwise, the decimal equivalent will be nonterminating and repeating. For instance, **(13)**

$$\frac{4}{9}\cdot 1 = \frac{4}{9}\cdot\frac{x}{x} = \frac{?}{10} \text{ or } \frac{?}{100} \text{ or } \frac{?}{1{,}000}$$

Observe there is no whole number x we can use that will produce a denominator that is a power of 10. The decimal equivalent of this rational number will thus be nonterminating, repeating.

Why decimals may be repeating

It usually seems reasonable to students that a decimal equivalent might not terminate. But why must a quotient repeat a cycle of digits indefinitely? In each division step, one of two things must happen: either **(14)**

*As shown above, the decimal equivalent of a fraction of this type may be indicated in either of two ways) (1) a horizontal bar (or line) may be placed over the repeating digits; or (2) after the pattern of digits is established, three dots may be used to indicate that the same digits repeat indefinitely.

the division ends or it does not end. If the division process ends, we have a terminating decimal. If the division process does not end, then we have a remainder. Now, we know that any remainder must always be less than the divisor; thus, there are only a limited (finite) number of possible remainders. (For instance, if the divisor is 7, then the only possible remainders are 1, 2, 3, 4, 5, or 6.) As the division process continues, eventually some remainder must repeat. When this happens, then the quotient also begins to repeat the same cycle of digits that occurred after the first time the remainder was obtained. This cycle then continues repeating infinitely and can never terminate.

As an example, consider the division used to convert $\frac{1}{7}$ to a decimal: **(15)**

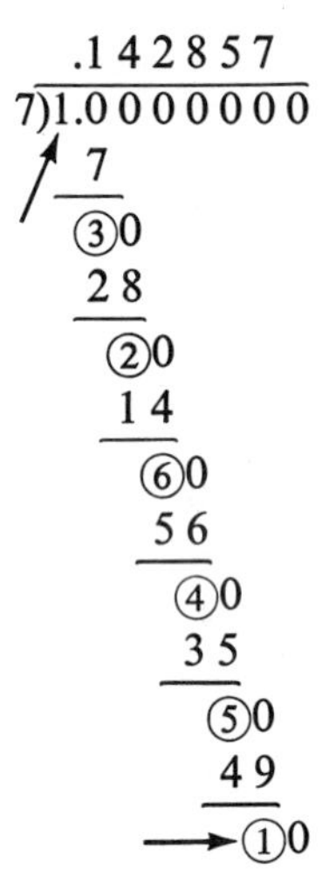

Notice that every possible remainder (1 – 6, as circled) has occurred. (This does not always happen, by any means.) As the arrows indicate, the same division process will now repeat a second time. So, the same sequence of digits (142857) begins repeating endlessly.

Rationals are (1) Terminating; or (2) Nonterminating, repeating

To repeat: Every rational number a/b can thus be expressed either as (1) a terminating decimal or (2) a nonterminating, repeating decimal. Conversely, every terminating decimal and every nonterminating repeating decimal denotes a rational number. **(16)**

B. DECIMAL FRACTIONS AS RATIONAL NUMBERS

In the section above we examined rational numbers in fractional form and converted each to a decimal equivalent—either a terminating decimal or a nonterminating, repeating decimal. We will now consider the pre- **(17)**

ceding process in reverse. That is, we will examine rational numbers in decimal form and convert them to their fractional equivalents. You have previously learned that terminating decimals may be restated in fractional form and reduced to lowest terms, as follows:

Objective 4: (1) Terminating decimals to fractions

◆ **Example 2** Express the following decimal fractions as common fractions and reduce each to lowest terms: (18)

(a) $0.4 = \frac{4}{10} = \frac{2}{5}$, reduced to lowest terms.

(b) $0.015 = \frac{15}{1{,}000} = \frac{3}{200}$

● **Practice 2** Rewrite each decimal as a fraction. Reduce to lowest terms: (19)

(a) 0.8 (b) 0.35 (c) 0.875
(d) 0.144 (e) 0.275 (f) 0.4375

See also Exercise 8.

Objective 5: (2) Nonterminating, repeating decimals to fractions

We will now consider the second decimal representation of rational numbers. That is, we will examine how nonterminating, repeating decimals (such as $0.83\bar{3}$) can be converted to their equivalent common fractions. This process requires some very simple equation solving, which we will review here. (General equation-solving techniques will be discussed in Chapter 9, Algebra). (20)

Simple equation solving

As normally used, *equation* means an equals relation that includes a *variable* (letter or symbol) to represent some unknown number. (For example, $2n = 6$ is an equation.) Both sides of an equation may be multiplied or divided by the same number. Also, the same number (or two expressions of equal value) may be added or subtracted on both sides of an equation. (21)

A *coefficient* is a number multiplied times a variable. (Given $2n$, the 2 is a coefficient). A coefficient and variable multiplied together (such as, $2n$) is called a *term*. Terms may be added or subtracted by combining their coefficients. For instance, (22)

$$\begin{array}{r} 5n \\ +2n \\ \hline 7n \end{array} \quad \text{and} \quad \begin{array}{r} 10n \\ -\ n \\ \hline 9n \end{array}$$

(Notice that "n" means "$1n$," and the coefficient 1 must be used in combining terms even if 1 is not written).

With this background, then, we are now ready to consider the procedure for converting a nonterminating, repeating decimal to its fractional equivalent: (23)

Procedure: Nonterminating repeating decimals to fractions

1. Let N equal the nonterminating, repeating decimal number. Multiply both sides of the equation by some power of 10 (that is, by 10, 100, 1,000, etc.) so that the decimal point passes *one complete cycle* of the repeating digits.
2. Subtract some number from the number obtained in step 1 so that the difference is a *whole number*. (Usually, it is only necessary to subtract N from the number obtained in step 1. However, it is sometimes necessary to multiply again by a power of 10, so that the digits of both decimals will repeat in the same order and the difference will be a whole number.)
3. Solve the difference of the two equations for N, thereby obtaining the fractional equivalent of the original decimal number. (Reduce to lowest terms, if necessary.)

This procedure is illustrated in the following examples. Notice that each step number of the procedure is indicated beside the corresponding step of the example.

◆ **Example 3** Find the fraction that corresponds to each of the following rational numbers (repeating, nonterminating decimals): (24)

(a) Let $N = 0.66\overline{6}$. Then,

$$\begin{array}{lrl} (1) & 10N = & 6.666\ldots \\ (2) & -\quad N = & -0.666\ldots \\ \hline & 9N = & 6.000\ldots \end{array}$$

(3) $9N = 6$

$$\frac{\cancel{9}N}{\cancel{9}} = \frac{6}{9}$$

$$N = \frac{2}{3}$$

(b) Let $N = 0.36\overline{36}$. Then,

$$\begin{array}{lrl} (1) & 100N = & 36.3636\ldots \\ (2) & -\quad N = & -\ 0.3636\ldots \\ \hline & 99N = & 36.0000\ldots \end{array}$$

(3) $\cancel{99}N = 36$

$$\frac{\cancel{99}N}{\cancel{99}} = \frac{36}{99}$$

$$N = \frac{12}{33}$$

(c) Let $N = 0.83\overline{3}$. Then,

$$\begin{array}{lrl} (1) & 100N = & 83.333\ldots \\ (2) & -\ 10N = & -\ 8.333\ldots \\ \hline & 90N = & 75.000\ldots \end{array}$$

(3) $90N = 75$

$$\frac{\cancel{90}N}{\cancel{90}} = \frac{75}{90}$$

$$N = \frac{5}{6}$$

(d) Let $N = 0.\overline{425}$. Then,

$$\begin{array}{lrl} (1) & 1000N = & 425.425425\ldots \\ (2) & -\quad N = & -\ 0.425425\ldots \\ \hline & 999N = & 425.000000\ldots \end{array}$$

(3) $999N = 425$

$$\frac{\cancel{999}N}{\cancel{999}} = \frac{425}{999}$$

$$N = \frac{425}{999}$$

● **Practice 3** Determine the fraction that is equivalent to each of the following nonterminating, repeating decimals: (25)

(a) 0.7777. . . (b) $0.6\overline{161}$
(c) 0.123123123. . . (d) $0.72\overline{2}$
(e) 0.41666. . . (f) $0.24\overline{24}$

See also Exercises 9 and 10.

C. THE SET OF IRRATIONAL NUMBERS

We have learned that every terminating decimal and every nonterminating, repeating decimal denotes a rational number. We know also that each rational number can be associated with some point on the number line. But we have said that there remain an infinite number of points on the number line which do not correspond to any number we have yet described. Let us now devise a new type of number and determine where it would be located on the number line. (26)

Objective Ia:

First consider the following rational numbers: (27)

	0.5151	Terminating*
(which also equals	0.5151000. . .	Nonterminating, repeating)
and	0.51515151. . .	Nonterminating, repeating

Next, study the following decimal number: (28)

0.51511511151111. . .

Notice that, after each succeeding five, the number of ones increases by one. This number is certainly nonterminating, since it establishes a pattern that we could continue indefinitely. However, the *same cycle* of digits does *not* repeat in the *same order*; therefore, this decimal number does not represent a rational number.

Objective 6:

On the other hand, this *is* a number, and it has a value. We can compare it with the other two numbers by considering the values of each corresponding digit: (29)

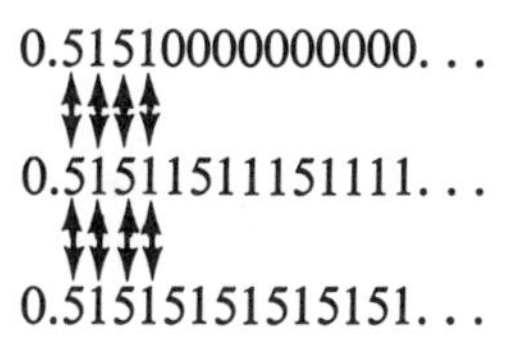

*We might even say that 0.5151 is a nonterminating, repeating decimal, since 0.5151 $= 0.5151\bar{0}$. Similarly, any terminating decimal could be considered to contain zeros that repeat infinitely. Thus, it is sometimes said that all decimal fractions are nonterminating, and that the set of rational numbers consists of those nonterminating decimals which repeat the same cycle of digits endlessly.

Notice that the first four digits correspond exactly. The fifth digits (0, 1, and 5) indicate the correct order of the numbers, which would be positioned on the number line approximately as follows: (30)

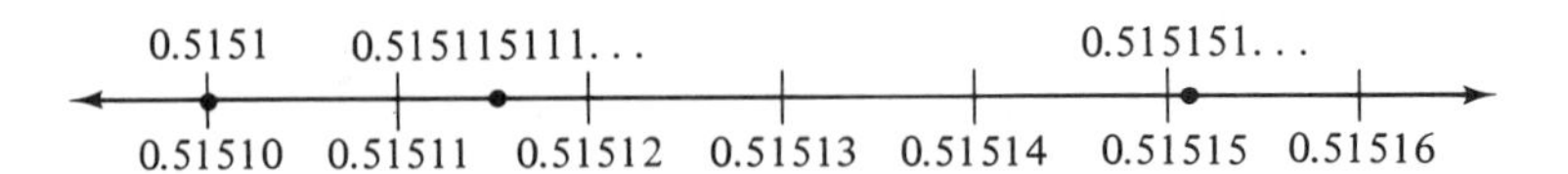

In a similar manner, other numbers we could construct could also be shown to correspond with points on the number line. Numbers such as these are known as irrational numbers, which are defined as follows: (31)

Definition of irrational numbers

Any number, a, belongs to the set, D, of irrational numbers if and only if a represents some nonterminating, nonrepeating decimal. That is, any number, a, is irrational if and only if a is a decimal number which *cannot* be expressed as the quotient of two integers. (32)

Irrationals are nonterminating, nonrepeating

The set, D, of irrational numbers may be represented as

$$D = \{a \text{ where } a \text{ is a nonterminating, nonrepeating decimal}\}$$

The very name "irrational" itself is derived from the definition of those numbers. Originally, the set of "rational" numbers were so named because each element could be expressed as a quotient (or *ratio*) of two integers. Thus, "irrational" does not imply that the elements of this set are unreasonable or illogical. Rather, the name "irrational" means simply "not rational"—which also defines the elements in the set of irrational numbers. (33)

Each member of the set of natural numbers can be associated with the irrational number 0.515115111... as follows: (34)

$$1 + 0.515115111\ldots = 1.515115111\ldots$$
$$2 + 0.515115111\ldots = 2.515115111\ldots$$
$$3 + 0.515115111\ldots = 3.515115111\ldots$$

Since the set of natural numbers is infinite, it follows that the set of irrational numbers also contains an infinite number of elements.

Other nonterminating decimals similar to 0.515115111... could also be divised, such as (35)

$$0.24224222422224\ldots$$

or

$$0.709700970009700009\ldots$$

The following example shows that an irrational number can be found between any two given rational numbers.

◆ **Example 4** Indicate an irrational number whose value falls between 0.437437 and $0.\overline{437}$. (36)

Objective 6: There are many irrational numbers that could be devised by many different patterns. The following irrational number is devised by establishing a pattern using the set of natural numbers. (Notice here that the pattern does not begin until after the second "437." The underline is there just to help you see the pattern.)

$$0.437437000000000\ldots$$

$$\text{Irrational} = 0.437437\underline{1}437\underline{2}437\underline{3}\ldots$$

$$0.437437437437437\ldots$$

● **Practice 4** Represent an irrational number whose value falls between each of the following pairs of rational numbers: (37)

(a) 0.6464 and $0.\overline{64}$ (b) 2.543543 and $2.\overline{543}$

(c) 0.1414 and $0.\overline{14}$ (d) 1.357357 and $1.\overline{357}$

See also Exercise 11.

D. SQUARE ROOTS AS IRRATIONAL NUMBERS

Irrational numbers do not necessarily contain digits that form some identifiable pattern, however; thus, we will now examine another form of irrational numbers—square roots. You know that a square root is not always a whole number. The fact is, however, that many square roots are actually irrational numbers, as the following paragraphs will verify. (38)

Objective 7: Square root

Suppose we wish to know the square root of two. There is no rational number, n, such that $n \times n = 2$. How, then, is $\sqrt{2}$ determined? The answer is simply, "By experimentation." The $\sqrt{2}$ must fall between one and two on the number line, since $1^2 = 1$ and $2^2 = 4$. We therefore try successive decimal digits after one, and test to determine how nearly the square equals two, as follows: (39)

◆ **Example 5** Find the approximate square root of 2. (40)

$a^2 < 2$		$b^2 > 2$	
1^2	$= 1$	2^2	$= 4$
$(1.4)^2$	$= 1.96$	$(1.5)^2$	$= 2.25$
$(1.41)^2$	$= 1.9881$	$(1.42)^2$	$= 2.0164$
$(1.414)^2$	$= 1.999396$	$(1.415)^2$	$= 2.002225$
$(1.4142)^2$	$= 1.99996164$	$(1.4143)^2$	$= 2.00024449$
$(1.41421)^2$	$= 1.9999899241$	$(1.41422)^2$	$= 2.0000182084$
$(1.414213)^2$	$= 1.999998409369$	$(1.414214)^2$	$= 2.000001237796$

Thus $\sqrt{2} \approx 1.414213$. (The symbol "$\approx$" means "approximately equals.")

● **Practice 5** Determine an approximate square root (to the first two decimal places) for the following natural numbers: (41)

(a) 5 (b) 29

See also Exercise 12.

Why is $\sqrt{2}$ irrational?

You might think, "Perhaps, if $\sqrt{2}$ were carried out to enough decimal places, the square would be exactly 2.0000. . . . How can we know for certain that $\sqrt{2}$ is irrational?" In fact, it can be proved that $\sqrt{2}$ is an irrational number, as will be given below by an indirect proof. (42)

Background for indirect proof

Indirect proof (a proof by contradiction) is used when there can be only two possible situations, and the existence of one prohibits the other. In this case, $\sqrt{2}$ is either (1) a rational number or (2) an irrational number. (By the definition of irrational, it cannot be both.) We will assume $\sqrt{2}$ is rational, show that this is impossible, and therefore conclude indirectly that $\sqrt{2}$ must be irrational. (43)

Proof that $\sqrt{2}$ is irrational

Assume $\sqrt{2}$ is a rational number. If so, by definition there is a rational number $\frac{a}{b}$ in lowest terms (relatively prime) such that $\frac{a}{b} = \sqrt{2}$. Then $\frac{a^2}{b^2} = 2$, or $a^2 = 2b^2$, which represents an even number. From number theory, a^2 being even implies that a is even; thus, we can let $a = 2n$ and then $a^2 = 4n^2$. We have now said $a^2 = 2b^2$ and also that $a^2 = 4n^2$; thus, by substitution $2b^2 = 4n^2$ and hence $b^2 = 2n^2$, which is another even number. This now implies that b is also an even number. But if a and b are both even, then $\frac{a}{b}$ is not in lowest terms (relatively prime) as we had assumed. This contradiction indicates that $\sqrt{2}$ cannot be a rational number. Thus, we conclude that $\sqrt{2}$ is irrational. (44)

Since $\sqrt{2}$ is an irrational number, you might think that it would be difficult to locate $\sqrt{2}$ on the number line. To the contrary, it is quite easy to locate $\sqrt{2}$ exactly on the number line by applying the Pythagorean theorem, as demonstrated below. (45)

The Pythagoreans were a semireligious group of Greek mathematicians and philosophers that flourished before 400 B.C. It was originally their belief that every number could be expressed in fractional form. They therefore became very upset when it was discovered that $\sqrt{2}$ could represent an actual distance but could not be written as a quotient of natural numbers, $\frac{a}{b}$. In fact, it is said that the member who betrayed the fraternity by revealing the existence of irrational numbers to the outside world suffered a fatal sailing "accident" soon afterward. (46)

The leader of this group was Pythagoras, for whom the theorem is named. Although the facts of the theorem had been known in ancient (47)

Egypt and China, Pythagoras is credited with being the first person to actually prove the theorem. According to the *Pythagorean theorem*, the square of the longest side (hypotenuse) of a right triangle (a triangle with a 90° angle) equals the sum of the squares of the other two sides (legs). In general,

Objective 8: Pythagorean theorem

$$c^2 = a^2 + b^2$$

For example,

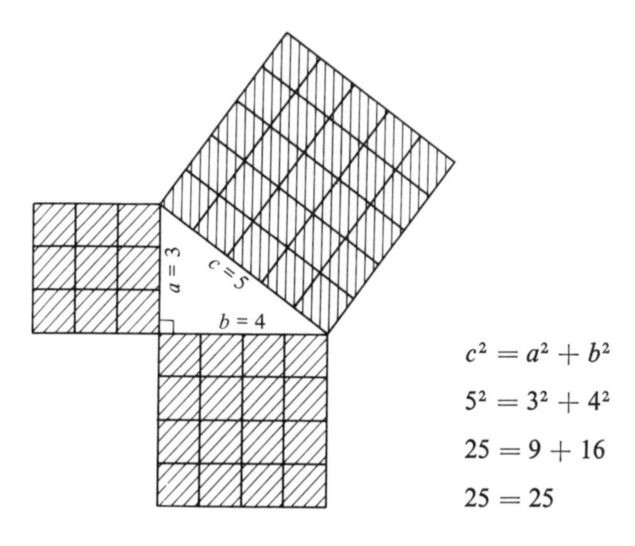

$$c^2 = a^2 + b^2$$
$$5^2 = 3^2 + 4^2$$
$$25 = 9 + 16$$
$$25 = 25$$

(48) The particular right triangle that puzzled the Pythagoreans may have been the one where $a = 1$ and $b = 1$. In this case,

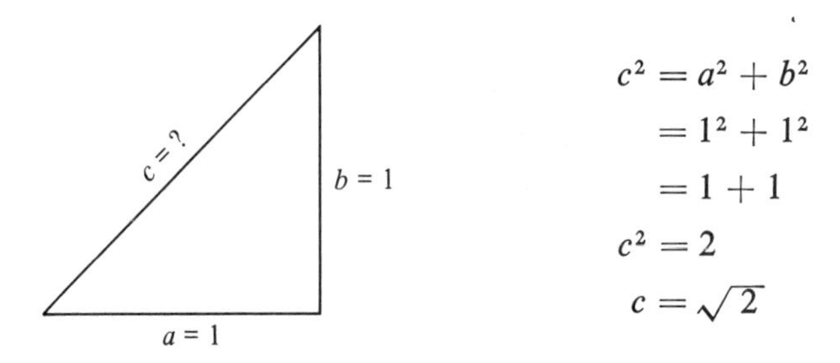

$$\begin{aligned} c^2 &= a^2 + b^2 \\ &= 1^2 + 1^2 \\ &= 1 + 1 \\ c^2 &= 2 \\ c &= \sqrt{2} \end{aligned}$$

(Note that only the positive square root is used here, since c represents a measurement.)

(49) We can construct this same triangle on a number line and use the hypotenuse to locate $\sqrt{2}$, as follows:

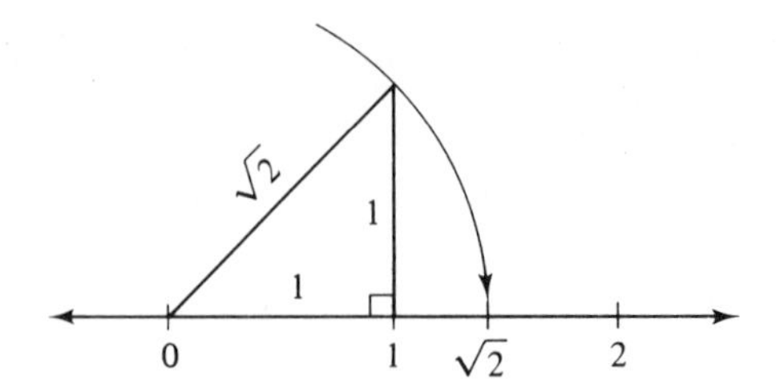

We know that the hypotenuse is $\sqrt{2}$ in length. By constructing a circle with the center at 0 and using the hypotenuse as radius, we then know the circle touches the number line at exactly $\sqrt{2}$. The location on the number line of other irrational numbers could be determined in a similar manner.

In no case above have we found the square root of a natural number to be a rational number that is not an integer (such as, 2.45 or 6.1412). It is interesting to observe that this always holds true: The square root of any natural number is always either an *integer* or else an *irrational* number (for example, $\sqrt{4} = 2$ or $\sqrt{2}$ is 1.414213. . .). (50)

◆ **Example 6** Use the Pythagorean theorem to find the unknown side of each triangle. Irrational lengths may be left in radical form. (The indicated value "$\sqrt{2}$" is in radical form.) (51)

(a) $a = 8$
$b = 15$
$c = ?$

(b) $a = ?$
$b = 5$
$c = 8$

Using the Pythagorean theorem,

(a)
$$\begin{aligned} c^2 &= a^2 + b^2 \\ c^2 &= 8^2 + 15^2 \\ c^2 &= 64 + 225 \\ c^2 &= 289 \\ c &= 17 \end{aligned}$$

(b)
$$\begin{aligned} a^2 + b^2 &= c^2 \\ a^2 + 5^2 &= 8^2 \\ a^2 + 25 &= 64 \\ a^2 &= 64 - 25 \\ a^2 &= 39 \\ a &= \sqrt{39} \end{aligned}$$

● **Practice 6** Apply the Pythagorean theorem in order to determine whether each unknown side is rational or irrational: (52)

(a) $a = 6$
$b = 8$
$c = ?$

(b) $a = 2$
$b = 3$
$c = ?$

(c) $a = ?$
$b = 16$
$c = 20$

(d) $a = 3$
$b = ?$
$c = 4$

See also Exercise 13.

Other irrationals

Nonterminating, pattern decimals and square roots are only two of the many types of irrational numbers. Another well-known irrational number is π (pronounced "pi"). The value of π is 3.14159265. . . . A very (53) good approximation of this value was discovered during the Greek efforts to express as a rational number, $\frac{C}{d}$, the ratio between the circumference of a circle and its diameter.

No true ratio (quotient) exists, of course, since the relationship is an irrational number. Further efforts to find a rational ratio comparing the area of a circle to its radius, $\frac{A}{r^2}$, also produced this same irrational number.* Other irrational numbers include e, the base of the natural logarithms ($e = 2.71828\ldots$), as well as many trigonometric functions and base 10 logarithms. Most irrational numbers, however, do not have any special symbol or name.

Objective 1a: Measurements, calculations for irrational numbers

(54) One point should be emphasized: any measurement or any calculation requires a rational number, since at some point a rational must be substituted for any irrational (nonterminating) number involved. Consider the substitution of $\frac{22}{7}$ for π:

$$\frac{22}{7} = 3.\overline{142857}$$

does not exactly equal

$$\pi = 3.14159265\ldots$$

but results close enough for practical purposes can be obtained. (It should also be noted that every measurement—no matter how accurate—is actually an approximation.)

Objective 9:

(55) Other roots besides square roots may also represent irrational numbers. For instance, since

$$2^3 = 2 \cdot 2 \cdot 2 = 8, \qquad \text{then} \qquad \sqrt[3]{8} = 2 \qquad \text{“the cube root of 8 is 2”}$$

Similarly, $\sqrt[4]{16}$ means “the fourth root of 16”; and $\sqrt[5]{56}$ means “the fifth root of 56.” For the general radical form, $\sqrt[n]{x}$, n is called the *index* and x is called the *radicand.* When no index is written, “$\sqrt{}$” always means square root.

(56) As in the case of square roots, higher roots of a natural number will either be *integers* or else *irrational* numbers. That is, a rational number that is not an integer (such as 6.4 or 2.15) may *not* be the root of an integer. Furthermore, although an *odd* root of a negative integer may be determined (such as $\sqrt[3]{-8}$) no even root of a negative integer (such as $\sqrt{-8}$) exists in any number system we have studied.

*When you think about it, isn't it amazing that such an irregular number as π would be the key to both the area and the circumference of a circle? When we have used formulas since childhood, however, we tend to take them for granted and fail to appreciate how amazing they really are.

◆ **Example 7** The following examples verify the indicated root of each integer. Since every radicand is an integer, we need only look for integers as roots. If none exists, then we know the root is either "irrational" or "not defined." (57)

(a) $\sqrt{16} = 4$ since $4^2 = 4 \cdot 4 = 16$

(b) $\sqrt[4]{16} = 2$ since $2^4 = 2 \cdot 2 \cdot 2 \cdot 2 = 16$

(c) $\sqrt[3]{8} = 2$ since $2^3 = 2 \cdot 2 \cdot 2 = 8$

(d) $\sqrt[3]{-8} = -2$ since $(-2)^3 = (-2)(-2)(-2) = -8$

(e) $\sqrt[3]{9}$ is irrational. There is no integer x for which $x^3 = x \cdot x \cdot x = 9$

(f) $\sqrt{-36}$ is not defined, since it would contradict rules for multiplication of positives and negatives. There is *no* x (rational or irrational) for which $x^2 = (x)(x) = -36$. (Numbers of this type will be considered in a later section.)

The preceding rules of odd and even roots may now be summarized as follows: (58)

Rules for even/odd roots

$\sqrt[\text{Even}]{(+)} = +$	Example: $\sqrt{4} = 2$
$\sqrt[\text{Odd}]{(+)} = +$	Example: $\sqrt[3]{8} = 2$
$\sqrt[\text{Odd}]{(-)} = -$	Example: $\sqrt[3]{-8} = -2$
$\sqrt[\text{Even}]{(-)} =$ not defined	Example: $\sqrt{-4}$ does not exist in the rationals or irrationals

● **Practice 7** Name all rational roots indicated below and verify that each is correct. State when an indicated root is "irrational" or "not defined." (59)

(a) $\sqrt{36}$ (b) $\sqrt{22}$ (c) $\sqrt[4]{81}$
(d) $\sqrt{81}$ (e) $\sqrt{-25}$ (f) $\sqrt[3]{27}$
(g) $\sqrt{27}$ (h) $\sqrt[3]{15}$ (i) $\sqrt[3]{-125}$
(j) $\sqrt[4]{-16}$ (k) $\sqrt{35}$ (l) $\sqrt{144}$

See also Exercise 14.

E. THE SYSTEM OF REAL NUMBERS

Objective 1a: In developing the set of irrational numbers, we have devised a completely new set of elements. We have seen that the digits of irrational numbers (60) may form a definite pattern or may occur in a random, unpredictable order. We have also found that every irrational number corresponds to some point on the number line.

The representation of rational and irrational numbers may not look so different from each other (for example, 0.515151... and 0.515115111...). But, by its very definition, a decimal number is an element of the set of irrational numbers only if it is *not* an element of the set of rational numbers.

Definition of real numbers

(61) Combing these two sets, we can now define the set of real numbers as follows:

Any number, a, belongs to the set, R, of real numbers if and only if a is either rational or irrational. In set form,

$$R = \{a \text{ where } a \text{ is either rational or irrational}\}$$

Development of number systems

(62) The relationship of the set of irrational numbers to other existing sets may be illustrated by a diagram similar to those given previously. That is, each new set of elements (in the lower box) was combined with the existing number system to produce a new number system.

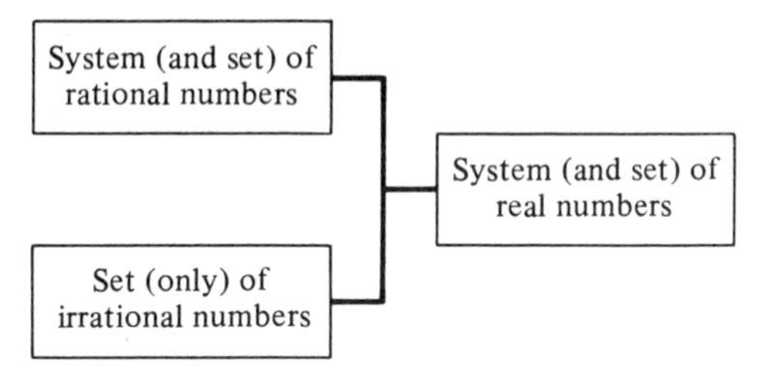

Development of real numbers

Similarly, the set of irrational numbers is a new *set* of elements that combines with the existing system of rational numbers in order to produce a new number *system* that is called the *system of real numbers.*

Irrationals not closed

(63) The set of negatives did not compose a number system, because the set was not closed under the operations that had been defined. Likewise, the set of irrational numbers does not compose a number system because this set also is not closed under the operations that have been defined. (Exercises will be given to verify that this is correct.)

(64) As defined above, however, the set of rational numbers combines with the set of irrational numbers to become the *set* of real numbers. The set of real numbers *is* closed under the operations; hence, the set of real numbers also composes a system.

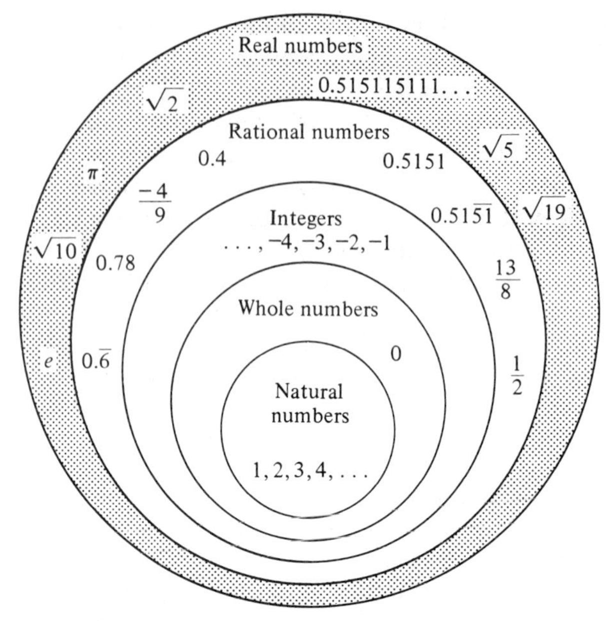

Systems (and Sets) of Numbers with Examples
(The shaded portion represents the set of irrational numbers.)

With our development of the real number system, we have finally achieved what is often called "the" number system. This system contains any number we might need to count objects, to compute a loss, to indicate one item broken into several pieces, or to measure any length in the physical (real) world. **(65)**

Completeness property

Whereas the rational numbers were said to be "dense" on the number line, the real numbers are now said to be *complete* or to possess the following property of *completeness*: Every real number corresponds to some point on the number line; and, finally, every point on the number line now corresponds to some real number. **(66)**

Field properties

Although this latest system adds nothing new to the list of fundamental number properties, the system does, as expected, maintain all properties already developed. Thus, the operations of addition and multiplication in the system of real numbers are closed, commutative, and associative; and multiplication is distributive over addition. Identity elements for addition and multiplication are included. There is also an additive inverse for each element, as well as a multiplicative inverse for each element except zero. Although it is impossible to write irrational numbers exactly, we can indicate them sufficiently to show that these field properties hold. The existence of these eleven basic properties means that the real number system, like the system of rational numbers, also composes a field. **(67)**

F. EQUALS, ORDER, AND EQUIVALENCE RELATIONS

The eleven field properties above are those properties that are considered fundamental to the number system, as is the equals relation. However, the system also includes a number of lesser properties and relations, some of which have already been introduced. For example, the order relation ($<$) was defined; and the trichotomy law was presented (that is, either $a = b$, $a < b$, or $a > b$). Other properties, which have simply been assumed, will now be listed for the sake of thoroughness. Proofs will not be given (although you can determine that the proofs would basically be derived from the closure and identity properties). (68)

Objective 10:

The following four properties form the basis for solving equations and inequalities in algebra: For each real number a, b, and c, (69)

Addition Property of Equals

$a = b$ if and only if* $a + c = b + c$

Multiplication Property of Equals

$a = b$ if and only if $a \cdot c = b \cdot c$ (where $c \neq 0$)

Addition Property of Less Than (70)

$a < b$ if and only if $a + c < b + c$

Multiplication Property of Less Than

$a < b$ if and only if $ac < bc$ (when $c > 0$)
$a < b$ if and only if $ac > bc$ (when $c < 0$)

◆ **Example 8** Use the given real numbers to illustrate the indicated properties: (71)

When $a = 12$, $b = (3 \times 4)$, and $c = 2$:

(a) $a = b$ implies $a + c = b + c$
$12 = (3 \times 4)$ implies $12 + 2 = (3 \times 4) + 2$
$12 = 12$ implies $14 = 14$

(b) $a = b$ implies $ac = bc$ (when $c \neq 0$)
$12 = (3 \times 4)$ implies $12 \cdot 2 = (3 \times 4)2$
$12 = 12$ implies $24 = 24$

*Recall that "if and only if" means that each part "implies" the other. The reverse portions of these properties are often called "cancellation" properties.

When $a = 3$, $b = 5$, and $c = 2$ (or $c = -2$):

(c) $a < b$ implies $a + c < b + c$
$3 < 5$ implies $3 + 2 < 5 + 2$
$3 < 5$ implies $5 < 7$

(d) $a < b$ implies $ac < bc$ (when $c > 0$)
$3 < 5$ implies $3 \cdot 2 < 5 \cdot 2$ (when $2 > 0$)
$3 < 5$ implies $6 < 10$

(e) $a < b$ implies $ac > bc$ (when $c < 0$)
$3 < 5$ implies $3(-2) > 5(-2)$ (when $-2 < 0$)
$3 < 5$ implies $-6 > -10$

● **Practice 8** Verify each of the following properties, using the given real numbers. (72)

When $a = (4 \times 2)$, $b = 8$, and $c = 3$:

(a) $a = b$ implies $a + c = b + c$
(b) $a = b$ implies $ac = bc$ (where $c \neq 0$)

When $a = -4$, $b = 2$, and $c = 3$:

(c) $a < b$ implies $a + c < b + c$
(d) $a < b$ implies $ac < bc$ (where $c > 0$)
(e) Let $c = -3$ and show:
$a < b$ implies $ac > bc$ (where $c < 0$)

See also Exercise 18.

Objective 11: Two relations have thus far been defined for the real number system—the equals relation (=) and the order relation (<). These relations have application in another type of relation, called an equivalence relation. To illustrate an equivalence relation, we may substitute the equals relation (=) for the relation R in the reflexive, symmetric, and transitive properties: (73)

Reflexive: $a = a$

Symmetric: If $a = b$ then $b = a$

Transitive: If $a = b$ and $b = c$, then $a = c$

Definition of equivalence relation

For all elements *a*, *b*, and *c*, any relation, R, is called an *equivalence relation* provided the following three properties hold: (74)

Reflexive: $a\text{R}a$

Symmetric: If $a\text{R}b$ then $b\text{R}a$

Transitive: If $a\text{R}b$ and $b\text{R}c$, then $a\text{R}c$

When the equals relation is substituted for R, the equivalence relation seems obvious. When other relations are tested, however, one or more of these three properties may fail to hold. Many relations that are not strictly mathematical relations can also be tested for equivalence, as follows: (75)

◆ **Example 9** Use persons A, B, and C to test the reflexive, symmetric, and transitive properties of the relation, "is a friend of." (76)

This is not an equivalence relation since it does not meet all three properties:

Relation	R	S	T	*Equivalence?*
"is a friend of"	✓	✓		No

To illustrate,

R: A is a friend of himself. (That is always true.)

S: If A is a friend of B,
then B is a friend of A.

T: If A is a friend of B,
and B is a friend of C,
we *cannot* conclude that A is a friend of C.
(A and C may not even know each other.)

● **Practice 9** Determine whether each relation below is an equivalence relation. Consider students A, B, and C when testing each property as illustrated above: (77)

(a) "has the same college major as"
(b) "studies more than"
(c) "scored within 5 points of." (Suppose they scored 90, 87, and 83, respectively, on a quiz.)

◆ **Example 10** Determine whether the relation "is a divisor (factor) of" is an equivalence relation by testing each property. Give an example to support your answer. (78)

Relation	R	S	T	*Equivalence?*
"is a divisor (factor) of"	✓		✓	No

R: 2 is a divisor of 2.

S: If 2 is a divisor of 4, it does *not* follow
that 4 is a divisor of 2.

T: If 2 is a divisor of 4
and 4 is a divisor of 8,
then 2 is a divisor of 8.

Thus, the relation "is a divisor of" is not an equivalence relation, because it is not symmetric.

● **Practice 10** Determine whether each of the following is an equivalence relation. (Write out the test of each property, as shown above.) (79)

(a) The relation "$<$" (is less than). Test using three different real numbers.

(b) The relation "$\subseteq$" (is a subset of—not necessarily a proper subset). Test using N, W, and I—the sets of natural numbers, whole numbers, and integers.

(c) The relation "is congruent to," for triangles. (Recall "congruent" triangles have the same shape and sides of the same length.)

(d) The relation "is next to" (or "beside"), for integers on the number line.

See also Exercise 19.

G. THE SYSTEM OF COMPLEX NUMBERS

As noted previously, the system of real numbers is often referred to as "the" number system. There is, however, one more extension which we shall study briefly for the sake of thoroughness. We found earlier that $\sqrt{-36}$ can be indicated, but there is *no* real number x for which $x \cdot x = -36$. However, we could say (80)

$$\sqrt{-36} = \sqrt{(36)(-1)} = \sqrt{36} \cdot \sqrt{-1} = 6\sqrt{-1}$$

In a similar manner, other negative numbers could be expressed as a positive real number times (-1). That is,

$$\sqrt{-x} = \sqrt{x(-1)} = \sqrt{x} \cdot \sqrt{-1}$$

And, $\sqrt{x}$ could then be determined as either rational or irrational.

Set of imaginary numbers

Now, since $\sqrt{-1}$ does not exist in the set of real numbers, this indicated number is called an *imaginary number*. (Actually, "imaginary" was a bad choice of names, because imaginary numbers are just as "real" as are negative numbers, for instance. But the name "imaginary" has been used for so long that it will undoubtedly remain with us.) For convenience, $\sqrt{-1}$ is indicated using the symbol, "i." Thus, substituting $i = \sqrt{-1}$, we have (81)

$$\sqrt{-36} = \sqrt{(36)(-1)} = \sqrt{36} \cdot \sqrt{-1} = 6i$$

In general, imaginary numbers are usually indicated in the form bi, where $i = \sqrt{-1}$ and b may represent any real number. Other examples of imaginary numbers would be the following: (82)

$$5i \qquad \frac{1}{2}i \qquad i\sqrt{3} \qquad \pi i$$

$$-9i \qquad -\frac{3}{11}i \qquad -i\sqrt{2} \qquad \frac{\pi}{2}i$$

(Notice that i is typically placed before the real number when a radical symbol is substituted for b; for example, $i\sqrt{5}$.) A number of the type bi is also frequently called a *pure imaginary* number. The following example further demonstrates imaginary numbers.

◆ **Example 11** Find each indicated root or power, or solve each equation, as directed. (83)

Each negative radicand may be rewritten as a positive value times (-1):

Objective 9:
Objective 12:

(a) $\sqrt{-16} = \sqrt{16(-1)} = \sqrt{16}\cdot\sqrt{-1} = 4\sqrt{-1} = 4i$

(b) $\sqrt{-7} = \sqrt{7(-1)} = \sqrt{7}\cdot\sqrt{-1} = \sqrt{7}\,i$ or $i\sqrt{7}$

The intermediate steps may be omitted as follows:

(c) $\sqrt{-25} = 5i$

The following problems demonstrate that the set of imaginary numbers is not closed:

(d) $i^2 = i\cdot i = (\sqrt{-1})(\sqrt{-1}) = -1$ This is the key to all other powers that include i.

(e) $(2i)^2 = 2i\cdot 2i = 4(-1) = -4$

(f) $(i\sqrt{2})^2 = i\sqrt{2}\cdot i\sqrt{2} = (-1)(2) = -2$

Notice that the first equation below* has a real-number solution, while the second equation has an imaginary solution:

(g) $x^2 = 5$
$x = \pm\sqrt{5}$

(h) $x^2 = -5$
$x = \pm\sqrt{-5}$
$x = \pm i\sqrt{5}$

● **Practice 11** Determine each root or power, or solve each equation, as indicated: (84)

(a) $\sqrt{9}$ (b) $\sqrt{-9}$ (c) $\sqrt{-64}$

(d) $\sqrt{-49}$ (e) $\sqrt{-3}$ (f) $\sqrt{-2}$

(g) $\sqrt{-11}$ (h) $\sqrt{-23}$ (i) i^2

(j) $(7i)^2$ (k) $(-4i)^2$ (l) $(i\sqrt{3})^2$

(m) $x^2 = 4$ (n) $x^2 = -4$ (o) $x^2 + 81 = 0$

(p) $x^2 + 9 = 3$

See also Exercises 14 and 20.

*Observe that every *number* (here 5 and -5) has both positive and negative square roots. But the symbol "$\sqrt{}$" denotes only the principal square root—the positive root. (That is, when $x^2 = 16$, then $x = \pm\sqrt{16}$, or $x = \pm 4$; but $\sqrt{16} = 4$.)

The set of imaginary numbers is similar to other new sets that have been developed in that it introduces a new set of elements that may be used in the operations that have been defined. We have demonstrated, however, that the set of imaginary numbers is not closed. Thus the set of imaginary numbers does *not* comprise a system of numbers. Rather, the set of imaginary numbers combines with the existing system of real numbers to form a new system called the *system of complex numbers*. The elements of the set of complex numbers are defined as follows: **(85)**

Objective 1b:

Definition of complex numbers

The set, C, of complex numbers consists of all elements

$$a + bi$$

where a and b are real numbers and i represents $\sqrt{-1}$.

Objective 13: $a + bi$

In the form $a + bi$ for complex numbers, a is designated the "real" portion and bi is called the "pure imaginary" part. The entire complex number, $a + bi$, may also be called a "mixed imaginary" number. Examples of complex numbers include such numbers as the following: **(86)**

$$3 + 2i \qquad \frac{1}{2} + i\sqrt{2} \qquad 7 - 5i$$

$$\sqrt{5} + 4i \qquad \pi + 3i \qquad \frac{3}{8} - i\sqrt{7}$$

(Notice that the "+" sign of $a + bi$ can become a "−" whenever b is negative, in accordance with the definition of subtraction.)

Observe further that either a or b can equal zero. Suppose $a = 4$ and $b = 0$; then **(87)**

$$a + bi = 4 + 0i = 4$$

Or, suppose $a = 0$ and $b = 2$; then

$$a + bi = 0 + 2i = 2i$$

From these examples we see that a complex number

$a + bi$ is a real number when $b = 0$

$a + bi$ is a (pure) imaginary number when $a = 0$

By this representation, we see that the set, R, of real numbers is a proper subset of the set, C, of complex numbers. That is,

$$\boldsymbol{R \subset C}$$

The formation of the system of complex numbers is illustrated by the two diagrams that follow. As illustrated by the diagram at the top of page 189, the set of complex numbers is formed by combining our newest set of elements (the imaginary numbers) with the previous set of real numbers. **(88)**

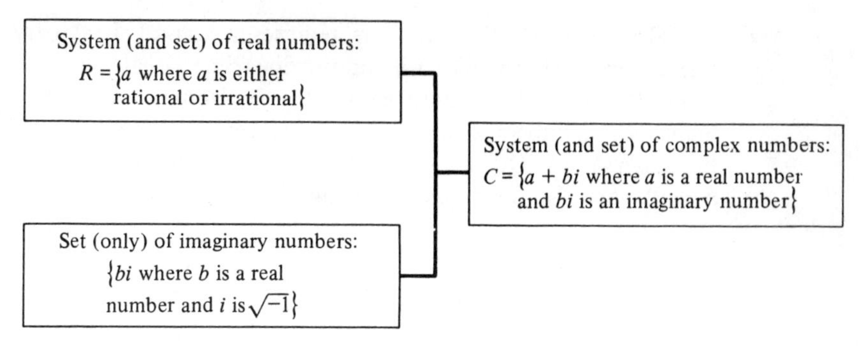

The diagram below depicts the relationship of all sets that we have studied previously. Observe that our new set of imaginary numbers is included in the shaded portion. As we have found, operations with elements of the set of imaginary numbers may produce results that are *not* imaginary numbers. Since the set of imaginary numbers is thus not closed, it does not compose a *system* of numbers. However, operations using imaginary numbers may produce a real number; thus, the system of real numbers (and all earlier systems) are subsets of the system of complex numbers.

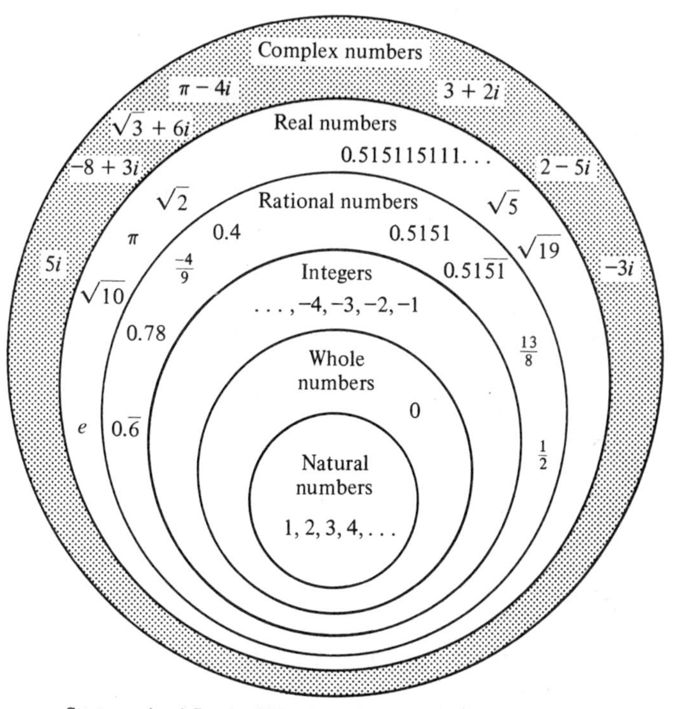

Systems (and Sets) of Numbers Showing Examples of Each
(The shaded portion includes the set of imaginary numbers.)

Graphing complex numbers

Graphing complex numbers requires a second, imaginary number line drawn perpendicular to our previous line. Real numbers (such as $-4 + 0i$, below) are graphed on the horizontal line as before. Pure imaginary numbers (such as $0 - 3i$) are graphed on the vertical (imaginary) line. Other complex numbers $a + bi$ (where $a \neq 0$ and $b \neq 0$) are also illustrated in the figure. (89)

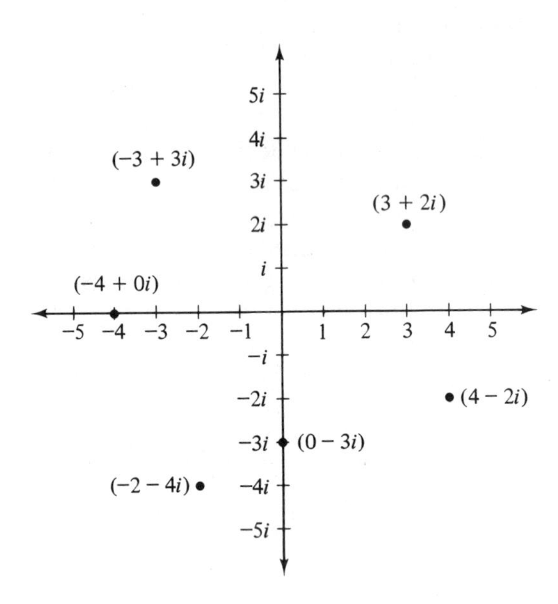

◆ **Example 12** Express each of the following as a complex number, $a + bi$: (90)

(a) $7 + \sqrt{-25}$ (b) $3 - \sqrt{-10}$ (c) 5

(d) -3 (e) $\sqrt{2}$ (f) $\sqrt{-3}$

The first two numbers are typical complex numbers of the form $a + bi$:

(a) $7 + \sqrt{-25} = 7 + 5i$ (b) $3 - \sqrt{-10} = 3 - i\sqrt{10}$

The following three numbers are real numbers; that is, $b = 0$:

(c) $5 = 5 + 0i$ (d) $-3 = -3 + 0i$

(e) $\sqrt{2} = \sqrt{2} + 0i$

The final number is pure imaginary; that is, $a = 0$:

(f) $\sqrt{-3} = 0 + i\sqrt{3}$

● **Practice 12** Rewrite each of the following in the form of a complex number, $a + bi$: (91)

(a) $3 + \sqrt{-16}$ (b) $-8 + \sqrt{-2}$ (c) $\frac{1}{2} - \sqrt{-81}$

(d) $\pi - \sqrt{-14}$ (e) 4 (f) -2

(g) 0 (h) $-\frac{3}{8}$ (i) $\sqrt{-36}$

(j) $\sqrt{15}$ (k) $\sqrt{-22}$ (l) $\sqrt{-7}$

See also Exercise 22.

The complex numbers are necessary in order to provide solutions to certain mathematical problems and theories (such as the theory of electrical circuits), as well as in the space program. (92)

Properties of complex numbers

The system of complex numbers contains the same eleven fundamental (field) properties as does the system of real numbers. (Our brief introduction will not include operations with complex numbers, however.) Surprisingly, the concept of order from the system of real numbers does *not* apply in the system of complex numbers. Whereas real numbers can be arranged in left-to-right order on the number line, complex numbers are scattered all around the plane of their graph. Thus, we cannot determine whether $6 + 2i$ is greater than or less than $3 + 7i$, in the same manner that we establish 3 is less than 6. (93)

No order relation

H. CONCLUSION

The number system we use has now been traced in building-block fashion from the simplest counting numbers to the more intricate real and complex numbers. Every mathematical problem that has been indicated now has a solution in the set of complex numbers. (94)

Other number systems?

A natural question is, "Do other number systems exist?" The answer is, "Yes." Mathematicians have identified other systems of numbers (such as the transfinite numbers used to represent the cardinal number of infinite sets). A mathematical system may exist with only a finite number of elements. Some abstract mathematical systems have been devised that have limited application. Mathematical systems without numbers have also been developed (such as in the study of geometry). (95)

And, of course, number systems can be expressed in bases other than base ten. (These other bases, however, are fundamentally just a translation of the real number system—much as a book translated from English into (96)

Spanish would still say the same thing). We will consider a few of these other systems in the following chapter. However, the system of complex numbers represents the maximum extension of our number system for most practical (as well as theoretical) purposes.

The chart on page 434 summarizes the entire system of real numbers; you may also refer to the diagram on page 435. Throughout our development of the real numbers, it is hoped that you have gained a new understanding and appreciation of the amazing system that the real numbers compose. (97)

EXERCISES

Refer to: decimal fractions

1. (a) The number system we use is the base ____ system.
   (b) Decimal fractions may also be called ____.
   (c) Any decimal is really an abbreviated fraction in which the denominator is not written but is always some ____.
   (d) A rational number may be defined as the ____ of two integers.
   (e) This definition implies that a rational number in fractional form may be converted to its decimal equivalent by ____.
   (f) The decimal representation of any rational number is either a ____ decimal or a ____ decimal.
   (g) If the denominator of a fraction has some multiple that is a power of ten, then the fraction will be a ____ decimal.
   (h) If the denominator of a fraction (in lowest terms) has prime factors other than two or five, then the decimal equivalent will be a ____ decimal.
   (i) Every nonterminating, repeating decimal represents some ____.
   (j) A terminating decimal may be expressed as a nonterminating decimal by ____.
   (k) Express 0.35 as a nonterminating, repeating decimal.

Irrational numbers

2. Answer the following. Explain each "no" response:
   (a) Every irrational number corresponds to a decimal that is ____.
   (b) Is the set of irrational numbers infinite?
   (c) Does each irrational number correspond to some point on the number line?
   (d) Does each point on the number line correspond to some irrational number?
   (e) Is the set of rational numbers a subset of the set of irrational numbers?
   (f) Can every irrational number be expressed as a nonterminating decimal that contains an identifiable pattern of nonrepeating digits?

More irrationals

3. (a) Can the square root of a natural number be an integer? a rational number that is not an integer? an irrational number?
(b) Is the square root of every prime number an irrational number? Explain.
(c) Is the square root of every composite number an integer? Explain and give examples.
(d) Can the length of any distance in the physical world be an irrational number?
(e) Is it possible to measure a line and know whether it corresponds to an irrational number?
(f) Since the exact value of an irrational number can never be written, how is it possible to locate an irrational number exactly on the number line?
(g) To denote the value of an irrational number, we must substitute some ____.
(h) If q represents some irrational number (say 1.41421. . .), then does $3q$ represent an irrational number?
(i) Can we compute $3q$? Explain.
(j) If $\pi d = C$, then, by the definition of division, $\pi = \frac{C}{d}$. Suppose we obtain C and d by measuring. Explain why $\frac{C}{d}$ does not really equal π.

Real numbers

4. Answer the following. Explain each "no" response:
(a) Is every integer a real number?
(b) Is every integer an irrational number?
(c) Is every rational number a real number?
(d) Is every irrational number a real number?
(e) Is every real number an irrational number?
(f) Is every real number either a rational number or an irrational number?
(g) Do the real numbers comprise a number system?
(h) Do the irrational numbers comprise a number system?
(i) Can every real number be expressed as the quotient of two integers $\left(\frac{a}{b} \text{ where } b \neq 0\right)$?
(j) Does every point on the number line correspond to some real number? What property answers this question?
(k) If F = the set of rational numbers and D = the set of irrational numbers, what is $F \cup D$?
(l) What is $F \cap D$?

5. Answer the following. Explain (or give an example for) each "no" response:

Equals, order, and equivalence relations

(a) Name the field properties.
(b) If $a = b$, does $a + c = b + c$ when $c < 0$?
(c) If $a = b$, does $ac = bc$ for all values of c?
(d) If $ac = bc$, does $a = b$ for all values of c?
(e) If $a < b$, is $ac < bc$ when $c < 0$?
(f) Relations that are reflexive, symmetric, and transitive are called ____ relations.
(g) Is the equals relation an equivalence relation?
(h) Is the order relation ($<$) reflexive?
(i) Is the order relation ($>$) transitive?

6. Answer the following. Explain each "no" response:

Complex numbers

(a) The $\sqrt{-1}$ is called an ____ number.
(b) Is every real number an imaginary number?
(c) Is every real number a complex number?
(d) Can every rational number be expressed in the form $a + bi$?
(e) Is every imaginary number a complex number?
(f) Is every complex number an imaginary number?
(g) Is every imaginary number an irrational number?
(h) Is every irrational number an imaginary number?
(i) Is every complex number an irrational number?
(j) Is every irrational number a complex number?
(k) Is $\sqrt{5}$ a pure imaginary number?
(l) Is $\sqrt{-4}$ a pure imaginary number?
(m) If $R =$ the set of real numbers and $P =$ the set of pure imaginary numbers, what is $\{R + P\}$?
(n) If $a = 0$, then $a + bi$ represents a ____ number.
(o) If $b = 0$, then $a + bi$ represents a ____ number.
(p) Do the imaginary numbers comprise a number system?
(q) Do the complex numbers comprise a number system?
(r) Do the field properties hold in the complex numbers?
(s) Does the trichotomy law hold in the complex numbers?
(t) Is $2 + 5i < 7 - 3i$?

7. Use division to convert each of the following fractions to its decimal equivalent:

Objective 3; Example 1

(a) $\frac{3}{5}$ (b) $\frac{7}{40}$ (c) $\frac{5}{8}$ (d) $\frac{11}{32}$

(e) $\frac{22}{30}$ (f) $\frac{5}{13}$ (g) $\frac{29}{37}$ (h) $\frac{4}{33}$

8. Rewrite each decimal as a fraction. Then reduce each to lowest terms:

Objective 4; Example 2

(a) 0.4 (b) 0.375 (c) 0.64
(d) 0.048 (e) 0.1125 (f) 0.4375

Objective 5; Example 3

9. Determine the fraction that corresponds to each of the following repeating decimals:

(a) $0.88\overline{8}$ (b) $0.5666\ldots$ (c) $0.72\overline{72}$
(d) $0.324324\ldots$ (e) $0.27\overline{27}$ (f) $0.58333\ldots$

Example 3

10. (a) Express 0.72 as a common fraction. Reduce to lowest terms.
(b) Rewrite 0.72 as a nonterminating, repeating decimal.
(c) Use your answer from part (b) and apply the procedure for converting a nonterminating, repeating decimal to its fractional equivalent. Compare this answer with part (a).

(d) Notice

$$\begin{array}{rcl} \dfrac{1}{3} & = & 0.3333\ldots \\ +\dfrac{2}{3} & = & +0.6666\ldots \\ \hline \dfrac{3}{3} & = & 0.9999\ldots \end{array} \qquad \begin{array}{rcl} \dfrac{1}{3} & = & 0.3333\ldots \\ \times\ 3 & \times & 3 \\ \hline \dfrac{3}{3} & = & 0.9999\ldots \end{array}$$

This implies $0.99\overline{9} = 1$. Convert $0.99\overline{9}$ to its fractional equivalent to find out.

Objective 6; Example 4

11. Express an irrational number whose value would fall between each given pair:
(a) 0.8686 and $0.\overline{86}$
(b) 3.952952 and $3.\overline{952}$
(c) 2.4747 and $2.4\overline{7}$
(d) 0.248248 and $0.\overline{248}$

Objective 7; Example 5

12. By trial and error, find an approximate square root for the following (to the first two decimal places):
(a) 12 (b) 38

Objective 8; Example 6

13. Determine the unknown side of each right triangle by applying the Pythagorean theorem:

(a) $a = 12$, $b = 9$, $c = ?$
(b) $a = 3$, $b = 5$, $c = ?$
(c) $a = 24$, $b = ?$, $c = 25$
(d) $a = ?$, $b = 2$, $c = 3$

Objective 9; Examples 7 and 11

14. The indicated roots below are either rational, irrational, or imaginary. Find each root, if possible, and verify your answer. (Irrational roots may be left as radicals. Express imaginary roots in the form *bi*.) Tell whether each root is "rational," "irrational," or "imaginary":

(a) $\sqrt{1}$ (b) $\sqrt[3]{1}$ (c) $\sqrt[4]{1}$ (d) $\sqrt[5]{1}$
(e) $\sqrt{-1}$ (f) $\sqrt[3]{-1}$ (g) $\sqrt{64}$ (h) $\sqrt[3]{64}$
(i) $\sqrt{-64}$ (j) $\sqrt[3]{-64}$ (k) $\sqrt[5]{64}$ (l) $\sqrt[6]{64}$
(m) $\sqrt{17}$ (n) $\sqrt{-81}$ (o) $\sqrt[3]{27}$ (p) $\sqrt{-14}$

(q) $\sqrt[3]{-125}$ (r) $\sqrt[4]{625}$ (s) $\sqrt{-15}$ (t) $\sqrt{100}$
(u) $\sqrt[3]{18}$ (v) $\sqrt{6}$ (w) $\sqrt{-16}$ (x) $\sqrt{-13}$

Miscellaneous Operations

15. (a) Consider the two irrational numbers 0.212112111. . . and 0.454554555. . . . Find the sum of these numbers. Is the sum an irrational number?
(b) Does $\sqrt{4} \cdot \sqrt{9} = \sqrt{4 \cdot 9}$? Does $\sqrt{9} \cdot \sqrt{16} = \sqrt{9 \cdot 16}$?
(c) Using inductive reasoning from part (b), find the product of the irrational numbers $\sqrt{2}$ and $\sqrt{8}$. Is the product an irrational number? (Note: the product $\sqrt{a} \cdot \sqrt{b} = \sqrt{ab}$ does *not* hold true when both a and b are negative numbers.)
(d) Why does the set of irrational numbers not compose a number system?
(e) Part (b) introduces an interesting characteristic about the product of square roots. Determine whether a similar situation holds for addition of square roots. Does $\sqrt{9} + \sqrt{16} = \sqrt{9 + 16}$? Does $\sqrt{36} + \sqrt{64} = \sqrt{36 + 64}$? Does $\sqrt{4} + \sqrt{9} = \sqrt{4 + 9}$?

Objective 2a

16. Use real numbers p, q, and r (if needed) to express the following properties of the system of real numbers:
(a) Closure property of addition. Verify this property using 4 and 2.502500250002. . . .
(b) Closure property of multiplication. Verify this property using 0.313113111. . . and 2.
(c) Commutative property of addition. Verify this property using 5 and 0.771772773. . . .
(d) Commutative property of multiplication. Verify this property using $\sqrt{2}$ and $\sqrt{18}$.
(e) Associative property of addition. Verify the property using 3 and $0.33\bar{3}$ and 0.040040004. . . .
(f) Associative property of multiplication. Verify this property using $\frac{1}{2}$ and 4 and 0.121221222. . . .
(g) Distributive property of multiplication over addition. Verify this property using 3 and 0.313113111. . . and 0.020220222. . . .
(h) Additive identity property. Verify this property using 3.14159265. . . .
(i) Multiplicative identity property. Verify this property using 1.414213. . . .
(j) Additive inverse property. Verify this property using 0.898898889. . . and its inverse.
(k) Multiplicative inverse property. Verify this property using π and its inverse.

17. You should now be able to identify the number (field) properties that hold in each system. Copy the following chart; beside each property,

Objective 2b

put a check under each number system to which the property applies:

Properties	*Natural numbers*	*Whole numbers*	*Integers*	*Rational numbers*	*Real numbers*
Closure, +	____	____	____	____	____
Closure, ×	____	____	____	____	____
Commutative, +	____	____	____	____	____
Commutative, ×	____	____	____	____	____
Associative, +	____	____	____	____	____
Associative, ×	____	____	____	____	____
Distributive, × over +	____	____	____	____	____
Identity, +	____	____	____	____	____
Identity, ×	____	____	____	____	____
Inverse, +	____	____	____	____	____
Inverse, ×	____	____	____	____	____
Also:					
Density	____	____	____	____	____
Completeness	____	____	____	____	____

Objective 10; Example 8

18. Verify the following addition and multiplication properties of equals and of "less than."

(a) When $a = (2 \times 2)$, $b = 4$, and $c = -9$:
$a = b$ implies $a + c = b + c$.

(b) When $a = (-2 \times 4)$, $b = -8$, and $c = -3$:
$a = b$ implies $ac = bc$.

(c) When $a = -7$, $b = -3$, and $c = -5$:
$a < b$ implies $a + c < b + c$.

(d) When $a = -4$, $b = -1$, and $c = 3$:
$a < b$ implies $ac < bc$ (when $c > 0$).

(e) When $a = -5$, $b = -2$, and $c = -3$:
$a < b$ implies $ac > bc$ (when $c < 0$).

Objective 11; Examples 9 and 10

19. Determine whether each of the following relations is an equivalence relation, by testing the reflexive, symmetric, and transitive properties:

(a) The relation "$\leq$" for real numbers. (Let a, b, and c each represent different real numbers.)

(b) The relation "$\subset$" (is a proper subset of). Test using N, I, and R—the sets of natural numbers, integers, and real numbers.

(c) The relation "$\approx$" (approximately equals) for irrational numbers a, b, and c.

(d) The relation "$\perp$" (is perpendicular to) for lines A, B, and C.

(e) The relation "is a multiple of." Test using real numbers.

20. Find each power or solve each equation:

Objectives 9 and 12; Example 11

(a) $(3i)^2$ (b) $(-6i)^2$

(c) $\left(\frac{2}{3}i\right)^2$ (d) $(i\sqrt{5})^2$

(e) $(\pi i)^2$ (f) $\left(\frac{-3}{4}i\right)^2$

(g) $x^2 = 64$ (h) $x^2 = -36$

(i) $x^2 - 4 = 0$ (j) $x^2 + 3 = 0$

(k) $x^2 + 2 = -7$ (l) $x^2 + 5 = 3$

21. You should now be able to identify the set (or sets) to which any given number belongs. Copy and complete the following chart. Beside each number, put a check under each set of which the number is an element:

Objective 1c

Number	*Natural numbers*	*Whole numbers*	*Integers*	*Rational numbers*	*Irrational numbers*	*Real numbers*	*Pure imaginary numbers*	*Complex numbers*
(a) -7	___	___	___	___	___	___	___	___
(b) 0	___	___	___	___	___	___	___	___
(c) $\frac{3}{8}$	___	___	___	___	___	___	___	___
(d) $\sqrt{5}$	___	___	___	___	___	___	___	___
(e) $\sqrt{-9}$	___	___	___	___	___	___	___	___
(f) $3 + 2i$	___	___	___	___	___	___	___	___
(g) $\sqrt[3]{7}$	___	___	___	___	___	___	___	___
(h) 12	___	___	___	___	___	___	___	___
(i) $\frac{-7}{8}$	___	___	___	___	___	___	___	___
(j) $0.\overline{12}$	___	___	___	___	___	___	___	___
(k) π	___	___	___	___	___	___	___	___
(l) 0.8	___	___	___	___	___	___	___	___
(m) $0.020020002\ldots$	___	___	___	___	___	___	___	___
(n) $3i$	___	___	___	___	___	___	___	___
(o) $\sqrt[3]{-27}$	___	___	___	___	___	___	___	___

Objective 13; Example 12

22. Express each of the following in the form of a complex number, $a + bi$:

(a) $2 + \sqrt{-36}$ (b) $-8 + \sqrt{-49}$ (c) $\frac{3}{8} - \sqrt{-7}$
(d) $\sqrt{5} - \sqrt{-3}$ (e) 9 (f) -1
(g) $\frac{-4}{7}$ (h) π (i) $\sqrt{17}$
(j) $\sqrt{16}$ (k) $\sqrt{-64}$ (l) $\sqrt{-5}$

OBJECTIVES

After completing this chapter, you should be able to

Characteristics of the real and complex numbers

1 General objectives

(a) Identify the characteristics of the set of real numbers, and tell the distinctions between the set of real numbers and previous sets studied. (Characteristics include description of elements, subset relationships, decimal representation, number line representation, etc.) **(8–10; 26–35; 53–54; 60–67)**

(b) Repeat part (a) for the set of complex numbers. (Characteristics include subset relationships, representation of complex numbers, properties of earlier systems that do or do not hold, etc.) **(80–82; 85–93)**

(c) Given a list of numbers, indicate to which set(s) of numbers each belongs. **(Exercise 21)**

Properties of the real numbers

2 (a) Express with variables and illustrate with real numbers the eleven field properties. **(Exercise 16)**

(b) Given a list of number properties, indicate in which number system(s) each property applies. **(Exercise 17)**

A. Rational numbers as decimal fractions

3 Convert any fraction to its decimal equivalent, either (1) terminating or (2) nonterminating, repeating. **(5–11)**

B. Decimal fractions as rational numbers

4 Express any given decimal as a fraction; reduce the fraction to lowest terms. **(17–19)**

5 Convert any nonterminating, repeating decimal fraction to its equivalent common fraction. (20–25)

6 Given any two rational numbers, indicate an irrational number whose value falls between the given values. (27–30; 36–37)

D. Square roots as irrational numbers

7 By trial and error, find the approximate square root of any natural number. (38–41)

8 Use the Pythagorean theorem to find the unknown side of any right triangle. (45–52)

9 Given a list of indicated roots, find and verify each rational root and each imaginary root. Tell whether each indicated root is rational, irrational, or imaginary. (55–59; 83–84)

F. Equals, order, and equivalence relations

10 Use real numbers to verify the addition and multiplication properties of "equals" and of "less than." (68–72)

11 Given any relation, determine whether it is an equivalence relation by testing the reflexive, symmetric, and transitive properties. (73–79)

G. The system of complex numbers

12 Find indicated roots or powers, or solve equations involving imaginary numbers. (80–84)

13 Express any given number in the form of a complex number, $a + bi$. (85–91)

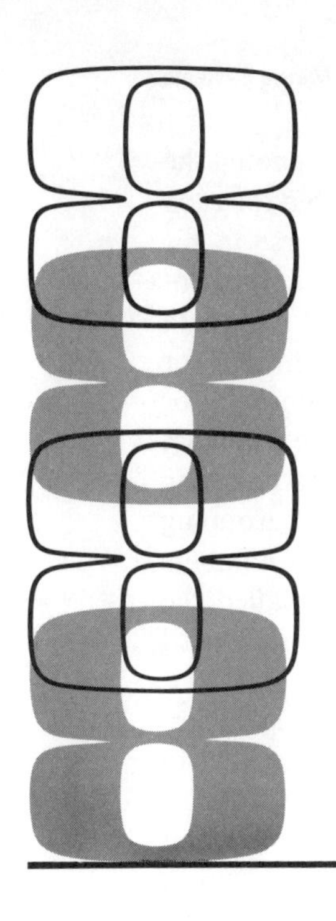

Numeration and Metric Systems

Historical background

Suppose you awoke one morning in a dream world—a world without numbers. How would your day go? Being accustomed to numbers, you would probably be fairly disoriented. Our days are filled with numbers: we cook, and measure two cups of this and three teaspoons of that; we bank, and deposit $35.83 and write a check for $15.37; we read, and finish a chapter of 24 pages; we travel, and go 3.8 miles to reach a friend's house. Our list could go on and on. We depend greatly on numbers in our day-to-day communication with others. (1)

Fortunately for us, we have a convenient system for denoting numbers, called the *Hindu-Arabic System*. It was invented by the Hindus and brought to Europe by the Arabs—hence the name, Hindu-Arabic. Although the system was introduced into Europe before the year 1000 A.D., it was not until the thirteenth century that a European author wrote a comprehensive discussion of the system. People were slow to recognize the advantages of this system, but it was widely used by the seventeenth century. (2)

The Hindu-Arabic System uses ten symbols (or digits): 0, 1, 2, 3, 4, 5, 6, 7, 8, 9. The numerals in this system are expressed in powers of ten and have place value. The concepts of powers and place value will be discussed later as we explore different systems of numeration. For now let it suffice to say that, when you see the number 183, you interpret it as one hundred plus eighty plus three, or $(1 \times 100) + (8 \times 10) + (3 \times 1)$. (3)

How did numbers begin? A primitive shepherd may have denoted the quantity of his sheep by using a bag in which he placed a pebble for each sheep, or a horseman may have matched a stick for each horse in his corral. The number of pebbles placed in a bag to represent the number of sheep was not found by counting; a pebble was merely matched with a sheep. A pebble was a symbol (object used to represent something) for the sheep. *Symbols* are used to express ideas, complex relationships, or abstractions. If pebbles or some other object were not available to match with the objects being counted, man could use his fingers. As counting developed through the centuries, man learned the advantages of grouping objects for counting. Because man has ten fingers, the system of numbers that developed was based on the number ten. This system is called the *decimal system*, the word "decimal" being derived from the Latin word *decem*, meaning ten. (4)

Man thus first became aware of the need for numbers when he recognized his need to count. He eventually recognized that, by giving names to numbers of objects, he could make the quantity known to others. Thus, names developed with certain meanings that became generally known and accepted. Today, for example, the meanings of words such as "one", "two", "fifteen", and "million" are accepted universally. A system for symbolizing these numbers is called a *system of numeration*. We will now investigate several such systems. (5)

A. DECIMAL SYSTEM

Decimal system of numeration

Let us look more closely at the *decimal system of numeration*. The decimal system is based on numbers arranged in groups of ten. There are ten digits or symbols in the system: 0, 1, 2, 3, 4, 5, 6, 7, 8, 9. Of course, we want to continue counting after we reach the numeral nine. In order to do this, we must invent a number which will be a combination of the first ten numbers. This is done by establishing a pattern of grouping by tens. That is, we write "10" (read "one, zero" or "ten") to indicate that we have used one group of the ten and are now starting on another group. Similarly, "11" means one group of ten plus one group of one and is read "one, one" or eleven. (6)

Each position in a decimal system numeral is associated with a certain *power of ten* (a value indicated by ten with some exponent, such as 10^4.) When we write 10^4, this is called *exponential notation*. Recall that ten would be called the *base* of the number and 4 is called the *exponent*. The use of exponents provides a convenient way of writing very large numbers. For example: (7)

$$10^5 = 10 \times 10 \times 10 \times 10 \times 10 = 100{,}000$$

$$10^7 = 10 \times 10 \times 10 \times 10 \times 10 \times 10 \times 10 = 10{,}000{,}000$$

A decimal system is thus called a *positional system of numeration* because the position of the various parts in a given symbol is important. By itself, a digit stands for one of the numbers from zero to nine, and its value is constant. However, a digit also represents some other number because of its position in a numeral; this latter value is called its *place value*. The place value which a digit represents is obtained by multiplying the digit times the power of ten assigned to the position which the digit occupies. (8)

Place value

By way of explanation, we shall begin with a dot, called the decimal point (see the following table for an illustration of this discussion). All the numbers to the right of the point would be fractional parts of a whole number. The numbers to the left of the point are whole numbers. For our present purposes, we shall be concerned with the whole numbers only. (9)

The various positions in a whole number are *associated* with powers of ten as follows: The first position to the left of the decimal is the "ones" or "units" position. When the positions are written in successive powers of ten, this first position would be 10^0 (read "ten to the zero power"). A number raised to the zero power equals 1; hence, 10^0 is the "ones" position. The second position to the left of the decimal represents ten to the first power (10^1). The first power of any number is that same number; therefore, 10^1 equals 10. The third position to the left of the decimal is ten to the second power, which is 10^2, or 100. Continuing to the left, the next position is ten to the third power (10^3), or 1,000. This continues for all successive ascending powers of 10.* (10) (11)

Place Values of Base Ten

Hundred thousands	Ten thousands	Thousands	Hundreds	Tens	Ones or units	Decimal point	
10^5	10^4	10^3	10^2	10^1	10^0	·	← Powers of ten
2	4	3	6	8	1	·	← Number example

The numeral "zero" is of particular interest in working with place value. If the digit 0 occurs in a number, its place value is zero times the value of its position in the number. In this respect, zero "holds" a position which would otherwise be left empty. In the numeral 206, for example, the (12)

*Place values to the right of the decimal point would be associated with descending powers of ten, indicated with negative exponents. Left to right, these powers of ten would be 10^{-1}, 10^{-2}, 10^{-3}, etc.

zero not only shows no tens, but it also maintains the symbols 2 and 6 in their proper positions.

(13) As stated above, the place value of a digit in a number is obtained by multiplying the digit times the power of ten assigned to the position the digit occupies. Thus, 5 in the 10^3 position would be $(5 \times 10^3) = (5 \times 10 \times 10 \times 10) = 5{,}000$. Therefore, any complete base ten number may be written as the sum of the place values of its digits, each associated with its respective, successive power of 10. This form of writing is called *expanded notation.*

Objective 1: Expanded notation

(14) ◆ **Example 1** Write 7,043 in expanded notation form:
First determine the position of each numeral. Then multiply each digit times the power of ten assigned to its position:

7 is in the 10^3 position;	(7×10^3)
0 is in the 10^2 position;	(0×10^2)
4 is in the 10^1 position;	(4×10^1)
3 is in the 10^0 position;	(3×10^0)

Finally, add the products:

$$7{,}043 = (7 \times 10^3) + (0 \times 10^2) + (4 \times 10^1) + (3 \times 10^0)$$

Summary of the decimal system

(15) In summary, there are four important characteristics of the decimal system. We will also find shortly that these same principles apply to numeration systems having bases other than ten:

1. Base ten: the decimal system arranges or groups numbers by tens and powers of ten.
2. Ten symbols: There are ten separate symbols in this system of notation. All base ten numerals are written using these symbols (digits): 0, 1, 2, 3, 4, 5, 6, 7, 8, 9.
3. Place value: By itself, a digit represents one of the numbers from zero to nine and its value is constant. However, a digit also represents some other value because of its position in a number. The positional values are successive, ascending powers of ten.
4. Place holder: Zero serves as a place holder because it occupies a position in a numeral that would otherwise be left empty.

(16) ● **Practice 1** Write the expanded notation name for the following base ten numerals:

(a) 8,346 (b) 1,327
(c) 98,234 (d) 470

See also Exercise 1.

B. NONDECIMAL SYSTEMS

In the decimal system, objects are grouped and counted in tens and powers of ten. In the decimal system 32 items would be grouped as follows: **(17)**

********** ********** ********** **
3 groups of ten plus 2 more

We could, however, just as easily group these items in other ways. In the following designs, we see how 32 asterisks can be grouped in three different ways: **(18)**

**********	*******	*********
**********	*******	*********
**********	*******	*********
**	*******	*****
	****	
3 groups of ten and 2 more	4 groups of seven and 4 more	3 groups of nine and 5 more

If we use a subscript to indicate our manner of grouping, we may write many different numerals (names) for the number of items in the same collection: **(19)**

3 groups of ten and 2 more becomes 32_{ten}

4 groups of seven and 4 more becomes 44_{seven}
(read "four, four, base seven")

3 groups of nine and 5 more becomes 35_{nine}
(read "three, five, base nine")

The study of positional numeration systems using bases other than ten should not be difficult, if the characteristics of the decimal system discussed on the previous page are kept in mind. Note the following points concerning nondecimal systems: **(20)**

1. For each base, b, the digits used are $0, 1, 2, \ldots, b-1$. That is, a numeration system whose base is a natural number b, $(b > 1)$, has b distinct digits:

   Example
   Base *ten* has *10* digits: 0, 1, 2, 3, 4, 5, 6, 7, 8, 9.
   Base *two* has *2* digits: 0, 1.
   Base *five* has *5* digits: 0, 1, 2, 3, 4.
   Base *three* has *3* digits: 0, 1, 2.

2. A number symbol in a base other than ten will include a subscript to identify the base (such as the base-eight numeral, 42_{eight}). Each subscript will be the decimal name of the base number. You might sometimes see a number written in the form 42_8 rather than 42_{eight}. We shall use the form 42_{eight} to avoid confusion, because, in base eight, there exists no such symbol as "8." The number 42_{eight} is read "four, two, base eight" *not* "forty-two base eight."
3. When no subscript is used, the numeral is understood to be a base ten number unless otherwise noted. However, the subscript "ten" will be used when we wish to emphasize that a number is in base ten. Thus, the numerals 54 and 54_{ten} both refer to the same decimal number.
4. In a numeration system with base b, a number written in the expanded notation name is expressed in terms of successive powers of b. Therefore, $4131_b = (4 \times b^3) + (1 \times b^2) + (3 \times b^1) + (1 \times b^0)$.
5. Conversely, given the expanded notation name of any number in base b, the number can easily be written in the usual number-symbol form. In the example above, $(4 \times b^3) + (1 \times b^2) + (3 \times b^1) + (1 \times b^0)$ should be recognized as the number 4131_b. In base ten, $(9 \times 10^2) + (3 \times 10^1) + (4 \times 10^0)$ may be written immediately as 934.

(21) The following general table indicates place values associated with any base, b:

General Table of Place Values

b^4	b^3	b^2	b^1	b^0	←Powers of base
3	5	4	1	2	←Base example

Base five or quinary system of numeration

(22) The system of numeration that groups numbers by fives is called the base five or the *quinary system*. The features of the quinary system are the same as for the decimal system except that the base is five. That is, the characteristics of the quinary system are:

1. Base five: Grouping by fives and powers of five.
2. Five basic symbols (digits): 0, 1, 2, 3, 4. The symbol "5" does not exist in base five. The number representing the group of symbols ***** is denoted 10_{five} and read "one, zero, base five."
3. Place value: In base ten, the place values for whole numbers starting with the ones position are ones (10^0), tens (10^1), hundreds (10^2), thousands (10^3), etc. In base five, the corresponding place values are ones (5^0), fives (5^1), twenty-fives (5^2), etc.—all successive powers of five.

4. Place holder: Zero serves as a place holder in the quinary system in a similar manner as in the base ten system.

Therefore, 431202_{five} means (23)

four groups of 5^5
three groups of 5^4
one group of 5^3
two groups of 5^2
zero groups of 5^1
two groups of 5^0

The following table illustrates the place values of base five:

Place Values of Base Five

Three thousand one hundred twenty-fives	Six hundred twenty-fives	One hundred twenty-fives	Twenty-fives	Fives	Ones	
5^5	5^4	5^3	5^2	5^1	5^0	← Powers of base
4	3	1	2	0	2	← Base example

◆ **Example 2** Write in expanded notation form: (24)

(a) 234_{five}

First, determine the position of each numeral.
Next multiply each digit times the power of five assigned to its position:

2 is in the 5^2 (twenty-fives) position (2×5^2)

3 is in the 5^1 (fives) position (3×5^1)

4 is in the 5^0 (ones) position (4×5^0)

Finally, add the products

$$234_{\text{five}} = (2 \times 5^2) + (3 \times 5^1) + (4 \times 5^0)$$

(b) 1304_{five}

$$1304_{\text{five}} = (1 \times 5^3) + (3 \times 5^2) + (0 \times 5^1) + (4 \times 5^0)$$

● **Practice 2** Write the expanded notation name for the following: (25)

(a) 3450_{five} (b) 10023_{five} (c) 10204_{five}

See also Exercise 1.

Base two or the binary system of numeration

The *binary system* has the natural number two as its base and is of special interest because it has only two digital symbols—0 and 1. The advantage of this system is that it can be used in electronic computers and electric circuits. There is a disadvantage in that a large number of digits are required in order to write large numbers in base two. In spite of this, computations in base two are relatively easy to perform because only two digits are involved. (26)

The characteristics of the binary system are identical to those of the decimal system and the quinary system, except that the base is two. Thus, base two has the following features: (27)

1. Base two: grouping by two's and powers of two.
2. Two basic symbols: 0 and 1. The symbol "2" does not exist in base two. The group of symbols ** would be designated by 10_{two} and read "one, zero, base two."
3. Place value: Place values include ones (2^0), twos (2^1), fours (2^2), eights (2^3), sixteens (2^4), etc.—all successive powers of two.
4. Place Holder: Zero is a place holder as in the preceding numeration systems.

The following table illustrates the place values and powers of base two: (28)

Place Values of Base Two

Thirty-twos	Sixteens	Eights	Fours	Twos	Ones or units	
2^5	2^4	2^3	2^2	2^1	2^0	← Powers of base
1	0	1	1	0	1	← Base example

Numbers in base two are written in expanded notation in the same manner as were numbers in base five and base ten, except that place values are successive powers of two.

◆ **Example 3** Write in expanded notation form: (29)

(a) 1011_{two}

First, determine the position of each numeral. Next, multiply each digit times the power of two assigned to its position:

1 is in the 2^3 (eights) position; (1×2^3)

0 is in the 2^2 (fours) position; (0×2^2)

1 is in the 2^1 (twos) position; (1×2^1)

1 is in the 2^0 (ones) position; (1×2^0)

Finally, add the products:

$$1011_{\text{two}} = (1 \times 2^3) + (0 \times 2^2) + (1 \times 2^1) + (1 \times 2^0)$$

(b) By the same procedure,

$$10101_{\text{two}} = (1 \times 2^4) + (0 \times 2^3) + (1 \times 2^2) + (0 \times 2^1) + (1 \times 2^0)$$

● **Practice 3** Write the expanded notation name for the following: **(30)**

(a) 11101_{two} (b) 11001_{two} (c) 1001_{two}

See also Exercise 1.

Base three numeration system

Base three follows the same pattern that the decimal, quinary, and binary systems follow, except that the base number is three. The following characteristics thus hold: **(31)**

1. Base three: grouping by threes and powers of three.
2. Three basic symbols (digits): 0, 1, 2. The symbol "3" does not exist in base three; the group of symbols *** would be denoted by 10_{three}, pronounced "one, zero, base three."
3. Place value: Place values are ones (3^0), threes (3^1), nines (3^2), twenty-sevens (3^3), eighty-ones (3^4), etc.—all successive powers of three.
4. Place holder: Zero is also a place holder in base three.

The following table represents place values and powers of three: **(32)**

Place Values of Base Three

Two hundred forty-threes	Eighty-ones	Twenty-sevens	Nines	Threes	Ones or units	
3^5	3^4	3^3	3^2	3^1	3^0	← Powers of base
2	1	0	2	1	1	← Base example

◆ **Example 4** Write the expanded notation form of the following: (33)

(a) 1221_{three}

First, determine the position of each digit (left to right). Next, multiply the digit times the power of three assigned to its position:

1 is in the 3^3 (twenty-sevens) position (1×3^3)

2 is in the 3^2 (nines) position (2×3^2)

2 is in the 3^1 (threes) position (2×3^1)

1 is in the 3^0 (ones) position (1×3^0)

Finally, add the products:

$$1221_{three} = (1 \times 3^3) + (2 \times 3^2) + (2 \times 3^1) + (1 \times 3^0)$$

(b) In a similar manner,

$$1012_{three} = (1 \times 3^3) + (0 \times 3^2) + (1 \times 3^1) + (2 \times 3^0)$$

● **Practice 4** Write the expanded notation name for the following: (34)

(a) 1102_{three} (b) 20112_{three} (c) 2001_{three}

See also Exercise 1.

When working with numeration systems, it is helpful to know how to change a nondecimal number to a decimal number. This is done by writing the base number in the expanded notation form, computing the notation in base ten, and then finding the sum of the resulting terms. (35)

Objective 2: Translating into a base ten numeral

◆ **Example 5** Translate the following to base ten numbers: (36)

(a) 4321_{five}

First, write the expanded notation name:

$$4321_{five} = (4 \times 5^3) + (3 \times 5^2) + (2 \times 5^1) + (1 \times 5^0)$$

Then, compute the notation in base ten and find the sum:

$$\begin{aligned} 4321_{five} &= (4 \times 125) + (3 \times 25) + (2 \times 5) + (1 \times 1) \\ &= 600 + 75 + 10 + 1 \\ &= 686_{ten} \end{aligned}$$

Therefore, $4321_{five} = 686_{ten}$ (Two names for the same number of elements.)

(b) 10110_{two}

First, write the expanded notation name:

$$10110_{two} = (1 \times 2^4) + (0 \times 2^3) + (1 \times 2^2) + (1 \times 2^1) + (0 \times 2^0)$$

Next, compute the notation in base ten and find the sum:

$$10110_{two} = (1 \times 16) + (0 \times 8) + (1 \times 4) + (1 \times 2) + (0 \times 1)$$
$$= 16 + 0 + 4 + 2 + 0$$
$$= 22_{ten}$$

Therefore, $10110_{two} = 22_{ten}$ (Two names for the same number of elements.)

(c) 21021_{three}

Following the procedures in (a) and (b) above:

$$21021_{three} = (2 \times 3^4) + (1 \times 3^3) + (0 \times 3^2) + (2 \times 3^1) + (1 \times 3^0)$$
$$= (2 \times 81) + (1 \times 27) + (0 \times 9) + (2 \times 3) + (1 \times 1)$$
$$= 162 + 27 + 0 + 6 + 1$$
$$= 196_{ten}$$

Therefore, $21021_{three} = 196_{ten}$ (Two names for the same number of elements.)

● **Practice 5** Change the following numerals to base ten numeration: **(37)**

(a) 321_{five} (b) 3242_{five}

(c) 101011_{two} (d) 212_{three}

(e) 120_{three}

See also Exercise 2.

C. METRIC SYSTEM

The need for a single worldwide coordinated measurement system has been recognized for hundreds of years. Gabriel Mouton, in 1670, proposed a complex system based on measurements of the earth. Other proposals for standard measurements were also made over the years. Finally, in 1790, a commission of the French Academy of Sciences was appointed by the National Assembly of France to create a uniform system. The system developed was simple and scientific. A portion of the earth's circumference was established as the unit of length, and measures for volume and weight were also derived from the unit of length, thus relating the basic units of the system to each other and to nature. Multiplication and division of the basic units by ten and its multiples established the larger and smaller versions of each unit. **(38)**

Basic units of the metric system

Thus, the metric system of measurement was established with the *meter* as the standard unit for length, the *gram* as the standard unit for **(39)** weight, and the *liter* as the standard unit for measuring volume. As mentioned above, the superior advantage of this system is that its measures are multiples of ten and are, therefore, related to the decimal system of numeration.

Objective 3: Metric prefixes

The metric system is based on the number ten as is the decimal system; thus, it is a base ten system. A metric unit can be increased into a unit that is a *power* of ten or decreased into a unit that is a *division* of ten. (40)

Powers of ten	*Divisions by powers of ten*
$10^1 = 10$	$1 \div 10 = 1/10$ or 0.1
$10^2 = 100$	$1 \div 100 = 1/100$ or 0.01
$10^3 = 1000$	$1 \div 1,000 = 1/1,000$ or 0.001

The metric system uses prefixes that are based on ten in order to indicate powers or divisors of the basic units of length, weight, and volume. (41)

Powers of any unit of measurement are preceded by prefixes with Greek origin:

Prefix	*Example*
deca* (10)	decameter = 10 meters
hecto (100)	hectometer = 100 meters
kilo (1,000)	kilometer = 1,000 meters

Divisions by powers of ten of any unit of measurement are preceded by prefixes derived from Latin:

Prefix	*Example*
deci** (0.1)	decimeter = 0.1 meter
centi (0.01)	centimeter = 0.01 meter
milli (0.001)	millimeter = 0.001 meter

The tables on page 213 illustrate how the metric prefixes are used in relation to the basic units of length (meters), weight (grams), and volume (liters).

As the metric system was further refined, the need developed for powers larger than 1,000 and divisions smaller than $\frac{1}{1000}$; thus, some additional prefixes were introduced to the system. The prefix "mega" meaning 1,000,000 was added for very large measurements; the prefix "micro" meaning $\frac{1}{1,000,000}$ was included for very small measurements. (45)

*The "c" is pronounced as "*k*."
**The "c" is pronounced as "*s*."

(42)

Length: Length is a measure of distance or extent.
Meter is the metric unit of measurement.

	Unit	*Number of meters*	*Abbreviation*
Powers of ten:	kilometer	1,000. meters	km.
	hectometer	100. meters	hm.
	decameter	10. meters	dkm.
	meter	1. meter	m.
Division by powers of ten:	decimeter	0.1 meter	dm.
	centimeter	0.01 meter	cm.
	millimeter	0.001 meter	mm.

(43)

Weight: Weight is a measure indicating heaviness.
Gram is the metric unit of measurement.

	Unit	*Number of grams*	*Abbreviation*
Powers of ten:	kilogram	1,000. grams	kg.
	hectogram	100. grams	hg.
	decagram	10. grams	dkg.
	gram	1. gram	g.
Divisions by powers of ten:	decigram	0.1 gram	dg.
	centigram	0.01 gram	cg.
	milligram	0.001 gram	mg.

(44)

Volume: Volume is a measure of space occupied.
Liter is the metric unit of measurement.

	Unit	*Number of liters*	*Abbreviation*
Powers of ten:	kiloliter	1,000. liters	kl.
	hectoliter	100. liters	hl.
	decaliter	10. liters	dkl.
	liter	1. liter	l.
Divisions by powers of ten:	deciliter	0.1 liter	dl.
	centiliter	0.01 liter	cl.
	milliliter	0.001 liter	ml.

◆ **Example 6** Write the correct number or prefix: (46)

(a) 1 kilogram = ____ grams

The prefix kilo means 1,000; therefore, the answer is *1,000*.

(b) 1 ____ meter = 0.1 meter

The decimal fraction 0.1 means deci; therefore, the answer is *deci*meter.

● **Practice 6** Complete the following statements by writing the correct number or prefix: (47)

(a) 1 decigram = ____ gram

(b) 1 ____ meter = 0.01 meter

(c) 1 ____ liter = 0.001 liter

(d) 1 hectometer = ____ meters

(e) ____ grams = 1 decagram

(f) 1 millimeter = ____ meter

(g) 1 ____ liter = 0.1 liter

(h) 1 kilometer = ____ meters

(i) 1 ____ gram = 100 grams

(j) 1 ____ meter = 10 meters

See also Exercise 3.

Objective 4: Converting between Fahrenheit and Celsius temperature

The standard unit of temperature measurement in the metric system is called *Celsius* after Anders Celsius, the man who devised the Celsius scale (formerly called the centigrade scale). Prefixes are not commonly used with temperature measurement. The following chart depicts some of the more common temperatures: (48)

Temperature: Temperature is a measure of hotness or coldness. *Celsius* is the metric unit of measurement.

Fahrenheit	*Celsius*	
32°	0°	Freezing point of water
50°	10°	Warm winter day
68°	20°	Mild spring day
86°	30°	Quite warm—almost hot
98.6°	37°	Normal body temperature
104°	40°	Heat wave conditions
212°	100°	Boiling point of water

There are two formulas that are useful when converting between Fahrenheit (F) and Celsius (C) temperature measurements: (49)

$$C = \frac{5}{9}(F - 32) \qquad F = \frac{9}{5}C + 32$$

The illustration at the right shows a comparison between the two scales.

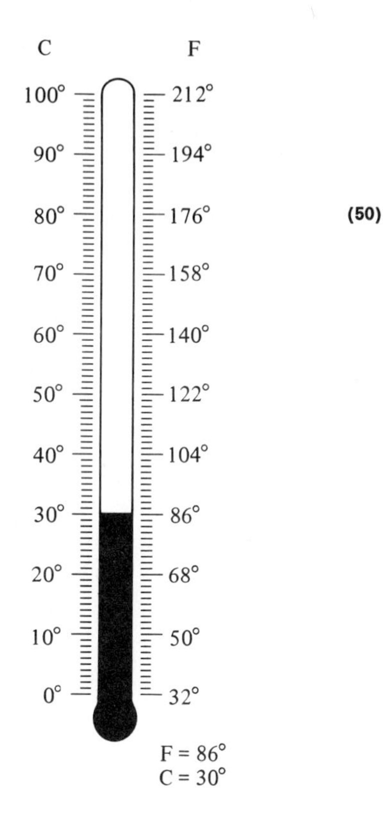

(50)

◆ **Example 7**

(a) Convert 30°C to a Fahrenheit reading:

$$\begin{aligned} F &= \frac{9}{5}C + 32 \\ &= \frac{9}{5}(30) + 32 \\ &= 54 + 32 \\ &= 86° \end{aligned}$$

(b) Convert 176°F to a Celsius reading:

$$\begin{aligned} C &= \frac{5}{9}(F - 32) \\ &= \frac{5}{9}(176 - 32) \\ &= \frac{5}{9}(144) \\ &= 80° \end{aligned}$$

● **Practice 7** Convert the following to Fahrenheit temperature: (51)

(a) 35°C (b) 40°C
(c) 22°C (d) 81°C

Convert the following to the Celsius temperature:

(e) 59°F (f) 95°F
(g) 158°F (h) 221°F

See also Exercise 4.

Objective 5: Converting from English to metric system

Most of our everyday measurements are made in the English system. The metric system has been officially adopted by most other countries; however, it is used generally in scientific work in the United States. Furthermore, since the metric system is almost certain to be adopted in the U.S., it is helpful to know both systems and to be able to change measurements from one system to another. (52)

Because there are different ways of expressing the same measurement, it is important to state the unit of measurement that is being used. For instance, a person's height may be 70 inches, 6 feet, or 182.88 centimeters. (53)

Describing the person's height as "70," or "6," or "182.88" is meaningless unless the units of measurement are stated also.

In order to convert given English measurements to their equivalent metric measurements, the table of equivalent measures on page 438 may be used. (54)

The following examples illustrate the use of the conversion factors.

◆ **Example 8**

(a) Change 60 inches to centimeters. (55)

From the table we know that 1 inch = 2.54 centimeters; therefore, we multiply both sides of the equation by 60 to make the conversion:

$$\begin{array}{rcr} 1 \text{ in.} & = & 2.54 \text{ cm.} \\ \underline{\times 60} & & \underline{\times\ 60} \\ 60 \text{ in.} & = & 152.4 \text{ cm.} \end{array}$$

(b) 5 pounds = ____ grams

From the table we know that 1 pound is 454 grams; therefore,

$$\begin{array}{rcr} 1 \text{ pound} & = & 454 \text{ grams} \\ \underline{\times 5} & & \underline{\times\ \ 5} \\ 5 \text{ pounds} & = & 2{,}270 \text{ grams} \end{array}$$

(c) 2 gallons = ____ liters

$$\begin{array}{rcr} 1 \text{ gallon} & = & 3.8 \text{ liters} \\ \underline{\times 2} & & \underline{\times\ 2} \\ 2 \text{ gallons} & = & 7.6 \text{ liters} \end{array}$$

● **Practice 8** Convert the following English system measurements to the metric system measurements given. (56)

(a) 2 inches = ____ centimeters (b) 6 pounds = ____ grams

(c) 2 quarts = ____ liters (d) 18 ounces = ____ grams

(e) 10 feet = ____ meters (f) 14 feet = ____ centimeters

(g) 3 inches = ____ millimeters (h) 14 pounds = ____ decagrams

(i) 3.5 gallons = ____ liters (j) 132 ounces = ____ grams

See also Exercises 5, 6, 7.

Objective 6: Converting from Metric to English system

Likewise we can convert given metric units of measurements to the English system of measurement. On page 438 is given the equivalent measures from the metric system to the English system. (57)

◆ **Example 9**

(a) 10 meters = ____ feet (58)

As before, the table equivalents are multiplied by the specific measurement given:

$$\begin{aligned} 1 \text{ meter} &= 3.28 \text{ feet} \\ \underline{\times 10} \quad &= \underline{\times\ 10} \\ 10 \text{ meters} &= 32.8 \text{ feet} \end{aligned}$$

(b) 32 grams = ____ ounces

$$\begin{aligned} 1 \text{ gram} &= 0.035 \text{ ounce} \\ \underline{\times 32} \quad &= \underline{\times\ 32} \\ 32 \text{ grams} &= 1.120 \text{ ounces} \end{aligned}$$

(c) 12 liters = ____ quarts

$$\begin{aligned} 1 \text{ liter} &= 1.06 \text{ quarts} \\ \underline{\times 12} \quad &= \underline{\times\ 12} \\ 12 \text{ liters} &= 12.72 \text{ quarts} \end{aligned}$$

● **Practice 9** Convert the following metric system measurements to the given English system measurements: (59)

(a) 10 kilometers = ____ miles

(b) 1295 centimeters = ____ inches

(c) 50 kilograms = ____ pounds

(d) 58 liters = ____ quarts

(e) 32 liters = ____ pints (remember 2 pints make a quart)

(f) 9 centimeters = ____ inches

(g) 55 grams = ____ ounces

(h) 716 centimeters = ____ inch

(i) 116 liters = ____ quarts

(j) 2 kilograms = ____ pounds

See also Exercises 8, 9.

Many rulers today are marked in millimeters and centimeters as well as in inches. If a meter stick and a yard stick were compared, the meter stick would be a little over three inches longer than the yard stick. The meter is divided into 10 decimeters, 100 centimeters, and 1,000 millimeters. (60)

Objective 7: Measuring with a metric ruler

The *millimeter* is the smallest unit distance marked off on a metric ruler. Ten millimeters equal one *centimeter* (10 mm. = 1 cm.); conversely, a millimeter is one-tenth of a centimeter (1 mm. = 0.1 cm.). A *decimeter* is a distance equal to one-tenth of a meter (1 dm. = 0.1 m.). Therefore, a centimeter is one-tenth of a decimeter (1 cm. = 0.1 dm.). A millimeter is one-hundredth of a decimeter (1 mm. = 0.01 dm.). The following table illustrates the measurement of the metric ruler. (61)

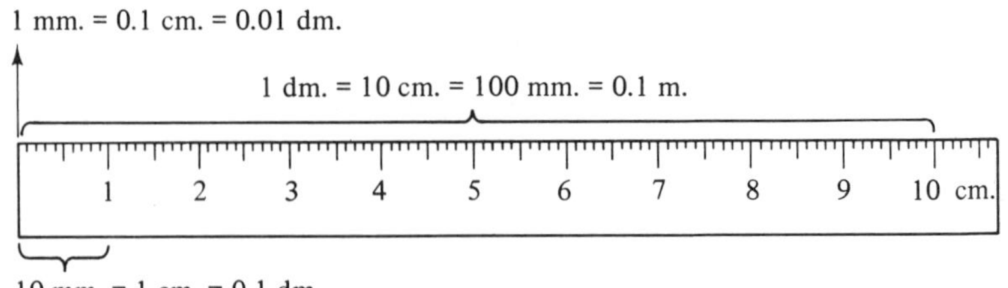

◆ **Example 10** Complete the following statements, using the diagram of the metric ruler shown below. All measurements are made from the left end of the "ruler." Write your answers in the space provided. **(62)**

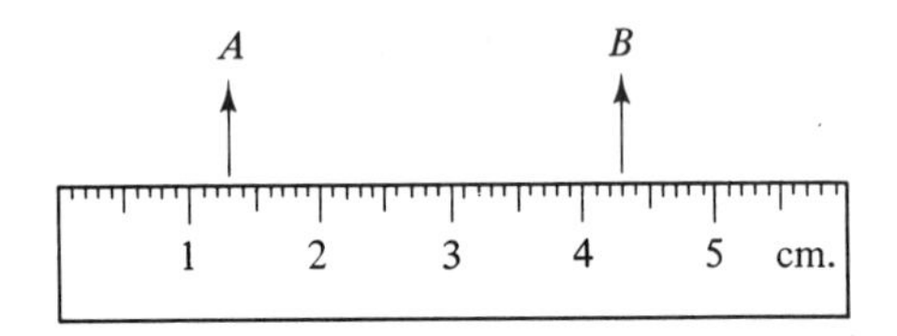

(a) *A* is ____ mm. or ____ cm.

Starting at the left, count the number of small lines to the arrow; these are millimeters. Thus, there are *13* millimeters. Since each centimeter is equal to 10 millimeters, then there are *1.3* cm.

(b) *B* is ____ cm. or ____ dm.

Again, starting at the left, count the number of millimeters. There are *43* millimeters or *4.3* centimeters or *0.43* decimeters.

● **Practice 10** Complete the following statements using the diagram of the metric ruler shown. **(63)**

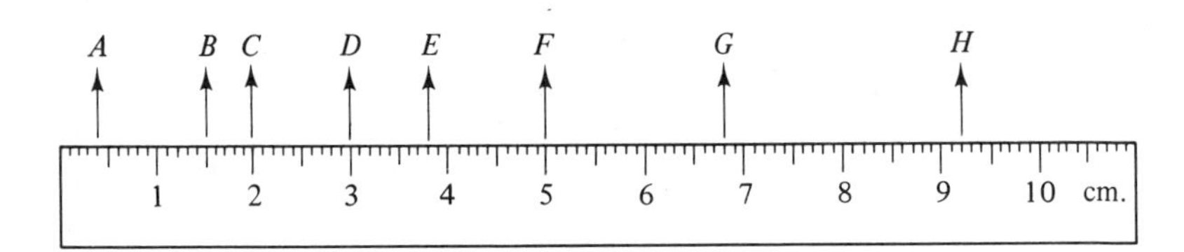

(a) *A* is ____ mm.

(b) *B* is ____ cm.

(c) *C* is ____ dm.

(d) *D* is ____ mm. or ____ cm.

(e) *E* is ____ cm. or ____ dm.

(f) *F* is ____ cm. or ____ dm.

(g) *G* is ____ mm. or ____ cm. or ____ dm.

(h) *H* is ____ mm. or ____ cm. or ____ dm.

See also Exercise 10.

EXERCISES

Objective 1; Examples 1, 2, 3, 4

1. Give the expanded notation name for each of the following:

(a) 1001_{two} (b) 120_{three} (c) 41_{five}
(d) 1983 (e) 1011_{two} (f) 389
(g) 1002_{five} (h) 112_{three} (i) 22_{three}
(j) 51037 (k) 111_{two} (l) 321_{five}

Objective 2; Example 5

2. Translate the following into the corresponding base ten numeral:

(a) 210_{three} (b) 1101_{two}
(c) 134_{five} (d) 100_{two}
(e) 1221_{three} (f) 441_{five}

Objective 3; Example 6

3. Complete the following statements:

(a) 1 milligram = ____ gram (b) 1 ____ gram = 100 grams
(c) 1 decameter = ____ meters (d) 1 ____ meter = 0.01 meter
(e) 1 deciliter = ____ liter (f) 1 decaliter = ____ liters
(g) 1 kiloliter = ____ liters (h) 1 meter = ____ millimeters
(i) 100 centimeters = ____ meters (j) 1 hectogram = ____ grams
(k) 10 liters = 1 ____ liter (l) 1000 meters = 1 ____ meter

Objective 4; Example 7

4. Determine the following:

(a) Convert 25°C to a Fahrenheit reading.
(b) Convert 221°F to a centigrade reading.
(c) Water boils at ____°C or ____°F.
(d) Water freezes at ____°F or ____°C.
(e) Normal body temperature is 98.6°F. What is this temperature on the centigrade scale?
(f) Convert −10°C to a Fahrenheit reading.
(g) Convert −13°F to a Celsius reading.
(h) The average February temperature in Anchorage was 4° below zero on the Fahrenheit scale. What is the Celsius reading?

Objective 5; Example 8

5. (a) The Washington Monument is 555 feet high. How many meters is this?
(b) If you are 65 inches tall, how many centimeters tall are you?
(c) How many millimeters long is an inch worm?

Objective 5; Example 8

6. (a) A camping water cooler holds 5 quarts. How many liters does it hold?
(b) A small fish aquarium holds 10 gallons. How many liters does it hold?
(c) A school is having a class picnic for 30 children. The teacher orders 10 pints of ice cream for the occasion. How many liters did the teacher order?

Objective 5; Example 8

7. (a) A bottle of perfume holds 3 ounces. How many grams does it hold?

(b) A shopper bought 5 pounds of fish. How many grams of fish were bought?

(c) Rabbit feed is bought in 50-pound sacks. How many grams does a sack of rabbit feed weigh?

Objective 6; Example 9

8. (a) Pikes Peak has an elevation of about 4,300 meters. How many feet is this?

(b) Your pen-pal writes that he lives 120 kilometers away from Paris. What is this distance in miles?

(c) A hospital recorded a baby's weight as being 4 kilograms. How many pounds did the infant weigh?

Objective 6; Example 9

9. (a) A bag of sugar is labeled as weighing 2,270 grams. How many pounds does the bag weigh?

(b) A room is 274.32 centimeters wide and 365.76 centimeters long. How many square feet of carpet would a person need to wall-to-wall carpet the room?

(c) Your car has a 76-liter gas tank. When full how many gallons does your tank hold?

Objective 7; Example 10

10. Using the illustration of the metric ruler below, name the following distances indicated by the arrows:

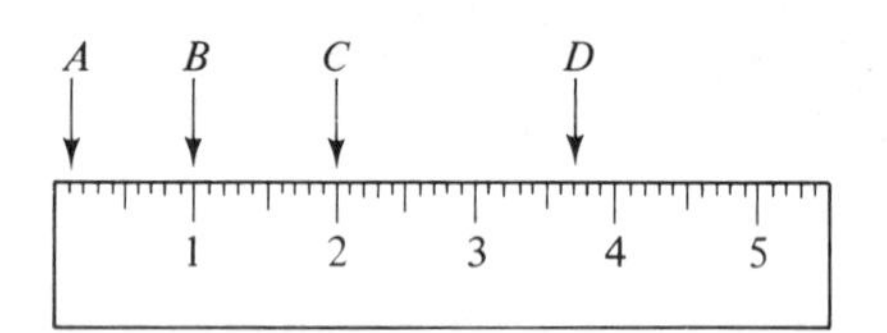

A is ____ mm or ____ cm or 0.01 dm

B is ____ mm or ____ cm or 0.1 dm

C is ____ mm or ____ cm or ____ dm

D is ____ mm or ____ cm or ____ dm

OBJECTIVES

After completing this chapter, you should be able to

A. Decimal system
B. Nondecimal system

1 Given a number in base two, three, five, or ten, write the expanded notation name. **(6–34)**

2 Given a number in base two, three, or five, translate it into the base ten numeral. **(35–37)**

C. Metric system

3 Demonstrate an understanding of the metric prefixes by completing statements similar to the following for weight, length, and volume. **(40–47)**

1 ____ gram = 1,000 grams

1 decameter = ____ meters

1 ____ liter = 0.01 liters

4 Use given conversion formulas to convert between Fahrenheit and Celsius temperature. **(48–51)**

5 Given a table of conversion (to metric measures) factors and given a measurement in the English system, convert the measurement to the metric system. **(52–56)**

6 Given a table of conversion (from metric measures) factors and given a measurement in the metric system, convert the measurement to the English system. **(57–59)**

7 Given a metric ruler (or a diagram of a metric ruler) with points marked, write the measurement indicated. **(60–63)**

Part Three

Introduction to Areas of Mathematics

Algebra: Equations and Inequalities

Mathematical sentences

The study of *algebra*, which we begin with this chapter, applies the operations and properties of the real number system in order to solve mathematical sentences involving variables (letters or symbols that represent numbers). Such *mathematical sentences* may be either equations or inequalities. In previous units we have solved some simple equations, primarily by intuition. In this unit we will learn to solve them in an organized manner. (1)

A. SOLVING EQUATIONS

Some mathematical sentences may be written entirely with numbers. For example, (2)

$$4 + 3 = 7 \qquad 3 + 7 < 2 \cdot 8$$

$$8 - 4 \neq 2 + 3 \qquad 3 \cdot 5 > 6 + 7$$

Open sentences

Such equations and inequalities may be classified as either true or false. (All the sentences above are true sentences.) Mathematical sentences that are written using variables are known as *open sentences*. For example,

$$x + 7 = 13 \qquad n - 3 < 5$$

$$2z = 36 \qquad \frac{3}{4}p > 6$$

Open sentences are not classified as true or false, since they may be true when the variable is replaced by some numbers and false for other replacements. The purpose in solving an open sentence is to find the value (or values) for which the sentence is true. This value is called the *solution set* or *truth set* or *root* of the sentence.

(3) There is, however, one special type of open sentence that holds true for any replacement of the variable. For instance,

$$y + 3 = 3 + y \qquad p < p + 1$$
$$4 \cdot c = c \cdot 4 \qquad r > r - 2$$

Identity sentences

Any open sentence such as these, which are true for any replacement, is called an *identity* or an *identity sentence*. Identities cannot be "solved" because the variable always "disappears," leaving only numbers (which express a true sentence).

Objective 1:

(4) Equations may be compared to old-fashioned balancing scales.

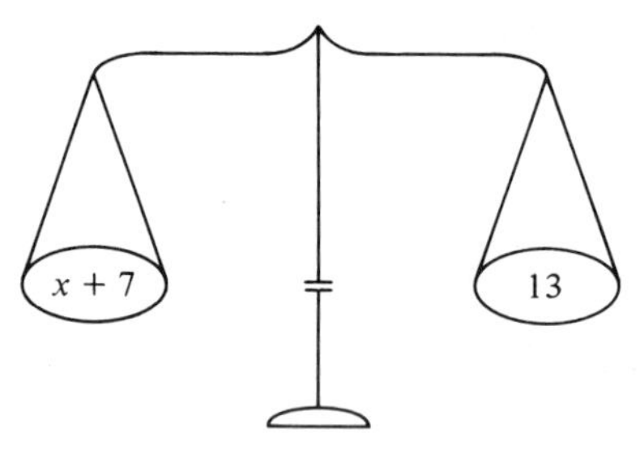

Weights may be added or removed from the pans of a scale and—as long as the same change is made in both pans—the scale will remain in balance. Similarly, numbers may be added, subtracted, multiplied, or divided in an equation and—as long as the same change is made on both sides of the equation—the equation will also remain in balance.

First-degree equations in one variable

(5) The equations we will study are *first-degree equations in one variable.* "First-degree" means the variable in each equation has an exponent of one (although the exponent may not actually be written). The "one variable" may appear more than one time, but there will be only one distinct variable (letter) in any equation. To illustrate:

$x + 4 = 11$	First-degree in one variable
$2n + n = 9$	First-degree in one variable
$r + s = 10$	First-degree in two variables
$q^2 - 4 = 0$	Second-degree in one variable
$m^2 - n^2 = 0$	Second-degree in two variables

(6) There are only four basic equation forms for first-degree equations in one variable—those whose solution requires a single operation of either

addition, subtraction, multiplication, or division. All other equations are variations or combinations of these four forms. The variable in a basic equation has one number associated with it on the left side of the equal marks. That is, a number is either added, subtracted, multiplied, or divided with the variable. To solve each equation, the *opposite* operation must be performed on each side of the equation.

◆ **Example 1** Solve the following equations: (7)

Variable plus number

(a) Solve for x: $x + 9 = 17$

Subtract 9 from both sides. $x + 9 - 9 = 17 - 9$

$9 - 9 = 0$; and $x + 0 = x$. $x \cancel{+ 9 - 9} = 17 - 9$

$x = 8$

The solution set is {8}

Recall that subtracting 9 is the same as adding (-9).

Variable minus number

(b) Solve for c: $c - 8 = 4$

Add 8 to both sides. $c - 8 + 8 = 4 + 8$

$-8 + 8 = 0$; $c + 0 = c$. $c \cancel{- 8 + 8} = 4 + 8$

$c = 12$

The solution set is {12}

Number times variable

(c) Recall that $3r$ means $3 \times r$ and that 3 is called the *coefficient* of the variable r. To solve the next equation, we divide by the coefficient: (8)

Solve for r: $3r = 15$

Divide 3 into both sides. $\frac{3r}{3} = \frac{15}{3}$

$\frac{3}{3} = 1$; $1 \cdot r = r$. $\frac{\cancel{3}r}{\cancel{3}} = \frac{15}{3}$

$r = 5$

The solution set is {5}

Again, recall that dividing by 3 is the same as multiplying by the reciprocal, $\frac{1}{3}$.

Variable divided by number

(d) Solve for v: $\frac{v}{4} = 7$

Multiply both sides by 4. $\frac{4}{1} \cdot \frac{v}{4} = 7 \cdot 4$

$\frac{4}{4} = 1$; $1 \cdot v = v$. $\cancel{4} \cdot \frac{v}{\cancel{4}} = 7 \cdot 4$

$v = 28$

The solution set is {28}.

Note: The various equations that appear on the succeeding lines of a solution are called *equivalent equations*, because each of those equations has the same root or truth set. Observe that the foregoing solutions are possible because of the addition property of equals (if $a = b$, then $a + c = b + c$) and the multiplication property of equals (if $a = b$, then $ac = bc$). (9)

Practice 1 Find the solution set for each equation: (10)

(a) $y - 15 = 17$ (b) $k + 6 = 3$

(c) $\frac{d}{3} = 12$ (d) $12s = 8$

See also Exercise 4.

Example 2 Some variations and combinations of the basic equations are given below. Determine each truth set: (11)

Fractional coefficient

(a) Just as $\frac{3 \times 10}{5}$ gives the same result as $\frac{3}{5} \times 10$, so $\frac{3t}{5}$ is the same as $\frac{3}{5}t$. In order to solve the equation below, we multiply both sides by the reciprocal of $\frac{3}{5}$:

Solve for t: $\frac{3t}{5} = 21$

Multiply by $\frac{5}{3}$, the reciprocal of $\frac{3}{5}$. $\frac{5}{3} \cdot \frac{3}{5} t = 21 \cdot \frac{5}{3}$

$\frac{5}{3} \cdot \frac{3}{5} = 1;\ 1t = t.$ $\frac{\not{5}}{\not{3}} \cdot \frac{\not{3}}{\not{5}} t = \overset{7}{\not{21}} \cdot \frac{5}{\not{3}}$

$t = 35$

The solution set is $\{35\}$

Combining variables

(b) A variable together with its coefficient is called a *term*. When an equation contains several variable terms (or "like" terms) we first combine them into a single variable term by combining the coefficients. (12)

Solve for x: $5x + x - 2x = 60$

Combine the coefficients $5 + 1 - 2 = 4$ $4x = 60$

Divide both sides by 4. $\frac{\not{4}x}{\not{4}} = \frac{60}{4}$

$x = 15$

The solution set is $\{15\}$.

Variable term and number on same side

(c) When one side of an equation contains both a variable term and a number term, it is customary to work with the number term first, in order to obtain an equivalent equation of the type (13)

$$\text{variable term} = \text{numbers}$$

Solve for n: $3n - 6 = 18$

Add 6 $3n - 6 + 6 = 18 + 6$

$3n = 24$

Divide by 3 $\left(\text{or multiply by } \frac{1}{3}\right)$ $\frac{\cancel{3}n}{\cancel{3}} = \frac{24}{3}$

$n = 8$

The solution set is $\{8\}$

Variable on right side

(d) The variable in an equation will sometimes appear on the right side. Although the equation could be reversed, it may also be solved just as it is written: (14)

Solve for m: $17 = \frac{2m}{3} + 5$

Add (-5) or subtract 5. $17 - 5 = \frac{2m}{3} \cancel{+ 5 - 5}$

$12 = \frac{2m}{3}$

Multiply by $\frac{3}{2}$. $\frac{3}{2} \cdot 12 = \frac{2m}{3} \cdot \frac{3}{2}$

$\frac{3}{\cancel{2}} \cdot \overset{6}{\cancel{12}} = \frac{\cancel{2}m}{\cancel{3}} \cdot \frac{\cancel{3}}{\cancel{2}}$

$18 = m$

The solution set equals $\{18\}$.

Variable term and number on both sides

(e) Variable terms or number terms can be combined only when they are on the *same side* of an equation. We thus "move" terms to the opposite side so that (15)

$$\text{variable terms} = \text{number terms}$$

It is customary (although not essential) to work with the variable term first and then the number term:

Solve for c: $6c + 3 = 4c + 19$

Subtract $4c$ from both sides. $6c - 4c + 3 = \cancel{4c - 4c} + 19$

$2c + 3 = 19$

Subtract 3 from both sides. $2c + 3 - 3 = 19 - 3$

$$2c = 16$$

Divide by 2. $\frac{2c}{2} = \frac{16}{2}$

$$c = 8$$

The solution set {8} makes this equation a true sentence.

Parentheses

(f) When an equation contains parentheses with a coefficient, the distributive property is first used in order to clear (remove) the parentheses: (16)

Solve for y: $5(y - 2) = 25$

Distributive property. $5y - 10 = 25$

Add 10. $5y - 10 + 10 = 25 + 10$

$$5y = 35$$

Divide by 5. $\frac{5y}{5} = \frac{35}{5}$

$$y = 7$$

The solution set is {7}

Minus before parentheses

(g) When an expression in parentheses is preceded by a minus sign, clear the parentheses by treating the coefficient (of the parentheses) as if it were a negative: (17)

Solve for r: $3r - 5(r - 4) = 4$

Distributive property:
$(-5)(-4) = +20$ $3r - 5r + 20 = 4$

$$-2r + 20 = 4$$

Subtract 20. $-2r + 20 - 20 = 4 - 20$

$$-2r = -16$$

Divide by (-2). $\frac{-2r}{-2} = \frac{-16}{-2}$

$$r = 8$$

The solution set is {8}

Fractional coefficients

(h) When one of the variable terms contains a denominator (and the other variable term does not), clear the denominator by mulitplying both sides of the equation by that denominator. That is, the distributive property is applied on both sides: (18)

Solve for d: $\frac{3d}{2} - 3 = d - 1$

Multiply both sides by 2. $2\left(\frac{3d}{2} - 3\right) = 2(d - 1)$

Distributive property. $\frac{2}{1}\cdot\frac{3d}{2} - 2\cdot 3 = 2d - 2\cdot 1$

$3d - 6 = 2d - 2$

Subtract $2d$. $3d - 2d - 6 = \cancel{2d - 2d} - 2$

$d - 6 = -2$

Add 6. $d \cancel{- 6 + 6} = -2 + 6$

$d = 4$

The solution set is $\{4\}$.

The solution set for any equation may be verified by substituting it for the variable in the original equation. For example, the solution set $\{4\}$ may be checked for the preceding equation as follows: (19)

Checking equations

Check: $\frac{3}{2}d - 3 = d - 1$

$\frac{3}{2}(4) - 3 = 4 - 1$

$6 - 3 = 4 - 1$

$3 = 3$

● **Practice 2** Find the solution set that makes each of the following a true sentence. (Check each solution if directed by the instructor). (20)

(a) $\frac{4d}{7} = 24$

(b) $3y + 4y - y = 18$

(c) $8k + 5 = 21$

(d) $14 = \frac{3t}{4} - 7$

(e) $9n - 5 = 6n + 7$

(f) $4(c + 3) = 32$

(g) $2p - 7(p - 2) = 9$

(h) $\frac{2s}{5} + 6 = s - 6$

(i) $6 = 7g - g + 3g$

(j) $3(b + 6) = 23$

(k) $\frac{5r}{2} = -10$

(l) $11z - 3 = 7z + 9$

(m) $3q + 10 = 4$

(n) $7 = \frac{2x}{9} + 3$

(o) $\frac{4c}{3} - 3 = c - 1$

(p) $2t - 4(t - 4) = 20$

See also Exercise 4.

B. SOLVING INEQUALITIES

Objective 2: Each first-degree equation that we have considered has had a solution set that contains a single element. The solution set for a first-degree inequality, however, is a set of many points that would themselves compose an inequality. A typical solution in set-builder notation would be $\{x \mid x < 3\}$. Any number specified by the solution inequality could be substituted in the original inequality and would produce a true sentence. (21)

Inequalities are solved using the same methods by which equations are solved, but with one distinction. Recall by the multiplication property of "less than" that multiplying by a negative reverses the sense (direction) of an inequality. That is, (22)

"Sense" of inequalities

$$a < b \quad \text{implies} \quad ac < bc \quad \text{when} \quad c > 0$$

but

$$a < b \quad \text{implies} \quad ac > bc \quad \text{when} \quad c < 0$$

A similar pattern holds when $a > b$. Thus, inequalities are solved the same as equations; however, you must be careful to use the correct inequality symbol, since the direction will change whenever you multiply (or divide) by a negative.

◆ **Example 3** Determine the truth set of each given inequality: (23)

(a) Using methods learned for equations, solve for k:

$$\frac{k}{3} - 7 < -3$$

$$\frac{k}{3} - 7 + 7 < -3 + 7$$

$$\frac{k}{3} < 4$$

$$3 \cdot \frac{k}{3} < 4 \cdot 3$$

$$k < 12$$

The solution set is $\{k \mid k < 12\}$

(Notice that since we multiplied by a positive value, 3, the sense of the inequality remained the same.)

To check: If 12 itself is substituted for k, the inequality does not hold. However, it is necessary to check 12 to verify that the starting point of the inequality is correct. The starting point checks when the same number is obtained on both sides of the mathematical sentence. Then, some value less than 12 must be checked to verify that the sense (direction) of (24)

Checking inequalities

the solution is correct. (Here $k = 9$ and $k = 6$ are used as examples. Any other integers or rational numbers less than 12 could also have been used to check the solution.)

$k = 12$	$k = 9$	$k = 6$
$\frac{k}{3} - 7 < -3$	$\frac{k}{3} - 7 < -3$	$\frac{k}{3} - 7 < -3$
$\frac{12}{3} - 7 < -3$	$\frac{9}{3} - 7 < -3$	$\frac{6}{3} - 7 < -3$
$4 - 7 < -3$	$3 - 7 < -3$	$2 - 7 < -3$
$-3 \nless -3$	$-4 < -3$	$-5 < -3$

(b) Solve for x: (25)

$$2x + 6(2 - x) \geq 4$$
$$2x + 12 - 6x \geq 4$$
$$-4x + 12 \geq 4$$
$$-4x + 12 - 12 \geq 4 - 12$$
$$-4x \geq -8$$
$$\frac{-4x}{-4} \leq \frac{-8}{-4}$$
$$x \leq 2$$

The solution set is $\{x \mid x \leq 2\}$

In this example, the direction of the inequality reverses, because dividing by -4 is the same as multiplying by $\frac{1}{-4}$.

Note: Since the inequality "$\leq$" includes "is equal to," a check when $x = 2$ would produce a true sentence $(4 \geq 4)$. Any value less than 2 should also produce a true sentence. (26)

● **Practice 3** Find the solution set for each inequality. (If directed, check each using two different values.) (27)

(a) $7t - 3 > 4t + 3$

(b) $x + 5 \geq 5x - 7$

(c) $5(y - 1) + 4 \leq 2y - 7$

(d) $\frac{3c}{2} - 2 < 2c - 4$

See also Exercise 5.

Objective 3: In many cases, it is necessary to restrict the possible solutions that open sentences might have. For example, it may be necessary that all candidates for a certain job must be at least 18 years old $(c \geq 18)$ but must be less than 30 years $(c < 30)$. It is possible to combine two such statements into a single statement, as illustrated below. (28)

◆ **Example 4**

(a) The statement $-2 < b \leq 3$ is a shortcut method of indicating $-2 < b$ *and* $b \leq 3$. (Observe that $-2 < b$ is the same as $b > -2$.) (29)

Reading double inequalities

Statements of this type are read by starting with the variable and reading out in both directions:

$$-2 < b \leq 3$$

is read: "b is greater than -2 *and* less than or equal to 3."

$$(b > -2 \quad \text{and} \quad b \leq 3)$$

● **Practice 4** Write the meaning of the following: (30)

(a) $-3 < y < 0$ (b) $-2 \leq k < 5$

(c) $1 < p \leq 4$ (d) $-4 \leq d \leq 2$

See also Exercise 6.

C. ENGLISH SENTENCES TO MATH SENTENCES

Objective 4:

Formulas are useful for solving problems that repeat frequently. However, many situations arise for which no established formula applies. Such "word problems" have traditionally been a plague to students. But this should not be the case. The key to solving word problems is to first express the situation in a concise English sentence. The verbal sentence will then translate directly into a mathematical sentence (equation or inequality). (31)

Verb represents equal marks

One fact that helps immensely in converting English sentences into mathematical sentences is that the *verb* of the English sentence corresponds to the *equal marks* (or inequality symbol) of the mathematical open sentence. The mathematical equivalents of several other words are illustrated in the following examples. (32)

◆ **Example 5**

(a) What number increased by 15 gives 37? (33)

What number	increased by	15	gives	37?
↓	↓	↓	↓	↓
n	$+$	15	$=$	37

"Increased by" means "plus"

Thus,

$$n + 15 = 37$$

$$n + 15 - 15 = 37 - 15$$

$$n = 22$$

"Of" means "times"

(b) An optometrist finds that $\frac{3}{5}$ of the new patients he examines need glasses. If 45 patients were found to have defective vision, how many patients did the optometrist examine? **(34)**

$$\underline{\tfrac{3}{5}}\ \underline{\text{of}}\ \underline{\text{patients}}\ \underline{\text{need}}\ \underline{\text{glasses}}$$

$$\downarrow\quad\downarrow\quad\downarrow\quad\downarrow\quad\downarrow$$

$$\frac{3}{5}\times\quad p\quad =\quad 45$$

$$\frac{3}{5}p = 45$$

$$\frac{5}{3}\cdot\frac{3}{5}p = 45\cdot\frac{5}{3}$$

$$p = 75$$

Note: let the variable in each equation be the first letter of the word the variable represents. With this habit, you will always know exactly what your final solution represents.

(c) Bob Watson is a student in a community college. Bob's clothing expenses ran only \$25 less than the cost of his living expenses. If Bob spent \$375 for clothes, how much were his living expenses? **(35)**

"\$ less than" means "\$ subtracted from"

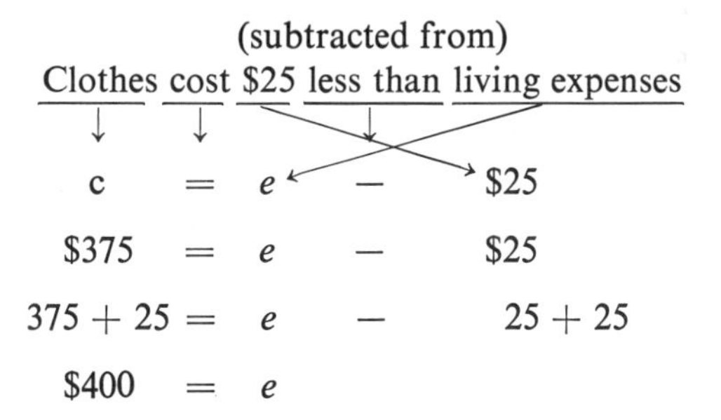

Observe here that "less than" does *not* translate "<." In this case, "less than" means "subtracted from" because the problem tells exactly *how much less* the amount was. That is, the clothing equals the "living expenses less \$25."

Given "*x* times the sum of . . ."

(d) The student above had a part-time job where his salary totaled three times the sum of his living expenses and books. If Bob earned \$1,500 from his part-time job, how much did his books cost? **(36)**

Salary	was	3	times	living expenses	and	books
↓	↓	↓	↓	↓	↓	↓
s	$=$	3	$\times$ (	e	$+$	b)
\$1500	$=$	3	$\times$ (	\$400	$+$	b)

Then,

$$1{,}500 = 3 \cdot 400 + 3b$$
$$1{,}500 = 1{,}200 + 3b$$
$$1{,}500 - 1{,}200 = 1{,}200 - 1{,}200 + 3b$$
$$300 = 3b$$
$$\frac{300}{3} = \frac{3b}{3}$$
$$\$100 = b$$

(e) During an art show, Adams and Baker together sold 45 paintings. If Adams sold twice as many paintings as Baker, how many paintings did each sell? (37)

Given joint total, when one variable is a multiple of the other

When a problem involves two amounts, it is usually better to let the main variable represent the *smaller* amount. Since Adams sold twice as many paintings as Baker, Baker is the smaller quantity:

Adams	sold	twice	Baker
A	$=$	$2 \times$	B

Now

Adams	and	Baker	together	sold	45 paintings
A	$+$	B		$=$	45
↓					
$2B$	$+$	B		$=$	45
		$\frac{3B}{3}$		$=$	$\frac{45}{3}$
	(Baker's sales)	B		$=$	15 paintings
	(Adams' sales)	A		$=$	$2B$
				$=$	$2(15)$
				$=$	30 paintings

(f) The total cost of running a community child care project was less than \$1,800.* Salaries for personnel were twice the cost of food supplies; and equipment ran three times as much as food. Determine the approximate cost of each item. (38)

*In this context, it is assumed that the total cost was almost \$1,800.

Type (e) repeated

Salaries	were	twice food		Equipment	was	3 times food
s	$=$	$2f$		e	$=$	$3f$

"Less than" means "<" when no amount less than is specified

Total cost was less than $1,800

$$s + e + f < \$1{,}800$$
$$2f + 3f + f < 1{,}800$$
$$\frac{6f}{6} < \frac{1{,}800}{6}$$
$$\text{(food) } f < \$300$$
$$\text{(salaries) } s = 2f$$
$$= 2 \times 300$$
$$s = \$600 \text{ (approximate)}$$
$$\text{(equipment) } e = 3f$$
$$= 3 \times 300$$
$$e = \$900 \text{ (approximate)}$$

Observe here that "less than" is translated "$<$" because we do *not* know *how much less* the cost was.

(g) A musical review was sponsored to raise money for charity. The sale of preferred ($4) and choice ($3) tickets totaled $3,400. If 1,000 tickets were sold altogether, how many tickets of each type were sold? (39)

Representing quantity of each

Regardless of the actual number of each type, 1,000 tickets were sold;

$$\text{Choice} + \text{Preferred} = 1{,}000$$
$$c + p = 1{,}000$$
$$c = 1{,}000 - p$$

Then,

Sum of "price times quantity" when only total quantity is known

Sale of choice plus sale of preferred brought $3,400

$$\$3 \times c + \$4 \times p = 3{,}400$$
$$\downarrow$$
$$3(1{,}000 - p) + 4p = 3{,}400$$
$$3{,}000 - 3p + 4p = 3{,}400$$
$$3{,}000 + p = 3{,}400$$
$$3{,}000 - 3{,}000 + p = 3{,}400 - 3{,}000$$
$$\text{(\# of preferred tickets) } p = 400 \text{ tickets}$$

$$\text{(\# of choice tickets) } c = 1{,}000 - p$$
$$c = 1{,}000 - 400$$
$$c = 600 \text{ tickets}$$

Practice 5 Express each of the following as an English sentence and then as an equation. Solve each. (40)

(a) What number increased by 27 gives 70?

(b) What number decreased by 27 gives 70?

(c) A number of citizens summoned for jury duty, decreased by 25 who are exempt, leaves 36 eligible jurors. How many people were summoned?

(d) The number of initial participants in a telephone survey, increased by 72, provides a total survey of 250 persons. How many persons did the survey originally include?

(e) Some $\frac{7}{8}$ of the letters received by a senator express support for his stand on election reform. If 280 writers supported the senator, how many letters did he receive?

(f) Superlative Foods found that $\frac{2}{3}$ of those surveyed preferred their new coffee over the other leading brands. How many people were surveyed, if 560 preferred the new brand?

(g) The number of students taking chemistry is 47 less than the enrollment for biology. If there are 168 chemistry students, how many students are studying biology?

(h) The budget for product development at Handy Gadgets is only $2,300 more than the amount allocated for advertising. If $14,800 is spent for product development, how much are advertising expenses?

(i) A homeowner finds his $250 mortgage payment is twice as much as his utilities and insurance combined. If his insurance payment is $25, how much are his utilities?

(j) The number of color television sets sold at Entertainment Associates was four times the combined total of stereos and radios. If their sales included 52 television sets and 4 radios, how many stereos were sold?

(k) Advertising and food for a fund raising dinner totaled $1,800. If the food cost four times as much as the advertising, how much was spent for each?

(l) Smith and Taylor together sold 144 tickets to the fund raising dinner. If Smith sold three times as many tickets as Taylor, how many tickets did each person sell?

(m) A family vacation cost the Ramseys less than $480. Transportation costs were about 3 times the cost of food; and lodging totaled about twice the amount spent for food. Find the approximate amount of each expense.

(n) The total cost of a hospital stay was more than \$420. Room expenses ran twice as much as x-rays, while doctors' fees were about four times as much as x-rays. What was the approximate cost of each item?

(o) History books cost \$9 each and English texts are \$12 each. A recent order for 27 books totaled \$279. How many of each book were purchased?

(p) Nonstudent tickets to a college football game cost \$3 for children and \$5 for adults. If the sale of 700 tickets totaled \$2500, how many tickets of each type were sold?

See also Exercise 7.

EXERCISES

Objective 5: Sentences

1. Answer the following. Explain each "no" response:

(a) The study which applies the number properties of arithmetic to sentences involving variables is called ____.

(b) Is $3x + 2 = 9$ a mathematical sentence? an open sentence?

(c) Is $4 + 3 \leq 6 + 2$ a mathematical sentence? an open sentence?

(d) Is $\frac{3n}{4} > n - 5$ a mathematical sentence? an open sentence?

(e) Is $3(b + 1) = 3b + 3$ a mathematical sentence? an open sentence? an identity sentence?

(f) Is $S = P(1 + rt)$ a mathematical sentence? an open sentence?

(g) An open sentence where the two sides exactly balance each other is called an ______.

(h) An open sentence where the two sides do not balance is called an ______.

(i) Is $2 + 3 \neq 8 - 4$ a true sentence?

(j) Is $n + 4 > 5$ a false sentence?

Equations

2. Classify each as true or false. Explain each "false" response:

(a) $4y - 8 = 16$ is a first-degree equation in one variable.

(b) $3d + 2 = 4d - 2$ is a first-degree equation in two variables.

(c) The simplest equations (such as $x * \# = \#$, where $*$ can be any arithmetic operation) are solved by performing the opposite operation on both sides of the equation.

(d) The truth set of the equation $2r + 3 = 15$ is the set, $\{r\}$.

(e) The reflexive, symmetric, and transitive properties are the basic properties used to solve equations.

(f) Given $2y - 6 = 10$, then $2y = 16$ would be an equivalent equation.

Inequalities

3. (a) Use set-builder notation to indicate all numbers greater than 2.
(b) Use set-builder notation to indicate all numbers between -3 and 5.
(c) Use set-builder notation to indicate all numbers that are at least as much as 0 but not as large as 4.
(d) Given $\{t \mid t < 3,\ t \text{ a whole number}\}$, would $t = 1$ make the sentence true? would $t = -2$ belong to the truth set?
(e) According to the multiplication property of $<$, multiplying by a negative changes the ______ of the inequality.
(f) In checking the solution to an inequality, what must be obtained in order to verify that the starting point of the solution inequality is correct?
(g) Why is it necessary to use a second point in order to verify an inequality solution?

Objective 1; Examples 1 and 2

4. Determine the solution set for each of the following equations. Then check each solution:

(a) $\frac{y}{4} = 15$
(b) $t + 3 = -5$
(c) $r - 6 = -2$
(d) $5g = -35$
(e) $\frac{3k}{8} = 9$
(f) $4s - 5 = 19$
(g) $3(h + 7) = 19$
(h) $2x - 9x + x = -18$
(i) $8 = \frac{2c}{7} + 4$
(j) $2d - 5(d + 1) = 10$
(k) $8a - 7 = 6a + 5$
(l) $\frac{5n}{6} - 2 = n - 5$
(m) $4z + 11 = 3$
(n) $\frac{5h}{3} = -20$
(o) $32 = 12p - 3p - 5p$
(p) $7r + 4 = 11r - 8$
(q) $6(q - 2) = 30$
(r) $3v - 7(v - 2) = 18$
(s) $4 = \frac{3t}{8} - 5$
(t) $\frac{3s}{2} + 1 = s - 2$

Objective 2; Example 3

5. Find the truth set for each inequality. Verify your solution using two different values:
(a) $5d - 4 \leq 3d + 2$
(b) $2y + 3 > 4y - 5$
(c) $g - 4(g + 3) \geq 3$
(d) $5r - 1 < 8r - 7$
(e) $3(w + 4) - 7 \leq w - 1$
(f) $\frac{4x}{3} - 1 > 2x + 3$

Objective 3; Example 4

6. Explain the meaning of the following statements:
(a) $2 < j < 6$
(b) $-3 \leq n < 7$
(c) $-5 < s \leq 0$
(d) $-4 \leq z \leq -1$

Objective 4; Example 5

7. Express each for the following as an English sentence and then as a mathematical sentence. Solve each equation.

(a) What number increased by 18 gives 64?
(b) What number decreased by 18 gives 64?
(c) A number of job applicants, decreased by 12 who did not meet the qualifications, leaves 27 remaining applicants. How many people originally applied?
(d) A survey revealed that $\frac{5}{8}$ of those interviewed felt an obligation to help conserve energy. If 750 persons favored energy conservation, how many persons were interviewed?
(e) River City Motors reports that sales of full-sized models totaled 14 less than compact sizes. How many compacts were sold, if 37 larger cars were sold?
(f) Exactly 10 more than the required $\frac{2}{3}$ of the legislature voted for an increase in taxes. If 74 members approved the increase, what is the total membership of the legislature?
(g) Contributions to the United Fund total $2,500 less than the projected goal. If total contributions are $87,300, what is the fund's goal?
(h) Preseason football ticket sales have already totaled slightly less than $60,000. If these sales represent $\frac{3}{4}$ of the total possible gross (sales), what will be the approximate income when all tickets have been sold?
(i) Modern Office Equipment experienced such a demand for electronic calculators that sales of the calculators ran three times the combined total of typewriters and dictaphones. If 54 calculators and 3 dictaphones were sold, how many new typewriters were bought?
(j) A total of 4,500 children and adult tickets were sold at the local zoo on a recent weekend. If the children's sales were four times the number of adult sales, how many tickets of each kind were bought?
(k) The boys and girls of Dover school sold 3,000 tickets to the pancake supper. If the boys sold twice as many tickets as the girls, how many tickets were sold by each group?
(l) The total cost of a community theater project was more than $1,100. The sets cost twice as much as the programs, and costumes cost 2.5 times as much as programs. Find the approximate cost of each item.
(m) The perimeter of a swimming pool (which has had tile installed) is twice the sum of the length and width of the pool. If the owner is charged for 96 ft of tile and he knows the pool is 32 ft long, what is the width of the pool?
(n) Given a choice of several items, eight less than half the passengers on an airliner selected steak for dinner. If 82 steak dinners were served, how many passengers were on board?

(o) A shipment of novels and travel guides cost $164. The novels cost $4 each and the travel guides were $7 each. If there were 32 books altogether, how many of each type did the shipment include?

(p) A total of 28 records and tapes cost $176. If the records were $5 each and the tapes were $8 each, how many of each item were purchased?

OBJECTIVES

After completing this chapter, you should be able to

A. Solving equations

1 Solve (and check, if directed) first-degree equations in one variable. **(1–20)**

B. Solving inequalities

2 Solve (and check, if directed) first-degree inequalities in one variable. **(21–27)**

3 Explain the meaning of a given statement that combines two inequalities into a single statement. **(28–30)**

C. English sentences to math sentences

4 Given a word problem, state the facts in an English sentence and convert this to a mathematical sentence that is used to solve the problem. **(31–40)**

General 5 Explain the basic concepts of sentences, equations, and inequalities. **(Exercises 1–3)**

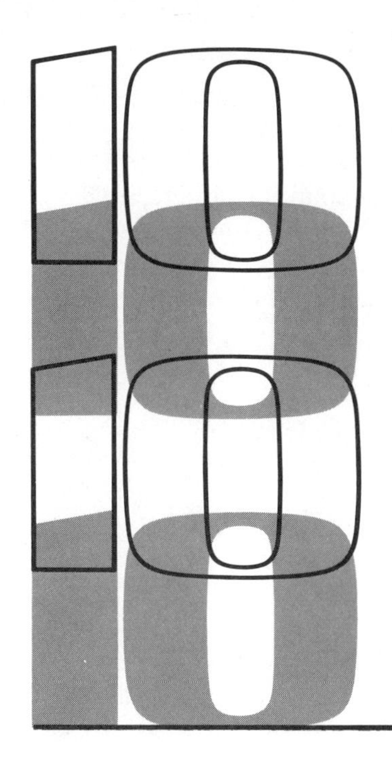

Algebra: Relations, Functions, and Graphs

Our study of algebra was introduced in the preceding unit, where we learned to solve first-degree equations and inequalities in one variable. In many cases, however, a correspondence exists in which two variables are related to each other in some way. The following topics of relations and functions concern the algebraic relationship between two variables: the equations and inequalities that express such relationships and the graphs that illustrate them. (1)

Many everyday situations are typical of functional relationships: the relationship between our light bill and the amount of electricity we use, between a weight loss and the number of calories we cut, between a moving bill and the number of miles the furniture is shipped, between a long-distance phone bill and the number of minutes we talk. (Do your grades also reflect how much time you spend studying?) (2)

A. RELATIONS AND FUNCTIONS

Recall from our study of sets that the Cartesian ("cross") product of two sets produces a set of ordered pairs. For example, given $A = \{1, 2\}$ and $B = \{x, y, z\}$, then (3)

$$A \times B = \{(1, x), (1, y), (1, z), (2, x), (2, y), (2, z)\}$$

Also recall that each pair is "ordered" because one component is considered the "first" component and the other is the "second" component. Thus, (a, b) is a different ordered pair from (b, a).

Definition of relation

Ordered pairs are particularly important in the study of algebra, where they are defined as follows: Any set of ordered pairs is called a *relation*. Thus, the Cartesian product $A \times B$ above is a relation. (4)

Domain

The set of all first components of a relation is called the *domain* of the relation. Thus, the (5)

$$\text{domain of } A \times B \text{ is} \quad \text{set } A = \{1, 2\}$$

Range

Further, the set of all second components of the ordered pairs of a relation is called the *range* of the relation. Therefore, the

$$\text{range of } A \times B \text{ is} \quad \text{Set } B = \{x, y, z\}$$

Objective 1: ◆ **Example 1a** Tell the domain and range of each relation, R: (6)

(a) $R = \{(a, d), (b, e), (c, f)\}$

Domain $= \{a, b, c\}$ Range $= \{d, e, f\}$

(b) $R = \{(a, e), (b, f), (c, f), (d, g)\}$

Domain $= \{a, b, c, d\}$ Range $= \{e, f, g\}$

(c) $R = \{(a, d), (b, e), (b, f)\}$

Domain $= \{a, b\}$ Range $= \{d, e, f\}$

Definition of function

Two of the sets of ordered pairs above form a special type of relation that is known as a function. A *function* is a relation where each first component defines exactly one ordered pair. That is, a function exists when there is exactly *one* ordered pair for each distinct first component. (7)

◆ **Example 1b** Tell whether each given set of ordered pairs represents a function: (8)

(a) $R = \{(a, d), (b, e), (c, f)\}$ is a function

(b) $R = \{(a, e), (b, f), (c, f), (d, g)\}$ is a function

Notice in (a) and (b) that there is *one* ordered pair for each first element; thus, both relations are also functions.

(c) $R = \{(a, d), (b, e), (b, f)\}$ is *not* a function

In this relation, the first element b forms two ordered pairs: (b, e) and (b, f); hence, this relation is *not* a function. To repeat: a function contains exactly *one* ordered pair for each distinct first component.

● **Practice 1**

(a) Name the domain and range of each relation, R: (9)

(1) $R = \{(r, u), (s, v)\}$

(2) $R = \{(r, x), (s, x), (t, y), (t, z)\}$

(3) $R = \{(p, t), (q, u), (r, v), (s, v)\}$

(4) $R = \{(x, m), (y, n), (y, o), (z, p)\}$

(5) $R = \{(t, r), (u, p), (v, q)\}$

(b) Indicate which relations are also functions.

See also Exercise 1.

Cartesian graphs on a plane (flat surface) are frequently used to determine whether a relation is a function. A graph represents a function provided a vertical line drawn anywhere on the plane would intersect the graph at only one point. Consider the following graphs: (10)

Objective 2: Vertical-line test

◆ **Example 2** Determine whether each graph below represents a function. (11)

(a)

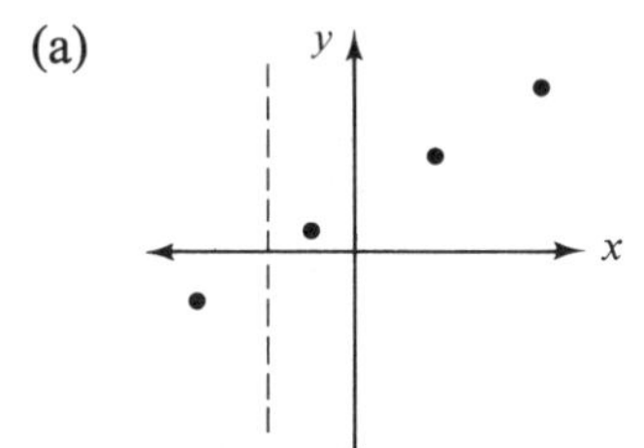

(b)

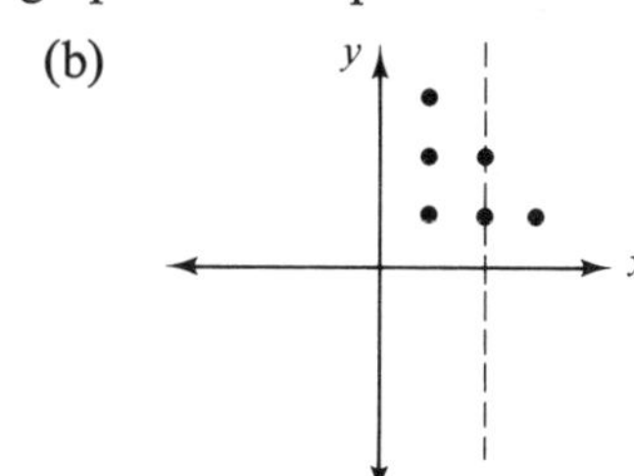

(c)

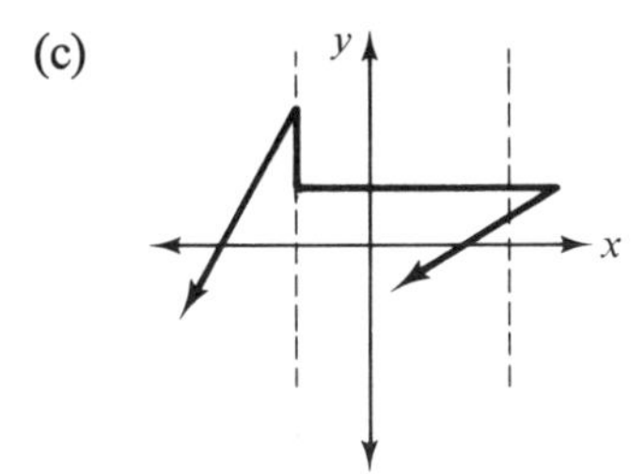

(d)

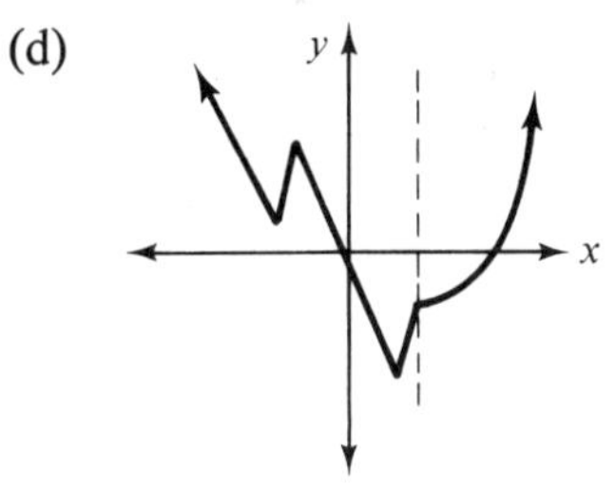

(a) and (b) are graphs of individual points. Graph (a) represents a function, since a vertical line drawn anywhere on the plane would intersect only one point. However, (b) is not a function, since it is possible to draw a vertical line on the plane that would intersect two or more points, as shown.

Similarly, the solid line of graph (c) does not represent a function, since vertical lines may be drawn that intersect the graph in at least two points. Graph (d) is a function, however, since any vertical line would intersect at only one point.

● **Practice 2** Examine the following graphs and tell which ones represent a function. (12)

(a)

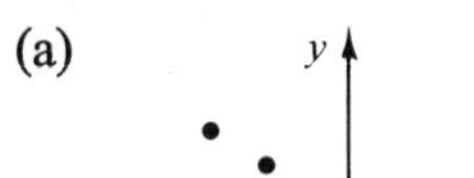
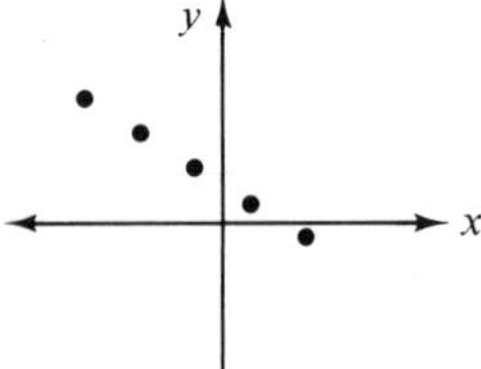

(b)

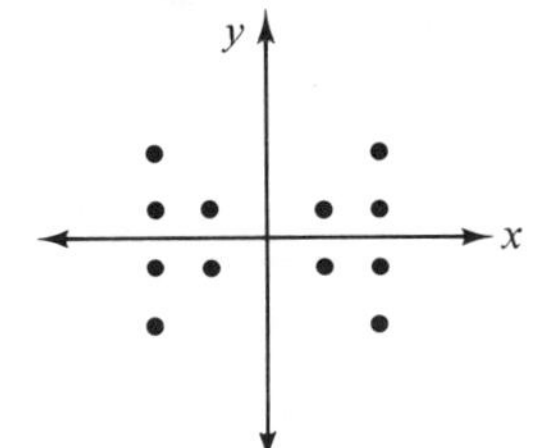

(c)

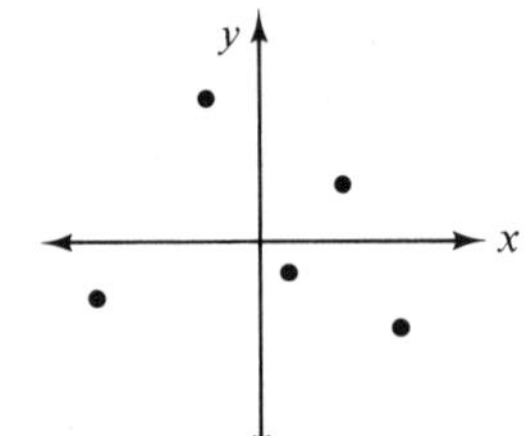

(d)

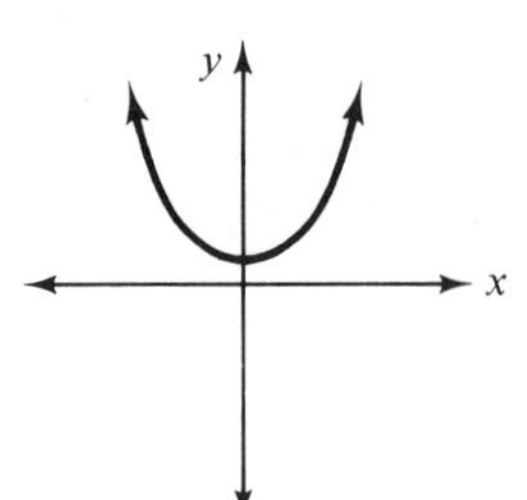

(e)

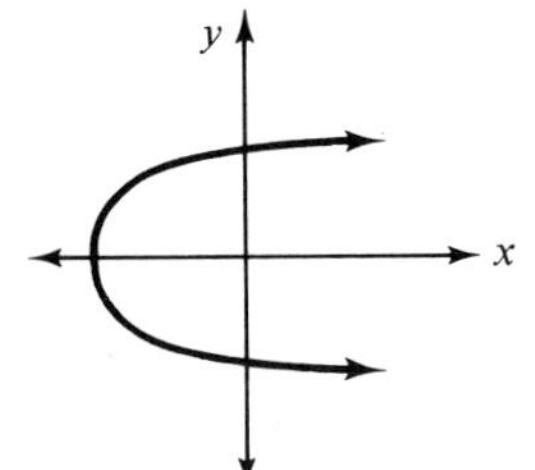

(f)

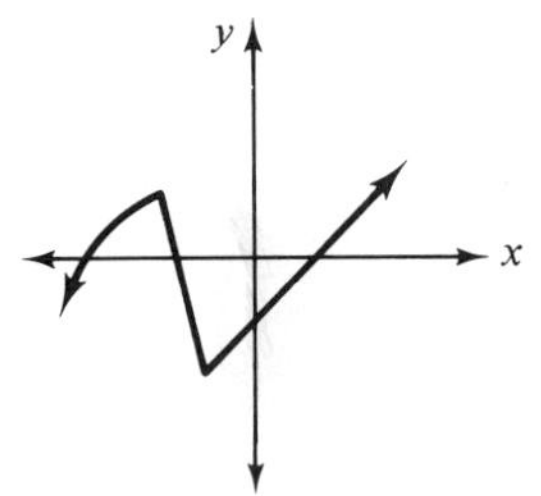

(g)

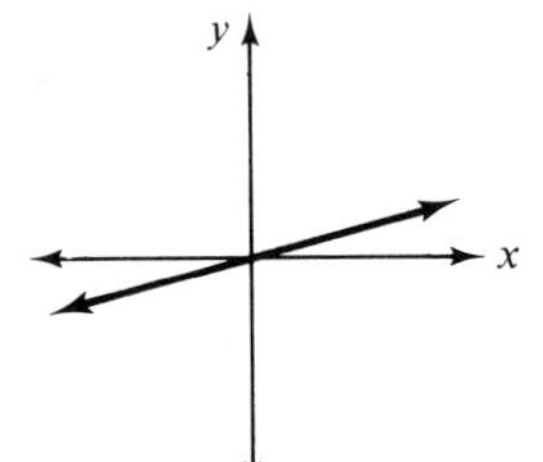

(h)

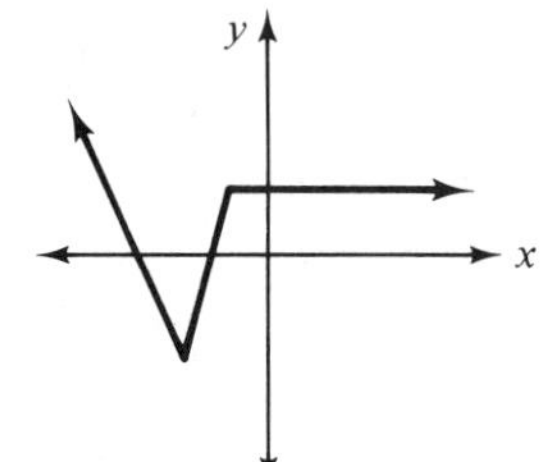

(i)

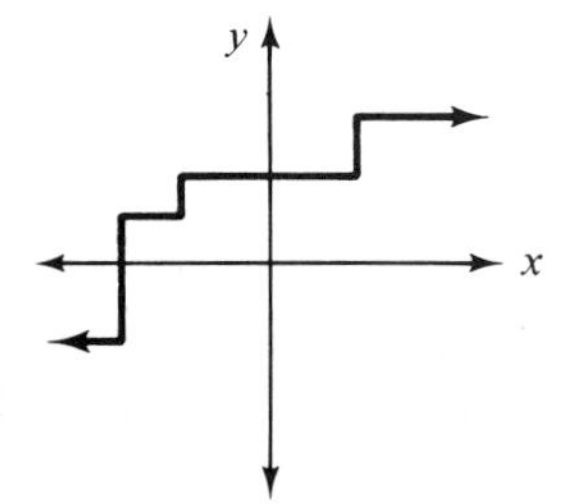

See also Exercise 5.

B. GRAPHING ON A PLANE

Earlier, we learned to solve first-degree equations in one variable, such as $\frac{3}{5}x + 4 = 13$. The following are examples of linear equations in two variables: **(13)**

(a) $y = 3x$ (b) $x + y = 0$

(c) $2x - 3y = 6$ (d) $y = 2x - 1$

Each of these equations is some variation of the basic equation form, $ax + by = c$, where a, b, and c represent some real numbers. The graph of each linear equation in two variables is always a straight line. Therefore, with the single exception of linear equations that graph as vertical lines, all other linear equations represent functions. Functions that graph as straight lines are thus called *linear functions*.

Linear functions

A linear equation, such as $y = 2x - 1$, is said to *define* a function.* That is, the equation is used to obtain the set of ordered pairs (x, y) which compose the function. This is done by substituting numerical values for x and computing the corresponding value of y. In the preceding equation, x is called the *independent variable*, and y is the *dependent variable*. Since y is dependent upon x, then y is said to be *a function of* x. The set of ordered pairs, (x, y), that make up a relation may be indicated in set-builder notation as **(14)**

Objective 3: Set-builder notation **(15)**

$$\{(x, y) \mid y = 2x - 1\} \qquad \text{read:}$$

"The set of all ordered pairs (or points), (x, y), such that $y = 2x - 1$."

(Observe that this type of notation can also be used to indicate a relation that is not a function. For example, $\{(x, y) \mid x = 3\}$ specifies a relation that is not a function.)

Functional notation

To show that an equation represents a function, we may use *functional notation*. The statement. **(16)**

"y is a function of x"

is translated $y = f(x)$

Thus, $y = 2x - 1$ is shown in functional notation as $f(x) = 2x - 1$

*Equations of higher degree (exponents) may also define functions. For example, $y = 3x^2 - 2x + 1$ and $y = x^3$ both define functions which graph as curved lines. However, we shall be concerned only with first-degree functions, which graph as straight lines.

The symbol "$f(x)$" strictly speaking means "f evaluated at x." For convenience, however, this is usually shortened to "f of x" or sometimes "f at x." Various numbers may then be substituted for x to obtain the corresponding value of y, as shown in the following example. (Other letters may also be used to indicate functions; the notations "$g(x)$" and "$h(x)$" are often used.) (17)

◆ **Example 3** Given the function $f(x) = 2x - 1$, evaluate the function when $x = -1, 0, 1$, and 2. (18)

x	$f(x) = 2x - 1$	(x, y)
0	$f(0) = 2(0) - 1 = -1$	$(0, -1)$
−1	$f(-1) = 2(-1) - 1 = -3$	$(-1, -3)$
1	$f(1) = 2(1) - 1 = 1$	$(1, 1)$
2	$f(2) = 2(2) - 1 = 3$	$(2, 3)$

● **Practice 3** Evaluate each of the following functions at $x = -2$, $x = 0$, $x = 2$, and $x = 4$: (19)

(a) $f(x) = x + 3$

(b) $\{(x, y) \mid y = 3x - 2\}$

(c) $f(x) = 3 - 2x$

(d) $y = \frac{1}{2}x + 5$

See also Exercise 6.

Objective 4: Graphing on a Cartesian plane

We will graph the preceding equations in order to verify that they do indeed represent linear functions. The plane used to denote (and to graph) the set of all ordered pairs of real numbers is called the *Cartesian plane.* The real numbers of any ordered pair that are used to identify a point on the plane are called the *Cartesian coordinates* of the point. (20)

The Cartesian plane is formed by two number lines (called *axes*) that are drawn perpendicular to each other. The point of intersection of the two axes is called the *origin* of the system and denotes the point $(0, 0)$. The horizontal axis is normally labeled as the *x-axis*, and the vertical axis is the *y-axis*. In any ordered pair (x, y), the x-coordinate is called the *abscissa*, and the y-coordinate is the *ordinate*. (21)

The point associated with any ordered pair (x, y) is located by first moving along the x-axis to the point named by x. From there, move vertically (either up or down, parallel to the y-axis) as many units as indicated by y. (Up is positive and down is negative.) This final position is the point indicated by the ordered pair, (x, y). (22)

◆ Example 4

(a) Use the ordered pairs of Example 3 to graph the function defined by $f(x) = 2x - 1$. **(23)**

(b) Graph $\{(x, y) \mid y = 2\}$.

(a) First, the points $(0, -1)$, $(-1, -3)$, $(1, 1)$, and $(2, 3)$ are located on the plane, as described above. However, these points do not compose the entire function. Any real number could have been substituted for x; therefore, the number of ordered pairs in the function is infinite. However, every point of the function would lie on the straight line that connects the given points and extends infinitely in both directions. Thus, it is the line itself (including the four given points) that represents the function. Notice that the graph in the figure below passes the vertical-line test for a function.

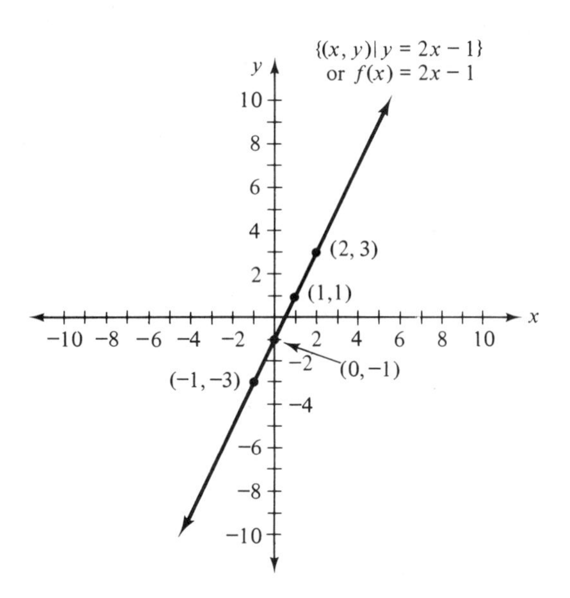

(b) $\{(x, y) \mid y = 2\}$ can also be expressed as $\{(x, 2)\}$. When an equation is stated in only one variable, (such as $y = 2$), this implies that the unspecified variable (x) may be *any* real number (but y will always be 2). Thus, $(-4, 2)$, $(-1, 2)$, $(0, 2)$, $(3, 2)$ would be examples of points on the graph: **(24)**

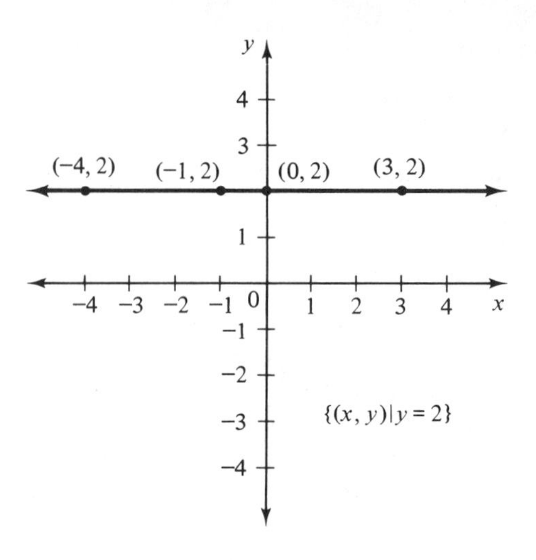

● **Practice 4**

(a)–(d) Graph the four functions of Practice 3, using the ordered pairs computed there. Observe that each passes the vertical-line test. (25)

(e) Graph $\{(x, -3)\}$. Is this a function by the vertical-line test?

(f) Graph $\{(x, y) \mid x = -2\}$. Is this a function?

See also Exercise 6.

C. INEQUALITIES ON A PLANE

Graphing inequalities

Inequalities are graphed in basically the same manner as equations. First, the line is located as though it were an equation. A broken line is drawn to indicate a simple inequality ($>$ or $<$). A solid line is drawn if the inequality includes equals ($\geq$ or $\leq$), in order to show that the line itself is included. In deciding which side of the line is indicated by an inequality, when y is dependent on x, the symbol "$>$" translates "above" the line, while "$<$" means "below" the line.* (26)

> is "above"
< is "below"

The following graphs demonstrate the various inequality combinations based on $y = 2$. *A shaded plane area indicates that every point in that area has coordinates which satisfy the inequality.* These areas are called *half-planes.* A solid line indicates the line itself is included in the half-plane and is called a *closed half-plane.* When the line itself is not included, the area is called an *open half-plane.* (27)

*When graphing inequalities that are solved for x rather than y (such as; $x > 2y + 1$), the symbol "$>$" translates "to the right of" the line, while "$<$" is interpreted "to the left of" the line. However, no inequalities of this type will be included in this survey unit.

◆ **Example 5** Graph the inequality variations based on $y = 2$. **(28)**

Objective 5:

(a)

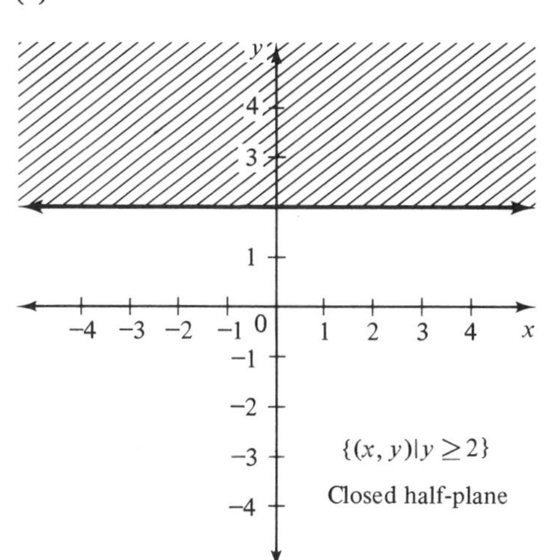

(b)

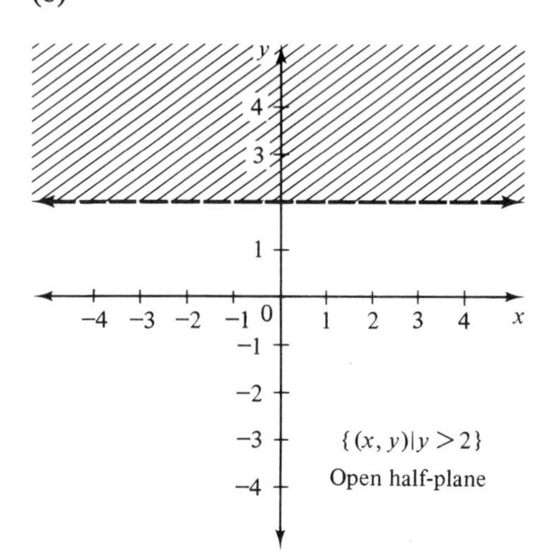

(c)

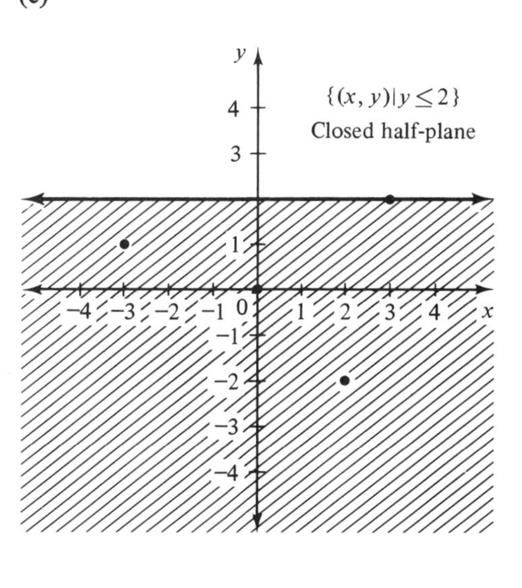

(d)

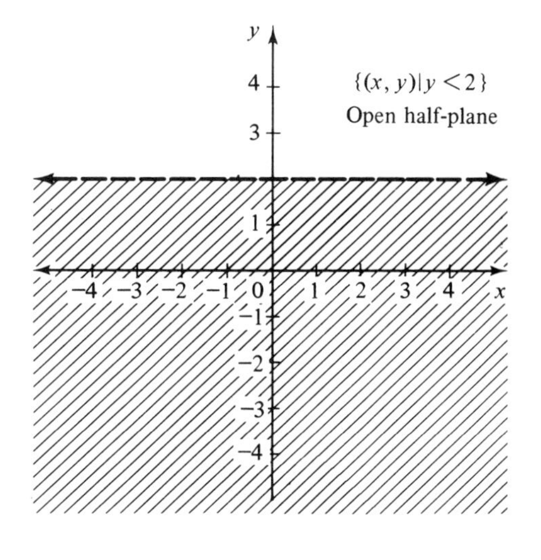

Random points are marked in graph (c) to illustrate that any point in the shaded area satisfies the inequality $y \leq 2$. Coordinates of the selected points would be as follows: (29)

$$(-3, 1) \qquad (2, -2)$$
$$(0, 0) \qquad (3, 2)$$

Notice that each ordered pair meets the specified condition: $y \leq 2$. (If the condition were $y < 2$, however, the pair (3, 2) would fall on a broken line and would not be included in the graph of $y < 2$.) Select some points in the graphs of $y \geq 2$ and $y > 2$ and verify for yourself that the ordered pairs you chose would meet the specified conditions.

Observe also that no graph of an inequality will pass the vertical-line test for functions. That is, any vertical line drawn on the graph of an inequality will intersect the inequality in an infinite number of points (the shaded area). Thus, no inequality can represent a function. (30)

Inequalities are *not* functions

● **Practice 5** Use a procedure similar to Example 3 to find coordinates (ordered pairs) for the following inequalities. (Let $x = -3$, $x = 0$, $x = 3$, and $x = 6$.) Then graph each inequality: (31)

(a) $\{(x, y) \mid y > x - 2\}$ (b) $\{(x, y) \mid y \leq 4 - 2x\}$

(c) $\{(x, y) \mid y \geq x\}$ (d) $\left\{(x, y) \mid y < \frac{1}{3}x + 7\right\}$

See also Exercise 6.

D. x- AND y-INTERCEPTS

We have graphed a number of equations to verify that they are linear functions. Thus, we are now ready to take a "short cut" for graphing such equations. This short cut is based on two facts: (32)

1. The graph of a first-degree equation in two variables* is a straight line; and
2. A straight line is determined by two points.

Objective 6: Thus, if we can locate any two points of the function, then the line that passes through these two points will be the graph of the entire function. Consider the following graph of $y = x + 2$.

*A "first-degree equation in two variables" as used here implies an equation that can be written in the form $ax + by = c$, where $a \neq 0$ and $b \neq 0$.

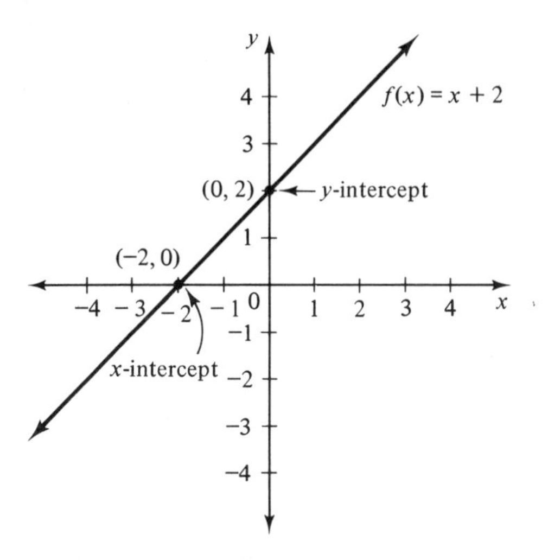

x-intercept

y-intercept

The point at which a line intersects the x-axis is called the *x-intercept.* (33) Similarly, the point at which a line intersects the y-axis is called the *y-intercept*. The x- and y-intercepts are usually the most convenient points to use in order to determine the line of a function. Notice at the point of the x-intercept that the value of y is zero. Also observe at the point of the y-intercept that x is zero. Thus, the two intercepts may be easily determined by first substituting $x = 0$ in the given function, and then substituting $y = 0$, as shown below.

◆ **Example 6** Determine the x-intercept and y-intercept for the function $f(x) = 2x - 1$. Use the intercepts to graph the function. (34)

When $x = 0$,

$$y = 2x - 1$$
$$= 2(0) - 1$$
$$y = -1$$

y-intercept: $(0, -1)$

When $y = 0$,

$$y = 2x - 1$$
$$0 = 2x - 1$$
$$0 + 1 = 2x - 1 + 1$$
$$1 = 2x$$
$$\frac{1}{2} = \frac{2x}{2}$$
$$\frac{1}{2} = x$$

x-intercept: $\left(\frac{1}{2}, 0\right)$

Thus, the intercept points are located and the function determined as illustrated below. (Compare this graph with Example 4(a), where the same equation was graphed by the previous method.)

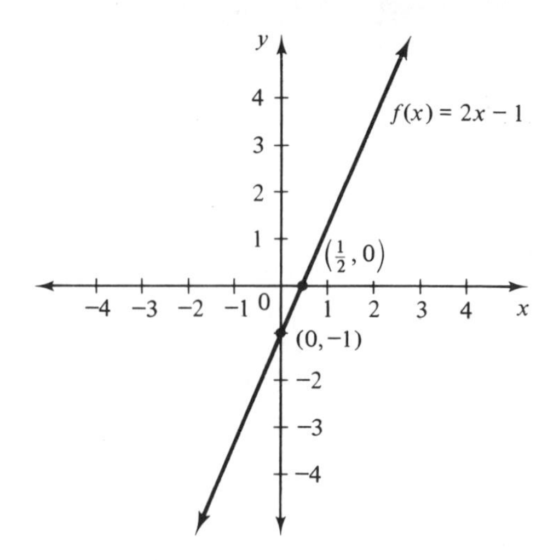

● **Practice 6** Determine the x-intercept and y-intercept for the following functions. Then graph each function by using the intercepts. (35)

(a) $y = 3x - 6$ (b) $f(x) = 4 - 2x$

(c) $\{(x, y) \mid y = x + 5\}$ (d) $y = \frac{1}{2}x + 3$

See also Exercise 7.

Cost equations

A typical example of the use of first-degree equations in two variables is *cost equations*, whereby a company can compute the cost of producing any given amount of merchandise. To illustrate, suppose a college club is making school banners as a money-raising project. First, they must buy some scissors, which cost \$5. This cost will be the same whether they make one banner or fifty banners. Materials for the banners themselves cost \$3 per banner. Thus, (36)

Total cost	equals	cost of banners	plus	cost of scissors
c	$=$	$\$3 \cdot b$	$+$	$\$5$

Hence, the total cost, $c = \$3 \cdot b + \$5$, can be expressed in an equation of the form,

$$y = 3x + 5$$

EXERCISES

Objective 7: Relations and functions

1. Answer the following. Explain each "no" response.
   (a) Let $A = \{1, 2\}$ and $B = \{3, 4\}$ and find $A \times B$. Is $A \times B$ a relation? a function?

(b) Find $A \times B$ when $A = \{2\}$ and $B = \{3, 4\}$. Is $A \times B$ a relation? a function?
(c) Find $A \times B$ when $A = \{1, 2\}$ and $B = \{3\}$. Is $A \times B$ a relation? a function?
(d) Which of the following relations are also functions?
(1) $\{(-2, 0), (0, -2), (1, 5)\}$
(2) $\{(-3, 2), (1, 2), (3, 0)\}$
(3) $\{(-4, 0), (-1, 2), (-4, 2)\}$
(4) $\{(-2, 0), (0, 1), (1, 3), (3, 1)\}$
(5) $\{(-1, -1), (0, 0), (1, 1), (3, 3)\}$
(6) $\{(3, -2), (1, -4), (3, 2), (-1, 4)\}$
(e) Does every $A \times B$ represent a relation? a function?
(f) If y is a function of x, then what is the function of 3 when $y = 2x + 1$? Express the indicated ordered pair.
(g) Does every linear function graph as a straight line?
(h) Does every function graph as a straight line?
(i) Does every straight line represent a function?
(j) Does every straight line represent a relation?
(k) Does every inequality (of the first degree in two variables) represent a relation?
(l) Does every inequality (as above) represent a function?

Equations as functions

2. Answer the following. Explain each "no" response.
(a) Given $g = 3h + 2$, is g a function of h? Would an ordered pair for this function be (g, h) or (h, g)?
(b) Given $x = 4y - 3$, is y a function of x? Use x and y to express an ordered pair for this function, either (x, y) or (y, x).
(c) Let $2c + d = 6$ and solve this equation for d. Is d a function of c? How would an ordered pair be stated using c and d?
(d) Solve (c) for c. Which variable is now a function of the other? Use c and d to express an ordered pair for the function as it is now expressed.
(e) Given $r = \frac{1}{3}s - 2$, which is the independent variable? State an ordered pair using r and s.
(f) Use set-builder notation to indicate the function $y = 7x - 2$.
(g) Use functional notation to indicate the function $y = 4x + 3$.
(h) Does $\{(x, y) \mid x = -2\}$ define a function?

Graphing

3. Think how you would graph the following and then answer each question. Explain *all* responses:
(a) Can a first-degree equation in one variable have an infinite solution set? (For example, $x + 3 = 8$.)
(b) Can a first-degree inequality in one variable have an infinite solution set?

(c) Can a first-degree equation in two variables have an infinite solution set? (For example, $y = 2x + 3$.)
(d) Can a first-degree inequality in two variables have an infinite solution set?

Answer the following:

(e) The Cartesian plane is formed by the perpendicular intersection of two ____, which are called axes.
(f) Name the abscissa of $(3, -2)$.
(g) Name the coordinates of the origin on the Cartesian plane.
(h) Which inequality symbol denotes a closed half-plane that is shaded above the graph line?

Objective 1; Example 1

4. (a) Identify the domain and range of each relation, R, below:
(1) $R = \{(a, x), (b, y), (c, y)\}$
(2) $R = \{(g, k), (h, j)\}$
(3) $R = \{(s, w), (t, w), (u, x), (v, y), (v, z)\}$
(4) $R = \{(a, n), (s, b), (d, r), (f, p)\}$
(5) $R = \{(p, a), (q, c), (r, b), (r, d)\}$
(6) $R = \{(f, t), (g, r), (h, s), (i, t)\}$
(b) Tell which of the relations are also functions.

Objective 2; Example 2

5. Use the vertical-line test to determine which of the graphs represent functions.

(a)

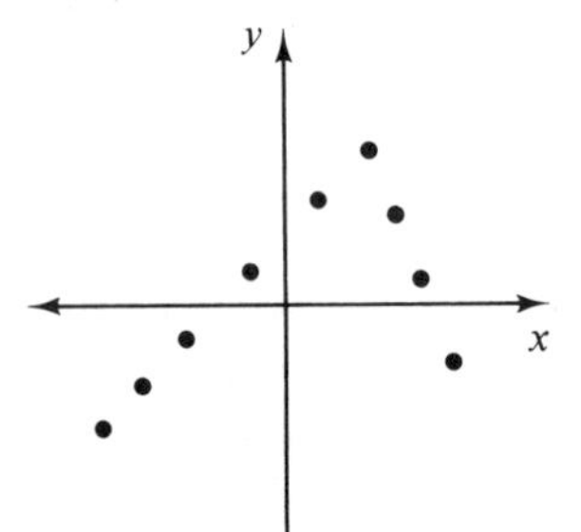

(b)

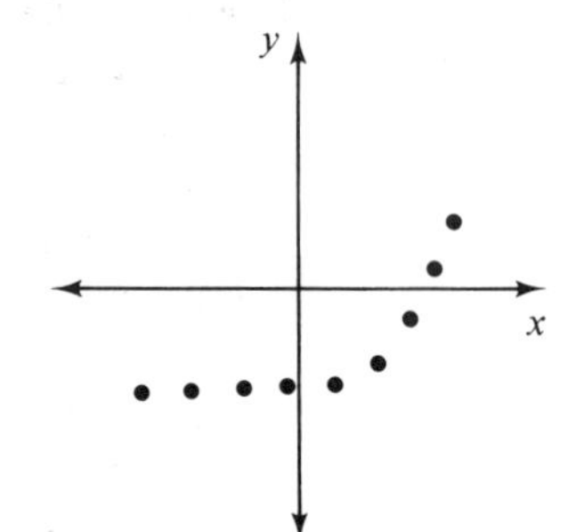

(c)

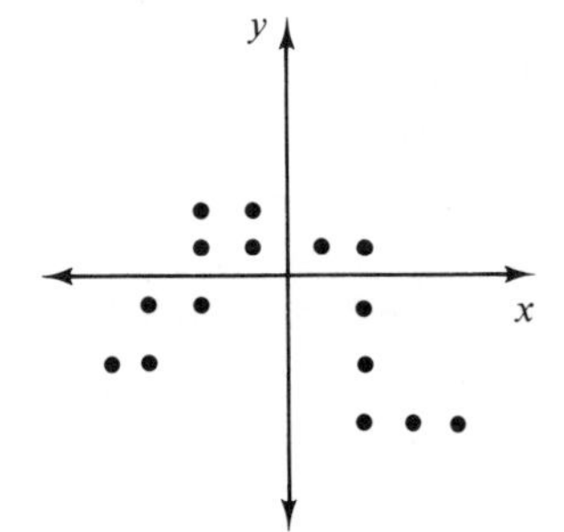

(d)

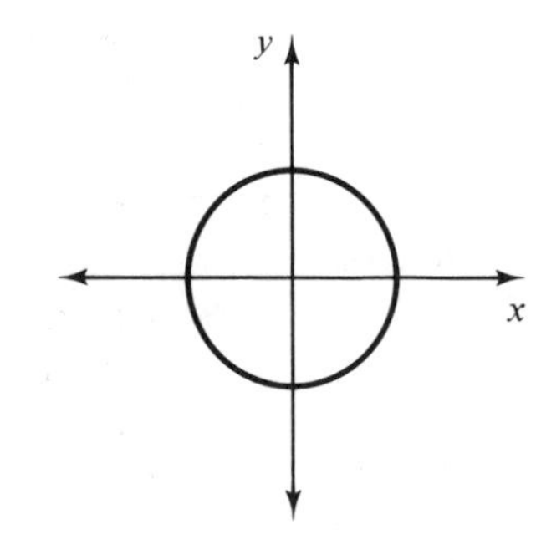

(e)

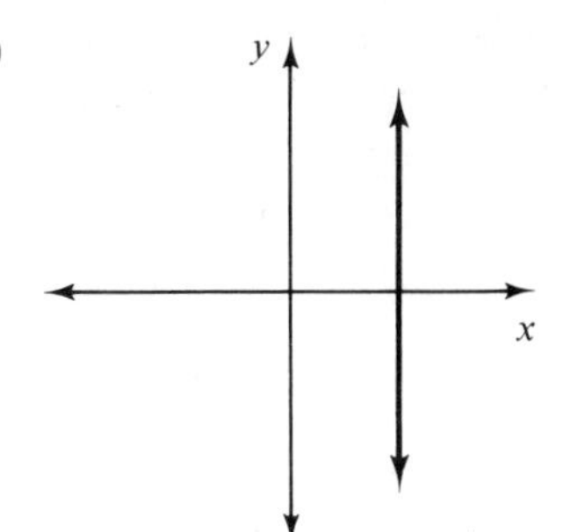

(f)

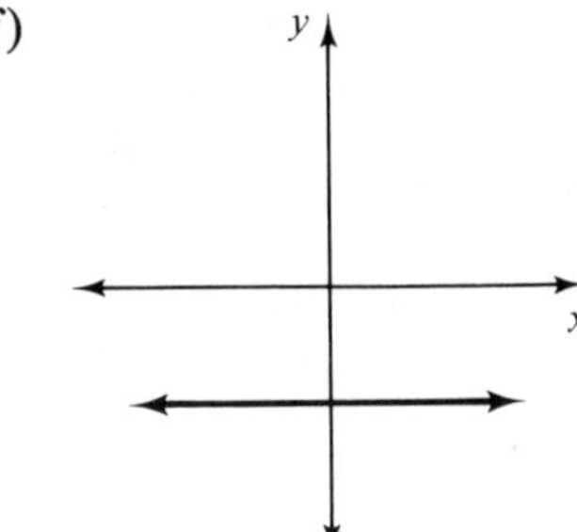

(g)

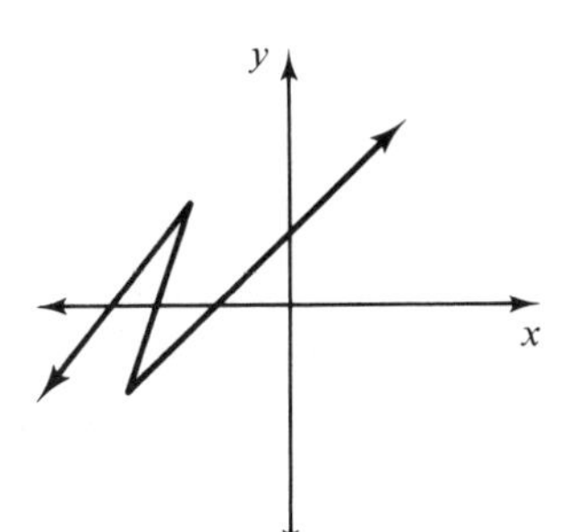

(h)

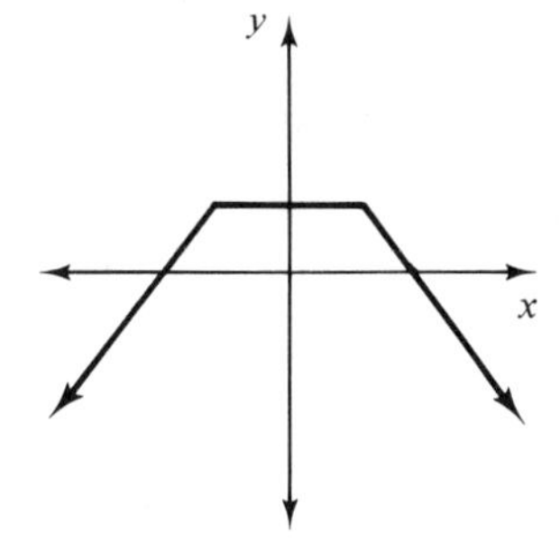

(i)

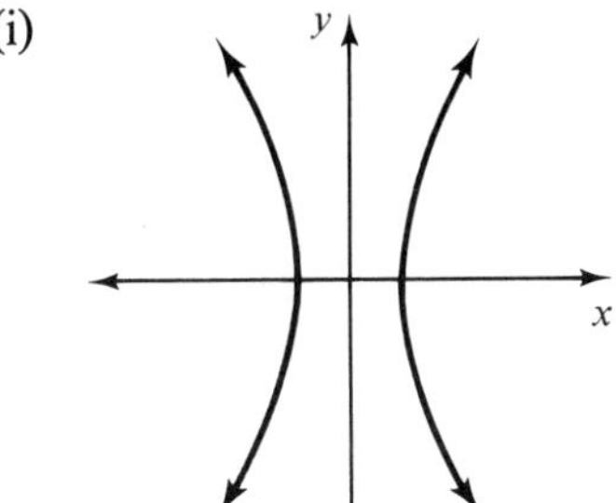

6. Compute ordered pairs for the following equations and inequalities using $x = -3$, $x = 0$, $x = 1$, and $x = 4$. Then graph each, and tell whether each is a function.

Objective 3, 4 and 5
Examples 3, 4 and 5

(a) $f(x) = x + 4$ (b) $\{(x, y) \mid y = -x\}$
(c) $f(x) = 1 - 2x$ (d) $y = 2x + 3$
(e) $\{(x, y) \mid y = 4\}$ (f) $\{(x, y) \mid x = -3\}$
(g) $\{(x, -1)\}$ (h) $\{(2, y)\}$
(i) $y \geq -3$ (j) $\{(x, y) \mid y < x - 1\}$
(k) $\{(x, y) \mid y \leq 3x\}$ (l) $\{(x, y) \mid y > 2x + 1\}$

7. Find the x-intercept and y-intercept for the following functions, and then graph each function using the intercepts. (Show (a) through (d) on one set of axes, and then (e) through (h) on a second set of axes.

Objective 6; Example 6

(a) $y = 4x + 2$ (b) $f(x) = x + 2$

(c) $f(x) = \frac{1}{2}x + 2$ (d) $\{(x, y) \mid y = -2x + 2\}$

(e) $f(x) = 2x - 8$ (f) $y = 2x - 2$

(g) $\{(x, y) \mid y = 2x + 4\}$ (h) $f(x) = 2x + 6$

OBJECTIVES

After completing this chapter, you should be able to

A. Relations and functions

1 Given a set of ordered pairs, **(3–9)**
(a) name the domain and range of each relation;
(b) tell whether each relation is also a function.

2 Use the vertical-line test to determine whether a given graph represents a function. **(10–12)**

B. Graphing on a plane

3 Given a function (or a relation) in set-builder notation, in functional notation, as a set, or as an equation, determine several ordered pairs, (x, y). **(13–19)**

4 Use ordered pairs to graph an equation on the Cartesian plane. **(20–25)**

C. Inequalities on a plane

5 Given an inequality, find several ordered pairs, (x, y), and graph the inequality on the Cartesian plane. **(26–32)**

6 Determine the x-intercept and y-intercept of a given function and use the intercepts to graph the function.

General 7 Explain the basic concepts of relations and functions, and graphing on the Cartesian plane. **(Exercises 1–3)**

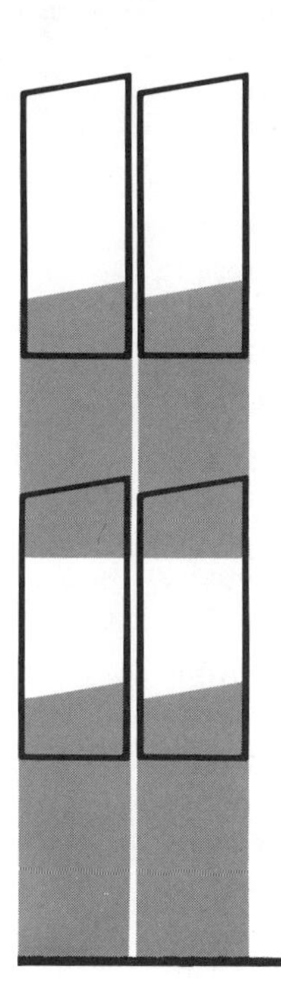

Statistics

Descriptive statistics

Inferential statistics

(1) The process of collecting, organizing, tabulating, and interpreting numerical data is called *statistics*. The information presented as a result of this process is also known collectively as *statistics*. Statistical methods are classified into two groups: *descriptive statistics*, which arranges data into a useful form to present a typical picture of past events; and *inferential statistics*, which examines a part of a larger group and generalizes to the whole group.

(2) In modern society, there are many reasons to describe the past experiences of persons, events, situations, and so on. Weather forecasters on a television news program, for example, describe the weather by giving the average temperature for a particular day or the range of the barometric pressure for a 24-hour period. Sportscasters report the average number of yards gained by a runner over the last ten football games or the most typical yardage punted by a given kicker. Similarly, technical reports contain informative facts like the average miles driven per gallon of gas for a particular automobile model or the typical cruising altitude for a DC-9 jetliner.

(3) It is also helpful to infer (predict) information about all persons, events, or situations in a specific classification by studying only a portion of them. For example, a scientist may test a small part of the corn in a cornfield and infer things about the whole cornfield from the part which was studied. This process of inferring characteristics about a whole cornfield by studying a part of the field is an application of inferential statistics.

Usually, as in this example, a small portion of the entire group is studied because it is not practical to test every member, and also because there is every reason to believe that the untested members would show results similar to the members which were studied. Another example of inferential statistics would be a study of 100 selected college freshmen that estimates their average height or weight and then infers (or states the likelihood) that the height or weight of all freshmen would be similar.

(4) Statistics is a way of producing numbers which make possible inferential, or relational, judgments about events. But why use statistics in the first place? Because statistics provides the logic for building objectivity into the processes of searching for new knowledge. Statistics provides objectivity by exploiting the laws of mathematics, thereby providing the logic and rules which, if followed, may generate useful and dependable information. In this chapter we will present a brief overview of some of the most important and most general facets of statistics.

All the information that is presented here is based on one or more *statistical models.* Keep this in mind throughout the chapter because, as we will see, it is of fundamental importance. Using a relatively precise set of rules, and with some experience, it is possible for the statistician to design mathematical models that will provide good estimates of specific aspects in which he is interested. From these estimates, he can then draw inferences about possible relationships that exist among the subjects in a larger group. This idea may seem obscure at this point, but we shall return to it at the end of the chapter when our grasp of statistical terminology has grown.

(5) Although statistics is a way in which a person attempts to make objective statements, the beginning student may wish to deal with this cautiously. We live in a scientific era where objective outcomes may be over valued. A decision or statement based on a valid system of logic and its resultant mathematical model will yield a logical result. However the objective data alone may not be enough to provide a reliable decision. For example, in psychology the subjective judgment of an experienced clinician may be more dependable than a decision based solely upon the more objective findings of any given psychometric or statistical analysis of other objective data. As will be seen in subsequent examples, even though the mathematics of a statistical process is all correct, it is possible for a situation to be greatly misrepresented, if invalid assumptions have been made, unusual factors about the data ignored, or inappropriate application made of a statistical model.

A. INFERENTIAL STATISTICS

Objective 1: Sample, population, statistic, parameter

(6) When a small portion of a total group of individuals is studied for statistical purposes, this portion is known as a *sample.* The total group or

universe from which the sample is drawn and about which inferences will be made is called the *population*. Whereas a population is all of some specified group (like all college sophomores in the United States), a sample would only be a small subset of the population. The term *statistic* (in the singular form) is used to describe a numerical characteristic that results from studying the numerical properties presented by a sample. On the other hand, when these numerical properties are derived from a population, they are called *parameters*. Usually population *parameters* are too difficult to determine directly and are inferred from a sample. Thus, statistic is to sample, as parameter is to population. That is, a statistic describes a sample; a parameter describes a population.

(7) A crucial logical connection can be made when it is seen that a statistic derived from a sample provides an *estimate* of the related population parameter. There are a number of factors which determine how good an estimate we are getting. For example, if a sample is selected *at random* in such a way that every person or thing in the population has an equal chance of being selected, then this *random sample* is said to be representative of the entire population. Accordingly, any statistical characteristic that is revealed about the representative sample is also believed to hold true for the whole population. Hence, from a statistic we obtain an estimate of the corresponding parameter; using a random sample, we are thus able to describe and characterize an entire population.

Random sample

(8) In order to draw valid conclusions from statistical information, it is necessary to follow two major prerequisites for selecting the groups of individuals to be measured: first, the population (all elements or individuals to be considered in a given situation) must be carefully defined. This can be a difficult task. However, information would be of little value unless the researcher specifies exactly what characteristics were used to identify groups such as "underachievers," "mentally retarded," or "gifted." Terms such as these could mean many different things to different people unless they are clearly defined, so that any researcher could determine whether a given individual belongs to the population being considered.

(9) Second, since it is almost always impossible to study entire populations, a random sample must be drawn from the population. That is, each member of the population must have an equal chance (probability) of being selected for the sample. A glaring example of failure to choose a random sample carefully (and in the context of the population defined) resulted in the confident assertion that Franklin Roosevelt would be defeated in a bid for the presidency. The population was defined (registered voters); the error was made by not choosing a valid sample of the voting public. In this particular case, the sample was drawn from telephone directories, while the population was defined as the "voting public." Apparently not all those who voted for President Roosevelt had telephones!

(10) Sampling is a subject of study in itself and will not occupy any more of our time, except to repeat that random sampling is necessary in order

to generalize conclusions about related populations. By analyzing statistics computed on random samples, we can find relationships, differences, or similarities revealed in those statistics, which we can often use to make accurate inferences (predictions) about the entire population. If samples are not drawn randomly, however, we must restrict our conclusions strictly to the subjects we studied. For example, suppose we wish to study characteristics of eighth graders in Texas. Failing to obtain a random sample in all areas of Texas, we would be required to limit our inferences to "eighth graders tested in the spring in Dallas, Ft. Worth, and Houston, Texas" instead of "Texas eighth graders." We will assume that all samples used in this unit are random samples. (Note: If a whole population is measured, then the population itself is a sample, and the measurement obtained is *both* a statistic and a parameter.)

Symbols for statistic and parameter

(11) Customarily, a statistic (that is, a measure computed from a sample) is denoted by some letter from the Latin (English) alphabet, whereas the corresponding parameter is designated by the counterpart letter in the Greek alphabet. Two statistical measures that we will study are the mean and the standard deviation. We indicate the mean of a sample with the (English) $\bar{x}$, and write μ (Greek letter "mu") to denote the mean for an entire population. Similarly, the (English) letter "s" denotes the standard deviation of a sample, while the Greek lowercase letter σ (sigma) refers to the population parameter of standard deviation.

(12) ◆ **Example 1** Given the following situation, relate the terms "population," "sample," "parameter," and "statistic" to the example:

> The average height (mean height) of all 10-year-old Samoan children is to be estimated from measures made on a few of them.

All 10-year-old Samoan children make up the *population*. The "few of them" we randomly choose is the *sample*. The average height (mean height) of the 10-year-old Samoan children we "sample" is a *statistic* ($\bar{x}$). The average height (mean height) of all 10-year-old Samoan children is the *parameter* (σ).

(13) ● **Practice 1** Choose the correct response:

(a) An actual value for a population is known as a
- (1) mean
- (2) parameter
- (3) hypothesized value
- (4) statistic

(b) A statistic is a(n) ____ value.
- (1) unknown
- (2) population
- (3) sample
- (4) universe
- (5) hypothesized

(c) In inferential statistics the statistic is to the sample as the ____ is to the population.
(1) parameter
(2) correlation
(3) limit
(4) validity

(d) If the average weight of 1-day-old babies in Cleveland, Ohio, is to be estimated from measures made on a few of them, all 1-day-old babies in Cleveland make up the
(1) sample
(2) range of values
(3) population
(4) frequency distribution

See also Exercises 1, 2.

B. FREQUENCY DISTRIBUTION: FREQUENCY POLYGON AND HISTOGRAM

Objective 2: Frequency distribution, histogram, frequency polygon

(14) Statistics are more understandable when they are presented in a way that allows the reader to see a coherent picture instead of reading listed masses of apparently unrelated numbers. A chart or table that indicates how many times each score (or interval of scores) occurs is called a *frequency distribution.* Two of the most straightforward graphical techniques for presenting statistical data in the form of a frequency distribution are the *frequency polygon* and the *histogram.*

(15) ◆ **Example 2** Arrange the following data as a frequency distribution, and then construct a frequency polygon and a histogram: Scores earned on a class project: 100, 200, 400, 100, 500, 300, 300, 500, 300, 200, 300, 500, 400, 200, 300, 300, 500.

Frequency Distribution

Score	*Frequency of score*
100	2
200	3
300	6
400	2
500	4
	17

(16) For both the frequency polygon and the histogram, the numbers along the vertical axis represent the number of times (*frequency*) that a given score* on the horizontal axis has occurred. Notice in the frequency

*A *score* in this chapter is defined as any number obtained by or assigned to the individuals or objects under study.

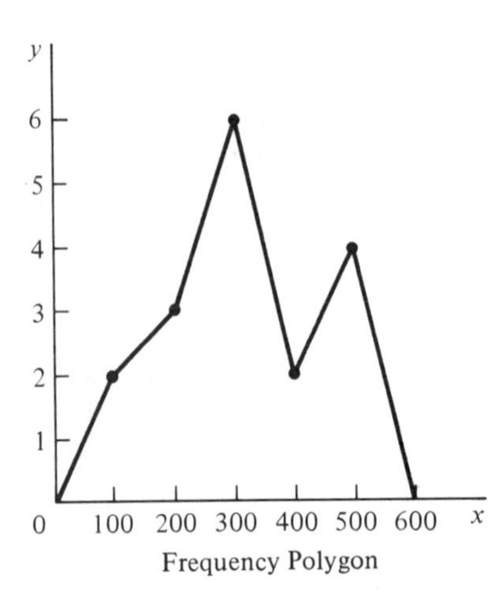

Frequency Polygon

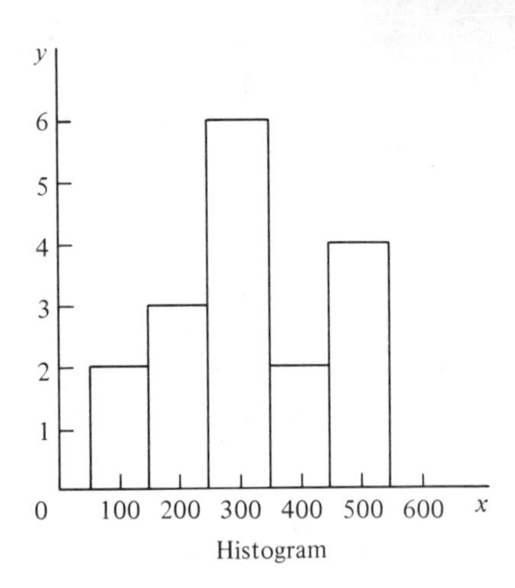

Histogram

polygon that the polygon starts at zero and also ends at zero in order to complete the polygon around the area defined by the given frequencies and scores. In the histogram a single score of 100 is defined as the center of a distance on the baseline from 50 through 150.

● **Practice 2** Arrange each set of data as a frequency distribution and construct a frequency polygon and a histogram: **(17)**

(a) Scores for a group of 8th grade students on an IQ test: 85, 120, 100, 105, 85, 90, 120, 120, 100, 100, 105, 90, 105, 120.

(b) Job applicants' scores on a 35-item math test: 21, 25, 15, 35, 25, 30, 32, 17, 30, 21, 17, 17, 25, 32, 35, 25, 30, 30.

See also Exercise 3.

C. MEASURES OF CENTRAL TENDENCY: MEAN, MEDIAN, MODE

The frequency polygon and histogram present *every* score from a frequency distribution. While these and similar methods are helpful, they often do not provide as much information as we need. Simple inspection of two frequency distributions, for example, may not enable us to make useful comparisons between them. Statisticians, therefore, have developed methods for obtaining a single number to represent a significant characteristic of an entire frequency distribution. This single number may or may not be directly related to actual scores in the distribution. Since such single numbers must represent vital characteristics of entire distributions, it is important that they be as representative as possible of each case in the distribution. Some single numbers which are used to describe primary **(18)**

Measures of central tendency

characteristics of a group of scores are called *measures of central tendency*, the statistical estimate of the central or most representative parameter describing one aspect of a population. Typical measures of central tendency include the mean, median, and mode.

Objective 3: Mode

(19) One score that is sometimes used to represent a group of scores is called the *mode*. When there are only a few scores in a sample, the mode may be only a very rough estimate of the most typical score. The crudeness of the mode for small samples thus makes it less useful than the other two measures of central tendency—the median and the mean. Practically, the mode is used primarily to describe the most "popular" score in a sample. For example, the mode might be used when determining the most popular record sold or the best-selling model of calculator.

◆ **Example 3** Determine the modal score for the following set of data: 60, 85, 89, 82, 85, 90, 92, 97, 93, 82, 87, 85. (20)

First, arrange the scores (from smallest to largest or largest to smallest). Then, record the frequency of each score:

Score	*Frequency*	*Score*	*Frequency*
60	1	90	1
82	2	92	1
85	3	93	1
87	1	97	1
89	1		

The score which appears most frequently is 85, with a frequency of 3. Therefore, 85 is the modal score of the sample.

● **Practice 3** For each set of data determine the mode: (21)

(a) The number of students enrolled in each of the 10 sections of freshman math is: 30, 32, 27, 25, 32, 35, 27, 32, 24, 23. Determine the modal number of students in freshman math classes.

(b) The number of phone calls received by the information desk at a college during each of the first 14 days of March is: 80, 97, 94, 72, 97, 81, 83, 97, 102, 78, 80, 81, 88, 97. Determine the modal number of calls at the college for the indicated time period.

See also Exercise 4.

(22) As the practice problems indicated, the mode may be a score at the extreme high or low end of the scores and, therefore, not very representative of the group as a whole. A more useful value that is more representative of the middle of a distribution is the *median*. The median is the middle value which divides the sample into two groups of equal size—one containing all the scores higher than the median and the other containing all the scores lower than the median.

Objective 4: Median

◆ Example 4

(a) The following achievement test sample contains an *odd number* (9) *of scores*. When the scores are *arranged in order*, we find that the median value would be 80 (the fifth score, counting either from the top or from the bottom of the ordered scores). Notice that four of the scores are above 80 and four are below this median value. **(23)**

Person	*Achievement test score*	*Counting from highest to lowest*	*Counting from lowest to highest*
Alice	100	1	9
Juan	90	2	8
Bill	90	3	7
George	90	4	6
Mary	80 (median score)	5	5
Sam	70	6	4
Bob	60	7	3
Sue	50	8	2
Oscar	45	9	1

(b) When there are even numbers of scores in samples, then there is no middle score which divides the ordered scores into two groups of equal size above and below that middle score. The median must therefore be computed, and it will be a new number which does not even appear in the sample. Finding this median value requires that we add the two middle scores of the sample and divide by two. **(24)**

Suppose we rework the example above but omit Oscar's score (leaving 8 scores). As illustrated below, we would first count up to the fourth (80) and fifth (90) scores, add the two scores together $(80 + 90 = 170)$, and divide by two $\left(\frac{170}{2}\right)$. We would thus obtain 85 as the median value:

Person	*Achievement test score*	*Counting from smallest to largest*
Alice	100	8
Juan	90	7
Bill	90	6
George	90	5
Mary	80	4
Sam	70	3
Bob	60	2
Sue	50	1

Fourth score = 80
Fifth score = +90
170 (sum)

$$170 \text{ divided by } 2 = \frac{170}{2} = 85 \quad \text{(median value)}$$

Half of the scores (four) fall above, and half of the scores fall below the median value of 85.

● **Practice 4** Calculate the median for each of the following samples: (25)

(a) 8, 7, 5, 3, 6, 2, 9, 3, 2
(b) 98, 84, 67, 75, 91
(c) 6, 6, 6, 6
(d) 3, 6, 8, 10, 12, 3, 7, 9

See also Exercise 5.

If one is working with small samples, the median is usually a more useful statistic for estimating the central population parameter than is the mode. In many cases, however, the median may not be as useful as the *arithmetic mean* (called simply the *mean*), which is the arithmetic average of all of the scores in a sample. (26)

Objective 5: Mean

The mean (denoted $\bar{x}$ for a sample) is the most commonly used measure of central tendency in the social sciences (history, sociology) and behavioral sciences (psychology).

◆ **Example 5** Suppose that a class had a member who had worked extremely diligently and sold $1,000 worth of greeting cards for a class project. Compute the mean of the distribution given with and without the $1,000 figure. (27)

(a) Computation of arithmetic mean without the $1,000:

Person	*Sale in dollars*
Alice	100
Juan	90
Bill	90
George	90
Mary	80
Sam	70
Bob	60
Sue	50
Oscar	45
Sum	675

$$\text{Arithmetic mean} \quad (\bar{x}) = \frac{\text{sum of scores}}{\text{number of scores}} = \frac{675}{9} = 75$$

Mathematical symbols are often used in the calculation of statistical tendencies. The same data above would be presented using mathematical symbols as follows: (28)

Summation symbols

$$
\begin{aligned}
x_1 &= 100\\
x_2 &= 90\\
x_3 &= 90\\
x_4 &= 90\\
x_5 &= 80\\
x_6 &= 70\\
x_7 &= 60\\
x_8 &= 50\\
x_9 &= 45\\
\hline
\sum_{i=1}^{n} x_i &= 675
\end{aligned}
\qquad
\bar{x} = \frac{\sum_{i=1}^{n} x_i}{n} = \frac{675}{9} = 75
$$

The Greek capital letter sigma (Σ) means "the summation of"; x_i refers to all the individual scores $x_1, x_2, x_3, x_4, \ldots, x_n$; "$n$" indicates the number of cases. Thus, the notation $\sum_{i=1}^{n} x_i$ means to sum all the scores, starting with the first ($i = 1$) score and stopping with the nth score (which in our example is the 9th score).

(b) Computation of the mean (arithmetic mean) including the $1,000. **(29)**

Person	*Sale in dollars*	or *Using mathematical symbols*		
Henry	1000	x_1	=	1000
Alice	100	x_2	=	100
Juan	90	x_3	=	90
Bill	90	x_4	=	90
George	90	x_5	=	90
Mary	80	x_6	=	80
Sam	70	x_7	=	70
Bob	60	x_8	=	60
Sue	50	x_9	=	50
Oscar	45	x_{10}	=	45
	Sum = 1675	$\sum_{i=1}^{n} x_i$	=	1675

$$
\begin{aligned}
\bar{x} &= \frac{\text{sum of scores}}{\text{number of scores}} \\
&= \frac{1675}{10} \\
\bar{x} &= 167.5
\end{aligned}
\qquad
\begin{aligned}
\bar{x} &= \frac{\sum_{i=1}^{n} x_i}{n} \\
&= \frac{1675}{10} \\
\bar{x} &= 167.5
\end{aligned}
$$

By including Henry's sale of $1,000, we have a sum of $1,675 instead of $675. Dividing by 10 (the new number of cases) instead of nine, we find the mean to be $167.50 instead of the previous $75.

The preceding example illustrates the great effect a single extreme score can have on the mean, which is the major disadvantage of the mean. In cases where extreme scores are present, the median is a more useful estimate of central tendency than is the mean. When a large number of cases is considered, however, this kind of distribution would be extremely rare, and a single extreme case would not have such a great effect on the mean. Actually, when the number of scores considered is large enough to form a normal curve (our next topic), then the mean, the median, and the mode would all equal (approximately) the same value. (30)

Among statisticians there is an expression that "statistics don't lie, but liars use statistics." In the preceding example, if the instructor wanted the entire class to look good, she could report the mean ($\bar{x}$) of $167.50. If she wanted people to be aware of the typical student's performance, she would report the median, which in this case would be $85. The mode in either case above would be $90. In other words, there can be many estimates of a particular parameter, all valid, but only one of which may be really relatively accurate. Because of this, the statistician or the user of statistics should decide which statistic, for his purposes, provides the most accurate estimate of the population parameter. In the example above, the teacher's legitimate purpose may be to make the class look as good as possible, or her purpose may be to represent typical student performance. As has already been shown, her answers would be very different.

● **Practice 5** Compute the arithmetic mean (using mathematical symbols) for the following sets of data: (It is not necessary to arrange the scores in order of size before summing them). (31)

(a) 65, 50, 45, 42, 60, 40, 55

(b) 45, 55, 20, 150, 80, 10

(c) 3, 8, 10, 14, 150

(d) 30, 200, 45, 330, 25, 20, 400

See also Exercise 6.

D. NORMAL CURVE

Suppose we were given the following frequency distribution of persons and weight categories: (32)

Weight (in pounds)	*Number of people (frequency)*
135–139	10
140–144	25
145–149	48
150–154	51
155–159	30
160–164	9

The above data could then be expressed as a histogram:

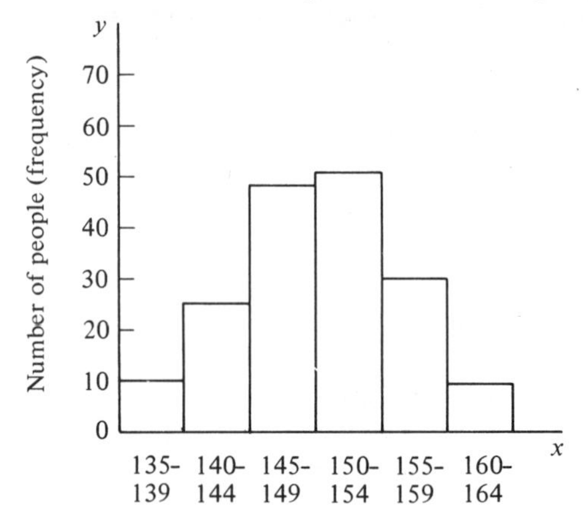

Weight in pounds

Instead of each category of weight being a cluster of five pounds, suppose we knew the exact number of people who weighed each separate pound between 135 pounds and 164 pounds, as follows:

Weight (in pounds)	*Number of people (frequency)*	*Weight (in pounds)*	*Number of people (frequency)*
135	1	150	11
136	1	151	11
137	1	152	10
138	2	153	10
139	2	154	9
140	3	155	8
141	4	156	7
142	5	157	6
143	6	158	5
144	7	159	4
145	8	160	3
146	9	161	2
147	10	162	2
148	10	163	1
149	11	164	1

This more complete data expressed as a histogram would be as shown at the top of page 271.

(33) If we sampled a very large number of persons and used very small weight categories (say one pound each), the resulting histogram would form a curve that resembles the shape of a bell. Such a smooth and continuous "bell-shaped" curve is called the *normal curve*. For most human characteristics, frequency distributions based on large samples take the shape of the normal curve, which is sometimes also called the *Gaussian*

Objective 6: Characteristics of the normal curve

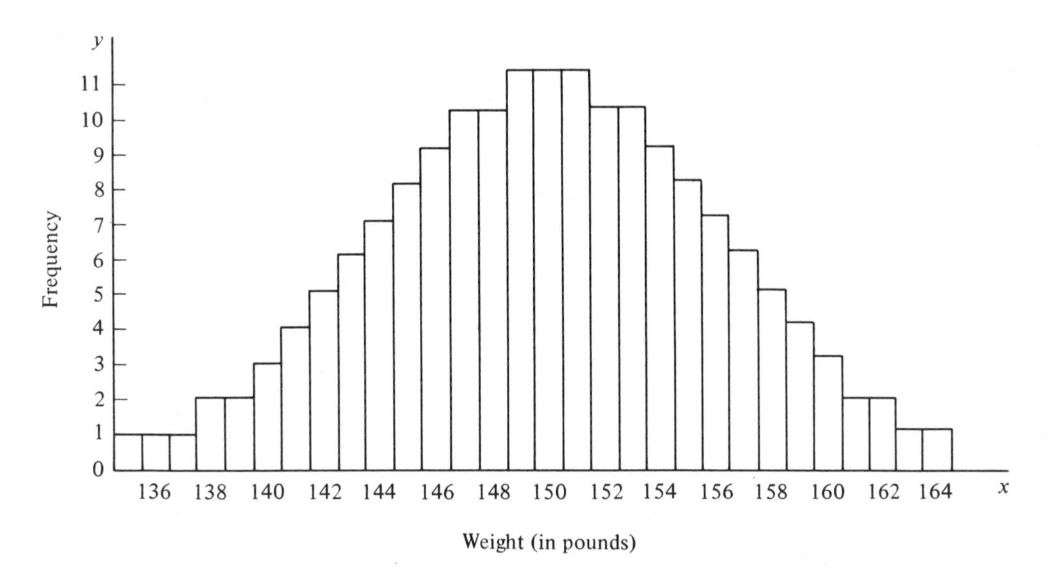

curve. The normal curve has definite mathematical characteristics, some of which are described below. For example, ten pennies shaken together and dropped an infinite number of times would most frequently produce 5 heads and 5 tails (the center of the curve) and least frequently produce 10 heads or 10 tails (the extreme ends of the curve) with other combinations distributed between.

All scores such as those above when taken together form the normal curve, also called the *probability* curve. It is *symmetrical* (balanced) about the mean. It is *asymptotic* to the baseline. (The term "asymptotic" means that the tails of the curve approach the baseline as it extends toward infinity but these tails never touch or cross the baseline.) The normal curve represents an infinite number of cases in any situation where the scores are assumed to be distributed by chance alone.

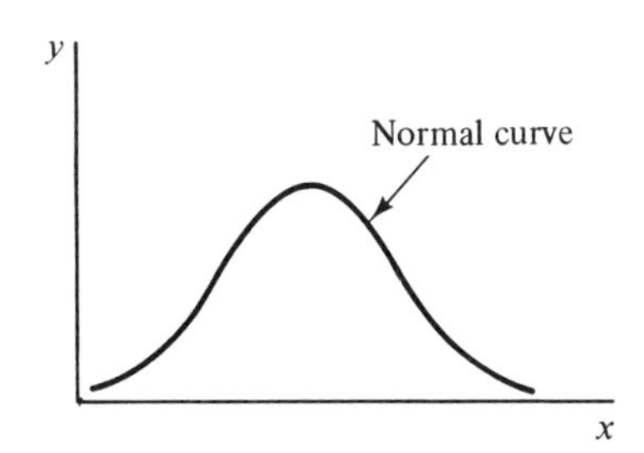

◆ **Example 6** Below is a normal curve with certain important characteristics noted. We will use this diagram in discussing four characteristics of the normal curve: (34)

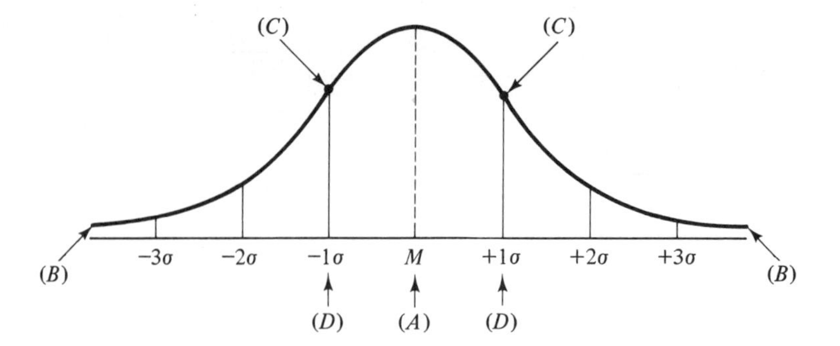

1. The normal curve is symmetrical about the mean, M, of the distribution. (Notice that the mean, median, and mode of the normal curve have the same value.) There are thus an equal number of scores on either side of the mean. The mean is the most frequently occurring score, the average score, and the middle most score.
2. The *tails* (extending ends, labeled B in the diagram) of a normal curve never touch the baseline (or line of abscissas) because the number of cases needed to form a normal curve is infinite. The tails of a normal curve are asymptotic to the baseline; that is, mathematically, both tails of the curve approach infinity without reaching points of intersection with the baseline.
3. There are two points on the normal curve where the curve changes direction (as viewed from above the curve) from convex to concave.* The two points (labeled C in the diagram) are called *points of inflection.*
4. Observe that perpendicular lines are drawn from the two points of inflection to the base line. This determines a unit of distance (or deviation) from the mean to the point where each perpendicular crosses the baseline. This unit of distance from the mean is called a *standard deviation,* which is denoted by the Greek letter sigma (σ) or the Latin letter, s. This "standard" distance can be used in dividing the baseline into several equal segments. Few cases in a normal curve occur more than plus-or-minus 3 standard deviations ($\pm 3\sigma$) beyond the mean. For all practical purposes, cases beyond $\pm 3\sigma$ are generally not considered.

*A convex line (∩) is rounded similar to the outside of the top of a circle, while a concave line (∪) is similar to the inside of the bottom of a circle.

Practice 6

(a) Which characteristics are *not* typical of the normal curve? (35)

(1) A point of inflection marks the $+1$ standard deviation.
(2) Mean, median, and standard deviation are identical.
(3) The curve is asymptotic.
(4) The curve is bell shaped.
(5) The curve represents the chance occurrence of an infinite number of cases.

(b) Draw a normal curve and label the following characteristics:

(1) The curve is symmetrical about the mean.
(2) The tails of a normal curve are asymptotic to the line of abscissas or baseline.
(3) There are two points of inflection. Label the approximate points.
(4) A perpendicular dropped from the points of inflection mark off $\pm 1\sigma$. Use these standard distances in marking off the curve into $+1\sigma$, $+2\sigma$, $+3\sigma$, and -1σ, -2σ, -3σ.

See also Exercise 7.

Objective 7: Area of the normal curve

(36) Theoretically, the total area under the normal curve approaches 100% (or 1 or unity). Between the mean and $+1\sigma$ from the mean is 34.13% of the total area under the curve.

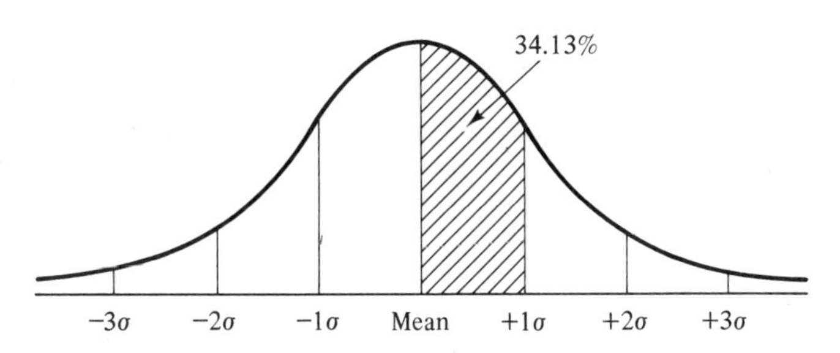

Since the normal curve is symmetric about the mean, then 34.13% of the area under the curve also lies between the mean and -1σ. Therefore, between $+1\sigma$ and -1σ lies 68.26% (34.13% + 34.13%) of the area of the total curve.

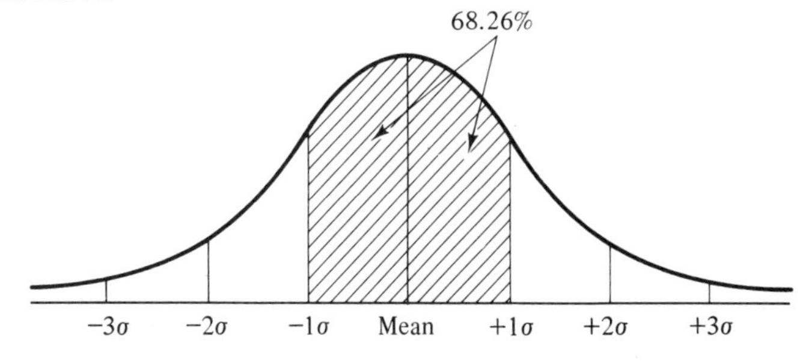

Although the normal curve extends infinitely to the left and to the right, the curve approaches the line of abscissas so closely that about 95.44% of the area under the curve is between -2σ and $+2\sigma$ from the mean. Also, 99.74% of the area under the curve lies between -3σ and $+3\sigma$ from the mean.

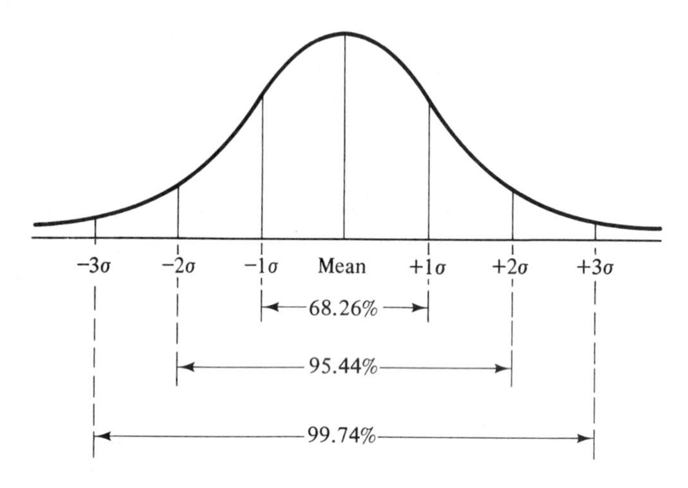

Since the portion of total cases that exist beyond $\pm 3\sigma$ is so slight (0.26%), data is often treated as if 100% of the total cases fall between $\pm 3\sigma$ Therefore, we could round off our percents as follows:

Between $\pm 1\sigma$ is 68.26% of the curve—this is rounded to 68%.

Between $\pm 2\sigma$ is 95.44% of the curve—this is rounded to 95%.

Between $\pm 3\sigma$ is 99.74% of the curve—this is rounded to 100%.

Objective 8: Percent of normal curve greater or less than a score or between two scores ◆

Example 7 In examining a set of normally distributed test scores, it is found that the mean is 60 and the standard deviation is 10. What percent of people equaled or exceeded the score of 70 on the test? (Draw the normal curve.) (37)

A general drawing of the normal curve would be as follows:

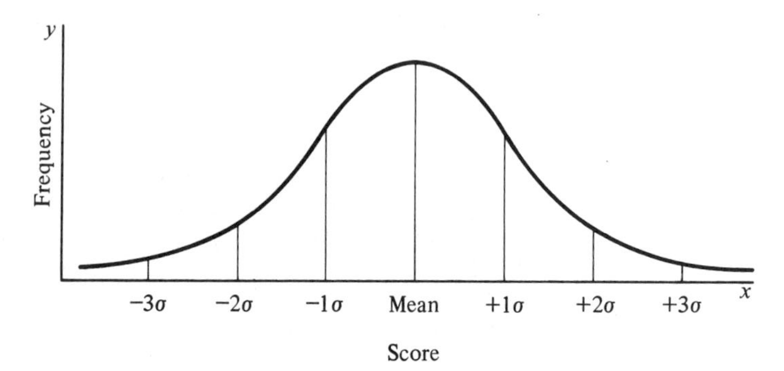

Since we know that the mean in our example is 60 and the standard deviation is 10, we can draw the curve more specifically. Since the σ is 10, the score associated with $+1\sigma$ is 70 (mean plus 1 standard deviation, or $60 + 10$). The score associated with -1σ is 50 (mean minus 1 standard deviation, or $60 - 10$). Similarly, the score associated with $+2\sigma$ is 80 (mean plus 2σ), with $+3\sigma$ is 90 (mean plus 3σ), with -2σ is 40 (mean minus 2σ), and with -3σ is 30 (mean minus 3σ).

We can thus draw our normal curve with this specific information.

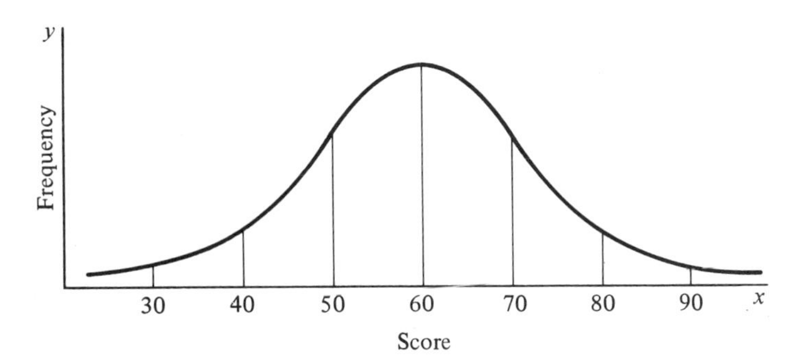

Shading in the normal curve area shows the people whose scores equaled or exceeded the score of 70.

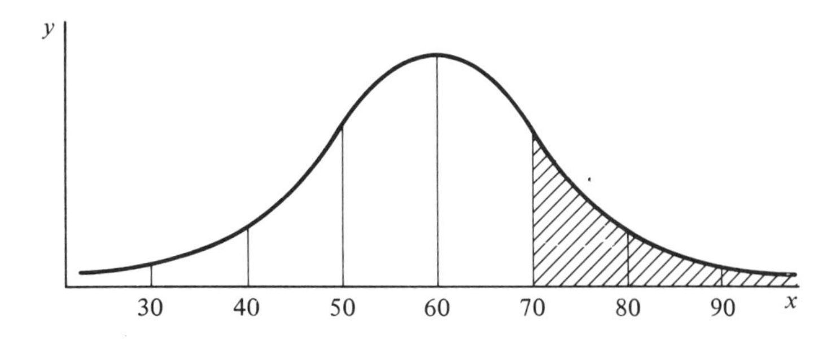

In our discussion of the normal curve, we have assumed that 100% of the total cases fall between $\pm 3\sigma$. We have also seen that the normal curve is symmetrical about the mean. Therefore, 50% of the cases lie to the right of the mean and 50% of the cases lie to the left of the mean. Between the mean and $+1\sigma$ of the mean (in our case a score of 70) is 34% of the total area. Therefore, the percent of people who equaled or exceeded the score of 70 in the test is: **(38)**

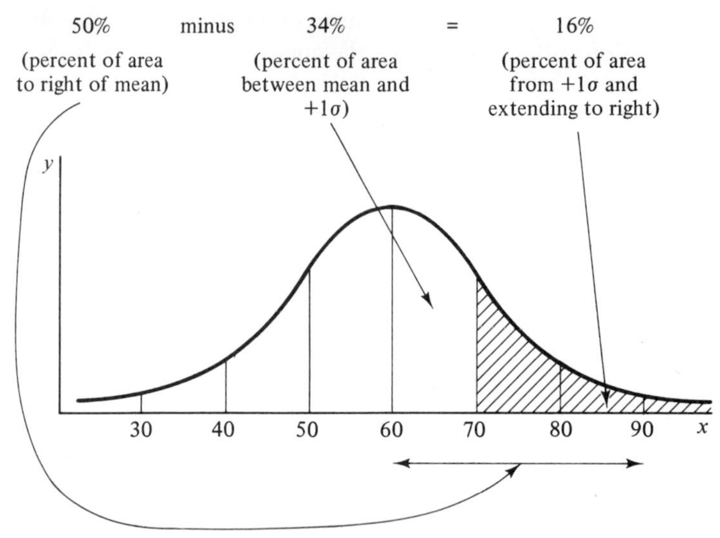

● **Practice 7**

(a) The math and verbal scales on the Scholastic Aptitude Test (SAT) each have a mean of 500 and a standard deviation of 100. A person with an SAT math score of 600 equals or exceeds what per cent of the test population? Illustrate your answer by drawing and labeling a normal curve. **(39)**

(b) A biology student received scores of 78, 75, and 85 on test I, test II, and the final exam, respectively. These tests had means and standard deviations as shown:

Test	*Mean*	*Standard deviation*
I	75	3
II	73	2
Final	80	5

In talking with a friend, this student said he did better on the final than on either of the other two tests. Do you agree? Illustrate your answer by drawing and labeling the three normal curves.

See also Exercises 8, 9, 10.

E. MEASURES OF VARIABILITY: RANGE, STANDARD DEVIATION, AVERAGE DEVIATION, VARIANCE

When a single score is used to represent a sample, we can add to our information about the entire sample by devising some measure of how scores scatter around (or vary from) the measure of central tendency. In all cases **(40)** here, we will consider how the scores scatter around the arithmetic mean. One of the simplest (and consequently one of the most limited) measures of variation (or dispersion) is the *range* (R). The range is simply the difference between the lowest and the highest score.

Objective 9: Range

◆ **Example 8** Calculate the range for the following set of data: (41)

Person	*Age*	*Person*	*Age*
R. B.	100	L. S.	66
S. T.	85	B. W.	65
E. W.	80	H. A.	60

The range (R) is simply the difference between the lowest score (60) and the highest score (100).

$$R = 100 - 60$$

$$R = 40$$

When the range is reported, the highest and the lowest scores are also usually stated, along with the measure of central tendency (the mean). For the data in this example we would thus say that the range is 40, the mean is 76, and the extreme scores are 60 and 100. (42)

● **Practice 8** Calculate the *range* for the following samples: (43)

(a) 6′5″; 5′6″; 5′8″; 6′6″; 5′10″

(b) 200 miles; 342 miles; 267 miles; 165 miles; 416 miles; 315 miles

See also Exercise 11.

A more useful measure of dispersion (or variation) gives an estimate of the average deviation from the mean that exists for the entire group of scores. In other words, the average deviation is a type of arithmetic mean of all the individual deviations from the mean of the samples. It tells how far, on the average, all single scores deviate from the mean score. This is an important concept and may be illustrated through the computation of a statistic called the *average deviation* (AD) or *mean deviation* (MD). (44)

Objective 10: Average deviation

◆ **Example 9** Suppose we had the following set of scores: 2, 3, 5, 6. The arithmetic mean ($\bar{x}$) of this set of scores is: (45)

$$\bar{x} = \frac{\sum_{i=1}^{4} x_i}{n} = \frac{2 + 3 + 5 + 6}{4} = \frac{16}{4} = 4$$

Let us find the distance (or deviation) which each score varies from the mean:

(I) Score	(II) Score minus the mean	(III) Deviation (D) of score from mean	
6	6 − 4	+2	This column contains the distance in
5	5 − 4	+1	score units that each
3	3 − 4	−1	score is above (+) or below
2	2 − 4	−2	(−) the mean.

Note: As Example 9 illustrates, the algebraic sum of all the deviations from the mean is always zero. This is an important characteristic of the mean. When we find the arithmetic mean of column III, *ignoring* the signs, this determines the average deviation: (46)

$$\text{AD} = \frac{2 + 1 + 1 + 2}{4} = 1.5$$

We may now interpret our information thus far in the following fashion: "The sample of scores has an arithmetic mean of 4; and each score deviates, on the average, 1.5 units from the mean."

● **Practice 9** Calculate the arithmetic mean for the following data. Set up a chart with columns labeled "score," "score minus mean," and "deviation of score from mean." Then calculate the average deviation (AD) for the data. (47)

(a) 7, 9, 6, 10, 8, 7, 5, 4
(b) 10, 70, 60, 60, 30
(c) 27, 20, 13, 5, 35, 15, 25
(d) 10, 2, 18, 13, 7

See also Exercise 12.

Objective 11: Standard deviation and variance

A simple modification of the computation in the example above provides us with the most important of the measures of dispersion, the *standard deviation* (s or σ), which we used briefly in analyzing the normal curve. As noted in the preceding example, the sum of all "deviations from the mean" is zero. This inconvenience can be avoided by squaring each deviation, as an intermediate step, and later taking the square root of their sum-divided-by-n. This process "gets us back," as it were, to an average of the deviations. If we did *not* take the square root of the sum-divided-by-n, we would have another statistic called, appropriately enough, the *variance* (s^2 or σ^2). (48)

◆ **Example 10** Calculate the variance and the standard deviation of the following scores: 2, 3, 5, 6. (49)

(a) First, we calculate the arithmetic mean as before:

$$\bar{x} = \frac{\sum_{i=1}^{4} x_i}{n} = \frac{2+3+5+6}{4} = \frac{16}{4} = 4$$

(b) Then, we set up the data in table form:

(I) Score	(II) *Score minus the mean*	(III) *Deviation (D) of score from mean*	(IV) *Square of deviation from mean (D^2)*
6	6 − 4	+2	+4
5	5 − 4	+1	+1
3	3 − 4	−1	+1
2	2 − 4	−2	+4

(c) The variance is the sum of the squares of all the individual deviations from the mean (the sum of column IV), divided by the number of scores. The general formula for variance is:

$$\sigma^2 = \frac{\sum_{i=1}^{n} D_i^2}{n}$$

Where σ^2 means variance,
D_1 is the deviation of each score from the mean, and
n is the number of cases (or scores).

Therefore, $$\sigma^2 = \frac{\sum_{i=1}^{4} D_i^2}{n} = \frac{4+1+1+4}{4} = \frac{10}{4}$$

$$= 2.5$$

(d) The standard deviation is the square root of the variance. General formula:

$$\sigma = \sqrt{\frac{\sum_{i=1}^{n} D_i^2}{n}}$$

Therefore,

$$\sigma = \sqrt{\frac{\sum_{i=1}^{4} D_i^2}{n}} = \sqrt{\frac{4+1+1+4}{4}}$$

$$= \sqrt{\frac{10}{4}} = \frac{\sqrt{10}}{2} \approx \frac{3.16}{2} \approx 1.58^*$$

$$\text{or} = \sqrt{2.5}^*$$

*Square roots that are not perfect square roots can be looked up in a table of square roots and your answer given as an approximate (≈) square root. However, for purposes of these exercises, we suggest that you leave approximate roots in radical form ($\sqrt{2.5}$).

● **Practice 10** Calculate the variance and the standard deviation for the following data. Set up the data in four columns as illustrated in Example 10. (50)

(a) 10, 20, 15, 25, 5.
(b) 6, 6, 6, 6.
(c) 4, 3, 3, 6, 5, 5, 2.

See also Exercise 13.

Relationship between range and standard deviation

If a distribution of sample scores is approximately normal (resembles a normal curve), the range may be used to obtain an approximation of the standard deviation. The relationship between the range and the standard deviation is: one sixth of the range approximately equals the standard deviation. Or, expressed in symbols: (51)

$$\frac{1}{6}R \approx \sigma$$

F. RELATIONSHIPS OF VARIABLES: CORRELATION AND REGRESSION

Objective 12: Correlation

Previous material in this chapter dealt with descriptive statistics in which mean, median, mode, or standard deviation were used to describe only a single parameter. We are now concerned with describing the relationship between two statistics, for example, between two sets of test scores. Thus, the major question of concern is, "How do two sets of scores on the same individuals relate to each other; that is, how do they *correlate*?" Many variables or events in nature are related to each other. As the sun rises, the day warms up; as children age, they think in more complex ways. Such relationships are called correlations. The relationship of one variable to another is known as a correlation. (52)

Positive and negative correlation

If the river rises when it rains, the two events are said to have a positive correlation. That is, when an *increase* in one variable coincides with an *increase* in another variable, the two variables have a *positive correlation.* By contrast, altitude and air pressure illustrate negative correlation: the greater the altitude, the less the air pressure. When an *increase* in one variable coincides with a *decrease* in another variable, the two variables have a *negative correlation.* (53)

In mathematics, a positive relationship or correlation is also called a *direct relationship* or *direct correspondence.* A negative correlation—as one variable increases (altitude) another variable decreases (air pressure)—is called *inverse correspondence* (relationship). Each relationship—direct or inverse correspondence—can be expressed as a linear equation in two variables, which graphs as a straight line. This ability to depict an algebraic (54) representation of the relationship between two variables enables statisticians to engage in a variety of processes useful in advancing knowledge. For purposes of this text, we will examine only the "Pearson product-moment correlation coefficient" (Pearsonian *r*) and its use in estimating relationships between scores or events. This process is called regression analysis, which we will also consider briefly.

Correlation and causation

(55) When there is a high correlation between two variables, we can estimate the values for one variable from those of the other. If there is a high positive correlation between drownings at the beach and ice cream sales, we can estimate that as ice cream sales increase the number of drownings will increase. We can estimate the occurrence of one event from another event, but we *cannot* say that one event causes the other event. There is a positive correlation between the number of drownings per day and ice cream sales, but drownings do *not* cause the ice cream sales or vice versa. A third variable, heat, may be the cause of both events.

(56) The distinction between correlation and causation is important. Correlation is a measure estimating relationship, association, or correspondence. As one variable changes, the other changes. Correlation is *not* the same as causation. Thus, if we set two accurate clocks (A and B) and start them simultaneously, then lists of minutes ("scores") taken at the same time from each clock would correlate perfectly. That is, we could estimate the time shown on one clock by knowing the time shown on the other (regardless of whether or not they had been set to read the same time) since these clocks bore, by our definition, a perfect relationship with each other. We could not, however, say that the action of clock A *caused* the time shown on clock B. The failure to make this distinction probably leads to much superstitious behavior, such as a child's thinking that he'll have bad luck if he walks under a ladder or if a "black cat" crosses his path. Thus, causation (or reason for correlation) must be determined logically, rather than simply being assumed because a correlation exists. In short, you must have a reasonable explanation and demonstrate it before you can claim "cause and effect" from a correlation between two variables.

(57) A final comment about correlation is that there must be a common link between the variables being correlated. If two tests are to be correlated, the same person or persons (or random samples drawn from the same population) must take both tests, and their corresponding scores must be matched for the variables which you want to compare.

(58) ◆ **Example 11** Indicate for which of the following pairs of variables one would ordinarily compute a correlation:

(a) IQ and achievement test scores for a class of students.

One could compute a correlation for these data—the assumption being that there is a relationship between a person's IQ and achievement test scores.

(b) The amount of rainfall in Sidney, Australia, and the number of babies born in Houston, Texas.

One would not ordinarily compute a correlation for these data—the assumption being that there is no relationship between these two sets of data.

Practice 11 Indicate for which of the following pairs of variables one would ordinarily compute a correlation: (59)

(a) Air temperature and number of persons in a swimming pool at various times.
(b) Advertising costs of a company and sales of a product.
(c) The number of letters in a person's name and the person's height.

See also Exercise 14.

A number of statistical indices (terms or measures) have been developed which quantify the relationship between two sets of data. Pairs of measurements (such as two test scores, height and weight, etc.) which yield two sets of related data are called *bivariate* (or "two-variable") data. The most commonly used measure of correlation is the *Pearson product-moment coefficient of correlation*, designated r when computed for samples. This correlation coefficient is commonly called the *Pearsonian r*. The value of r indicates the degree to which the relationship between two variables can be represented by a straight line. (60)

The higher a correlation, the closer a graph of the scores approaches a single line. The variable X in the diagrams below has the highest r with the variable A, because the scores in that graph (part II) more nearly describe a straight line. Perfect correspondence or relationship would define a single line, which is formed by graphing each pair of corresponding variables on a two-dimensional (Cartesian) plane. If the correlation is positive (direct), the line will rise as it moves toward the right (╱); if there is a negative (inverse) correlation, the line will fall as it moves toward the right (╲).

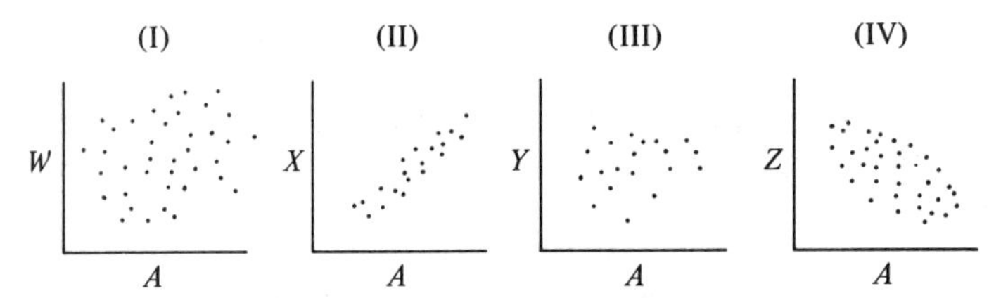

A Pearsonian r is a correlation coefficient (or index or value) bounded by the mathematical limits (values) of -1 and $+1$. In other words, the greatest magnitude of r is $+1$ or -1. If r equals either plus or minus one (61)

(a situation which would seldom, if ever, occur with psychological or educational data), then the correlation is said to be either a perfect positive correlation (direct or $+1$) or else a perfect negative correlation (inverse or -1). If r equals 0, there is no linear—direct or indirect—relationship or correspondence between the two sets of data. This lack of correspondence between variables is shown in parts I and III of the preceding illustration.

Simply speaking, when correlation coefficients approach plus one, we may state that the individuals who obtained *high* values on one set of scores probably also obtained *high* values on the other set of scores or measurements (positive correlation); see part II in previous illustration. Conversely, when r approaches minus one we know that, in general, individuals who score *high* on one of the two measurements probably scored *low* on the other (negative correlation). See part IV in the previous illustration for a negative correlation.

(62) When two variables are related, their corresponding correlation coefficient, as stated above, will exist between the limits of ± 1 or $(-1 \leq r \leq +1)$. A correlation of -1.00 is as great a correlation as $+1.00$. The algebraic sign ($+$ or $-$) of the correlation coefficient indicates the *direction* of the relationship (whether direct or inverse). It is the absolute value of r that indicates the closeness of the relationship. An r of -0.80 is a higher correlation than an r of $+0.65$.

(63) Coefficients of correlation also may be said to be "high, moderate, or low." For example, coefficients of correlation of the size .70 to 1.00 (either plus or minus) are often said to be "high," while coefficients of .20 to .50 are "low," and coefficients of .50 to .70 are "moderate."

Another point concerning the interpretation of a coefficient of correlation is required. Correlation coefficients are *not* measures on a scale of equal units. Thus, the distance from .40 to .50 is *not* the same distance as the distance from .80 to .90. One cannot say that a correlation of $+0.90$ is three times as close a relationship as $+0.30$, but merely that it indicates a much higher degree of relationship. Similarly, two tests having an r of $+0.40$ are not twice as related as two tests having an r of $+0.20$.

Objective 13: Degree and amount of relationship ◆

(64) **Example 12** What statements about the *amount* and about the *direction* of the correlation or the relationship would you make with respect to the following set of data?

Subject	*Test X*	*Test Y*	*Subject*	*Test X*	*Test Y*
A	90	30	F	40	80
B	80	60	G	40	65
C	80	50	H	30	75
D	75	60	I	20	80
E	60	60	J	20	100

Amount?	*Direction?*
high?	positive?
low?	negative?
zero?	

(a) *Interpreting data by noting relationship or degree of correspondence.* The relationship above tends to show the correspondence to be inverse or reciprocal. That is, when X is high, Y is low; or when X is low, Y is high. You can see how the inverse relationship works by observing that high values on Test X have corresponding low values on Test Y.

(b) *Interpreting data by plotting.* Plot points *A* through *J* using Test X and Test Y scores as the coordinates for each subject. For example, subject A graphs as the point (90, 30):

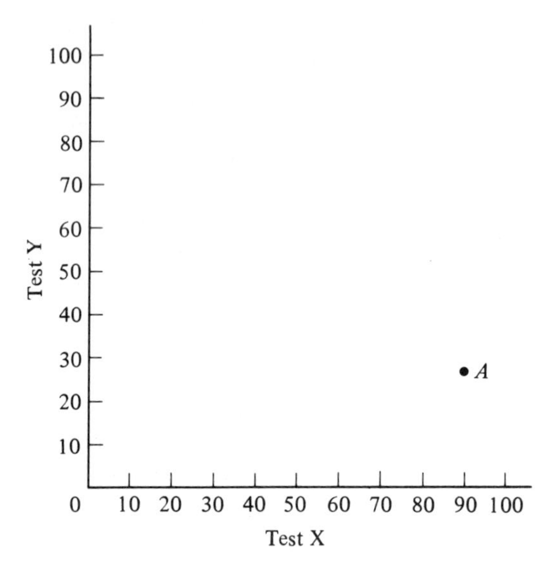

Amount of relationship: high
Direction of relationship: negative (or inverse)

● **Practice 12** What statements about the amount and about the direction of the relationship would you make with respect to the following set of data? (Plot the points.) **(65)**

A group of salespersons were given a test of sales ability and their scores were compared with their sales on the job:

Salesperson	*Test score*	*Sales*
Anderson	30	$ 60
Brown	50	80
Cook	60	75
Davis	70	90
Eastman	90	100

Amount?	*Direction?*
high?	positive?
low?	negative?
zero?	

See also Exercise 15.

Although coefficients of correlation for curvilinear data (data which graphs as a curved line) can be computed, their complexity and interpretation are beyond the scope of this chapter. All of the coefficients mentioned previously assume a straight-line (linear) relationship when graphed; that is, as one set of scores increases, the other set also increases (or else decreases). In some psychological studies (such as correlations between age and some types of memory tests), curvilinear relationships apparently exist. More sophisticated methods of correlation are required when the probable relationship is such that, say, both variables increase together up to a point, and then one decreases as the other increases. On occasion a sizable relationship between sets of data is overlooked because a low linear relationship was observed, when the data are actually curvilinearly related. **(66)**

Regression

Another topic closely allied to correlation is regression. Whereas correlation is concerned primarily with the degree and direction of relationship based on given data for two variables, *regression analysis* involves using a known correlation and known data for one variable in order to estimate a value of a corresponding variable. One might be interested in predicting an individual's score on some variable Y, using knowledge of another variable X and the correlation between X and Y. The prediction of this Y value might be expressed by the following formula: **(67)**

$\hat{Y} = r$ times (number of standard deviations from the mean of a known test score)

$\hat{Y}$ (read "Y hat") is the estimate of a person's deviation from the mean on the second test; and r is the known Pearson product-moment correlation coefficient depicting the degree of correspondence between the two tests being considered.

Objective 14: Estimation and correlation

◆ **Example 13** If we knew that a vocabulary test and an intelligence test correlated .80 and that Joe Jones obtained a vocabulary score which was 2 standard deviations above the mean, determine the best estimate we could make of what Joe's intelligence test score would be. **(68)**

General formula: $\hat{Y} = r$ times (number of standard deviations from the mean of the known test score)

Specific formula: $\hat{Y}$ of intelligence test $= .80(2)$

$= 1.60$

The answer, 1.60, is interpreted as being our best prediction (knowing the vocabulary score) of how many standard deviations Joe's intelligence test score will be from the intelligence test mean.

The preceding example illustrates a general statistical principle, that scores of psychological and educational variables tend to regress (move toward) the mean score of future, correlated tests. Thus, high scores tend to become relatively lower, and low scores tend to become relatively higher on repeated, correlated measurements. This phenomenon is called the *law of mean regression.*

● **Practice 13** If we know that a math test and a job-satisfaction test for computer programmers correlates .75 and that Alice Rubio obtained a math test score of three standard deviations above the mean, what is the best estimate we could make of her job-satisfaction test score? (69)

See also Exercises 16, 17.

G. STATISTICAL SIGNIFICANCE

Objective 15: Statistically significant findings

A primary purpose of *inferential* statistics is to determine the probability that certain relationships or differences exist between samples and general populations. The following statements are similar to those made in inferential statistics: (70)

1. I am 99% sure that men have greater mathematical ability than women.
2. There is better than a 95% chance that intelligence test scores and vocabulary test scores measure approximately the same trait. That is, they are related.
3. I am 99% sure that teaching statistical concepts by a self-paced unit is more effective than teaching the same concepts by the discussion method.
4. I am 80% sure that even brief educational-vocational counseling improves the counselee's perceptions of self.

When statistical results are obtained which could occur by chance alone only 5 times (or less) out of 100, these results are held to be *statistically significant.* The process and data used to determine the results must also be valid; that is, appropriate mathematical models must be used and followed to produce reliable, consistent results. (71)

Generally speaking, results are said to be statistically significant at a level which is the complement of the amount of confidence one can have in those results. Specifically, if we are 99% confident that the results are at least as accurate as they appear to be, then we can say the results are "significant at the 1% level." (This 1% level is more customarily expressed as a decimal: at the .01 level.) It means that the value of the difference would only occur *one* time out of 100 by chance alone. If, on the other hand, we are only 95% sure that the results are what they seem to be, then the results are said to be "significant at the .05 level"; if 70% confident, "significant at the .30 level"; if 80% confident, then ".20 level of significance," etc. Thus, the *higher* the degree of confidence, the *lower* the deci- (72)

mal "level of significance." As noted above, general practice is to accept significance only at the .05 level or below.

Statistical results are, strictly speaking, always significant at some level. Before results are considered "statistically significant," however, researchers in the social and behavioral sciences customarily require 95% confidence (.05 level of significance) or "better" (lower significance such as the .02, .01, or .001 levels). While statistical significance is a vitally important concept, it should be mentioned here that results may be statistically significant at an acceptable level and still be worthless for practical purposes, and vice versa. For example, height and intelligence may be significantly correlated at the .05 level. This finding is virtually meaningless in practical terms, however, since very little can be done to change either a person's height or intelligence. By contrast, intelligence test scores for individuals in certain occupational groups may produce statistically significant differences at a level of only .10. Yet, these results may be of definite practical value when applied in large-scale, personnel-selection programs. Primarily for this reason, researchers often report the actual level of significance obtained, rather than simply dismissing the findings as "not significant" (N. S.) when those results fail to reach the .05 level or some "better" (lower) level of significance. **(73)**

By this time, you may well ask, "Well, this is all fine; but, if I have obtained some statistical results, how would I know what degree of confidence to attach? 95% confidence? 99%? 80%?" The answer to this question involves mathematical probability. If we compute the mathematical probability of obtaining results as extreme as our actual results, and if this computation reveals that, on the average, these results would occur strictly by chance only one time in a hundred, then we can be 99% confident of our results. The actual computation of this probability, however, is beyond the scope of this book, although we will consider some aspects of probability in the following chapter. For our present purposes, we will only interpret levels of significance that have already been computed. **(74)**

To better understand the meaning of statistical-significance levels, consider the following situation: Suppose we found a difference of 10 points between the mean intelligence test scores for "blue-collar" and "white-collar" workers. Assume also that this result is statistically significant at the .01 level (meaning 99% confidence in the results). This means that 99 out of every 100 times when we tested a random sample of such blue- and white-collar workers, we could expect to obtain the same difference between the two groups of intelligence test scores. Notice that it would still be possible for us to be mistaken if we made the unqualified statement that "blue- and white-collar workers differ in intelligence." However, since we are 99% confident of our results (significant at the .01 level), we would be wrong in making such a statement only one time in **(75)**

100, on the average. Similarly, if the results had been significant at only the .05 level, we should expect to be wrong, on the average, five times in a hundred. At the .07 level, we would probably be wrong seven times in one hundred.

◆ **Example 14** While reading the report of a psychological experiment, you notice the following statement: (76)

> "The difference between Group A and Group B was found to be significant at the .05 level."

What does this mean?

The statement concerning significance would mean that this much difference between the scores of Group A and Group B would have occurred five times out of 100 simply by chance. Stated differently, if the study were repeated 100 times, this great a difference could occur only five times by chance alone.

● **Practice 14** Write the meanings of the following statements: (77)

(a) An investigator reported research findings of significance at the .01 level.
(b) The findings were significant at the .02 level.
(c) The results of the study were labeled N. S.

See also Exercise 18.

H. MODEL BUILDING

Objective 16: Model building process

Science is a process whereby models of phenomena (events) in the world are conceived and tested. The process is given below. (78)

1. Through observation, logic, previous knowledge, etc., an individual develops a view or idea concerning a phenomenon, which leads to certain expectations that are stated as hypotheses.
2. These logical structures are then operationalized, which means ultimately stating the phenomenon in numerical form. This then produces a mathematical model.
3. Data are gathered and an attempt is made to see how clearly the data fit the model.
4. As a result of step 3, the hypotheses are rejected or they are not, depending upon how well the model and the data fit together.

Statistical tests, which are themselves mathematical models concerned with decision making, help to make the process of model rejection or nonrejection more objective. The attempt is to find the simplest model to represent the phenomenon which is not rejected. This unrejected model can then be used, but by no means is it necessarily the only or the "best" model of the phenomenon; however, it is "better" than any rejected model.

(79) As was stated at the beginning of the chapter, all of the statistical properties and quantities we have been examining are based upon mathematical, statistical *models*. These models attempt to provide us with good *estimates* of population parameters in any specific case we happen to be examining at the time.

(80) This last statement may be clarified by an example. Recall from our discussion of correlation and regression that a "perfect correlation" (+1.00 or −1.00) produced a graph which resembled a straight line. Using our knowledge of algebra, the equation for a straight line is $Y = aX + b$. Y here is a reference point on the ordinate; X is the reference point on the baseline; a is the *slope* of the line (amount of change in Y per unit X); and b is the intersection of the line with the ordinate when $X = 0$ (commonly called the *y-intercept*). The formula $Y = aX + b$ is thus a specific mathematical and statistical model when applying correlation and regression.

The straight line which would result if there were perfect correlation between two variables is called a *regression line*. This line gives us the best estimate we can make of the population parameter we hope to find, given the one we already know. Even if the correlation between two variables is not perfect, the regression to the mean (described in Example 13) indicates that we can determine where this regression line would lie, and if we have a given value for one variable we can plot the other variable on the regression line.

(81) The most important fact to be realized is that the model above is only one special case of a whole family of models called *linear models* on which a great deal of statistics is based. These models when developed properly and accurately provide a means for making judgments about relationships between sample statistics. By mathematically comparing two similar models, one of which is somewhat simpler than the other, we can decide which is more accurate. Having done this, we can then infer that the more accurate model in some way represents a "more true" picture of the area we are examining.

(82) The specific processes for making these comparisons and inferences are beyond the scope of this chapter. These processes may be examined in detail from two basic points of view in more advanced texts. The classical point of view provides relatively simple computational formulas for the reader to use in making comparisons. The linear-model point of view provides the reader with a logical system for designing his own comparisons, but leaves the computational work to a modern high-speed computer.

EXERCISES

Objective 1: Population, sample, parameter, statistic

1. Discuss the relationships among the following terms: population, sample, parameter, statistic.

Objective 1: Example 1

2. Relate the terms "population," "sample," "parameter," and "statistic" to the following situations:
   (a) The mean weight of all U.S. professional football players is to be estimated from measures made on the N.Y. Jets.
   (b) In Atlanta three hundred random persons over 65 years of age are interviewed in November about their income, in order to aid in establishing for the following year a senior citizens' program based on average income.
   (c) There are 40 students in the ninth grade at Central High School with a tested IQ of 125 or over. These students were all given a questionnaire regarding the average number of hours per week which they spend studying.

Objective 2: Example 2

3. Arrange the following data as a frequency distribution and then construct a histogram and a frequency polygon:
   (a) Scores for math students on a 10-point quiz: 10, 6, 6, 2, 7, 8, 9, 5, 4, 8, 8, 10, 10.
   (b) Scores for persons on an IQ test: 110, 100, 110, 125, 95, 85, 130, 105, 100, 90, 110, 130.
   (c) Number of "heads" when flipping 10 coins at once: 1, 3, 9, 6, 6, 4, 5, 3, 7, 4, 5, 5, 2, 8.

Objective 3: Example 3

4. For each set of data determine the mode:
   (a) The number of persons responding to a newspaper ad the first 10 days the ad was in the paper: 35, 22, 25, 15, 10, 22, 32, 15, 22, 23. Determine the modal number of responses for the indicated period.
   (b) The number of calls requesting room service at 8 a.m. at a hotel for a 14-day period: 30, 31, 17, 23, 28, 18, 17, 28, 27, 8, 31, 12, 26, 31. Determine the modal number of calls for the indicated period.
   (c) The number of persons flying first class on all airlines from Houston to Dallas in the first 15 days of April: 200, 305, 178, 280, 240, 197, 280, 240, 210, 240, 301, 225, 240, 250, 280. Determine the modal number of first-class passengers for the indicated period.

Objective 4: Example 4

5. Calculate the median for the following sets of data:
   (a) 4, 6, 10, 11, 7, 5, 6, 3, 12, 13, 18.
   (b) 23, 43, 21, 29, 27, 41, 37, 33.

(c) 5, 100, 105, 210, 300, 350, 840.
(d) 1, 2, 2, 2, 3, 4, 5, 8, 9, 9, 10.

Objective 5: Example 5

6. Calculate the mean for the following sets of data (use the summation symbols in your calculation):
(a) 6, 1, 3, 7, 9, 10, 11, 2, 4 5, 8.
(b) 6, 6, 9, 11, 14, 14, 13, 12, 8, 7, 7, 13.
(c) 1, 1, 1, 1, 2, 2, 2, 8, 54.
(d) $6,000; $6,900; $5,000; $7,000; $500; $34,000.

Objective 6: Example 6

7. List (and illustrate on a normal curve) four characteristics of the normal curve.

Objective 7: Area under normal curve

8. Draw a normal curve and show the percent of area between:
(a) $+1\sigma$ and -1σ.
(b) $+2\sigma$ and -2σ.
(c) $+3\sigma$ and -3σ.

Objective 8: Example 7

9. Suppose the distribution of ages at which normal children walk approximates a normal curve, with a mean of 15 months and a standard deviation of 2 months. A parent is worried about a child because the child did not walk until he was 17 months old. You could reassure the parent by saying that the portion of normal children who learn to walk when they are 17 months old or older is about what percent? (Illustrate answer with a drawing of the normal curve.)

Objective 8: Example 7

10. The following is a partial list of scores which the same group of students made on two different math exams. Indicate for each student the test on which he scored "higher" (that is, scored better than the larger percent of the other students). (Draw the normal curve.)

Student	*Test A*	*Quiz*
A	53	8
B	60	5
C	47	17
D	79	13
E	81	7
F	39	9
G	56	6
H	66	9

	Test A	*Quiz*
Mean:	60	8
Standard deviation:	7	2

Objective 9: Example 8

11. Calculate the range for the following sets of data:
(a) $380, $550, $275, $840, $670, $565, $950.
(b) 27 years, 23 years, 18 years, 32 years, 30 years, 42 years, 27 years.
(c) 3, 3, 3, 3, 7, 14, 6, 8, 5.
(d) 180 pounds, 175 pounds, 210 pounds, 135 pounds, 200 pounds.

Objective 10: Example 9

12. For each of the following sets of data, calculate the arithmetic mean and the average deviation. (Set up the data in table form.)
 (a) 3, 6, 9, 12, 15, 18, 21.
 (b) 3, 3, 4, 5, 8, 7, 9, 7, 5, 9.
 (c) 2, 2, 2, 2.

Objective 11: Example 10

13. Calculate the variance and the standard deviation for each of the following sets of data. (Set up the data in table form.)
 (a) 50, 45, 60, 85.
 (b) 10, 10, 15, 20, 60, 70, 95.
 (c) 2, 2, 8, 7, 6, 3, 3, 9.

Objective 12: Example 11

14. Indicate whether or not you would ordinarily compute a correlation for each of the following pairs of variables and indicate the reason for your answer.
 (a) A person's IQ score and her grade point average.
 (b) The amount of insulation in the roof of a home and the yearly heating costs for the home.
 (c) The number of letters in a person's first name and his age.

Objective 13: Example 12

15. What statements about the amount and about the direction of the relationship would you make with respect to the following sets of data? (Plot the points on a two-dimensional graph.)
 (a) A correlation was made between the number of books read in one year by individual members of a civic club and the number of days which each of these members spent in outdoor recreation:

Member	*Books read*	*Days of outdoor recreation*
A	1	150
B	4	100
C	12	79
D	27	65
E	70	3
F	35	15
G	24	74
H	6	80
I	30	25
J	28	37

Amount: high? low? zero? *Direction:* positive? negative?

 (b) A correlation was made between the weight of male members of a local YMCA who joined in May and the number of days each used the gym facilities during June, July and August:

Member	*Weight*	*Days person used gym*
A	150	15
B	200	30
C	135	10
D	180	20
E	240	27
F	175	20
G	160	13
H	210	45

Amount: high? low? zero? *Direction:* positive? negative?

Objective 14: Example 13

16. If we knew that a reading test and a math test correlated .60 and that a person obtained a reading score 3 standard deviations above the mean, what is the best estimate we could make of the person's math test score?

Objective 14: Example 13

17. If we knew that a math test and an aptitude test for creative writing correlated −.35 and that a person obtained a math score 2 standard deviations below the mean, what is the best estimate we could make of the person's creative writing aptitude score?

Objective 15: Example 14

18. Write what is meant by the following statements regarding statistical significance.

(a) The difference between the two groups was reported to be significant at the .01 level.

(b) At the end of an article reporting some research findings, you read the following statement: "Results of study—N.S."

(c) Height and weight of 5-year-old girls showed a .75 correlation at the .05 level.

Objective 16

19. List and describe the four steps of the model building process.

OBJECTIVES

After completing this chapter, you should be able to

A. Inferential statistics

1 Define the relationships among the following terms: population, sample, parameter, and statistic. **(6–13)**

B. Frequency distribution: frequency polygon and histogram

2 Given a set of data, arrange the data as a frequency distribution and then construct a histogram and a frequency polygon. (14–17)

C. Measures of central tendency: mean, median, mode

3 Given a set of data, determine the mode. (18–21)

4 Given a set of data, calculate the median. (22–25)

5 Given a set of data, calculate the arithmetic mean (use the summation symbols in your calculation). (26–31)

D. Normal curve

6 List (and illustrate on a normal curve) the four basic characteristics of the normal curve. (32–35)

7 Draw a normal curve and show the percent of area between $\pm 1\sigma$, between $\pm 2\sigma$, and between $\pm 3\sigma$. (36)

8 Given specific test data, calculate the per cent of persons scoring: (37–39)
(a) from a given standard deviation unit "and greater."
(b) from a given standard deviation unit "and less".
(c) between two given standard deviations.

E. Measures of variability: range, standard deviation, average deviation, variance

9 Calculate the range for a given set of data. (40–43)

10 Given a set of data, calculate the arithmetic mean and the average deviation. (44–47)

11 Given a set of data, calculate the standard deviation and the variance. (48–51)

F. Relationships of variables: correlation, regression

12 Given a pair of variables, indicate whether or not you would ordinarily compute a correlation for this pair.

13 Given a set of data, depicting varying values of two related variables, indicate whether there is or appears to be a high or low degree of relationship or correspondence (correlation) and whether the relationship, correspondence, or correlation is positive (direct) or negative (inverse). Plot the points of intersection on a two dimensional graph. (60–66)

14 Given (1) the correlation of two variables and (2) the standard deviation of a given score from the mean for one variable, estimate how many standard deviations from the mean another score would be for the second variable. **(67–69)**

G. Statistical significance

15 Explain the meanings of statistically significant findings. **(70–77)**

H. Model building

16 List and describe the four steps of the model building process. **(78–82)**

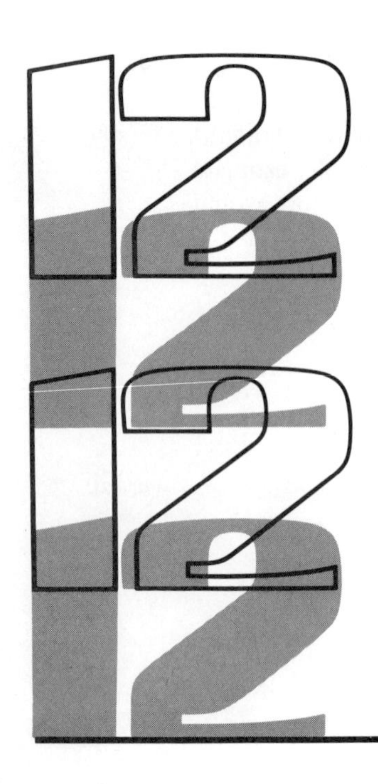

Probability

Even though you might not be planning an extended trip to Las Vegas, you will find that you will use probability in many everyday situations! Indeed, probabilities are still used to understand games of chance, such as the drawing of cards, throwing dice, or playing roulette. However, probabilities also have many practical applications in daily life. For instance, probabilities are used to determine the cost of insurance (life expectancy or car accident probability) and in conducting opinion surveys (the probability of a certain candidate being elected or of a bond issue being passed). (1)

A. GENERAL PRINCIPLE OF COUNTING

Before a specific probability can be calculated, it is frequently necessary to enumerate all the possible outcomes that might occur. Thus, counting becomes an important part of our study of probability. (2)

The *general principle of counting* states that, if one task can be performed in m different ways and a second task can be performed in n different ways, then the first and second tasks together can be performed in mn different ways. For instance, if we have two different breads $(m = 2)$ and three different kinds of meat $(n = 3)$, then a sandwich having one kind of bread and one slice of one kind of meat can be put together in $mn = 2 \cdot 3$ or 6 different ways.

The general principle of counting can also be extended indefinitely when there are additional tasks: $m \cdot n \cdot p \cdot \cdots \cdot z$. That is, for each additional task, one must multiply the number of different ways the new task can be performed times the combined total of the different ways the previous tasks could be performed.

(3) Tree diagramming is a convenient means by which to visualize the general principle of counting. In order to make a tree diagram, we first list all the different ways in which task A can be performed (all possible outcomes for that variable). Then list the different ways task B can be performed. By connecting a line (branch) from each possible outcome of task A to each possible outcome of task B, we thus determine the total possible ways in which tasks A and B together can be performed. The branches on the tree identify all the possible outcomes. Therefore, through tree diagramming one can visualize the process that is used to obtain a list of all possibilities for any given situation.

Objective 1: General principle of counting

(4) ◆ **Example 1** A panel for a class discussion, consisting of one man and one woman, is to be chosen from set S of students in a history class. Apply the general principle of counting and tree diagramming to determine how many different panels are possible.

$$S = \{\text{Charles, Sally, Frank, Ann, Betty}\}$$

General Principle of Counting

Let: m = number of men   Then: $m = 2$
n = number of women $n = 3$

Therefore, there are $mn = 2 \cdot 3$ or 6 possible panels consisting of one man and one woman. This can be illustrated as follows.

Tree Diagram

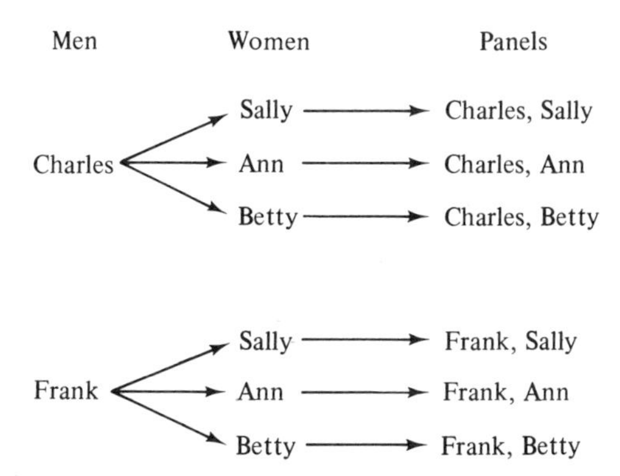

As before, there are 6 possible panels consisting of one man and one woman. In this situation, order is not considered; that is, the panel

{Charles, Sally} would be the same as {Sally, Charles}. Therefore, the same six panels would be formed if the women were taken first in the tree diagram.

● **Practice 1** Apply the general principle of counting and tree diagramming to solve the following: (5)

(a) How many two-digit numbers may be formed from the set of natural numbers $N = \{3, 4, 5\}$ if the same digit may be repeated?

(b) How many three-digit numbers may be formed from the set of natural numbers $N = \{3, 4, 5\}$ if repetitions of digits are *not* allowed?

(c) How many two-person teams of one man and one woman can be formed from the set of persons, $P = \{$Ann, Jack, Sarah, George$\}$?

See also Exercise 1.

B. COMBINATIONS AND PERMUTATIONS

Objective 2: *N* factorial

Before we consider combinations and permutations, let's practice the factorial calculation which will be used in calculating combinations and permutations. (6)

$N!$ is read "N factorial" and is equal to the product

$$(N)(N-1)(N-2)(N-3)\ldots(1)$$

Zero factorial, $0!$, is defined as equal to 1.

◆ **Example 2** (7)

(a) Calculate $N!$ for $N = 6$.

General formula

$N! = (N)(N-1)(N-2)(N-3)\ldots(1)$

Specific example

$6! = (6)(5)(4)(3)(2)(1) = 720$

(b) Determine a numerical answer for the following: $\dfrac{9!}{6!\,3!}$

$$\frac{9!}{6!\,3!} = \frac{9\cdot8\cdot7\cdot6\cdot5\cdot4\cdot3\cdot2\cdot1}{6\cdot5\cdot4\cdot3\cdot2\cdot1\cdot3\cdot2\cdot1}$$

At this point you observe it would be rather tedious to do all this multiplying and then dividing, so either:

(1) look for common factors and "cancel"

$$\frac{9\cdot 8\cdot 7\cdot \cancel{6}\cdot \cancel{5}\cdot \cancel{4}\cdot \cancel{3}\cdot \cancel{2}\cdot \cancel{1}}{\cancel{6}\cdot \cancel{5}\cdot \cancel{4}\cdot \cancel{3}\cdot \cancel{2}\cdot \cancel{1}\cdot 3\cdot 2\cdot 1} = \frac{\overset{3}{\cancel{9}}\cdot \overset{4}{\cancel{8}}\cdot 7}{\underset{1}{\cancel{3}}\cdot \underset{1}{\cancel{2}}\cdot 1} = 84$$

or (2) write the original problem as:

$$\frac{9!}{6!\,3!} = \frac{9\cdot 8\cdot 7\cdot \cancel{6}!}{\cancel{6}!\,3\cdot 2\cdot 1}$$ and "cancel" the common factor of 6!

$$= \frac{\overset{3}{\cancel{9}}\cdot \overset{4}{\cancel{8}}\cdot 7}{\cancel{3}\cdot \cancel{2}\cdot 1} = 84$$

● **Practice 2** Determine a numerical answer for each of the following: (8)

(a) $N!$ for $N = 4$, $N = 1$, $N = 2$, and $N = 6$.

(b) $2!3!$ (c) $\frac{7!}{5!\,2!}$ (d) $\frac{10!}{7!\,3!}$

See also Exercise 14.

Objective 3: Combinations

In Example 1 the panel {Charles, Sally} was considered to be the same as the panel {Sally, Charles} —that is, order was not considered. If the panel consists of Charles and Sally, the order in which they are selected makes no difference. That is, the panel formed by first selecting Charles and then Sally is the same as the panel formed by selecting Sally and then Charles. When order is *not* considered, we are calculating combinations. A *combination* is the number of *different* subsets containing s elements from a set of N elements $(N \geq s)$. The general symbol for combinations is ${}_NC_s$. This is read the "combination of N things taken s at a time." The formula for ${}_NC_s$ is: (9)

$${}_NC_s = \frac{N!}{s!\,(N-s)!}$$

◆ **Example 3**

(a) How many ways can a hand of 13 cards for a bridge hand be selected from a bridge deck of 52 cards? (10)

This example involves 52 things taken 13 at a time. It is a combination because order is not important—that is, order of the cards dealt is not considered.

In this case $N = 52$ and $s = 13$. Therefore, $N - s = 39$

$${}_{52}C_{13} = \frac{52!}{13!\,39!}$$

$$= \frac{52\cdot 51\cdot 50\cdot 49\cdot 48\cdot 47\cdot 46\cdot 45\cdot 44\cdot 43\cdot 42\cdot 41\cdot 40\cdot 39!}{13\cdot 12\cdot 11\cdot 10\cdot 9\cdot 8\cdot 7\cdot 6\cdot 5\cdot 4\cdot 3\cdot 2\cdot 1\cdot 39!}$$

$$= 635{,}013{,}559{,}600$$

There are 635,013,559,600 ways that a hand of 13 cards for a bridge hand can be selected from a bridge deck of 52 cards!

(b) How many ways can a committee consisting of nine members be selected from a club having nine members?

This example involves nine things taken nine at a time. It is a combination because order is not important.

In this case $N = 9$ and $s = 9$. Therefore, $N - s = 0$.

$$_9C_9 = \frac{9!}{9!\,0!} = \frac{9!}{9!\cdot 1} = \frac{9!}{9!} = 1$$

This result might seem obvious to you. Later compare this result to the result in Example 4(a), when order is considered.

● **Practice 3**

(a) Your history class has a reading list of 10 books from which you must choose 3 to read during the semester. In how many ways can you choose the 3 books? **(11)**

(b) You are planning your schedule for the spring semester. You have listed 8 courses you could take. How many ways can you plan your schedule to take 3 of these courses (assume that all 8 are offered at different times and you can take them in any order).

(c) You are one of 11 equally qualified persons being considered for a part-time job. Five persons are to be selected out of the eleven applicants. How many ways can the choice of the five persons be made?

See also Exercise 15.

Objective 4: Permutations

If order were considered in Example 1—that is, if set {Charles, Sally} were *not* considered to be the same as the set {Sally, Charles} —then we would be calculating permutations. A *permutation* is an *ordered* subset (of s elements) of a given universal set (of N elements). The general symbol for permutations is $_NP_s$. This is read the "permutation of N things taken s at a time." The formula for $_NP_s$ is: **(12)**

$$_NP_s = \frac{N!}{(N - s)!}$$

◆ **Example 4**

(a) There are nine horses registered for a racing event. Each jockey draws a number (from 1 to 9) which indicates his or her position in reference to the post at the starting gate. In how many ways can the horses be arranged at the starting gate? **(13)**

This example involves nine things taken nine at a time. It is a permutation because order is considered.

$$_NP_s = \frac{N!}{(N-s)!}$$

In this case $N = 9$ and $s = 9$. Thus, our formula becomes:

$$_9P_9 = \frac{9!}{(9-9)!}$$

$$= \frac{9!}{0!} = \frac{9 \cdot 8 \cdot 7 \cdot 6 \cdot 5 \cdot 4 \cdot 3 \cdot 2 \cdot 1}{1} = 362{,}880$$

There are 362,880 ways that the nine horses can be arranged at the starting gate.

This is a special case of our general formula, $_NP_s = \frac{N!}{(N-s)!}$. When $N = s$, as in the example, the formula becomes:

$$_NP_N = \frac{N!}{(N-N)!}$$

$$= \frac{N!}{0!}$$

$$= \frac{N!}{1} \quad \text{or} \quad N!$$

$_NP_N$ is read "the permutation of N things taken N at a time." Since there are nine horses and nine starting gate positions, the first position can be filled with any of the nine horses. After that space is filled, the second place can be filled with any of the eight remaining horses. Since there are 9 choices for the first place and eight choices for the second place, there are 9×8 or 72 ways the first two places can be filled (general principle of counting). Continuing our same reasoning, we see that the nine starting gate positions can be filled by the nine horses in

$$9 \times 8 \times 7 \times 6 \times 5 \times 4 \times 3 \times 2 \times 1 = 362{,}880 \text{ ways.}$$

Or, written in factorial notation, the nine starting gate positions can be filled by the nine horses in $9!$ ways. Compare Example 3(b), where there is only one way a combination of nine people can be taken nine at a time (disregarding order).

(b) How many five-letter "words" can be formed from the set of letters $L = \{d, a, c, b, e, o, g\}$?

This example involves seven things taken five at a time. It is a permutation because order is considered—that is, the "word" *dacbe* is not the same as the "word" *abcde*.

In this case $N = 7$ and $s = 5$. Thus, our formula becomes:

$$_7P_5 = \frac{7!}{(7-5)!}$$

$$= \frac{7!}{2!} = \frac{7 \cdot 6 \cdot 5 \cdot 4 \cdot 3 \cdot \not{2}!}{\not{2}!} = 2{,}520$$

There are 2,520 five-letter "words" that can be made from the set of letters, $L = \{d, a, c, b, e, o, g\}$.

● **Practice 4**

(a) You need to make 6 phone calls. In how many orders can you make the calls? (14)

(b) A telephone extension system at a company is installed. Each extension number has 4 digits (1 through 9). How many extension numbers can be formed if a digit cannot be repeated?

(c) Six people are forming a small company, $P = \{$Adams, Carver, Fugal, Hammer, James, Mosley$\}$. From this set, four persons are to be chosen as president, vice-president, secretary, and treasurer, respectively. In how many ways can this be done?

See also Exercise 16.

Objective 5: Distinguishing between a combination and a permutation

Basically, the distinction between calculating a permutation and a combination involves a judgment as to whether order is considered important. (15)

◆ **Example 5** Determine whether the following situations involve a combination or a permutation. Calculate the permutation or combination. (16)

(a) How many ways can a family of six sit in a row at a movie theatre?

Permutation—because order is considered. $N = 6$; $s = 6$:

$$_6P_6 = \frac{6!}{(6-6)!} = \frac{6!}{0!} = \frac{6!}{1} = 6 \cdot 5 \cdot 4 \cdot 3 \cdot 2 \cdot 1 = 720$$

There are 720 ways a family of six can sit in a row in a movie theatre.

(b) A family of six received two free tickets to a baseball game. How many ways can the two who go to the game be chosen?

Combination—because order is not considered. $N = 6$; $s = 2$:

$$_6C_2 = \frac{6!}{2!\,4!}$$

$$= \frac{6 \cdot 5 \cdot 4!}{2 \cdot 1 \cdot 4!} = \frac{30}{2} \quad \text{or} \quad 15$$

There are 15 ways two family members can be chosen to use the baseball tickets.

● **Practice 5** Determine whether the following events involve a combination or a permutation. Calculate the combination or permutation. (17)

(a) A family is showing home movies. They have 20 reels of movie film of which they plan to show 4. In how many ways can their program be shown?

(b) A child is playing a game by stacking five different colored blocks one on top of the other. In how many ways can the child stack the blocks?

(c) You are at a movie theatre that has displayed 11 types of candy. In how many ways can you choose three different types of candy to eat in the movie?

(d) There are 16 members of a bridge club assembled to play bridge. In how many ways can the pairs of bridge partners be selected?

(e) A movie theatre has a newsreel, feature, preview of next movie, and a cartoon to run. In how many different ways can they arrange these film pieces to be shown?

See also Exercise 17.

C. DEFINITION OF PROBABILITY

Objective 6: Ratio

The probability of success for a given event is defined here as the ratio of the number of possible successes to the total number of possible outcomes.* For example, the probability of drawing a heart is the ratio of all hearts to the total number of cards in a deck. Before working directly with calculating the probability of events, however, we will consider the proper use of ratio. (18)

A *ratio* expresses a numerical comparison or relationship. We may indicate the ratio of men to women in your math class; or the ratio of dogs to cats in the city pound; or the ratio of bad apples to good ones in the grocery store. (19)

The ratio of the number a to the number b can be expressed by the fractional number, $\frac{a}{b}$. That is, the ratio of a to b is $\frac{a}{b}$. The ratio of men to women in a class, if there were 14 men and 10 women in the class,

*Our definition of probability is confined to a finite number of outcomes each of which is equally likely to occur.

would be $\frac{14}{10}$ $\left(\text{or } \frac{7}{5}\right)$. The first element of the "ratio relationship" is the numerator in the fraction, and the second element of the relationship is the denominator. Therefore, the ratio

$$\text{men to women} = \frac{\text{number of men}}{\text{number of women}} = \frac{14}{10} = \frac{7}{5}.$$

(In computing ratios, reduce fractions to lowest terms.)

Ratio notation

A ratio can be expressed by any of three methods of notation. The ratio of 1 to 2 may be written as the fraction $\frac{1}{2}$ (as above), as (1, 2), or as 1:2. Each is read, "one is to two." (20)

◆ **Example 6**

(a) If a box contains 9 yellow marbles, 5 blue marbles, 4 green marbles, and 1 red marble, what is the ratio of blue marbles to yellow marbles? green marbles to red marbles? red marbles to blue marbles? Show each ratio notation in all three forms: (21)

The ratio of blue marbles to yellow marbles:

$$\frac{5}{9} \quad \text{or} \quad (5, 9) \quad \text{or} \quad 5:9$$

The ratio of green marbles to red marbles:

$$\frac{4}{1} \quad \text{or} \quad (4, 1) \quad \text{or} \quad 4:1$$

The ratio of red marbles to blue marbles:

$$\frac{1}{5} \quad \text{or} \quad (1, 5) \quad \text{or} \quad 1:5$$

(b) In a standard deck of cards (52 cards), what is the ratio of hearts to black cards?

There are 13 hearts, 13 diamonds, 13 spades, and 13 clubs in a standard deck of cards. Spades and clubs are both black suits, thus the ratio of hearts to black cards is

$$\frac{\text{hearts}}{\text{spades} + \text{clubs}} = \frac{13}{13 + 13} = \frac{13}{26} \text{ or } \frac{1}{2}.$$

The ratio $\frac{1}{2}$ is read "1 to 2" (for every one heart there are 2 black cards in a standard deck of cards).

● **Practice 6** Express the following relationships in each of the three forms of ratio notation. Reduce the ratio to lowest terms: (22)

(a) In a standard deck of 52 cards, what is the ratio of aces to face cards?

(There is an ace in each of the four suits, and each suit contains face cards of jack, queen, and king.)

(b) If a child's bank contains 9 quarters, 17 dimes, 12 nickels, and 50 pennies, what is the ratio of nickels to pennies? dimes to total coins? quarters to nickels and pennies?

See also Exercise 2.

Objective 7: Definition and formula for probability

As stated above, the probability of success for a given event is defined as the ratio of the number of possible successes to the total number of possible outcomes of that event. The formula for the probability of the success of an event is **(23)**

$$P(A) = \frac{n_A}{n}$$

where n_A is the number of possible successes of the event A, and n is the total number of all possible outcomes.

Range of probability values

The probability of success (known simply as probability), $\frac{n_A}{n}$, is greater than or equal to zero and smaller than or equal to 1: **(24)**

$$0 \leq \frac{n_A}{n} \leq 1 \qquad (n \neq 0)$$

Probability is equal to 1 when success is inevitable; that is, when n_A equals n. Toss a coin. What is the probability of tossing either a head or a tail? Since the coin has both a head and a tail, the probability of tossing either a head or a tail is one chance in one toss or five chances in five tosses, etc. Since you will get either a head or a tail any time you toss the coin, the probability of success thus equals one.

Probability is equal to 0 when success of an event is impossible, or when n_A equals zero. For example, if a bag contained 3 yellow balls and 2 red balls, what is the probability of drawing a green ball? Since there are no green balls in the bag, the probability of drawing a green ball is zero.

Objective 8: Solving for probability

◆ **Example 7** Given a standard deck of cards (52 cards), what is the probability of selecting an ace from the deck? **(25)**

Let $P(A)$ represent the probability of selecting an ace.

$$P(A) = \frac{n_A}{n}$$

where n_A is the number of possible successes for the event and n is the total number of all possible outcomes. In this case, $n_A = 4$. (There are 4 aces in a standard deck.) The total number of possible outcomes is $n = 52$. (There are 52 cards in a standard deck.)

Therefore, $$P(A) = \frac{n_A}{n} = \frac{4}{52} = \frac{1}{13}.$$

The chances of drawing an ace are 1 to 13, represented by the fraction $\frac{1}{13}$.

● **Practice 7**

(26)

(a) What is the probability of selecting a diamond from a deck of 52 cards on a single draw?

(b) What is the probability of selecting a black card from a deck of 52 cards on a single draw?

(c) What is the probability of obtaining a number different from one, given a single toss of one die? ("Die" is the singular of dice. You have one "die" or two or more "dice.")

(d) If a box contains 2 dimes and 4 nickels, what is the probability of drawing a nickel out of the box, given one chance?

See also Exercises 3 and 4.

D. SAMPLE SPACE

Objective 9: Sample space

(27)

With our understanding of tree diagrams and the definition of probability, we are now able to answer such questions as: What is the probability of tossing three coins and having two of them come up heads? Let us construct a *sample space* (list of all possible outcomes) in order to answer this question.

(28)

◆ **Example 8** What is the probability of tossing three coins and having two of them come up heads?

Tree diagram:

First Coin	Second Coin	Third Coin	Coins on Table
H	H	H →	H, H, H (3 heads)
		T →	H, H, T (2 heads)
	T	H →	H, T, H (2 heads)
		T →	H, T, T (1 head)
T	H	H →	T, H, H (2 heads)
		T →	T, H, T (1 head)
	T	H →	T, T, H (1 head)
		T →	T, T, T (0 heads)

$$P\text{ (two heads)} = \frac{n_H}{n}$$

As before, n_H = number of possible successes of the event. In our example, $n_H = 3$. (Three times we obtain two heads.) Also, n = total number of all possible outcomes. In our example, $n = 8$. (There was a total of 8 possible outcomes in throwing three coins.) Therefore,

$$P(\text{two heads}) = \frac{n_H}{n} = \frac{3}{8}.$$

● **Practice 8** What is the probability of tossing two dice and having each given result? (Draw a sample space for each.) (29)

(a) The probability of tossing two of the same number (like two 3's or two 4's, etc.)? (A die contains the numbers 1 to 6.)

(b) The probability of obtaining a sum greater than 7 when you add both dice?

See also Exercise 5.

E. PRINCIPLES OF PROBABILITY

Objective 10: Probability of an event not occurring

Let the dots in the box below represent 25 students in a class. Also, let thc dots in the area labeled M represent the set of men students. (30)

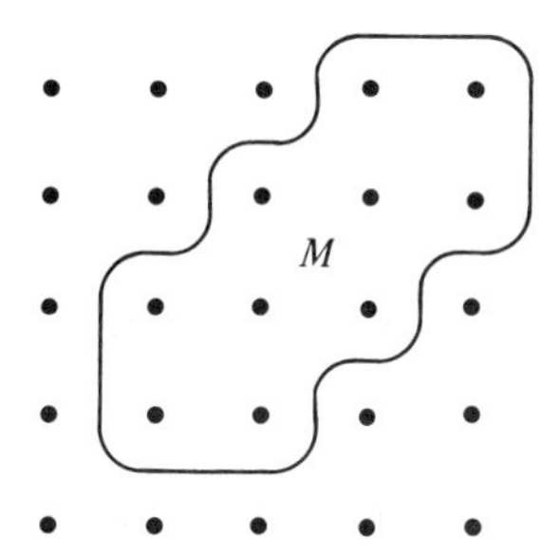

The 15 dots *not* in the M-area thus represent students in the class who are not members of the M-group, or the women students. If we let $P(M)$ represent the probability of picking a man, then $P(M)$ denotes the number of outcomes in the M-area, divided by the total number of outcomes in the class (box). That is,

$$P(M) = \frac{n_M}{n} = \frac{10}{25} \text{ or } \frac{2}{5} \text{ or } .4.$$

The probability of *not* picking a man is written as $Q(M)$. If $P(M)$ is .4, then $Q(M)$ is .6, because the probability of an event occurring plus the probability of an event not occurring must equal 1. In general,

$$P(X) + Q(X) = 1$$

◆ **Example 9**

(a) If $P(B) = .7$, what number would $Q(B)$ equal? **(31)**
Since in general,

$$P(X) + Q(X) = 1$$

Then, $$Q(X) = 1 - P(X)$$

Therefore, $$Q(B) = 1 - P(B)$$

$$Q(B) = 1 - .7 = .3$$

(b) If the probability of a baseball player getting a hit is .45, what is the probability of the player not getting a hit?

$$P(\text{hit}) = .45$$

$$P(\text{no hit}) \text{ or } Q(\text{hit}) = 1 - P(\text{hit}) = 1 - .45 = .55$$

● **Practice 9**

(a) If $P(D) = \frac{3}{7}$, what does $Q(D)$ equal? **(32)**

(b) If $P(P) = .45$, what does $Q(P)$ equal?

(c) There are 30 students in a class. If there are 24 men, what is the probability of picking a woman (of not picking a man)?

See also Exercise 6.

Objective 11: Mutually exclusive outcomes

Mutually exclusive outcomes are outcomes in which two or more of them can *not* occur together. Drawing hearts or clubs out of a card deck are mutually exclusive outcomes, since one could not draw both a heart and a club in a single draw. (By contrast, drawing aces and hearts are not mutually exclusive events, since one could draw both an ace and a heart on the same draw.) **(33)**

Let the outcomes in categories X and Y below represent mutually exclusive outcomes, since there are no instances in which an outcome is both an X and a Y.

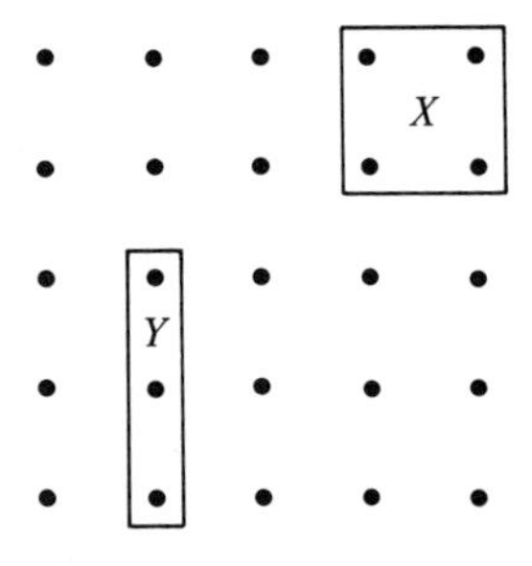

If n_X and n_Y represent the number of elements in categories X and Y respectively, *and* if these categories are mutually exclusive, then $n_X + n_Y$

denotes the number of outcomes that are *either* X or Y. That number in our illustration is 7 (number in category $X +$ number in category $Y = 4 + 3$, or 7).

If X and Y represent mutually exclusive outcomes then:

$$P(X \text{ or } Y) = \frac{n_X + n_Y}{n} = \frac{n_X}{n} + \frac{n_Y}{n} = P(X) + P(Y)$$

◆ **Example 10**

(a) Given real situation first, where A and B are mutually exclusive outcomes and $n = 20$, $n_A = 5$, and $n_B = 6$, what does $P(A \text{ or } B)$ equal? **(34)**

General formula

$$P(X \text{ or } Y) = \frac{n_X + n_Y}{n} = \frac{n_X}{n} + \frac{n_Y}{n} = P(X) + P(Y)$$

Specifically

$$P(A \text{ or } B) = \frac{5 + 6}{20} = \frac{11}{20}$$

(b) If C and D are mutually exclusive outcomes and $P(C) = .6$ and $P(D) = .1$, what does $P(C \text{ or } D)$ equal?

General formula

$$P(X \text{ or } Y) = \frac{n_X + n_Y}{n} = \frac{n_X}{n} + \frac{n_Y}{n} = P(X) + P(Y)$$

Specifically

$$P(C \text{ or } D) = P(C) + P(D)$$
$$= .6 + .1 = .7$$

(c) A deck of cards has 52 cards. Determine the probability of picking a club or a red ace.

$n = 52$

$n_C = 13$ (There are 13 clubs in a deck of 52 cards)

$n_{RA} = 2$ (There are two red aces in a deck of 52 cards—the ace of diamonds and the ace of hearts)

$$P(C \text{ or } RA) = \frac{n_C + n_{RA}}{n} = \frac{n_C}{n} + \frac{n_{RA}}{n}$$
$$= \frac{13}{52} + \frac{2}{52} = \frac{15}{52}$$

The probability of picking a club or a red ace out of a standard deck of 52 cards is $\frac{15}{52}$.

● **Practice 10** Given that all the following situations involve mutually exclusive outcomes, use the formula: (35)

$$P(X \text{ or } Y) = \frac{n_X + n_Y}{n} = \frac{n_X}{n} + \frac{n_Y}{n} = P(X) + P(Y)$$

to calculate the indicated portion.

(a) If $P(A) = .42$ and $P(B) = .16$, what does $P(A \text{ or } B)$ equal?

(b) If $n_A = 3$, $n_B = 7$, and $n = 28$, what does $P(A) + P(B)$ equal?

(c) If $P(A) = \frac{3}{20}$ and $P(B) = \frac{7}{20}$, what does n_A equal? n_B equal? n equal?

(d) A woman has 10 blouses, 3 of which are white and 2 of which are green. Determine the probability she will have on a white blouse or a green blouse.

(e) From a deck of 52 cards you pick a card. Determine the probability of picking a face card or a 7.

See also Exercise 7.

Objective 12: Outcomes not mutually exclusive

In the figure below the outcomes in categories X and Y are *not* mutually exclusive: (36)

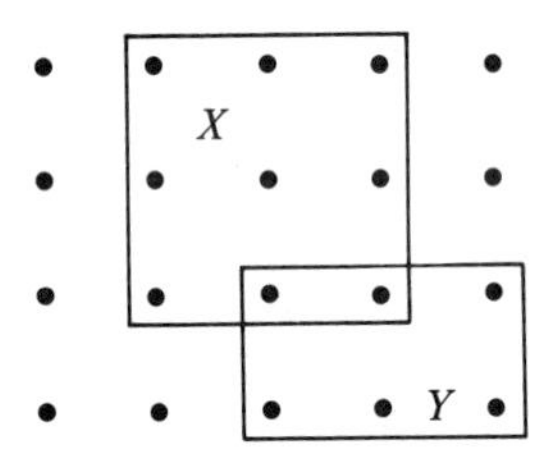

There are two instances in which an outcome is both an X and a Y.

The number of instances in which an outcome is both an X and a Y is written in symbols as n_{XY}. The probability that an outcome will be *both* an X *and* a Y is written $P(XY)$.

If X and Y are *not* mutually exclusive outcomes, then:

$$P(X \text{ or } Y \text{ or both}) = \frac{n_X + n_Y - n_{XY}}{n} = P(X) + P(Y) - P(XY)$$

Observe that adding $P(X) + P(Y)$ counts the common elements twice; therefore, $P(XY)$ is subtracted to eliminate this duplication.

◆ **Example 11** Referring to the given diagram, there are nine possible X outcomes, six possible Y outcomes, and two outcomes that are both X and Y. What does $P(X \text{ or } Y \text{ or both})$ equal? (37)

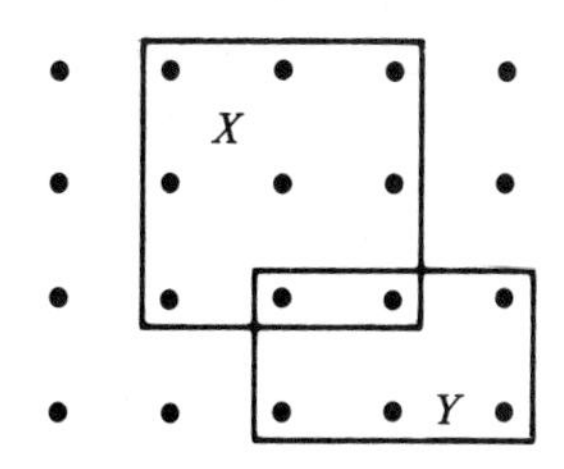

General formula

$$P(X \text{ or } Y \text{ or both}) = \frac{n_X + n_Y - n_{XY}}{n} = P(X) + P(Y) - P(XY)$$

Specific problem

$n_X = 9$ $n_{XY} = 2$

$n_Y = 6$ $n = 20$

Therefore, for our problem:

$$P(X \text{ or } Y \text{ or both}) = \frac{9 + 6 - 2}{20} = \frac{13}{20}$$

● **Practice 11** Given that all of the following situations are *not* mutually exclusive outcomes, use the formula: (38)

$$P(X \text{ or } Y \text{ or both}) = \frac{n_X + n_Y - n_{XY}}{n} = P(X) + P(Y) - P(XY)$$

to calculate the indicated portion:

(a) If $P(A) = .2$, $P(B) = .4$ and $P(AB) = .15$, what does $P(A$ or B or both) equal?

(b) If $n_A = 12$, $n_B = 10$, $n_{AB} = 3$ and $n = 30$, what does $P(A$ or B or both) equal?

(c) If $P(A$ or B or both) $= .8$ and $P(A) = .4$ and $P(B) = .6$, what does $P(AB)$ equal?

(d) You draw a card from a deck of 52 cards. What is the probability that you will draw a queen, a diamond, or the queen of diamonds?

(e) A family has 7 children—4 have red hair, 5 have brown eyes, and 3 have red hair and brown eyes. Determine the probability of randomly selecting a child in the family and picking one that has red hair, brown eyes or both.

See also Exercise 8.

Objective 13: Exhaustive categories

Categories of outcomes are called *exhaustive* if every possible relevant outcome is included in one of these categories. On a coin, "heads" and "tails" represent exhaustive categories of coin sides, since any side of a coin (39) must be either a "head" or a "tail." Also, the categories "heads" and "tails" are mutually exclusive.

Objective 14: Relationship between exhaustive and mutually exclusive outcomes

(40) The categories of events X, Y, and Z below are mutually exclusive but not exhaustive:

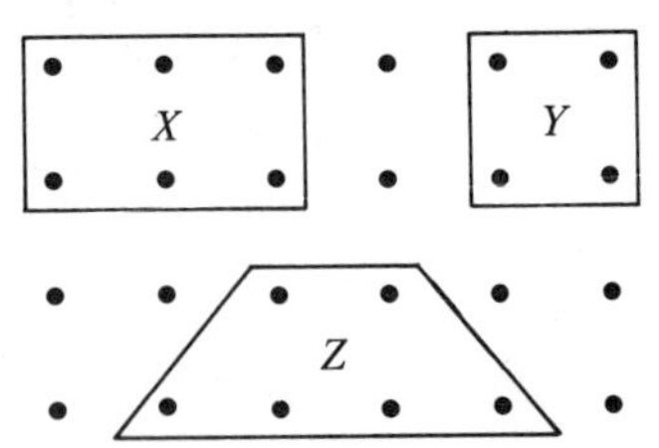

However, the categories of events A, B, and C below are both mutually exclusive *and* exhaustive:

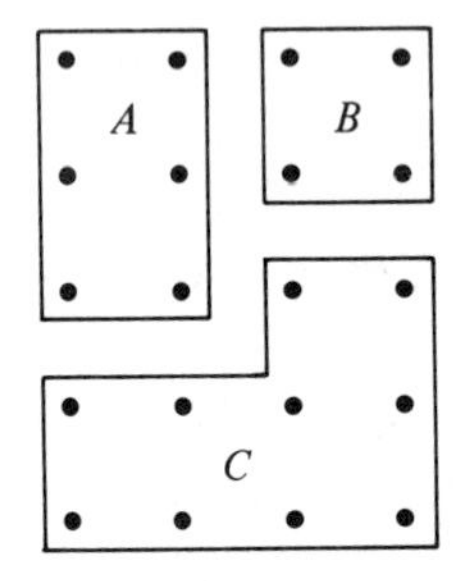

(41) ◆ **Example 12** Classify the following into the categories of "mutually exclusive but not exhaustive," or as "mutually exclusive and exhaustive."

(a) People who weigh 150 to 160 pounds; people who weigh 170 to 180 pounds:
Mutually exclusive but *not* exhaustive.
These categories do not overlap, but they do not cover all the possibilities of the weight of people.

(b) People under the age of 40; people 40 years old and older:
Mutually exclusive and exhaustive.
A person is *either* under 40 years old *or* 40 years old and older. Every possible person is included in one of these categories.

(42) ● **Practice 12** Classify the following as "mutually exclusive but not exhaustive" or as "mutually exclusive and exhaustive":

(a) A roll of 4 on a die; a roll of 3 on a die.

(b) Choosing a heart or a diamond from a deck of cards; choosing a club or spade from a deck of cards (assume a standard deck).

(c) Choosing a face card from a deck of cards; choosing a card with a number on it from a deck of cards (assume a standard deck).

See also Exercises 9 and 10.

Objective 15: Calculating probabilities of mutually exclusive and exhaustive outcomes

If categories of events X, Y, and Z are mutually exclusive and exhaustive, then (43)

$$P(X) + P(Y) + P(Z) = 1$$

◆ **Example 13**

(a) Given that A, B, and C are mutually exclusive and exhaustive categories, the $P(A) = .6$ and $P(B) = .3$, what is $P(C)$? (44)

$$P(A) + P(B) + P(C) = 1$$

because A, B, and C were given to be mutually exclusive and exhaustive categories.
Therefore,

$$P(C) = .1 \quad \text{because} \quad .6 + .3 + .1 = 1$$

(b) There are 100 colored balls of gum in a machine—green, red, yellow. The probability of getting a green ball is $\frac{3}{10}$. The probability of getting a red ball is $\frac{9}{20}$. How many yellow gum balls are there in the machine?

$P(G) + P(R) + P(Y) = 1$ because the outcomes are mutually exclusive and exhaustive. We know that $P(G) = \frac{3}{10}$ and $P(R) = \frac{9}{20}$.

Therefore,

$$P(Y) = 1 - P(G) - P(R)$$

$$= 1 - \frac{3}{10} - \frac{9}{20}$$

$$= 1 - \frac{15}{20}$$

$$= \frac{5}{20} \quad \text{or} \quad \frac{1}{4}$$

The probability of getting a yellow gum ball, $P(Y)$, is $\frac{1}{4}$. Since $P(Y) = \frac{n_Y}{n}$ and $n = 100$, then $P(Y) = \frac{n_Y}{100}$. Also, $P(Y) = \frac{1}{4}$.

Therefore,

$$P(Y) = \frac{n_Y}{n}$$

$$\frac{1}{4} = \frac{n_Y}{100}$$

$$4n_Y = 100$$

$$n_Y = 25$$

There are 25 yellow gum balls in the machine.

● **Practice 13** Given that the categories in the following are mutually exclusive and exhaustive, calculate the unknown probability. (45)

(a) $P(X) = \frac{3}{10}$; $P(Y) = \frac{1}{10}$; $P(Z) = ?$

(b) $P(A) = .35$; $P(B) = ?$; $P(C) = .47$

(c) Chicken, ham, or steak dinners are served on a flight from Atlanta to Dallas. The probability of getting a ham dinner is .3; the probability of getting a chicken dinner is .6. What is the probability of getting a steak dinner? If there are 300 people on the plane, how many will get steak?

See also Exercise 11.

Objective 16: Probability of independent events

(46) Two or more events may be defined as *independent* when the events are not related. That is, events are independent when the outcome of any single event does not affect the outcome of any of the other events. For example, consider two urns: the first contains a red and a black marble; the second contains a white, a green, and a yellow marble. What is the probability of drawing first a black and then a white marble? A sample space can be used to illustrate this probability:

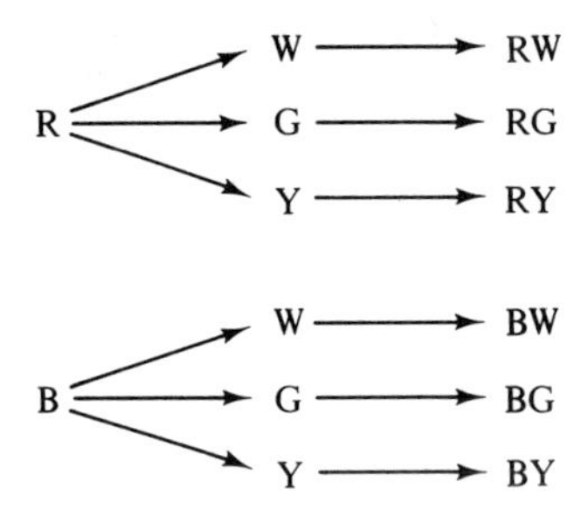

Thus, the probability of drawing a black and then a white marble is $\frac{1}{6}$; or $P(BW) = \frac{1}{6}$.

(47) Observe also the individual probabilities. On the first draw, $P(B) = \frac{1}{2}$. On the second draw, $P(W) = \frac{1}{3}$. The probabilities of these independent events are significant in determining the joint probability. Notice that

$$P(BW) = P(B) \cdot P(W)$$
$$= \frac{1}{2} \cdot \frac{1}{3} = \frac{1}{6}$$

This illustrates the general formula for independent events. That is, if X and Y are independent events,

$$P(XY) = P(X) \cdot P(Y)$$

Conversely, if $P(XY) = P(X) \cdot P(Y)$, then X and Y are independent. This formula can be extended to any number of independent events. For example, if A, B, C, and D are independent, $P(ABCD) = P(A) \cdot P(B) \cdot P(C) \cdot P(D)$.

◆ **Example 14** Assume that the rolling of two dice are independent events. Compute the probability of getting "snake eyes" (two 1's). (48)

The probability that the first die comes up 1 (event X) is:

$$P(X) = \frac{n_X}{n} = \frac{1}{6}$$

The probability that the second die comes up 1 (event Y) is:

$$P(Y) = \frac{n_Y}{n} = \frac{1}{6}$$

The probability of getting "snake eyes" is:

$$P(XY) = P(X) \cdot P(Y)$$

$$= \frac{1}{6} \cdot \frac{1}{6} = \frac{1}{36}$$

● **Practice 14** Assume that flipping a coin repeatedly represents independent events. Calculate the indicated probabilities. (49)

(a) Probability of flipping 2 heads in a row?
(b) Probability of getting a head, then a tail, and then a head?
(c) Probability of getting a tail, then a tail, and then a head?
(d) Probability of flipping 2 tails in a row?

See also Exercise 13.

Objective 17: Conditional probability

Consider the illustration below of a class of 25 students where 60% of the students are women (W), and 80% of the students did their assignment (A). (50)

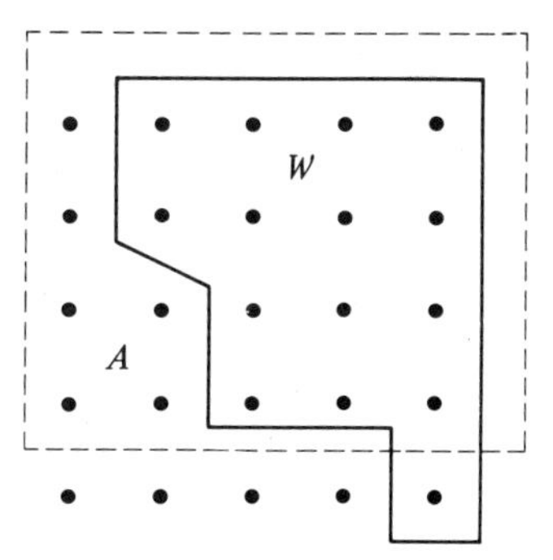

From our illustration: 15 of the students are women; 20 students did the assignment; 14 women students did the assignment.

Suppose n_{WA} (or n_{AW}) stands for the number of students who are both W (women) and A (did assignment). Of the students who did the assignment, the proportion who are women is $\frac{n_{WA}}{n_A} = \frac{14}{20}$ or $\frac{7}{10}$. The probability that a student is a woman if it is known that the person's assignment was done is indicated as $P(W/A) = \frac{7}{10}$. This is an example of a conditional probability. The symbol $P(W/A)$ is read "the probability of being W, known (or given) that it is an A." Conditional probability always includes an "if" that is known.

In the following example, we present this same situation again and use the general formula of conditional probability to formally compute $P(W/A)$ in two ways.

◆ **Example 15** A class has 25 students in which: 15 of the students are women (W), 20 of the students did the assignment (A), 14 of the women students did their assignment. Calculate the probability that a student did the assignment, $P(A)$; the probability that a student is a woman who did the assignment, $P(WA)$; and the probability that a student is a woman, if it is known that the student did the assignment, $P(W/A)$. (51)

General formula

$$P(G/H) = \frac{n_{GH}}{n_H} = \frac{\frac{n_{GH}}{n}}{\frac{n_H}{n}} = \frac{P(GH)}{P(H)}$$

Specific example

$$P(W/A) = \frac{n_{WA}}{n_A} = \frac{\frac{n_{WA}}{n}}{\frac{n_A}{n}} = \frac{P(WA)}{P(A)}$$

where $n_{WA} = 14$ women who did the assignment

$n_A = 20$ students who did the assignment

$n = 25$ students in the class

Now $$P(A) = \frac{n_A}{n} = \frac{20}{25} = \frac{4}{5}$$ (probability that any given student completed the assignment)

Also, $$P(WA) = \frac{n_{WA}}{n} = \frac{14}{25}$$ (probability that a student is a woman who completed the assignment)

Then, substituting in the general formula,

$$P(W/A) = \frac{P(WA)}{P(A)}$$

we have

$$= \frac{\frac{14}{25}}{\frac{4}{5}}$$

$$= \frac{14}{25} \cdot \frac{5}{4}$$

$$= \frac{7}{10}, \quad \text{as before}$$

However, the computation may be shortened by simply using the n_{WA} and n_A, as follows.

In short, $P(W/A) = \frac{n_{WA}}{n_A} = \frac{14}{20} = \frac{7}{10}$ (probability that a student is a woman, if it is known that the assignment was done)

- **Practice 15**

(a) A group of 100 people has: 60 people who are men (M); 85 people who have brown eyes (B); and 50 men who have brown eyes (MB). Calculate: the probability that a person is a man if it is known the person has brown eyes, $P(M/B)$; the probability a person is a man who has brown eyes, $P(MB)$; and the probability a person has brown eyes, $P(B)$. (52)

(b) Use the following diagram to calculate $P(X/Y)$, $P(XY)$, and $P(Y)$:

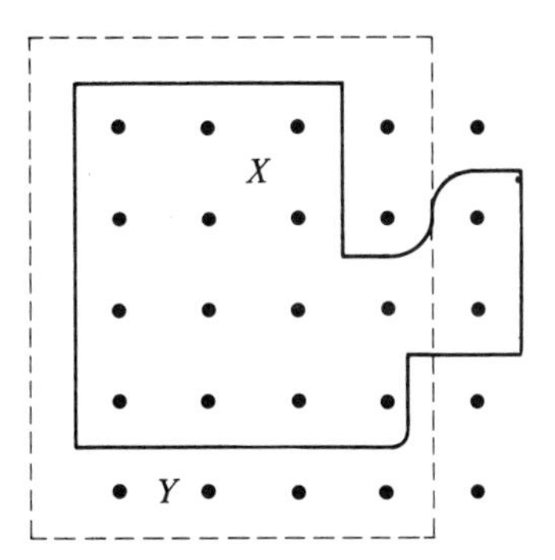

See also Exercise 12.

In the statistics chapter we found that if two variables are related, knowledge of one would help in the prediction of the other. Refer again to our illustration of a class of 25 in which 15 of the students are women, 20 students did an assignment, and 14 women students did their assignment: (53)

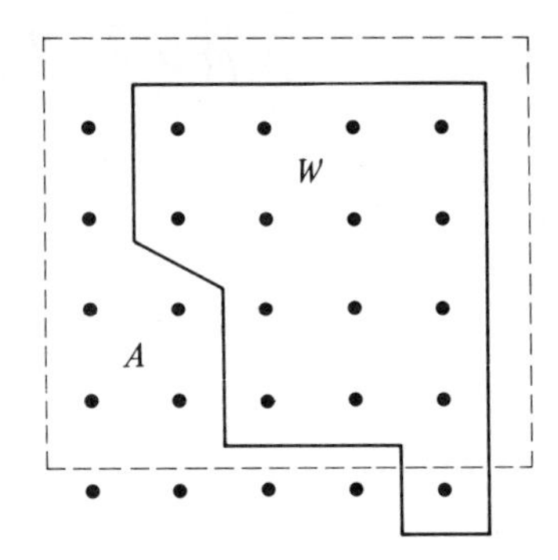

If we did not know whether or not the person's assignment was done, our best guess of any random student's sex would be "female," and our probability of being correct would be $\frac{15}{25}$ or .6. In symbols, this probability would be denoted $P(W)$. If we know that the person's assignment was done, our best guess of the person's sex would still be "female," but this time our probability of being correct would be $\frac{14}{20}$ or .7, denoted $P(W/A)$.

In this example, knowledge of whether a person's assignment was done helps us predict the person's sex. That is, completion of the assignment and sex of the person are related, or correlated. Therefore, for these two related variables, $P(W)$ did not equal $P(W/A)$; the former was .6, the latter .7. One would expect to find "the proportion of students whose assignment was done and who were women" to be the same as "the proportion of all students who were women" *if and only if* completing one's assignment and being a woman were independent events.

EXERCISES

Objective 1; Example 1

1. Apply the general principle of counting and tree diagramming to solve the following:
   (a) A man has three jackets, two pair of pants and two pair of shoes to choose from. How many different outfits can he form? (Assume all go together. Label jackets J1, J2, J3; pants P1, P2; shoes S1, S2.)
   (b) How many two-digit numbers may be formed from the set of natural numbers $N = \{1, 2, 3, 4\}$, if repetition of digits is allowed? If repetition of digits is not allowed?
   (c) How many three-digit numbers may be formed from the set of natural numbers $N = \{2, 3, 4\}$, if repetition of digits is allowed? If a digit can only appear once in the three-digit number?
   (d) How many two-letter "words" (the two letters do not have to form a word you can find in a dictionary) may be formed from the set of letters $L = \{t, e, p, m\}$, if repetition of letters is allowed? If repetition of letters is not allowed?

Objective 6; Example 6

2. Express the following as a ratio in the three standard forms for ratio:

(a) Seven books to 18 books.

(b) Hearts in a standard deck to all 52 cards.

(c) Number of fingers on one hand to total number of fingers.

(d) A person has 10 goldfish, 3 cats, 2 dogs and 20 rabbits. What is the ratio of goldfish to rabbits? Cats to total number of animals? Cats and dogs to goldfish and rabbits?

Objective 7: Definition and formula for probability

3. Define probability and relate the formula for probability, $P(A) = \frac{n_A}{n}$, to your definition.

Objective 8; Example 7

4. What is the probability of:

(a) selecting a jack from a deck of 52 cards on a single draw?

(b) throwing a 3 on a single toss of one die?

(c) drawing a green ball out of a box of 10 green balls, 2 red balls, and 7 white balls, given one chance?

(d) selecting a card other than a heart, diamond or spade from a deck of 52 cards on a single draw?

(e) throwing a number other than a six on a die on a single toss of one die?

Objective 9; Example 8

5. Draw a sample space for each of the following and determine the indicated probability:

(a) What is the probability of tossing three coins and having either one or two of them come up heads?

(b) What is the probability of throwing two dice and having one or both of them be a six?

(c) A box has 1 red ball, 1 green ball, 1 white ball and 1 orange ball. Two balls are drawn in succession without replacement. What is the probability that one of the balls drawn will be green?

Objective 10; Example 9

6. Calculate the probability of an event *not* occurring, given the following probabilities of an event occurring:

(a) If $P(A) = \frac{3}{10}$, what does $Q(A)$ equal?

(b) If $P(C) = .6$, what does $Q(C)$ equal?

(c) A box has 10 green balls and 15 red balls. What is the probability of not picking a red ball? Of not picking a green ball?

(d) A class of 25 students has 6 students six feet tall or over. What is the probability of picking a student under six feet tall?

Objective 11; Example 10

7. Given that all the following situations involve mutually exclusive outcomes, use the formula

$$P(X \text{ or } Y) = \frac{n_X + n_Y}{n} = \frac{n_X}{n} + \frac{n_Y}{n} = P(X) + P(Y)$$

to calculate the indicated portion.

(a) If $P(A) = \frac{3}{10}$ and $P(B) = \frac{4}{10}$, what do n_A, n_B, and n equal?
(b) If $n_X = 6$, $n_Y = 12$, and $n = 35$, what do $P(X)$, $P(Y)$, and $P(X \text{ or } Y)$ equal?
(c) If $P(A) = .3$ and $P(B) = .42$, what does $P(X \text{ or } Y)$ equal?
(d) $P(X \text{ or } Y) = \frac{13}{35}$ and $n_X = 9$ and $n_Y = 4$, what do n, $P(X)$, and $P(Y)$ equal?
(e) There are 32 children on a class picnic. Six chose to eat chicken, 15 chose to eat hamburgers. What is the probability a child eats chicken or a hamburger?
(f) A grocery store has eleven brands of washing detergent. Twenty percent of the customers choose brand A. Thirty percent of the customers choose brand B. Five hundred customers bought washing detergent today. What is the probability that a customer bought brand A or brand B?
(g) A store is giving away (at random) a turkey, a ham or a case of canned fruit with the purchase of a new refrigerator. The store has 150 turkeys, 200 hams, and 400 cases of fruit. What is the probability a customer who buys a refrigerator will get a turkey or a ham?

Objective 12; Example 11

8. Given that all of the following situations are not mutually exclusive outcomes, use the formula

$$P(X \text{ or } Y \text{ or both}) = \frac{n_X + n_Y - n_{XY}}{n} = P(X) + P(Y) - P(XY)$$

to calculate the indicated portion:

(a) Calculate $P(X \text{ or } Y \text{ or both})$.

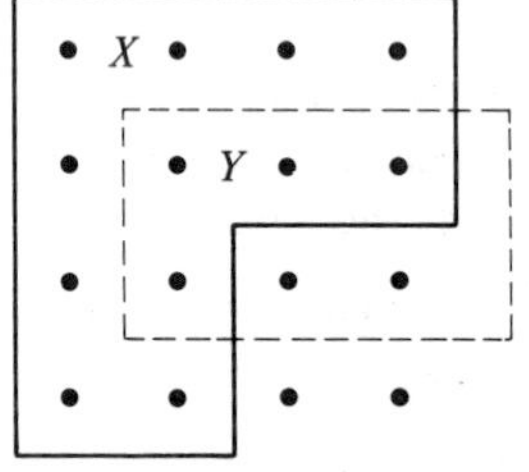

(b) If $P(A) = .7$ and $P(B) = .2$ and $P(AB) = .08$, what does $P(A \text{ or } B \text{ or both})$ equal?

(c) Calculate $P(AB)$.

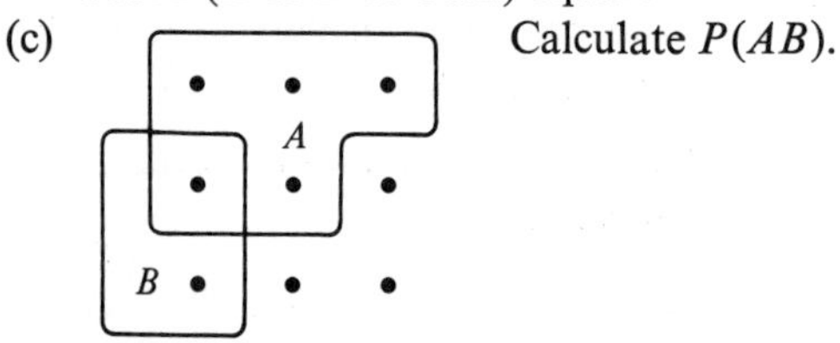

(d) If $P(A \text{ or } B \text{ or both}) = .7$ and $P(A) = .2$ and $P(B) = .85$, what does $P(AB)$ equal?

(e) The probability a person will order green beans at a restaurant is .8. The probability of ordering corn is .45. The probability of a person ordering both green beans and corn is .3. What is the probability a person will order green beans, corn or both?

(f) A church had a fall bazaar. Eight hundred people attended. The number of people who ate lunch at the bazaar is 200. The number of people who bought something from the booths is 450. The number of people who ate lunch and bought something is 150. What is the probability a person ate lunch and bought something or both?

(g) The probability of a person buying a car with an automatic transmission and an air conditioner is .45. The probability that a person will buy a car with an automatic transmission only is .8. The probability that a person will buy a car with an automatic transmission or an air conditioner or both is .9. What is the probability a person will buy a car with an air conditioner only?

Objective 13: Exhaustive categories

9. Define and give an example of exhaustive categorics.

Objective 14; Example 12

10. Classify the following into categories of "mutually exclusive but not exhaustive" or "mutually exclusive and exhaustive":

(a)

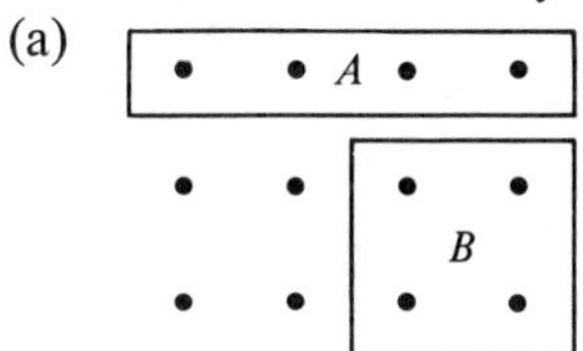

(b) Choosing a 7 from a deck of cards; choosing a face card from a deck of cards.

(c)

(d) People 5′5″ tall or shorter; people over 5′5″ tall.

Objective 15; Example 13

11. Given that the following categories are mutually exclusive and exhaustive, calculate the unknown probability.

(a) $P(A) = \frac{7}{9}$; $P(B) = \frac{1}{9}$; $P(C) = ?$

(b) $P(A) = .32$; $P(B) = ?$; $P(C) = .58$

(c) $P(A) = P(B) = P(C)$; $P(C) = \frac{1}{3}$; $P(A) = ?$; $P(B) = ?$

(d) $1 - P(C) = .8$; $P(A) = .6$; $P(B) = ?$; $P(C) = ?$

(e) A grab bag game at a school Halloween carnival has 3 toys: balls, whistles, and rubber spiders. The probability of drawing a spider is $\frac{7}{10}$. The probability of drawing a ball is $\frac{1}{10}$. What is the probability of drawing a whistle?

(f) The probability of drawing a face card from a standard deck of cards is $\frac{3}{13}$. The probability of drawing an ace is $\frac{1}{13}$. What is the probability of drawing a numbered card?

(g) A little league team is selling raffle tickets. There will be awarded one first prize and six second prizes. Two thousand tickets have been sold. What is the probability of winning the first prize? second prize? of winning nothing?

Objective 17; Example 15

12. Use the given formula for conditional probability to calculate the indicated probabilities:

$$P(G/H) = \frac{n_{GH}}{n_H} = \frac{\frac{n_{GH}}{n}}{\frac{n_H}{n}} = \frac{P(GH)}{P(H)}$$

(a) A class of 30 has 18 men (M) and 12 women (W). Ten of the men and 9 of the women have bought the book for the course (B). Draw a diagram to illustrate the situation and calculate the probability that: person bought a book, $P(B)$; person is a woman, $P(W)$; person is a man, $P(M)$; person is a woman, if it is known that the person bought a book, $P(W/B)$; person is a man, if it is known that the person bought a book, $P(M/B)$; person is a woman who bought a book, $P(WB)$; person is a man who bought a book, $P(MB)$.

(b) Use the following diagram to calculate: $P(X/Y)$, $P(XY)$, $P(Y)$, $P(X)$.

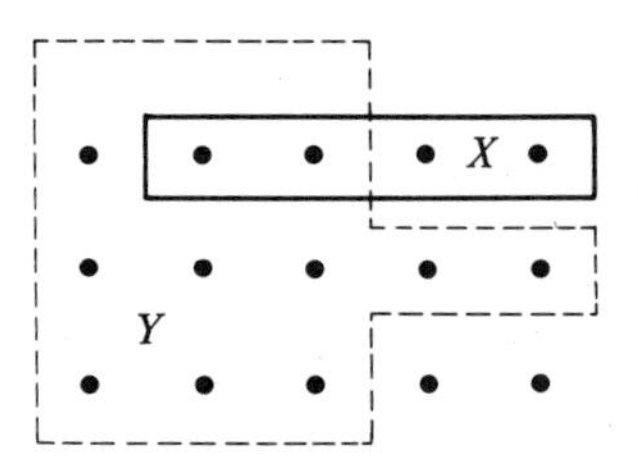

(c) The city animal shelter has 50 animals (A); 32 are dogs (D); 42 of the animals have identification tags (T); 30 dogs have identification tags. Draw a diagram to illustrate the situation and calculate the probability that: animal is a dog, $P(D)$; animal has an identification tag, $P(T)$; animal is a dog, if it is known that animal has a tag, $P(D/T)$; animal has a tag, if it is known that the animal is a dog, $P(T/D)$.

Objective 16; Example 14

13. Assume that the following are independent events. Calculate the indicated probabilities.

(a) Probability of throwing three 6's in a row on a die?

(b) Probability of flipping a coin and getting 3 heads in a row?

(c) Probability of picking an ace from a deck of cards 2 times in a row (assume that you replace the ace and shuffle the deck after each choice).

(d) Probability of flipping a coin and getting a head, then a tail, then a head and so on for 6 flips?

Objective 2; Example 2

14. Determine a numerical answer for each of the following:

(a) $N!$ for $N = 6$, $N = 3$, $N = 1$, $N = 0$ (b) $6!2!$

(c) $\frac{4!}{3!}$ (d) $\frac{8!}{5!\,3!}$ (e) $\frac{10!}{4!\,6!}$ (f) $\frac{10!}{9!\,1!}$

Objective 3; Example 3

15. Calculate the combination of N things taken s at a time using the formula

$$_{N}C_{s} = \frac{N!}{(s!)(N-s)!}$$

(a) How many ways can a hand of 5 cards bc selected from a standard 52-card deck?

(b) If there are 10 questions and you get 8 correct and 2 incorrect, what are the number of arrangements of 8 correct and 2 incorrect answers?

(c) How many ways can 4 balls be selected from a jar having 12 balls of different colors?

(d) There are 15 students in a class. In how many ways can the students be selected in a group of 9 to play baseball?

Objective 4; Example 4

16. (a) You have to read five books for an English literature class during this semester. In how many orders can you read the books?

(b) You are dressing a small child to go out in the rain. You are putting on a right boot, a left boot, a rain hat, and a rain coat. In how many ways can you dress the child?

(c) A radio announcer has fifteen commercials, 5 of which he will play between noon and 1 p.m. In how many ways can he play 5 commercials?

(d) A person has eight items in a shopping cart. In how many ways can the grocery checker ring up on the cash register the cost of five of these items?

Objective 5; Example 5

17. Determine if the following involves a combination or a permutation. Calculate the combination or permutation.

(a) Your son has a list of 17 boys he wants to invite to a slumber party. You say he has to limit his party to 11 boys. In how many ways can he invite the 11 boys?

(b) An elementary school is licensing bicycles with the letters A, B, C, D, E, F, G. A license has three letters. How many different licenses can it issue?

(c) There are eight candidates running for a civic office. In how many ways can the local television studio arrange for each candidate to talk for 3 minutes during a 30-minute broadcast?

(d) You are at a pet store buying baby chicks. The store has 30 and you want 2. In how many ways can you select your 2 chicks?

(e) A school club has eight members of an executive council. The council will elect one person as the president and another as secretary/treasurer. In how many ways can this be done?

OBJECTIVES

After completing this chapter, you should be able to

A. General principle of counting

1 Apply the general principle of counting in order to solve for all possible outcomes, using both numerical computation and tree diagramming. **(2–5)**

B. Combinations and permutations

2 Calculate expressions using $N!$ **(6–8)**

3 Calculate the combination of N things taken s at a time using the formula: **(9–11)**

$${}_NC_s = \frac{N!}{s!\,(N-s)!}$$

4 Calculate the permutation of N things taken s at a time using the formula: **(12–14)**

$${}_NP_s = \frac{N!}{(N-s)!}$$

5 Given situations involving "N things taken s at a time," determine whether it is a combination or permutation and perform the appropriate calculation. **(15–17)**

C. Definition of probability

6 Solve problems of ratio, writing the notation for a ratio in the three recognized forms. **(18–22)**

7 Define probability and relate the formula for probability, $P(A) = \frac{n_A}{n}$, to your definition. (23–24)

8 Solve for probability for a situation where the conditions are given. (25–26)

D. Sample space

9 Produce a sample space during the process of solving problems of probability. (27–29)

E. Principles of probability

10 Given the probability of an event occurring, determine the probability of the event not occurring. (30–32)

11 Define and give an example of mutually exclusive events. Use the following formula to calculate the probability that event X or event Y will occur (if X and Y represent mutually exclusive outcomes): (33–35)

$$P(X \text{ or } Y) = \frac{n_X + n_Y}{n} = \frac{n_X}{n} + \frac{n_Y}{n} = P(X) + P(Y)$$

12 Given that event X and event Y are not mutually exclusive outcomes, calculate the probability that event X or event Y or both will occur, using the following formula:

$$P(X \text{ or } Y \text{ or both}) = \frac{n_X + n_Y - n_{XY}}{n} = P(X) + P(Y) - P(XY)$$

13 Define and give an example of exhaustive categories. (39)

14 Classify categories of events as either (a) mutually exclusive but not exhaustive or (b) mutually exclusive and exhaustive. (40–42)

15 Use the following formula to calculate probability problems involving mutually exclusive and exhaustive categories: (43–45)

$$P(X) + P(Y) + P(Z) = 1$$

16 Compute the probability of independent events using the formula: (46–49)

$$P(XY) = P(X) \cdot P(Y)$$

17 Solve problems involving conditional probability using the formula: (50–53)

$$P(G/H) = \frac{n_{GH}}{n_H} = \frac{\frac{n_{GH}}{n}}{\frac{n_H}{n}} = \frac{P(GH)}{P(H)}$$

13 Business/Consumer Mathematics

There are very few, if any, professions that do not require some business transactions. Any kind of professional office must make purchases, pay bills and taxes, and keep records of its business affairs. Budgets for the coming year and statements of financial condition must be prepared. Even teachers or other persons who do not work in an office must still conduct their own private financial affairs. The following business mathematics topics are therefore presented in order to meet these fundamental needs of most professional persons. (1)

A. SALES AND PROPERTY TAXES

Sales tax provisions

A large majority of state legislatures have approved taxes on retail sales as a means of securing revenue. Retail merchants are responsible for collecting these taxes at the time of each sale and relaying them to the state. In a number of states, cities also have the option of adopting a municipal sales tax in order to obtain local funds. Sales taxes ordinarily apply only within the jurisdiction of the taxing body. Thus, merchandise sold outside a state (such as by mail or by telephone) is not subject to that state's sales tax. In some cases, sales tax may apply to only part of a purchase. For example, food and prescription drugs are often exempt, whereas other purchases in a grocery or pharmacy would be subject to sales tax. (2)

Objective 1: Computing sales tax

Sales tax is usually designated as a percentage rate. In general, (3)

$$\text{Rate} \times \text{Price} = \text{Tax}$$

(Sales tax is always rounded to the next cent if the third decimal place is 5 or larger.)

◆ **Example 1** Determine the sales tax on a purchase of \$15.89, if the tax rate is 4%. (4)

$$4\% \times \text{price} = \text{tax}$$
$$0.04 \times \$15.89 = \text{tax}$$
$$\$0.6356 = \text{tax}$$
$$\$0.64 = \text{tax}$$

The sales tax on this purchase adds \$0.64, making the total price \$15.89 + \$.64 = \$16.53.

● **Practice 1** Find sales tax and total price on the following purchases: (5)

(a) \$9.32 with 5% tax rate (b) \$31.79 with 6% tax rate

See also Exercise 5.

On occasion, only the total price (including tax) may be known and it would be helpful to know the marked price of the merchandise, as follows: (6)

◆ **Example 2**

Objective 2: Finding marked price

(a) The total price of a pocket calculator, including a 4% sales tax, was \$46.80. What was the marked price of the calculator? (7)

We know that sales tax (at 4% of marked price) must be added to marked price to obtain the total cost. Thus,

$$\begin{aligned}
\text{Marked price} + \text{Sales tax} &= \text{Total} \\
\text{price} + 4\% \text{ of price} &= \text{Total} \\
p + 0.04p &= \$46.80 \\
1.04p &= \$46.80 \\
p &= \frac{46.80}{1.04} \\
p &= \$45.00
\end{aligned}$$

The marked price* of the calculator was \$45.00.

*Observe that the sales tax is 4% of the marked price, rather than 4% of the total price. It is thus *not* possible to solve transactions of this type by finding 4% of the total price and subtracting.

● **Practice 2** Calculate the marked price for each purchase below: (8)

(a) After a 6% sales tax, the total price was $55.12.
(b) The total price was $30.45, including a 5% sales tax.
(c) After a 4% sales tax, a total of $70.72 was paid.

See also Exercise 5.

The typical means by which cities and counties secure revenue is the property tax. This method is also used by school districts, if they are organized as an independent, taxing body. Property tax applies to real estate (land and the building improvements on it), as well as to personal property (cash, cars, household furnishings, appliances, jewelry, etc.) (9)

Property tax provisions

Two things affect the amount of tax that will be paid: (1) the assessed value of the property, and (2) the tax rate. Property tax is normally paid on the *assessed value* of property, which is usually determined by taking a designated percent of its estimated fair market value. Thus, the tax rate alone is not a dependable indication of how high taxes are in a community; one must also know how the assessed value of property is determined. There are several ways in which the property tax *rate* of an area might be expressed; however, rates are now most often expressed as a per cent or as an amount per $100 of assessed valuation. (10)

Computation of property tax, like sales tax, is a variation of the basic per cent formula. That is, (11)

Objective 3: Computing property tax

$$\text{Rate} \times \text{Valuation} = \text{Tax}$$

◆ **Example 3**

Tax

(a) A property tax rate (R) of $2.43 per $100 (or "per C") is applied to a home assessed at $18,500 ($V$). Compute the property tax (T) due on the home: (12)

$R = \$2.43$ per C $\qquad R \times V = T$

$V = \$18{,}500$ $\qquad 2.43 \times 185 = T$

$= 185$ hundred $\qquad \$449.55 = T$

$T = ?$

Rate

(b) A lawyer paid $642 in property taxes on his office building, which was assessed at $26,750. Find the tax rate (as a per cent correct to one decimal place). (13)

$R = ?$ $\qquad R \times V = T$

$V = \$26{,}750$ $\qquad R(\$26{,}750) = \$642$

$T = \$642$ $\qquad 26{,}750R = 642$

$$R = \frac{642}{26{,}750}$$

$$R = 0.024 \quad \text{or} \quad 2.4\%$$

Assessed value

(c) In a city where the school tax rate was \$1.18 per C, a homeowner paid \$177 in school taxes. What is the assessed value of this home? (14)

$R = \$1.18$ per C $\qquad R \times V = T$

$V = ?$ $\qquad 1.18V = \$177$

$T = \$177$ $\qquad V = \frac{177}{1.18}$

$V = 150$ hundred or \$15,000

Notice that the initial value obtained for V represents that many *hundreds* of dollars of valuation.

● **Practice 3** Find the property tax, the rate, or the assessed value, as indicated below: (15)

	Rate	*Assessed value*	*Tax*
(a)	1.7%	\$10,500	——
(b)	\$1.75 per C	9,400	——
(c)	2.5%	——	\$475
(d)	\$1.85 per C	——	259
(e)	?%	\$27,500	440
(f)	? per C	18,000	477

(g) San Gabriel County has a property tax rate of \$2.36 per \$100 valuation. Determine the obligation of a citizen whose property is assessed at \$12,500.

(h) Residents of Wooten Heights are subject to a 2.4% property tax. Find the assessed value of an unimproved lot on which the tax is \$84.

(i) A property tax of \$572 was paid on a commercial building. If the tax rate was \$1.76 per C, what was the assessed value of the building?

(j) A property owner in Yates County paid \$240 in property tax on a home assessed at \$16,000. What per cent is the county tax rate?

(k) A businessman in Greenville paid \$517 in property taxes. Compute the tax rate per \$100 if the assessed value of his property was \$27,500.

See also Exercise 6.

B. FINANCIAL STATEMENTS

The bookkeeping records of every business are audited periodically (usually quarterly or annually) to determine the financial condition of the business: the volume of business, the profit (or loss), how much these things have changed, how much the business is actually worth, and so on. Similar analysis is often done on the financial affairs of private citizens. As a means of presenting this important information, two financial state- (16)

ments are normally prepared: the income statement and the balance sheet.

Essentials of the income statement

(17) The *income statement* covers business operations during a certain *period of time*. The basic calculations of an income statement are these:

Net sales
−Cost of merchandise
Gross profit
−Operating expenses
Net profit (or net loss)

(18) Notice that *gross profit* (frequently called *margin*) is the amount remaining after the merchandise has been paid for. (An organization that sells only services—such as a lawyer or counseling agency—would have no merchandise to pay for; thus, all of its income would represent gross profit.) Out of this gross profit must be paid all the *operating expenses* or *overhead*. (Overhead consists of expenses such as salaries, rent, utilities, supplies, property taxes, insurance, etc. *Depreciation*, or loss in value of business equipment or buildings, is also an allowable business expense.)

(19) Whatever amount remains after these operating expenses is the clear or spendable profit—the *net profit*. (It should be noted that the government will require some of this net profit to be spent for taxes! Thus, many firms add two additional items to their income statements in order to show the taxes due on the net profit and the net profit after taxes.) Furthermore, the computation of net profit does *not* imply that the firm ends the year with this much cash on hand. Some of this profit may still be owed to the business by credit buyers. Some of it may also have already been spent for real estate, equipment, or other investments that are not considered operating expenses.

Percents on the income statement

(20) In order to effectively analyze expenses or to compare this year's income statement with preceding ones, it is customary to convert the dollar amounts to *percents of the total net sales*. These computations are based on the basic percent equation, as shown in the following example:

WISE MANUFACTURING, INC.

Income Statement for Year Ending December 31, 19xx

Net sales	\$50,000	100%
Cost of goods sold	− 37,500	−75
Gross profit	\$12,500	25%
Operating expenses	− 7,500	−15
Net profit	\$ 5,000	10%

The fact that 25% of the net sales (income) was gross profit is computed (21)

$$? \% \text{ of Net sales} = \text{Each item}$$
$$? \% \times \text{Net sales} = \text{Gross profit}$$
$$? \% \times \$50{,}000 = \$12{,}500$$
$$50{,}000r = 12{,}500$$
$$r = \frac{12{,}500}{50{,}000}$$
$$r = 25\%$$

The remaining percents can be computed in a similar manner. Notice that the percents can be checked by subtracting from net sales (which represents 100%).

Objective 4: Completing income statement

The preceding example of an income statement was highly condensed. The following example presents an itemized income statement: (22)

◆ Example 4

OLD WORLD IMPORTS

Income Statement for Year Ending December 31, 19xx

Income from sales:				
Total sales	$102,500		102.5%	
Less: Sales returns and allowances	−2,500		−2.5	
Net sales		$100,000		100.0%
Cost of goods sold:				
Inventory, January 1	$ 25,700			
Purchases	+70,800			
Goods available for sale	$ 96,500			
Inventory, December 31	−26,500			
Cost of goods sold		−70,000		−70.0
Gross profit on sales		$ 30,000		30.0%
Operating expenses:				
Salaries	$ 12,500		12.5%	
Rent	3,500		3.5	
Utilities	700		.7	
Office supplies	200		.2	
Depreciation	1,500		1.5	
Insurance	900		.9	
Miscellaneous	+ 700		+.7	
Total operating expense		−20,000		−20.0
Net profit		$ 10,000		10.0%

Of the various topic breakdowns, only "cost of goods sold" requires particular mention. To the value of the inventory on hand at the beginning of the year is added the total of all merchandise bought during the year. This gives the total value of all merchandise which the company had available for sale during the year. Since some of this merchandise remains unsold, however, the value of the inventory on hand at the end of the year must be deducted, in order to obtain the value of the merchandise which was actually sold. (23)

Of the percents indicated in the right-hand column, most important by far to the owners, to possible investors, or to potential creditors is the final per cent: What per cent of sales was the net profit? This is usually the first question asked by anyone considering investing money in a business. (24)

● **Practice 4** Copy and complete the following income statement, computing percents to the nearest tenth: (25)

WEST PRODUCTS, INC.

Income Statement for Year Ending December 31, 19xx

Income from sales:				
Gross sales	$406,000		--- %	
Sales returns and allowances	6,000		---	
Net sales		$ ---		--- %
Cost of goods sold:				
Inventory, January 1	$100,000			
Purchases	266,000			
Goods available for sale	$ ---			
Inventory, December 31	106,000			
Cost of goods sold		---		---
Gross profit on operations		$ ---		--- %
Expenses:				
Salaries	$ 75,000		--- %	
Rent and utilities	25,000		---	
Supplies	4,000		---	
Insurance	6,000		---	
Advertising	7,500		---	
Depreciation	2,000		---	
Miscellaneous	500		---	
Total operating expenses		---		---
Net profit on operations		$ ---		--- %
Income taxes		4,000		---
Net income		$ ---		--- %

See also Exercise 9.

Essentials of the balance sheet

The purpose of the *balance sheet* is to give an overall picture of what a business is worth at a specific *point in time*. On a balance sheet are listed all of a business's *assets* (possessions and money owed to the business) and all its *liabilities* (money which the business owes to someone else). The balance remaining after the liabilities are subtracted from the assets is the *net worth* of the business. Thus, (26)

$$\text{Assets} = \text{Liabilities} + \text{Net worth}$$

The general headings of a balance sheet are ordinarily broken down as follows: Assets are divided into *current assets* and *fixed assets*. Among the current assets are cash, bank accounts, accounts receivable (accounts owed to the business by customers who have bought merchandise "on credit"), notes receivable (money owed to the business by customers who have signed a written promise to pay by a certain date), and the merchandise inventory. Fixed assets are the buildings, furnishings, machinery, equipment (all listed at their current book value after depreciation), and land (which may not be depreciated). (27)

The liabilities of a business are separated according to *current liabilities* and *fixed liabilities*. Current liabilities include the accounts payable (accounts owed elsewhere for goods purchased on credit), notes payable, and interest and taxes due within a short period of time. Fixed liabilities are such long-term debts as mortgages and bonds. (28)

Other common terms used instead of net worth are *net ownership*, *net investment*, *proprietorship* (for an individually-owned business), *owner's equity*, *stockholders' equity* (for a corporation), and *capital*.

Percents on the balance sheet

The example on page 334 is a typical balance sheet. Each asset is represented as a *per cent of the total assets*. That is, (29)

$$?\% \text{ of Total assets} = \text{Each item}$$

Recall that total liabilities plus net worth is the same as the total assets. Thus,

$$\text{Net worth} = \text{Total assets} - \text{Total liabilities}$$

Income statement vs. balance sheet

Although both the income statement and the balance sheet are important indicators of a firm's financial condition, neither is sufficient by itself to show a complete picture. For instance, the income statement might show a nice profit. It is the balance sheet, however, that indicates to a large extent what the business is worth and whether there is a "financial cushion" to fall back on to meet obligations if sales decline. (30)

Thus, both statements are needed in order to assess how financially sound a business is. (Actually, such statements covering several years' time need to be available.)

◆ **Example 5** (31)

Objective 5: Completing Balance Sheet

R. B. SMITH DISTRIBUTORS, INC.

Balance Sheet, December 31, 19xx

Assets				
Current assets:				
Cash on hand	$ 200		.2%	
Cash in bank	4,200		3.5	
Accounts receivable	32,400		27.0	
Inventory	+13,200		+11.0	
Total current assets		$ 50,000		41.7%
Fixed assets:				
Plant site	$12,600		10.5%	
Building	33,000		27.5	
Equipment	19,000		15.8	
Delivery trucks	+5,400		+4.5	
Total fixed assets		+70,000		+58.3
Total assets		$120,000		100.0%
Liabilities and Net Worth				
Current liabilities:				
Accounts payable	$26,400		22.0%	
Notes payable	+4,800		+4.0	
Total current liabilities		$ 31,200		26.0%
Fixed liabilities:				
Mortgage		+34,800		+29.0
Total Liabilities		$ 66,000		55.0%
Net Worth		+54,000		+45.0
Total Liabilities and Net Worth		$120,000		100.0%

Practice 5 Copy and complete the following balance sheet, computing percents to the nearest tenth. (32)

HERRSCHER ANTIQUES

Balance Sheet, August 31, 19xx

Assets				
Current assets:				
Cash	$ 5,000		--- %	
Accounts receivable	15,000		---	
Notes receivable	3,000		---	
Merchandise inventory	22,000		---	
Total current assets		$ ---		--- %
Fixed assets:				
Building (less depreciation)	$18,000		--- %	
Furnishings and equipment (less depreciation)	9,000		---	
Trucks (less depreciation)	13,500		---	
Land	4,500		---	
Total fixed assets		---		---
Total assets		$ ---		--- %
Liabilities and Proprietorship				
Current liabilities:				
Accounts payable	$13,500		--- %	
Taxes payable	800		---	
Note payable	1,000		---	
Total current liabilities		$ ---		--- %
Fixed liabilities:				
Mortgage		22,500		---
Total liabilities		$ ---		--- %
Proprietorship:				
S. B. Herrscher, capital		---		---
Total liabilities and proprietorship		$ ---		--- %

See also Exercise 10.

C. MARKUP: BASED ON COST AND ON SELLING PRICE

Merchants offer their goods for sale with the intention of recovering expenses and making a profit. On the other hand, prices must also meet competition from other businesses. It is therefore important that business people have a thorough understanding of the factors involved in determining the selling price of their merchandise. (33)

The process and the terminology used in pricing individual merchandise is similar to that used for the income statement that has already been introduced. That is, the *cost* of a piece of merchandise includes not only its catalog price but also insurance, freight, and any other charges incurred prior to delivery of the merchandise. The difference between this cost and the selling price is the *gross profit* (or *markup* or *margin*). As noted previously, this gross profit includes the *overhead* expenses such as salaries, rent, office supplies, utilities, advertising, etc. Whatever remains after these expenses have been paid constitutes the clear or *net profit*. The net profit remaining is often quite small; indeed, expenses often exceed the margin, in which case there is a loss. (34)

The following diagram illustrates the breakdown on selling price: (35)

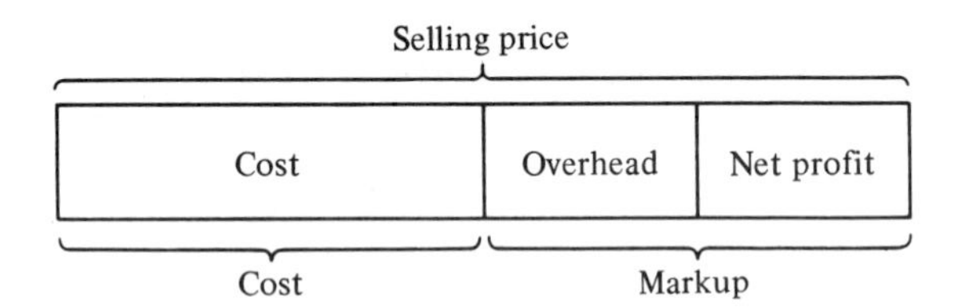

Cost plus *Markup* (or margin or gross profit) always equals *Selling price*. From this relationship is obtained the fundamental selling price formula:

$$C + M = S$$

where C = cost, M = markup, and S = selling price. Since the Margin (markup) is composed of overhead plus net profit, the same formula can be restated

$$C + (OH + P) = S$$

where OH = overhead and P = net profit.

Before calculating selling price, let us consider the following situation: Suppose a store bought a dress for \$20 and sold it for \$25. The "gross profit" is \$5; however, if overhead expenses are \$8, then there is actually a \$3 net *loss* on the sale. This illustrates a vital fact about selling—the simple fact that an item sells for more than it cost does not necessarily mean that any profit is made. Now let us consider some examples where selling price is computed to include a specific markup. (36)

◆ **Example 6** A hardware manager has found that expenses plus the net profit he wishes to make usually run $33\frac{1}{3}\%$ of the cost of his goods. For what price should he sell a fireplace set which cost him $24? (37)

Objective 6a
Markup based on cost

Since the markup equals $33\frac{1}{3}\%$ or $\frac{1}{3}$ of the cost, and $\frac{1}{3}$ of \$24 = \$8, then the selling price is \$24 + \$8 = \$32.

It is customary in business, however, to price merchandise by the following, more efficient method: Since the markup equals $\frac{1}{3}$ of the cost $(M = \frac{1}{3}C)$,

$$C + M = S$$

$$C + \frac{1}{3}C = S$$

$$\frac{4}{3}C = S$$

$$\frac{4}{3}(\$24) = S \qquad \text{(since cost equals \$24)}$$

$$\$32 = S$$

If the business above uses a $33\frac{1}{3}\%$ markup on cost for all of its merchandise, then the basic calculation $\frac{4}{3}C = S$ would be used each time an item is priced. This coefficient $\frac{4}{3}$ (or 1.33) is called the selling price factor. The *selling price factor* (*Spf*) is a number which can be multiplied times the cost to obtain the selling price, as expressed by the formula (38)

Selling price factor

$$Spf \times C = S$$

Thus, in actual practice, a retail store would perform only the calculation based on $Spf \times C = S$. For example, if an item cost \$36, then

$$\frac{4}{3}C = S$$

$$\frac{4}{3}(\$36) = S$$

$$\$48 = S$$

● **Practice 6** Use the method illustrated in Example 6 in order to price the following merchandise. (Markup may be expressed either as a fraction or as a decimal, whichever is more convenient.) Also name the selling price factor that corresponds to each markup: (39)

Markup	*Cost*	*Selling price*	*Spf*
(a) 25% of cost	\$16	____	____
(b) 20% of cost	35	____	____
(c) 45% of cost	40	____	____
(d) 38% of cost	50	____	____

Have you observed an easy method for determining *Spf*?

See also Exercise 7.

Many business expenses are calculated as a percent of net sales (for example, salespeople's commissions or sales taxes). Many companies take inventory at its sales value. The income statement lists the firm's sales, all of its expenses, and its net profit; each of these items—including the gross profit or margin—is then computed as a percent of net sales. For these reasons, most business people prefer to price merchandise using a markup based on sales (selling price), as the following example demonstrates. (40)

◆ **Example 7**

(a) A store's income statement indicates that its expenses and net profit together are 30% of sales. On that basis, price an item which cost \$35. (41)

As before, we start with the equation $C + M = S$. We know that the markup (overhead plus net profit) equals 30% of sales, or $M = \frac{3}{10}S$. Thus we substitute $\frac{3}{10}S$ for M in the selling price formula:

Objective 6b: Markup based on selling price

$$C + M = S$$

$$C + \frac{3}{10}S = S$$

$$C + \frac{3}{10}S - \frac{3}{10}S = S - \frac{3}{10}S$$

$$C = \frac{7}{10}S$$

The \$35 is next substituted for the cost C, and the equation is solved for the selling price S:

$$C = \frac{7}{10}S$$

$$\$35 = \frac{7}{10}S$$

$$\frac{10}{7}(\$35) = \frac{10}{7}\left(\frac{7}{10}S\right)$$

$$\$50 = S$$

Thus the selling price must be \$50 so that the markup (\$15) will be 30% of this selling price (30% of \$50 = \$15). Notice in this example, $\frac{10}{7}$ was the selling price factor because $\frac{10}{7}$ multiplied times the cost, \$35, produced the selling price, \$50.

(b) If a firm's expenses are 24% of sales and the net profit is 8% of sales, price an item which cost \$85. (42)

This example illustrates a situation where decimals are more convenient than fractions. Observe that markup equals $24\%S$ plus $8\%S$, or $M = 32\%$ of sales:

$$C + M = S$$
$$C + 0.32S = S$$
$$C + \cancel{0.32S - 0.32S} = S - 0.32S$$
$$C = 0.68S$$
$$\$85 = 0.68S$$
$$\frac{85}{0.68} = \frac{0.68S}{0.68}$$
$$\$125 = S$$

(43) Observe above that since $\frac{\$85}{0.68} = \frac{1}{0.68}(\$85)$, the selling price factor would be $\frac{1}{0.68} = 1.4706$. We would then have $Spf \times C = 1.4706 \times \$85 = \$125$. You will not be asked to compute such selling price factors (since it involves tedious division). However, you should be able to *indicate* the selling price factor as $\frac{1}{0.68}$.

Spf with decimals

(44) ● **Practice 7** For each of the following items

1 Use the formula $C + M = S$ to compute the selling price where markup is a percentage of sales. (Express markup as either a fraction or a decimal, as convenient.)

2 Use a fraction to indicate the selling price factor that corresponds to each markup:

	Markup	*Cost*	*Selling price*	*Spf*
(a)	20% of sales	$32	____	____
(b)	25% of sales	45	____	____
(c)	48% of sales	91	____	____
(d)	Overhead: 32% of sales Net profit: 8% of sales	54	____	____
(e)	Expenses: 22% of sales Net profit: 6% of sales	90	____	____

See also Exercise 7.

(45) A merchant may wish to know what margin based on sales is equivalent to a margin based on cost, or vice versa. To find the percent of markup requires the formula

Objective 7: Finding percent of markup

$$?\% \text{ of } C = M \qquad \text{(based on cost)}$$

or

$$?\% \text{ of } S = M \qquad \text{(based on sales)}$$

where C = cost, S = selling price, and M = markup.

◆ **Example 8** An electric mixer cost $36 and required $4 freight. It sold for $50. What was the markup percent (a) based on cost? (b) based on selling price? (46)

The actual cost of the mixer was $\$36 + \$4 = \$40$; the markup $M = S - C = \$50 - \$40 = \$10$.

(a) *Markup on cost*

$$?\% \text{ of } C = M$$
$$?\% \times \$40 = \$10$$
$$40r = 10$$
$$r = \frac{1}{4}$$
$$r = 25\% \text{ on cost}$$

(b) *Markup on selling price*

$$?\% \text{ of } S = M$$
$$?\% \times \$50 = \$10$$
$$50r = 10$$
$$r = \frac{1}{5}$$
$$r = 20\% \text{ on sales}$$

● **Practice 8** For each of the following items, determine (1) What percent markup on cost was obtained? and (2) What was the percent of gross profit based on sales? (47)

(a) A book cost $4 and sold for $6.
(b) A fishing rod cost $18 and sold for $24.
(c) A phonograph cost $35 and freight was $5. It sold for $56.
(d) A jacket listed for $43 and tax and postage totaled $5. It retailed for $60.

See also Exercise 8.

D. SIMPLE INTEREST AND AMOUNT

Persons who rent buildings or equipment expect to pay for the use of another person's property. Similarly, rent paid for the privilege of borrowing another's money is called *interest*. The amount of money that was borrowed is the *principal* of a loan. A certain percentage of the principal is charged as interest. The per cent or *rate* is quoted on a yearly (per annum) basis, unless otherwise specified. The *time* is the number of days, months, or years for which the money will be loaned. (In order to make rate and time correspond, time must be converted to years, since rate is always given on a yearly basis.) (48)

Basic simple interest formulas

Simple interest is calculated on the *whole* principal for the entire length of the loan, using the formula (49)

$$I = Prt$$

where I = interest, P = principal, r = rate, and t = time. The *amount* due at the end of the loan (also called *maturity value*) is the sum of the principal plus the interest, expressed by

$$S = P + I$$

where $S =$ amount (or maturity value or sum), $P =$ principal, and $I =$ interest.

◆ **Example 9A**

(a) Find the simple interest and amount (maturity value) due on a loan of $500 at 8% for 9 months. (50)

$P = \$500$ | $I = Prt$ | $S = P + I$

$r = \frac{8}{100}$ | $= \$500 \times \frac{8}{100} \times \frac{3}{4}$ | $= \$500 + \$30$

$t = 9$ mos. | $I = \$30$ | $S = \$530$

$= \frac{9}{12}$ or $\frac{3}{4}$ yr.

Objective 8: Using basic simple interest formulas

(b) How long will it take for $2,400 to amount to $2,500 at a simple interest rate of 10%? (51)

First, the interest is determined from the formula $P + I = S$:

$I = S - P$

$= \$2{,}500 - \$2{,}400$

$I = \$100$

$P = \$2{,}400$

$r = \frac{10}{100} = \frac{1}{10}$

$t = ?$

$I = Prt$

$\$100 = \$2{,}400 \times \frac{1}{10}t$

$100 = 240t$

$\frac{100}{240} = \frac{240t}{240}$

$\frac{5}{12} = t$

$t = \frac{5}{12}$ yr. or 5 months

A person who borrows money usually signs a written promise to repay the loan; this document is called a *promissory note*, or just a *note*. The quantity of money that appears on a note is the *face value* of the note. If an interest rate is mentioned in the note, the face value of this simple interest note is the *principal* of the loan. Interest is computed on the whole principal for the entire length of the loan, and the *maturity value* (principal plus interest) is repaid in a single payment on the *due date* (or maturity date) stated in the note. If no interest is mentioned, the face value is the maturity value of the note. The person who borrows the money is called the *maker* of the note, and the one to whom the money will be repaid is known as the *payee*. (52)

NAME George D. Rogers DUE DATE Oct. 5 NO. 08335 $ 410.00

$400 PRIN | $10 INT | X INS

Austin, Texas, July 7, 19--

ON DEMAND, or if no demand is made, then ______ after date, without grace, for value received, I, we, and each of us, as principals, promise to pay to the order of NORTH AUSTIN STATE BANK, at Austin, Texas,

the sum of Four hundred and no/100-- DOLLARS,

with a **FINANCE CHARGE** of $10 which is an **ANNUAL PERCENTAGE RATE** of 10 % from July 7 until maturity, and if not then paid, at the rate of 10% per annum until paid.

In the event of default in the payment of this note, when due, or in the performance of any agreement contained in the security agreement if any is taken to secure payment hereof, or in the event the holder deems itself insecure, then the holder of this note shall have the option, without demand or notice, to declare the principal and interest at once due and payable and to exercise any and all other rights or remedies provided in this note and in the security agreement, if any, including the right to set off against this note and all other liabilities of the undersigned to the holder, all money or other property in its possession held for or owed to the undersigned.

Each maker, surety, endorser, and guarantor of this note hereby waives presentment for payment or acceptance, notice of non-payment or dishonor, protest, notice of protest, and diligence in the collection hereof or in filing suit hereon and agrees that liability for the payment hereof shall not be affected or impaired by any release of or change in the security, if any, or by any extension in the time for payment; and further agrees to pay all costs and expenses of collection incurred by the holder, and if this note is placed in the hands of an attorney for collection after maturity, or is collected by legal proceedings of any kind, to pay a reasonable attorney's fee, which shall not in any event be less than the sum of $50.00 and shall bear interest at the rate of 10% per annum from the date of its accrual.

Payment of this note is secured by all money or other property of the undersigned now or at any time hereafter in the possession of the holder in any capacity and also

by ______

INSURANCE AGREEMENT

CREDIT LIFE INSURANCE is not required to obtain this loan. No charge is made for credit insurance and no credit insurance is provided unless the borrower checks the appropriate statement below:

(a) The cost for Credit Life Insurance alone will be $______ for the term of credit.

☐ I desire Credit Life Insurance.

☒ I do not desire Credit Life Insurance.

______ Date

______ Signature for Insurance Acknowledgment

Address 4206 Running Brook Lane

Phone: Bus. 472-8911 Res. ______

Age ______ Filing Fee ______

George D. Rogers

CO-SIGNERS ARE EXPECTED TO PAY IN THE EVENT THE BORROWER DOES NOT.

Bell's - Austin

Interest-bearing Promissory Note

A simple-interest bank note appears on page 342. (The face value of this note is \$400. Notice that this principal and the interest of \$10 are entered in the margin for the bank's convenience.) Notes between individuals are also usually simple interest notes. Many banks write notes for a maximum time of one quarter (three months). At the end of that time, the borrower must pay the accrued interest and may, if desirable, "renew" the note by signing a new note for the same amount. The new note would then reflect any change in interest rates that had occurred during the previous quarter. (53)

Bankers' rule

In computing simple interest on loans, financial institutions determine the "time" factor according to the *bankers' rule*. This method uses the exact number of days in the term of the loan and assumes a 360-day year. That is, by the (54)

$$\text{Bankers' rule:} \qquad \text{time} = \frac{\text{exact days}}{360}$$

For example, by the bankers' rule, the time between March 1 and June 1, rather than being three months or $\frac{1}{4}$ year, would be $\frac{92}{360}$ year. This is the most profitable of the various ways in which time can be expressed.

◆ **Example 9B**

Objective 8: (cont'd.) Using basic simple interest formulas

(c) At what rate will \$600 earn \$13.80 in interest after 92 days? (55)

$P = \$600$

$r = ?$

$t = \dfrac{92}{360}$

$I = \$13.80$

$$I = Prt$$

$$\$13.80 = \$600r\frac{92}{360}$$

$$13.80 = \overset{5}{\cancel{600}} \cdot \frac{92}{\underset{3}{\cancel{360}}}r$$

$$13.80 = \frac{460}{3}r$$

$$\frac{3}{460}(13.80) = \frac{3}{460}\left(\frac{460}{3}r\right)$$

$$\frac{41.40}{460} = r$$

$$.09 = r$$

$$9\% = r$$

(d) What was the principal on a loan where interest of \$12.40 was due after 62 days at 8%? (56)

$$P = ? \qquad I = Prt$$

$$r = \frac{8}{100} \qquad \$12.40 = P\frac{8}{100}\cdot\frac{62}{360}$$

$$t = \frac{62}{360} \qquad 12.40 = P\frac{\overset{1}{\cancel{8}}}{\cancel{100}}\cdot\frac{.62}{\underset{45}{\cancel{360}}}$$

$$I = \$12.40 \qquad 12.40 = \frac{.62}{45}P$$

$$\frac{45}{.62}(12.40) = \frac{45}{.62}\left(\frac{.62}{45}P\right)$$

$$\$900 = P$$

● **Practice 9** Solve each of the following as indicated: (57)

Find the interest and amount
(a) \$600 at 8% for 5 months
(b) \$1,400 at 10.5% for 6 months
(c) \$800 at 9% for 62 days
(d) \$720 at 10% for 92 days

(e) What is the interest rate if \$30.80 interest is due on a \$480 loan after 7 months?

(f) How long will it take for \$800 to amount to \$820 at 7.5% simple interest?

(g) What was the principal of a loan, if \$27 interest is due after 9 months at 8%?

(h) After 61 days at 10%, the interest on a note is \$12.20. What was the face value of the note?

(i) At what rate will \$1,800 amount to \$1,881.45 after 181 days?

(j) If interest of \$91 is due on a \$3,000 note written at 12%, for how many days was the money borrowed?

See also Exercise 11.

E. INSTALLMENT PLANS

No doubt you already have some knowledge of that uniquely American tradition, the *installment plan*. Almost all retail stores, as well as charge-card companies, finance companies and banks, offer some form of install- (58)

ment plan whereby a customer may take possession of a purchase immediately and, for an additional charge, pay for it later by a series of regular payments.

(59) We will consider the typical installment plan where a single purchase (automobile, furniture, washing machine, etc.) is made. In this case, the account will be repaid in a *specified number of payments*; hence, finance charges for the entire time are computed and added to the cost at the time of purchase.

The typical procedure for an installment purchase is as follows: The customer makes a *down payment* (sometimes including a trade-in) which is subtracted from the cash price to obtain the *outstanding balance*. A *carrying charge* is then added to the outstanding balance. The resulting sum is divided by the number of payments (usually weekly or monthly) to obtain the amount of each installment payment.

(60) This type of time-payment plan is also offered by finance companies and by banks, with the exception that there is no down payment (the amount of the loan becomes the outstanding balance) and the interest corresponds to a carrying charge. This type of installment note is shown on page 346.

Truth in Lending Law

(61) The Truth in Lending Law, which took effect in 1969, enables the consumer to determine how much the privilege of credit buying actually costs. The creditor must disclose both the *finance charge* (the total amount of extra money paid above the cash price) and also the *annual percentage rate* to which this is equivalent. The finance charge includes any carrying charge, service charge, interest, insurance, or other special charges levied only upon credit sales. The equivalent annual percentage rate (correct to the nearest fourth of a percent) can easily be computed using tables available from the Federal Reserve System.

(62) It should be pointed out that the Federal Truth in Lending Law does not establish any maximum rates or maximum finance charges (although some state laws do). Also, the seller may determine the finance charge by any desired method—such as a set charge, a set percent, by simple interest, etc. The law only requires that the buyer be fully informed of what actually is paid for the privilege of credit buying. As noted, the principal items which must be disclosed are (1) the finance charge and (2) the annual percentage rate.

◆ **Example 10A**

Objective 9: Finding monthly installment payments

(63) (a) When Andy Gibson bought a \$3,500 compact car, he received a \$900 trade-in allowance on his old car and also paid \$200 cash. The dealer arranged a two-year loan computed at 8% simple interest. What is Gibson's (1) monthly payment and (2) the total cost of his car? Determine (3) the finance charge and (4) the annual percentage rate required under the Truth in Lending Law.

STATEMENT OF LOAN

TOTAL OF PAYMENTS		$1,080.00
FINANCE CHARGE		$80.00
AMOUNT FINANCED		$1,000.00
INSURANCE:		
Credit Life	$ -0-	
Credit A/H	$ -0-	
Property	$ -0-	
Less Total Insurance		$ -0-
Filing Fees	$ -0-	
Recording Fees	$ -0-	
Less Total Fees		$ -0-
Net Amount Financed		$1,000.00
Previous A/C No.	-	
Previous Bal.	$ -0-	
Less Refund on Charges	$ -0-	
Less Refund on Insurance	$ -0-	
Net Balance		$ -0-
Net Proceeds		$1,000.00
ANNUAL PERCENTAGE RATE		14.50 %

$ 1,000.00 Austin, Texas, April 15, 19 X1.

For value received, I, we, and each of us, as principals, promise to pay to the order of NORTH AUSTIN STATE BANK, at 7600 Burnet Road, Austin, Texas, the sum of One thousand eighty and no/100------------ **Dollars, in** 12 **equal installments of $** 90.00 **each and a final installment of $** -0- **, the first installment to be due and payable on the** 15 **day of** May **, 19** X1 **, and one installment to be paid on the same day of each month thereafter until full payment of this note, and maturing on the** 15 **day of** April **, 19** X2.

4-15-X1 — **Date**

John D. Grant — **Signature**

Home Address 408 W. Perry Lane

Business Address P. O. Box 1425

Phone: Bus. 483-1880 **Res.** 475-3421

Age 37

Installment Note

(1) Gibson's monthly payment will be \$116, which is determined as follows:

Cash price	\$3,500	
Down payment	−1,100	(\$900 trade-in + \$200 cash)
Outstanding balance	\$2,400	
Finance charge	+ 384	$\left(\begin{array}{l} I = Prt \\ \quad = \$2{,}400 \times \frac{8}{100} \times \frac{2}{1} \end{array}\right)$
Total of payments	\$2,784	

$$\frac{\$2{,}784}{24} = \$116 \text{ per month}$$

(2) The total cost of his new car will be \$3,884; this is shown as follows:

Down payment:	Trade-in	\$ 900
	Cash	200
Total of monthly payments		2,784
Total cost		\$3,884

Finance charge

(3) The total cost includes an extra \$384 interest (or carrying charge) which represents the finance charge disclosure required by law.

Computing annual Percentage rate

(4) The annual percentage rate which the lender must disclose is found using the table on page 436. The instructions indicate that the finance charge should first be divided by the amount financed (the outstanding balance) and the result multiplied by 100. This gives the *finance charge per \$100 of amount financed*: **(64)**

$$\frac{\text{Finance charge}}{\text{Amount financed}} \times 100 = \frac{\$384}{\$2400} \times 100 = .16 \times 100 = 16$$

Thus the buyer pays \$16 interest for each \$100 being financed.

Using annual percentage rate table

To use the annual percentage rate table, first find the column headed by the number of monthly installment payments—24, in this example. Then look down the column until you locate the amount nearest in value to the \$16 calculated above; the table shows 16.08. Following that same line across to either outside column, *the annual percentage rate is 14.75% or 14¾%*. **(65)**

For practical purposes, this means that if interest were computed at 14¾% on the remaining balance each month and the \$116 monthly payment were applied first to the interest due and then to reduce the principal, then 24 payments of \$116 would exactly repay the \$2,400 loan plus the \$384 interest. Thus it is the 14¾% rate that the merchant must state is being charged—not the 8% used in computing simple interest.

Nominal vs. effective rates

When simple interest is calculated on the *original balance* for the entire length of a loan, this rate is sometimes called a *nominal interest* rate. By contrast, an annual rate which is applied only to the *balance due at the time of each payment* is called an *effective interest* rate or an *actuarial* rate. The Truth in Lending rates are effective rates. (66)

◆ **Example 10B**

(b) What actuarial (or effective) interest rate is equivalent to a nominal (or simple) interest rate of 10%, if there are six monthly installments? (67)

Multiplying *rate* × *time* gives the value we look for in the table:

$$10\% \times \frac{1}{2}\text{ year} = 5$$

This means the interest due is 5% of the amount borrowed. It also means the borrower pays \$5 interest for each \$100 borrowed.

Using the annual percentage rate table we look in the 6-month column for the value nearest to 5.00. The nearest value is 5.02. We now see that a simple (nominal) interest rate of 10% for six months corresponds to an actuarial (effective) rate of 17.00%, correct to the nearest $\frac{1}{4}\%$.

Effective interest rate verified

We will demonstrate below that *the simple (nominal) interest rate of 10% above requires the same amount of interest as an effective rate of 17%.* Assuming a \$400 loan, $\$400 \times 10\% \times \frac{1}{2}$ year gives \$20 simple interest. Thus, $\frac{\$420}{6}$ means the payment is \$70 per month. In the following calculation, however, each \$70 monthly payment first pays the interest due, and the remainder of each payment is then used to reduce the principal on which interest at 17% is computed for the following month: (68)

Month	*Prt = I*	*Payment to principal*	
1st	$\$400.00 \times \frac{17}{100} \times \frac{1}{12} = \$5.67$	\$ 64.33	
2nd	$335.67 \times \frac{17}{100} \times \frac{1}{12} = 4.76$	65.24	
3rd	$270.43 \times \frac{17}{100} \times \frac{1}{12} = 3.83$	66.17	
4th	$204.26 \times \frac{17}{100} \times \frac{1}{12} = 2.89$	67.11	
5th	$137.15 \times \frac{17}{100} \times \frac{1}{12} = 1.94$	68.06	
6th	$69.09 \times \frac{17}{100} \times \frac{1}{12} = .98$	69.02	
	\$20.07* +	\$399.93	= \$420.00

*The extra 7¢ interest here results from the annual percentage rate table being rounded to the nearest quarter of a percent.

Practice 10 Find each installment payment and/or annual percentage rate, as indicated: (69)

(a) A portable color television sells for $240 cash; or a $\frac{1}{3}$ down payment, a carrying charge of $20, and the balance in 12 payments. What is the monthly payment?

(b) A sofa would cost Carole Dickerson $400 cash. On the time payment plan, she may pay 10% down and finance the balance at 8% simple interest over 36 months. Find the monthly payment.

(c) Determine the effective interest rate equivalent to a nominal rate of 10% over 24 months.

(d) What actuarial (or effective) interest rate is equivalent to a simple interest rate of 16%, if there are 6 monthly payments?

(e) A piano costs $1,000. This piano may also be purchased for a 10% down payment and 18 monthly payments. For convenience, the manager computes the finance charge at 12% simple interest. Determine (1) the finance charge that must be disclosed, (2) the monthly payment, and (3) annual percentage rate that the seller must reveal.

(f) A bedroom suite is priced at $800. The time-payment plan requires a 25% down payment and 30 monthly payments. For simplicity, the finance charge is computed at 10% nominal (simple) interest. What are (1) the carrying (finance) charge and (2) the amount of each payment? (3) What annual rate must be disclosed under the Truth in Lending Law?

See also Exercise 12.

F. COMPOUND INTEREST AND AMOUNT

Money invested at a bank or savings institution earns compound interest. At compound interest, "interest is earned on interest"; that is, interest is earned, not only on the original principal, but also on all previously accumulated interest. For a simple interest investment, by comparison, interest is paid on the original principal only. This comparison is demonstrated in the following illustration. (70)

Suppose investor A invests $1,000 for three years at 6% simple interest. This investment would earn as follows: (71)

$$I = Prt \qquad\qquad S = P + I$$

$$= \$1{,}000 \times \frac{6}{100} \times 3 \qquad\qquad = \$1{,}000 + \$180$$

$$I = \$180 \qquad\qquad S = \$1{,}180$$

Long investment vs. repeated short investments

Now suppose investor B invests \$1,000 for only six months, also at 6% interest. In that case, $t = 6$ months or $\frac{1}{2}$ year, and investor B would have \$1,030 at the end of this six-month investment. Then suppose investor B reinvests this \$1,030 for another six months at 6%; interest of \$30.90 would be earned on this second investment, making the total amount \$1,060.90. If this procedure were repeated each six months for three years, investor B would have made six investments and the computations would be as follows: (72)

(73)

1st 6 months

$$I = Prt$$
$$= \$1{,}000 \times \frac{6}{100} \times \frac{1}{2}$$
$$I = \$30$$
$$S = \$1{,}030$$

2nd 6 months

$$I = Prt$$
$$= \$1{,}030 \times \frac{6}{100} \times \frac{1}{2}$$
$$I = \$30.90$$
$$S = \$1{,}060.90$$

3rd 6 months

$$I = Prt$$
$$= \$1{,}060.90 \times \frac{6}{100} \times \frac{1}{2}$$
$$I = \$31.83$$
$$S = \$1{,}092.73$$

4th 6 months

$$I = Prt$$
$$= \$1{,}092.73 \times \frac{6}{100} \times \frac{1}{2}$$
$$I = \$32.78$$
$$S = \$1{,}125.51$$

5th 6 months

$$I = Prt$$
$$= \$1{,}125.51 \times \frac{6}{100} \times \frac{1}{2}$$
$$I = \$33.76$$
$$S = \$1{,}159.27$$

6th 6 months

$$I = Prt$$
$$= \$1{,}159.27 \times \frac{6}{100} \times \frac{1}{2}$$
$$I = \$34.78$$
$$S = \$1{,}194.05$$

Thus, after three years, investor B's original principal would have amounted to \$1,194.05. Since investor A had only \$1,180 after that single, three-year investment, investor B made \$1,194.05–\$1,180 or \$14.05 more interest by making successive, short-term investments.

Compound interest

The preceding situation illustrates the idea of *compound interest*: each time that interest is computed, the interest is added to the previous principal; that total then becomes the principal for the next interest period. All money invested in savings accounts earns interest in this way. (74)

How banks earn compound interest on loans

Lending institutions, however, are not permitted to *charge* compound interest from their loan customers. Rather, bankers earn compound interest by lending one borrower's simple interest to some other borrower (75) who will pay interest on it. This is the primary reason why banks issue promissory notes for no more than 90 days.

Terminology

Interest is said to be "compounded" whenever interest is computed and added to the previous principal. This is done at regular intervals known as *conversion periods* (or just *periods*). Interest may be compounded annually (once a year), semiannually (twice a year), quarterly (four times a year), or monthly. It has also become popular for financial institutions to pay "daily interest" (interest compounded daily), although the interest is not actually added to accounts that often. The total value at the end of the investment (original principal plus all interest) is the *compound amount*. The *compound interest* is thus the difference between the compound amount and the original principal. The length of the investment is known as the *term*. The quoted interest *rate* is always the nominal (or yearly) rate. (76)

Compound amount formula

Compound amount may be found using the formula (77)

$$S = P(1 + i)^n$$

where $S =$ compound amount, $P =$ original principal, $i =$ interest rate per period, and $n =$ number of periods. In the investment described earlier, interest at 6% was compounded each six months; thus, the 6% annual interest divided into two periods would give $i = 3\%$ per period. (Notice that "i" is also equivalent to "rate $\times$ time.") Since there were two periods each year for three years, $n = 6$ periods altogether, as shown earlier. The previous compound amount of $1,194.05 could also be obtained similarly to the next example.

Compound amount table

Computation of compound amount can be greatly simplified through the use of a compound amount table—a list of the values obtained when the parenthetical expression $(1 + i)$ is used as a factor for the indicated numbers of periods, n. The tabular value must then be multiplied by the appropriate principal. The tables on pp.432- 3 give the values for $(1 + i)$ used to compute compound amount when annual rates of 5% and 6% are compounded monthly. (78)

◆ **Example 11** Find the compound interest and compound amount when $1,000 is invested for $2\frac{1}{2}$ years at a rate of 5% compounded monthly. (79)

Objective 10: Finding compound amount using tables

The time of $2\frac{1}{2}$ years is equivalent to 30 months (or $n = 30$ compounding periods). Using line 30 of the compound amount table for 5% annual interest, we find $(1 + i)^n = 1.132\ 854\ 2177$.

$P = \$1{,}000$	$S = P(1 + i)^n$
$n = 30$	$= \$1{,}000(1.132\ 854\ 2177)$
(5%)	$S = \$1{,}132.85$

The compound interest earned is thus $1,132.85 − $1,000, or a total of $132.85. By comparison, simple interest on $1,000 at 5% for $2\frac{1}{2}$ years (80) would total \$125. Thus, the investment at compound interest would earn \$132.85 − \$125 or \$7.85 more interest.

Effect of longer and frequent periods

It should be noted that the difference between compound interest and simple interest becomes increasingly greater (in percentage difference as well as in dollars) as an investment remains on deposit longer. Similarly, more interest is earned when compounding periods are more frequent; for example, an investment at 6% compounded monthly would earn more than the same investment at 6% compounded semiannually. (81)

● **Practice 11** Use the formula $S = P(1 + i)^n$, along with the compound amount table, in order to determine compound amount and interest on the following investments: (82)

(a) \$1,000 at 5% compounded monthly for 5 years.

(b) \$1,000 at 6% compounded monthly for 3 years.

(c) Interest at 5% compounded monthly for 4 years on (1) \$100, (2) \$200, and (3) \$400. How does doubling principal affect the amount of interest earned?

(d) \$1,000 at 6% compounded monthly for (1) 2 years, (2) 4 years, and (3) 8 years. What general effect do you observe when time is doubled?

(e) Find the simple interest on \$1,000 at 6% for 1 year. Compare with the same investment at 6% compounded monthly.

(f) Repeat part (e) for 5-year investments.

(g) Repeat part (e) for 7-year investments.

See also Exercise 13.

EXERCISES

Answer the questions of Exercises 1 through 4. Explain any "no" responses:

Objective 11: Sales and property tax

1. (a) Sales and property tax computations are applications of what basic type of calculation?
   (b) Is merchandise ordered by mail from another state exempt from that state's sales tax?
   (c) If a sales tax rate of 4% is applied on a purchase of \$7.85, then is the tax due 4% × \$7.85 = \$.314 = \$.31?
   (d) If a \$16.80 total price includes a 5% sales tax, can the marked price be found by taking 5% of \$16.80 and subtracting?
   (e) The term "property tax" covers both ______ and ______ property.
   (f) Property tax is computed on the ______ value of property.

Financial statements

2. (a) Which financial statement reports a firm's net profit or loss?
(b) Which financial statement reveals how much a company is worth?
(c) Do the income statement and balance sheet both describe a business' condition during a period of time?
(d) What basic business formula is derived from the income statement?
(e) Buildings and real estate are called ______ assets.
(f) Salaries and utilities are examples of ______.
(g) The amount remaining after merchandise has been paid for is the ______ profit; this may also be called ______ or ______.
(h) Accounts payable and notes payable are ______ liabilities.

Markup

3. (a) Which item on an income statement corresponds to markup in the selling price formula, $C + M = S$?
(b) Does a net profit always result if an item is sold for more than its wholesale cost?
(c) What would be the selling price factor equivalent to a markup of 20% on cost? of 20% on sales?
(d) Which gives a higher markup, 40% on cost or $33\frac{1}{3}$% on sales?
(e) An item cost \$20 and a markup of 25% of cost is used for pricing. Another item is also marked using 25% on cost. Since these items have the same % markup based on cost, would their equivalent % markups based on sales also be the same? (Try your own examples, if necessary.)

Simple and compound interest; installments

4. (a) Interest may be compared to ______ paid for the use of someone else's property.
(b) Explain briefly the difference between simple interest and compound interest.
(c) At what simple interest rate will an investment double itself in 12 years? (Hint: Choose any principal; what is the total interest after 12 years?)
(d) How long would it take for an investment to increase by half at 6% simple interest?
(e) How long (approximately) would it take for an investment to increase by half at 6% interest compounded monthly?
(f) Explain how bankers can earn compound interest on the money they loan, although the borrowers pay simple interest.
(g) Name two reasons bankers would require promissory notes to be renewed (and the interest due paid) at least every quarter.
(h) Which would earn more interest, 5% compounded semiannually or 5% compounded monthly?
(i) Name the two principal disclosures required under the Truth in Lending Law.

(j) Explain the difference between a nominal interest rate many installment purchases include and an effective (or actuarial) interest rate.

Objectives 1 and 2; Examples 1 and 2

5. (a) Find the sales tax and total price on the following:
(1) \$7.75 purchase with 4% tax rate
(2) \$24.78 purchase with 5% tax rate
(b) Determine the marked price of the following purchases:
(1) A total price of \$26.50 included a 6% sales tax.
(2) After a 4% sales tax, the total price was \$17.68.

Objective 3; Example 3

6. Compute property tax, rate, or assessed value, as indicated:

Rate	*Assessed value*	*Tax*
(a) 2.2%	\$12,500	
(b) \$1.95 per C		\$312
(c) ? per C	28,000	490

(d) Manor County's property tax rate is \$1.42 per \$100 valuation. What is the tax due on an undeveloped lot assessed at \$5,000?
(e) The town of Tryon has a 1.8% property tax rate. What was the assessed value on a home for which taxes of \$495 were paid?
(f) A property tax of \$387 was assessed on a home in Gainesville. If the assessed value of the home was \$18,000, what was the tax rate per \$100 of valuation?
(g) Citizens of Whiteville are subject to a tax rate of \$1.84 per C. Find the assessed value of an office building on which the tax is \$1,150.
(h) A property owner in Grand Rapids paid \$900 in taxes on a home assessed at \$37,500. What per cent is the city's tax rate?

Objective 6; Examples 6, 7

7. Use the formula $C + M = S$ in order to (1) compute each selling price. Also, (2) indicate the selling price factor in each case, using either a decimal or a fraction:

Markup	*Cost*	*Selling price*	*Spf*
(a) 20% on cost	\$45	?	?
(b) 20% on sales	28	?	?
(c) Expenses: 10% of cost Net profit: 5% of cost	60	?	?
(d) Overhead: 18% on sales Net profit: 7% on sales	24	?	?

Objective 7; Example 8

8. For the following sales, determine (1) What percent of cost was the gross profit? and (2) What percent of sales was the markup?

(a) A coat cost $35 and retailed for $56.

(b) A watch cost $48 and sold for $80.

(c) A styling comb listed for $18 and shipping charges were $2. The retail price was $24.

Objective 4; Example 4

9. Copy and complete the following income statement, computing percents to the nearest tenth:

PRESTRIDGE FURNITURE CO., INC.
Income Statement for Year Ending December 31, 19xx

Income from sales:				
Gross sales	$510,000		--- %	
Sales returns and allowances	10,000		---	
Net sales		$ ---		--- %
Cost of goods sold:				
Inventory, January 1	$ 80,000			
Purchases	298,000			
Goods available for sale	$ ---			
Inventory, December 31	78,000			
Cost of goods sold		---		---
Gross profit on operations		$ ---		--- %
Expenses:				
Salaries	$ 80,000		--- %	
Rent and utilities	20,000		---	
Maintenance	9,000		---	
Office supplies	6,500		---	
Insurance	5,000		---	
Advertising	10,000		---	
Depreciation	7,000		---	
Other	2,500		---	
Total operating expenses		---		---
Net profit on operations		$ ---		--- %
Income taxes		35,000		---
Net income		$ ---		--- %

Objective 5; Example 5

10. Copy and complete the following balance sheet, computing percents to the nearest tenth:

FULLERTON WHOLESALERS, INC.
Balance Sheet, September 30, 19xx

Assets				
Current assets:				
Cash	$ 4,000		--- %	
Accounts receivable	22,000		---	
Notes receivable	2,000		---	
Merchandise inventory	20,000		---	
Total current assets		$ ---		--- %
Fixed assets:				
Building (less depreciation)	$16,000		--- %	
Furnishings and fixtures (less depreciation)	3,200		---	
Delivery truck	2,800		---	
Land	10,000		---	
Total fixed assets		---		---
Total assets		$ ---		--- %
Liabilities and Stockholders' Equity				
Current liabilities:				
Accounts payable	$20,000		--- %	
Taxes payable	1,600		---	
Note payable	2,400		---	
Total current liabilities		$ ---		--- %
Fixed liabilities:				
Mortgage		32,000		---
Total liabilities		$ ---		--- %
Stockholders' equity		---		---
Total liabilities and equity		$ ---		--- %

Objective 8; Example 9

11. Find the indicated information about each simple interest loan:
(a) Find the interest and amount on $800 at 9% for 7 months.
(b) Find the interest and amount on $600 at 12% for 121 days.
(c) How long will it take for $750 to amount to $765 at 8% simple interest?
(d) What is the interest rate if $36.40 interest is due on a $720 loan after 182 days?

(e) What was the principal on a loan, if \$84 interest was due after 7 months at 12%?
(f) After 92 days at 9%, the interest on a promissory note was \$46. What was the face value of the note?

Objective 9; Example 10

12. Determine each installment payment and/or annual percentage rate, as indicated:
(a) A sewing machine sells for \$200 cash. On the "easy payment plan," the buyer pays a $\frac{1}{4}$ down payment, a \$30 carrying charge, and makes 18 monthly payments. How much is the monthly payment?
(b) A metal sculpture costs \$750 cash. The art gallery will also accept 20% down and finance the balance at 8% simple interest over 24 months. Compute the monthly payment.
(c) Compute the effective interest rate that is equivalent to a nominal (simple) interest rate of 10% over 18 months.
(d) A stereo is priced at \$400. The time-payment plan requires a 25% down payment and 24 monthly payments. The finance charge is computed at 10% nominal interest. Find (1) the finance charge, (2) the monthly payment, and (3) the annual percentage rate that must be revealed under the Truth in Lending Law.
(e) Monthly payments of \$23 are required for one year in order to repay a loan of \$240. Determine (1) the finance charge, and (2) the effective annual percentage rate.
(f) A console color television costs \$800 cash. The buyer received an \$80 allowance on his old set and paid 36 equal installments of \$26. Compute (1) the total cost of the set, (2) the finance charge, and (3) the annual percentage rate that must be disclosed.

Objective 10; Example 11

13. Use the formula $S = P(1 + i)^n$ and the compound amount table in order to compute compound amount and interest:
(a) \$1,000 at 6% compounded monthly for 4 years.
(b) \$2,000 at 5% compounded monthly for $6\frac{1}{2}$ years.
(c) \$800 at 6% compounded monthly for $3\frac{1}{3}$ years.
(d) Determine (1) the simple interest on \$1,000 at 5% for 2 years. (2) Compare with the same principal invested at 5% compounded monthly.
(e) Repeat part (d) for a 5-year investment.
(f) Repeat part (d) for an 8-year investment.

OBJECTIVES

After completing this chapter, you should be able to

A. Sales and property tax

1 Compute sales tax, based on a percentage rate. **(2–5)**

2 Find the marked price of an item, given (a) the sales tax rate and (b) the total price. **(6–8)**

3 Using the property tax formula $R \cdot V = T$ and given two of the following items, determine the unknown: (a) property tax, (b) tax rate, or (c) assessed value of property. **(9–15)**

B. Financial statements

4 On an income statement with individual amounts given, (a) compute the net profit and (b) find the percent of net sales that each item represents. **(16–25)**

5 On a balance sheet with individual amounts given, compute (a) total assets, (b) total liabilities, (c) net worth, and (d) percent each item represents. **(26–32)**

C. Markup

6 Using the formula $C + M = S$, determine selling price (and selling price factor) when markup is (a) based on cost or (b) based on selling price. **(33–39)** **(40–44)**

7 Use the formula $?\%$ of $C = M$ to determine the percent of markup based on cost; and $?\%$ of $S = M$ to find the percent of markup based on selling price. **(45–47)**

D. Simple interest and amount

8 (a) Use the simple interest formula $I = Prt$ to find any variable, when the other items on a simple interest note are given.
(b) Use the amount formula $S = P + I$ to find amount or interest. **(48–57)**

E. Installment plans

9 Compute a monthly installment payment; also determine the finance charge and annual percentage rate required by the Truth in Lending Law. **(58–69)**

F. Compound interest and amount

10 Calculate compound amount using the formula $S = P(1 + i)^n$ and a compound amount table. **(70–82)**

General **11** Explain the concepts and procedures required to compute

(a) Sales and property tax. **(Exercise 1)**

(b) Financial statements. **(Exercise 2)**

(c) Markup. **(Exercise 3)**

(d) Simple and compound interest, and installment plans. **(Exercise 4)**

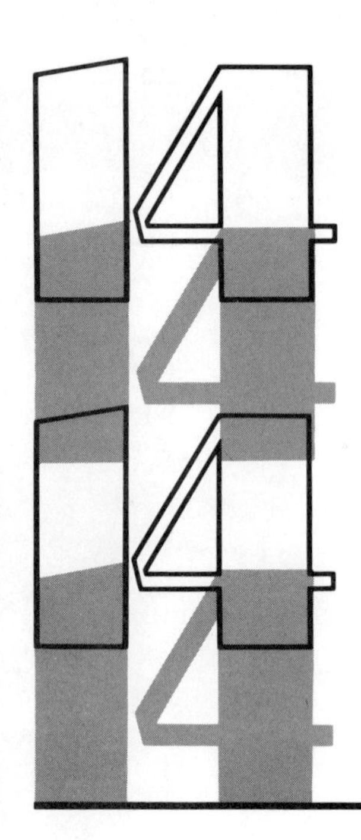

Geometry

The branch of mathematics which studies the relationships between points, lines, and space is called geometry. There are actually many kinds of geometry, but Euclidean geometry is the geometric system which is most easily applied to everyday concepts. Euclidean geometry was developed by the Greek mathematician, Euclid, about 300 B.C. and has remained essentially unchanged since that time. This chapter introduces some of those Euclidean properties. (1)

Proofs of theorems in this discussion of geometry are generally presented informally. In a more rigorous presentation of geometry a formal proof would be required. A formal proof consists of the figure, a statement of what is given and what is to be proved, and a list consisting of each step and the reason for the step. In this chapter you will be asked only to supply the reason for each step of a given proof. Acceptable reasons in a proof include "given" information, definitions, axioms, theorems, or any previously learned mathematical properties. Typical properties include the reflexive property (any element is equal to itself), the transitive property (if $a = b$ and $b = c$, then $a = c$), and substitution (any element may be substituted for its equal).

A. POINTS, LINES, AND PLANES

Point

(2) The basic elements of the Euclidean system are points, lines, and planes. Though these terms are not defined, each can be illustrated by a sketch in order to visualize the basic assumptions about the system. A *point* can be depicted by a dot, a fixed position which has no width, length or depth. Points are named by using capital letters. A and B below are two distinct points.

A B

Line

(3) A *line* is understood as a straight path which has length but no width or depth. It is drawn with arrows on each end to indicate that it continues indefinitely in both directions. A lowercase letter may be used to name a line.

$\overleftrightarrow{l}$

Plane

(4) A *plane* is considered to be a flat surface which extends infinitely in length and width but has no depth. The idea of a plane can be visualized as a sheet of paper or as the top of a table which has no thickness but extends indefinitely in length and width.

(5) To describe the system there are some basic assumptions, called *axioms* or *postulates*, about these three basic concepts of point, line and plane. The first two axioms describe the relationship between points and lines:

Axiom 1 *Every line is a infinite set of points that includes at least two distinct points.*

Objective 1: Six basic axioms of points, lines and planes

By this axiom, at least two positions on any given line may be associated with two specific points.

Axiom 2 *If A and B are two distinct points, then there is one and only one line through A and B.*

Stated differently, this second axiom is "Two points determine a line." This assumption provides another means of naming a line. That is, if A and B are two points on a line, the line may be called $\overleftrightarrow{AB}$. However, there may be several ways of naming a single line. For example, the line below could also be called $\overleftrightarrow{BC}$, $\overleftrightarrow{AC}$, $\overleftrightarrow{BA}$, $\overleftrightarrow{CB}$, or $\overleftrightarrow{CA}$.

The following two axioms describe the relationship between points and planes: (6)

Axiom 3 *A plane is a set of points which contains at least three points that are not all on the same line.*

As shown by the figure below, a plane may be named either by a lowercase letter (n) or by three points in the plane (ABC).

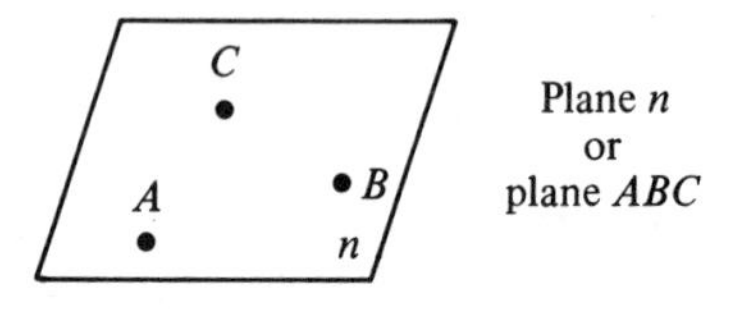

Plane n
or
plane ABC

Axiom 4 *If A, B, and C are three distinct points which do not lie on the same line, there is one and only one plane through A, B, and C.* (7)

This concept may also be expressed as "Three points determine a plane."

Axiom 5 *If two points of a line lie in a plane, then all the points of the line lie in the plane.* (8)

This axiom shows a relationship among all three concepts of point, line and plane. To illustrate, in the figure below the next axiom, every point of line $\overleftrightarrow{AB}$ lies in the plane ABC.

Axiom 6 *There are at least four points, not all of which are in the same plane.* (9)

This assumption assures that not all points lie in one plane. The figure below represents three points (A, B, and C) in one plane and a fourth point (D) not in the plane.

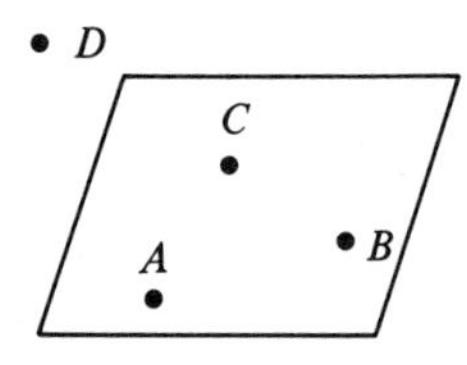

◆ **Example 1** Apply the six basic axioms to determine the following relationships among points, lines, and planes: (10)

(a) How many lines are determined by points A, B, and C?

C

$\bullet A \qquad \bullet B$

(b) Given a line and a point not on that line, how many planes contain the line and the point?

(c) If two lines intersect, explain why they intersect in only one point.

(a) Two points determine a line. Thus, there is a line through each pair of points. Draw a sketch of the lines determined by the points.

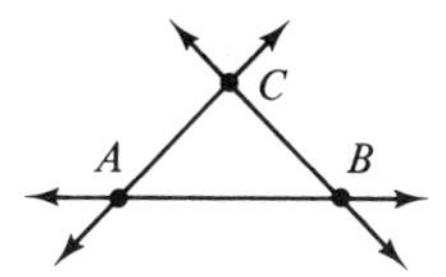

There are three lines determined by A, B, and C.

(b)

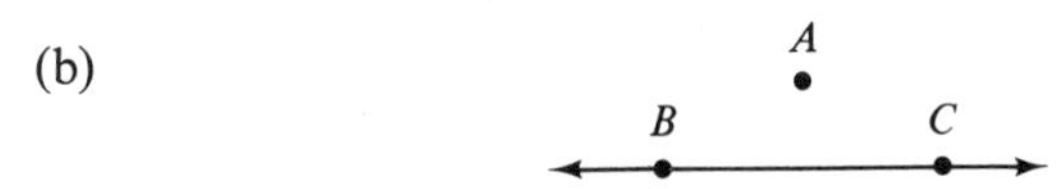

The line and the point determine exactly one plane. Pick two points, B and C, on a line. There is one and only one plane through 3 points. There is only one plane which contains the three *points*, A, B, and C. All of the *line* must also lie in that plane, since two points of the line lie in the plane.

(c) Sketch two intersecting lines $\overleftrightarrow{AB}$ and $\overleftrightarrow{BC}$:

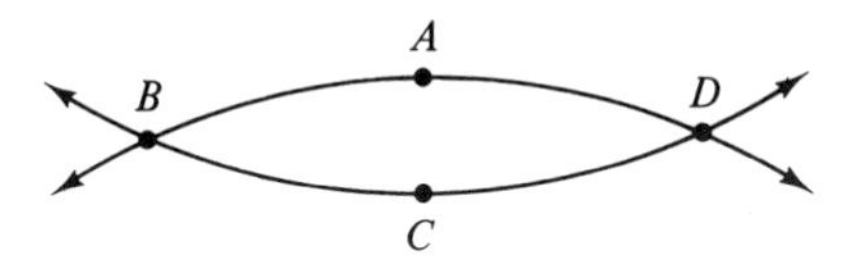

Suppose the two lines also intersected in some other point D. Then there would be two different lines through the points B and D. But there can be only one line through two given points. Thus, if two lines intersect, they intersect in one point.

● **Practice 1** Use the six basic axioms (also listed on page 431) in the following applications of points, lines, and planes: (11)

(a) How many lines are determined by the four points R, S, T, and U?

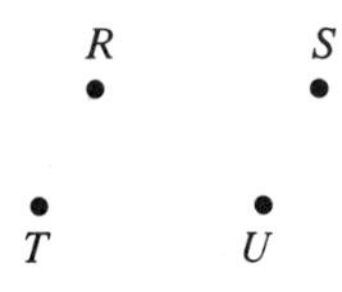

(b) Given a line, how many planes are there which contain that line?

(c) Which of the six axioms says that a line cannot "jump off" a plane as in the figure below?

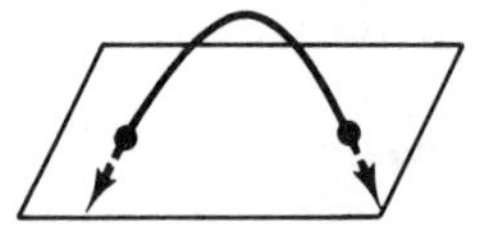

See also Exercises 1, 2.

B. RAYS, LINE SEGMENTS, AND ANGLES

Now that the undefined terms—point, line, and plane—have been given some properties (axioms), new terms can be introduced into the system by means of a statement called a *definition*. These axioms and definitions can then be used in order to prove other new properties of the geometric system. These proven statements are called *theorems*. For example, the statement in Example 1(c) could be written as a theorem. (12)

Theorem *If two lines intersect, they intersect in only one point.*

We will now expand our geometric system by considering several axioms and definitions. Another assumption which further describes a line is called *Dedekind's axiom*.

Dedekind's axiom

Axiom *If all points on a line are separated into two nonempty sets such that every point of the first set is to the left of every point of the second set, then there exists a point which is the boundary of this separation, and the point belongs to one of the two sets.* (13)

This axiom means that any point on a line divides the line into two parts, and the point will belong to one of the two parts.

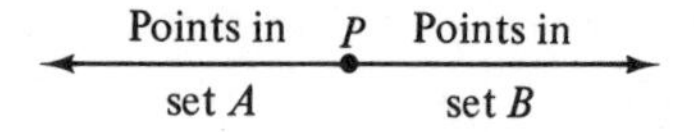

The concept of this axiom leads to the following definition:

Ray

A *ray* is a part of a line, consisting of a point on the line and all points either to the left or to the right of that point. A ray is named by the end (14)

point and any other point on the ray, with an arrow above the letters. The ray above denoted is $\overrightarrow{PA}$.

Line segment

Another subset of a line is called a line segment. A *line segment* is composed of two points on a line together with the infinite set of all points (15) on the line between these two endpoints. A line segment is named by its endpoints; $\overline{AB}$ below is a line segment:

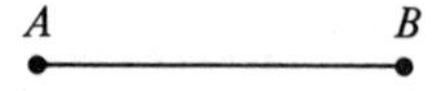

Observe the distinction between $\overline{AB}$, $\overrightarrow{AB}$, and $\overleftrightarrow{AB}$. $\overline{AB}$ is a line segment; $\overrightarrow{AB}$ is a ray; and $\overleftrightarrow{AB}$ is a line.

(16) The means of measuring a line segment has already been considered while studying real numbers and the number line. That is, we can establish a one-to-one correspondence between the set of real numbers and the set of points on a line, as follows:

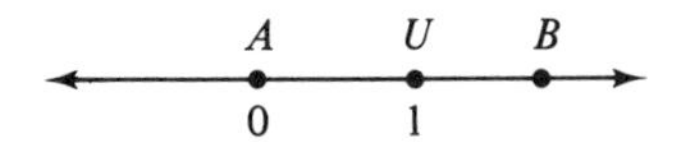

Given a line, one can arbitrarily choose some point A to be the point which corresponds to the number 0 and some point U to be the point which corresponds to the number 1. Once these two points have been chosen and a unit measure thus established, the real number which corresponds to any other point on the line is thereby determined. However, line segments are ordinarily measured using some common unit such as centimeters, meters, inches, or feet.

(17) Just as a point divides a line into two parts, a line divides a plane into two parts, as indicated by the following axiom.

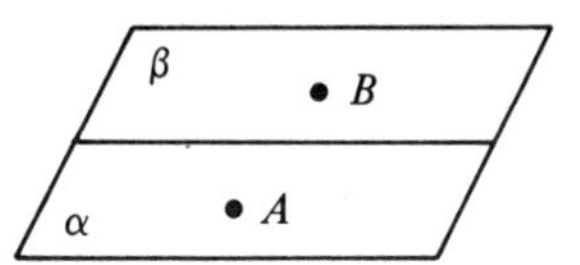

Axiom A line divides a plane into two sets, α and β, such that

(a) if two points belong to one of the sets, the line segment determined by those two points also belongs to the set.

(b) If a point A belongs to α and a point B belongs to β, then the line $\overleftrightarrow{AB}$ must intersect the boundary line.

Half-plane

The line is the boundary of the two sets and belongs to one of the two sets. This axiom leads us to the definition of a half-plane. A *half-plane* consists of a line and all the points in a plane which lie on one side of that line.

Angle

(18) When two rays have a common endpoint, an angle is formed. An *angle* is the union of two rays with a common endpoint. The common endpoint is called the *vertex*. The rays are called the *sides of the angle*.

Objective 2: Naming an angle by the vertex and two points

An angle is indicated by the symbol $\measuredangle$. This symbol is used to name an angle in any one of three ways described below.

(a) Using the vertex of the angle:

$\measuredangle A$

(b) Using the vertex and one point on each ray, with the vertex listed between the other two points:

$\measuredangle BAC$

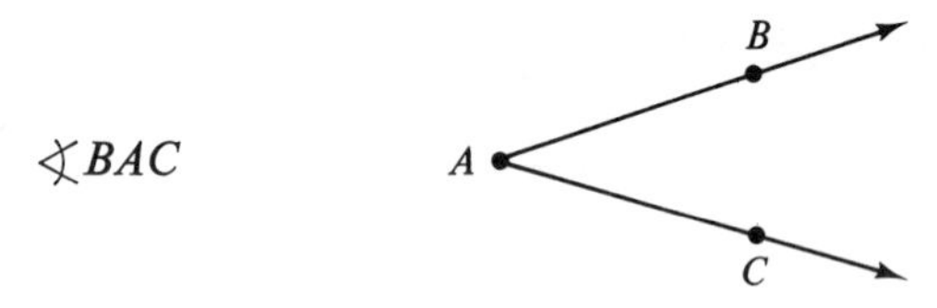

(c) Using an Arabic numeral:

$\measuredangle 1$

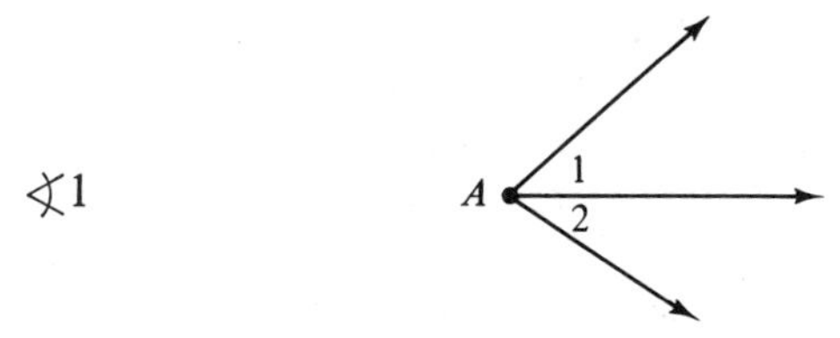

(Notice in this figure that one could not call the angle "$\measuredangle A$," because there is more than one angle with its vertex at point A.)

Objective 3: Adjacent angles

Adjacent angles are two angles with a common vertex and a common ray between them. In (c) above, $\measuredangle 1$ and $\measuredangle 2$ are adjacent angles. (19)

Straight angle

A *straight angle* is an angle whose two sides lie on the same line. $\measuredangle BAC$ below is a straight angle.

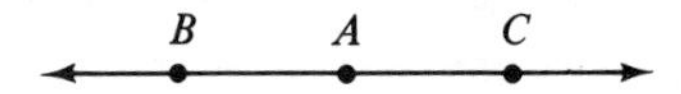

Interior of an angle

The *interior of an angle BAC* is the area where the half-plane on the B side of AC intersects with the half-plane on the C side of AB.

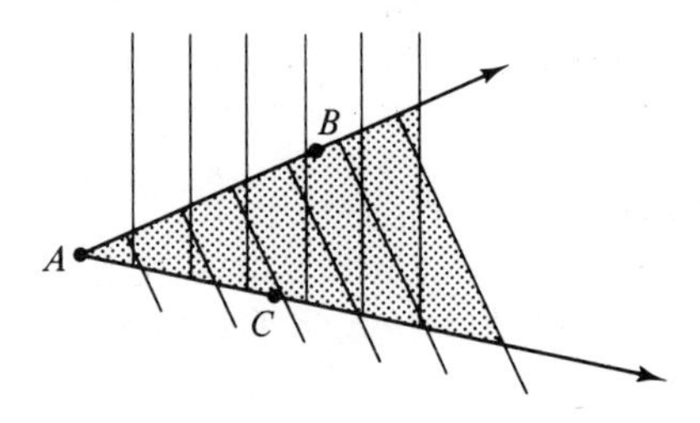

◆ **Example 2**

(a) Name all of the angles in the figure below. (20)
(b) Name a pair of adjacent angles.
(c) Name a pair of angles which have a common ray but are not adjacent angles.
(d) Name a straight angle.
(e) Shade in the interior of angle *COD*.

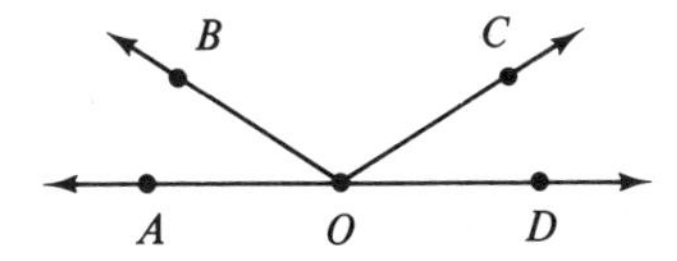

(a) The angles shown above are $\measuredangle AOB$, $\measuredangle AOC$, $\measuredangle AOD$, $\measuredangle BOC$, $\measuredangle BOD$, and $\measuredangle COD$.
(b) The pairs of adjacent angles are $\measuredangle AOB$ and $\measuredangle BOC$, $\measuredangle AOB$ and $\measuredangle BOD$, $\measuredangle BOC$ and $\measuredangle COD$, and $\measuredangle AOC$ and $\measuredangle COD$.
(c) $\measuredangle AOB$ and $\measuredangle AOD$.
(d) $\measuredangle AOD$ is a straight angle because $\overline{AO}$ and $\overline{OD}$ are part of the same line.
(e)

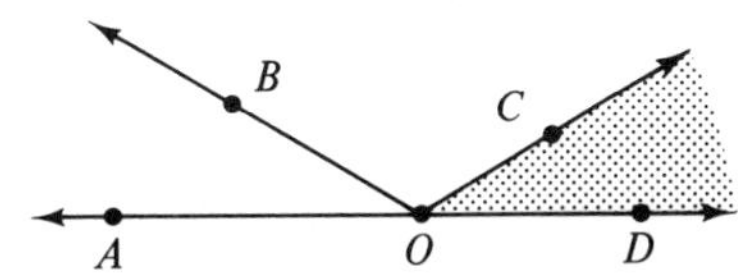

● **Practice 2** Draw three rays with a common endpoint, *B*. (21)

(a) Label the drawing by naming the vertex and a point on each ray.
(b) How many different angles are formed? Name them.
(c) How many pairs of adjacent angles are formed? Name them.
(d) Shade in the interior of one of the angles.

See also Exercise 3.

C. MEASUREMENT OF ANGLES

The unit of measurement most commonly used for angles is the degree. (22)
Degree — A *degree* is $\frac{1}{180}$ of a straight angle. Thus, a straight angle measures 180 degrees, denoted 180°. The notation commonly used to indicate degree measurement for a specified angle is $m(\measuredangle A) = 180°$, given a straight angle, $\measuredangle A$. By definition, *two angles are equal* when their degree measures are equal. That is, $m(\measuredangle A) = m(\measuredangle B)$ means that $\measuredangle A = \measuredangle B$.

Of special importance is the angle which measures one-half of a (23)
Right angle — straight angle, or 90°. A *right angle* is an angle whose measure is 90°. The symbol for right angle is $\perp$.

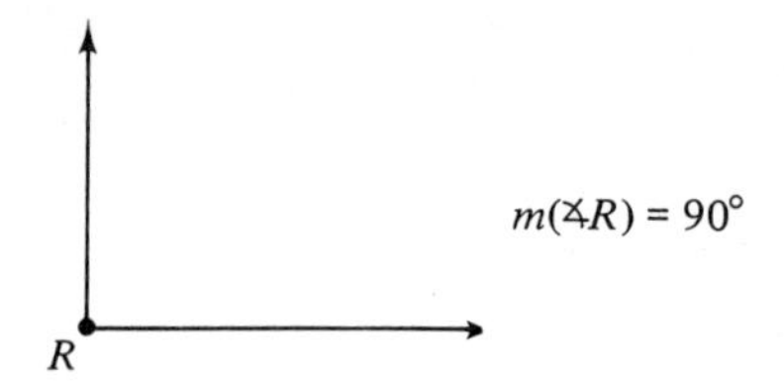

Every angle (other than a straight angle) may be classified according to its measured relationship to 90°. The trichotomy law states that, given two numbers a and b, one of three things must be true: either $a = b$, $a < b$, or $a > b$. Similarly, given any angle, one of three statements must be true: either $m(\measuredangle A) = 90°$, $m(\measuredangle A) < 90°$, or $m(\measuredangle A) > 90°$. This fact allows one to classify any angle as one of four types: right, acute, obtuse, or straight.

Acute angle

An *acute angle* is an angle whose measure is less than 90°.

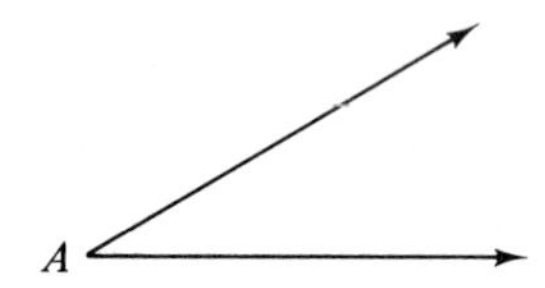

Obtuse angle

An *obtuse angle* is an angle whose measure is more than 90°.

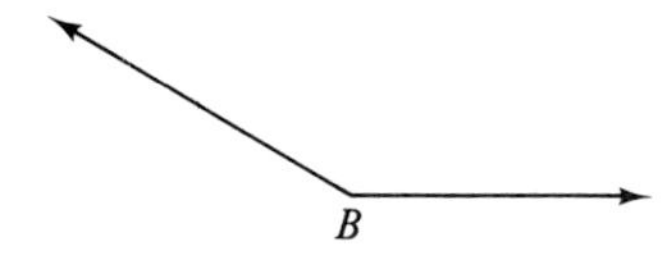

◆ **Example 3**

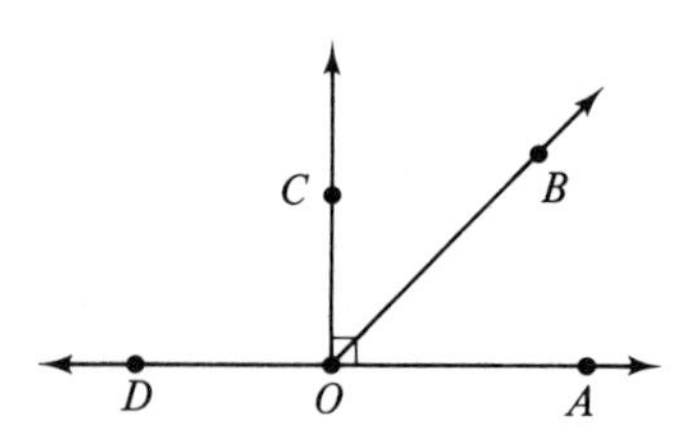

Objective 4: Acute, obtuse, and right angles

(24)

(a) Name two acute angles in the figure above.
(b) Name a right angle.
(c) Name an obtuse angle.
(d) Draw a sketch of two adjacent angles whose sum measures 90°. What kinds of angles must they be?

(a) $\measuredangle AOB$ and $\measuredangle BOC$ are acute angles.
(b) $\measuredangle DOC$ is a right angle.
(c) $\measuredangle DOB$ is an obtuse angle.

(d)

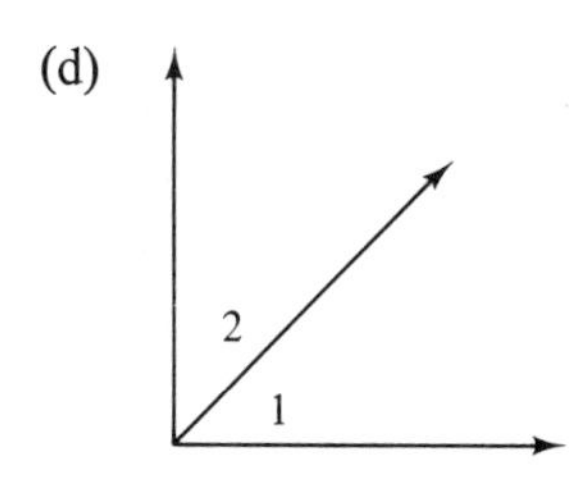

(The angles must be acute, since each angle measures less than 90°.)

● **Practice 3** Given the following figure, answer the questions below: (25)

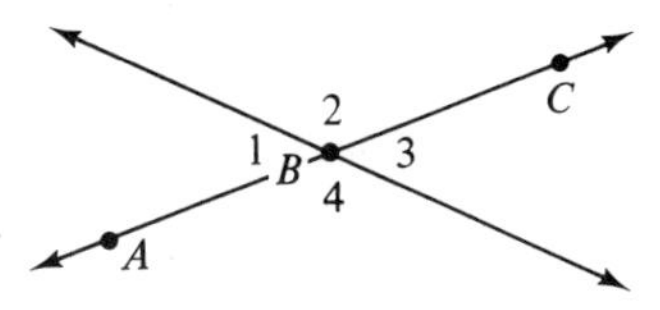

(a) Name the acute angles in the figure.
(b) Name the obtuse angles in the figure.
(c) Name the straight angle in the figure.
(d) What is the apparent relationship of the pair of acute angles?
(e) What is the apparent relationship of the pair of obtuse angles?
(f) What is the sum of the measures of any acute angle and either adjacent obtuse angle in the figure?

See also Exercise 4.

Complementary angles

In Example 3(d), two angles were drawn whose measures total 90°. (26) Two such angles are said to be complementary. By definition, two angles are *complementary* if the sum of their measures is 90°. Notice that complementary angles are not required to be adjacent angles; the 30° angle and the 60° angle below are also complementary:

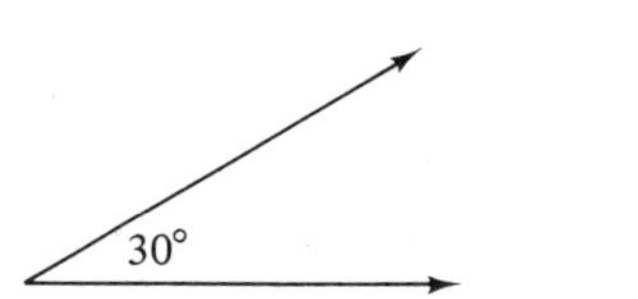

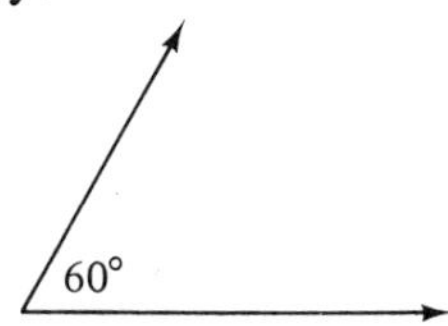

Supplementary angles

Two angles are *supplementary* if the sum of their measures is 180°. The 110° angle and the 70° angle below are supplementary. Observe that

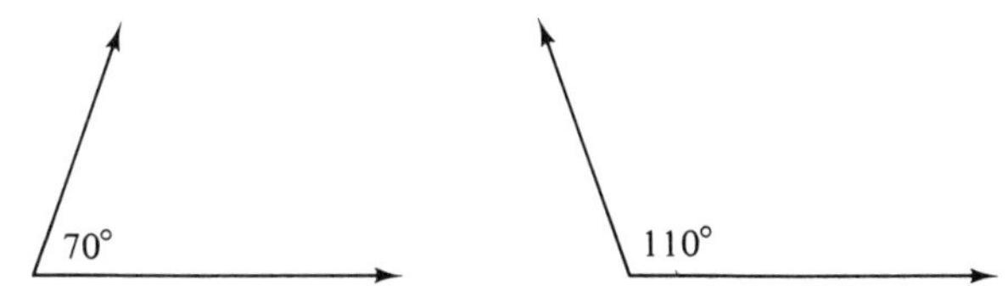

supplementary or complementary angles are considered in pairs. One does not say that $\angle A$ is complementary or that $\angle B$ is a supplement.

Example 4

Objective 5: Complement and supplement of an angle

(a) Find the measure of the complement of $\sphericalangle A$, if $m(\sphericalangle A) = 75°$. (27)

(b) Find the measure of the supplement of $\sphericalangle B$, if $m(\sphericalangle B) = 34°$.

(c) Draw a sketch of two adjacent supplementary angles which are equal in measure. What kind of angles are they?

(d) Suppose that $\sphericalangle A$ is complementary to $\sphericalangle C$ and that $\sphericalangle B$ is complementary to $\sphericalangle C$. What conclusions can be drawn about the angles A and B? Explain your answer.

(a) Call the complement $\sphericalangle B$.

$$m(\sphericalangle A) + m(\sphericalangle B) = 90°$$
$$75° + m(\sphericalangle B) = 90°$$
$$m(\sphericalangle B) = 90° - 75°$$
$$m(\sphericalangle B) = 15°$$

(b) Let the supplement be $\sphericalangle D$.

$$m(\sphericalangle B) + m(\sphericalangle D) = 180°$$
$$34° + m(\sphericalangle D) = 180°$$
$$m(\sphericalangle D) = 180° - 34°$$
$$m(\sphericalangle D) = 146°$$

(c)

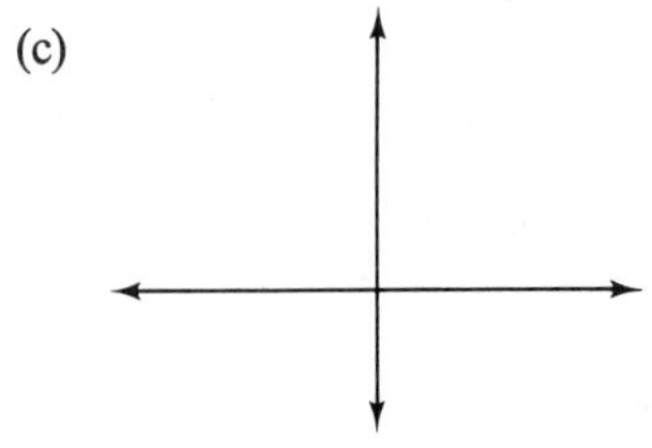

If two angles are equal and supplementary, then each angle measures 90°. The angles are thus right angles.

(d) $\sphericalangle A$ and $\sphericalangle C$ are complementary means that

$$m(\sphericalangle A) + m(\sphericalangle C) = 90°$$

$\sphericalangle B$ and $\sphericalangle C$ are complementary means that

$$m(\sphericalangle B) + m(\sphericalangle C) = 90°$$

By the transitive property it follows that

$$m(\sphericalangle B) + m(\sphericalangle C) = m(\sphericalangle A) + m(\sphericalangle C)$$

Thus, $$m(\sphericalangle B) = m(\sphericalangle A)$$

Practice 4

(a) Find the measure of the complement of a 70° angle. (28)

(b) Find the measure of the supplement of a 110° angle.

(c) If two angles are complementary, are they both acute angles? If your answer is no, explain.

(d) Suppose $\measuredangle A$ is supplementary to $\measuredangle D$ and that $\measuredangle B$ is supplementary to $\measuredangle D$. What can you conclude about $\measuredangle A$ and $\measuredangle B$? Explain your answer.

See also Exercises 5, 6, 7d, 7e.

D. VERTICAL ANGLES AND PERPENDICULAR LINES

The conclusions obtained in the preceding example and practice are the basis for two new theorems which will be useful in later proofs. (No proofs will be given here for these theorems.) (29)

Theorem *If two angles are complementary to the same angle or to equal angles, they are equal to each other.*

Theorem *If two angles are supplementary to the same angle or to equal angles, they are equal to each other.*

When two lines intersect, four angles are formed, as illustrated below. Those angles which are not adjacent to each other are called vertical angles, as follows. *Vertical angles* are a pair of nonadjacent angles formed by two intersecting lines. In the figure below, $\measuredangle 1$ and $\measuredangle 2$ are vertical angles; also $\measuredangle 3$ and $\measuredangle 4$ are vertical angles. Notice that, like complementary or supplementary angles, vertical angles are also discussed in pairs. (30)

Vertical angles

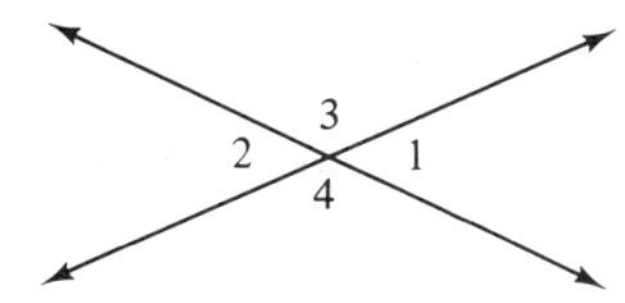

It appears from the sketch that the pairs of vertical angles are equal. This statement can be proved: (31)

Theorem *Vertical angles are equal.*

Observe that $\measuredangle 2$ is supplementary to $\measuredangle 3$ because the angles compose a straight angle, which is 180°. Similarly, $\measuredangle 1$ is supplementary to $\measuredangle 3$. But if two angles are supplementary to the same angle, they are equal to each other. Thus, $\measuredangle 1 = \measuredangle 2$. Similarly, $\measuredangle 3 = \measuredangle 4$.

If two lines intersect to form a right angle, it follows from this theorem that all of the angles formed will be right angles. Lines of this type have (32) special significance in the study of geometry and are defined as follows. If two lines intersect so that one of the angles formed is a right angle, then the lines are said to be *perpendicular*. The symbol for perpendicular is "$\perp$."

Perpendicular lines

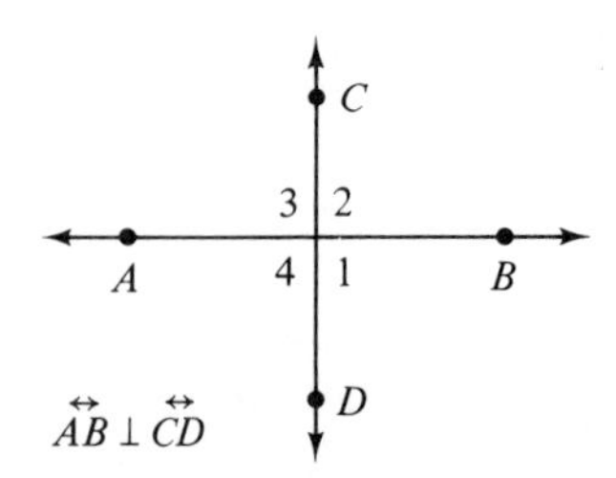

Pairs of rays (or pairs of line segments) are also said to be perpendicular when they are subsets of different perpendicular lines.

Below are an axiom and two theorems (that could both be proved):

Axiom *If P is a point not on a line, there is one and only one perpendicular from the point to the line.*

Theorem *All of the angles formed by perpendicular lines are equal right angles.* (33)

In the figure above, suppose $\measuredangle 1$ is a right angle. Then $m(\measuredangle 1) = 90°$ by the definition of a right angle. Now, vertical angles are equal; therefore, $\measuredangle 1 = \measuredangle 3$ and $\measuredangle 2 = \measuredangle 4$. Thus, $m(\measuredangle 3) = 90°$. But $\measuredangle 1$ and $\measuredangle 2$ are supplementary; therefore, since $m(\measuredangle 1) = 90°$ then also $m(\measuredangle 2) = 180° - 90° = 90°$. Furthermore, since $\measuredangle 2 = \measuredangle 4$, then also $m(\measuredangle 4) = 90°$. Thus, all of the angles formed by perpendicular lines are 90° in measure, or right angles.

Theorem *If two intersecting lines form equal adjacent angles, then the lines are perpendicular.*

◆ **Example 5** Use the figure below to answer the following questions: (34)

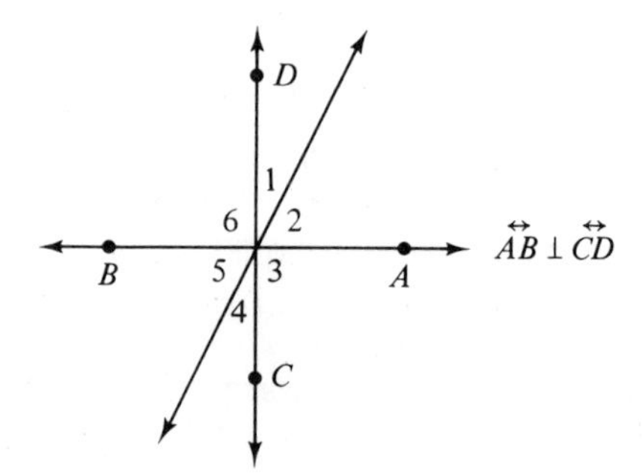

Objective 6: Finding unknown angles by using vertical angles

(a) Name the pairs of vertical angles.

(b) Name an angle which is complementary to ∢1 but not adjacent to it.

(c) Given a pair of intersecting lines which form one right angle, what can be said about the other angles formed by the intersecting lines?

Objective 7: Properties of perpendicular lines

(a) The pairs of vertical angles are ∢1 and ∢4, ∢2 and ∢5, ∢3 and ∢6.

(b) ∢1 is equal to ∢4, and ∢4 is complementary to ∢5. Thus, ∢1 is complementary to ∢5. ∢1 and ∢5 are not adjacent because they have no common rays.

(c) Since one right angle was formed, the lines are perpendicular and all of the angles are right angles.

● **Practice 5** Use this figure to answer the following questions, given that ∢1 = ∢5: (35)

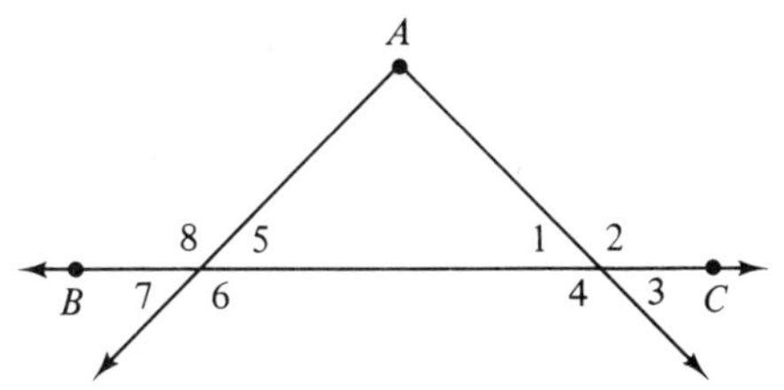

(a) Name all angles equal to ∢3.

(b) Name the angles supplementary to ∢3.

(c) If two lines are perpendicular, what is the relationship between each pair of the angles formed?

See also Exercises 7a, b, d.

E. PARALLEL LINES

"Parallel lines are lines which do not intersect." Most people would consider this a good definition of parallel lines. However, it is not a complete definition, as the following sketch of a room illustrates: (36)

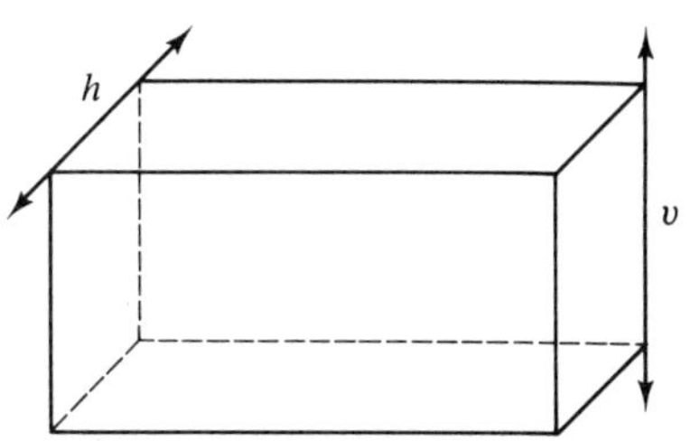

The vertical line, $\overleftrightarrow{v}$, at the corner of the room is parallel to any of the other vertical lines. The horizontal line, $\overleftrightarrow{h}$, is either parallel to or intersects any of the other horizontal lines. Now, the horizontal line, $\overleftrightarrow{h}$, and the vertical line, $\overleftrightarrow{v}$, do not intersect, but neither are they parallel. These two lines are positioned so that no plane can pass through both lines. Such lines are called skew lines. Thus, the key words left out of the preceding definition of parallel lines are "in the same plane." We, therefore, define "parallel" and "skew" lines as follows:

Objective 8: Parallel and skew lines

Parallel lines

(37) *Parallel lines* are lines in the same plane which do not intersect. That $\overleftrightarrow{AB}$ is parallel to $\overleftrightarrow{CD}$ may be denoted "$\overleftrightarrow{AB} \parallel \overleftrightarrow{CD}$." This definition may also be extended to include parallel rays and line segments which are subsets of parallel lines. Lines which are not in the same plane and which do not intersect are *skew lines*.

Skew lines

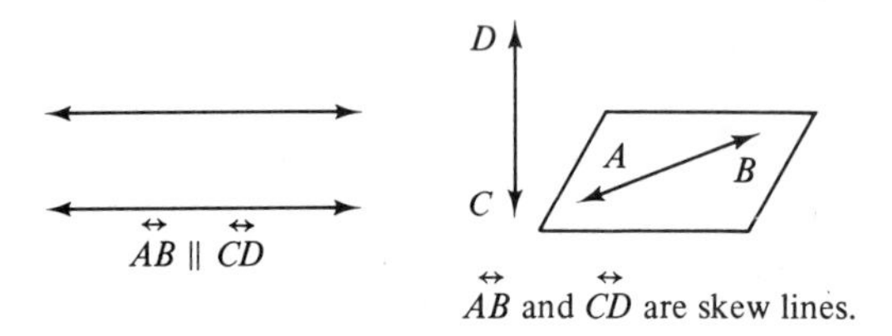

$\overleftrightarrow{AB} \parallel \overleftrightarrow{CD}$

$\overleftrightarrow{AB}$ and $\overleftrightarrow{CD}$ are skew lines.

(38) One of the basic assumptions of Euclidean geometry concerns parallel lines.

Axioms and theorems concerning parallel lines

Axiom *If P is a point not on a given line, there is one and only one line parallel to the given line which contains point P.*

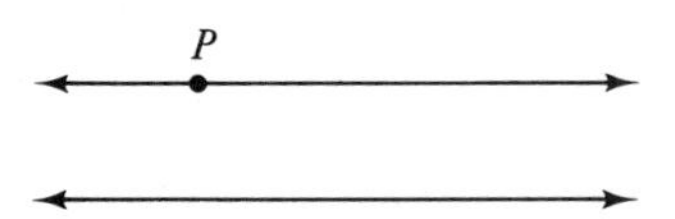

(39) ***Theorem*** *If two lines are perpendicular to the same line and if the two lines lie in the same plane, then the lines are parallel.*

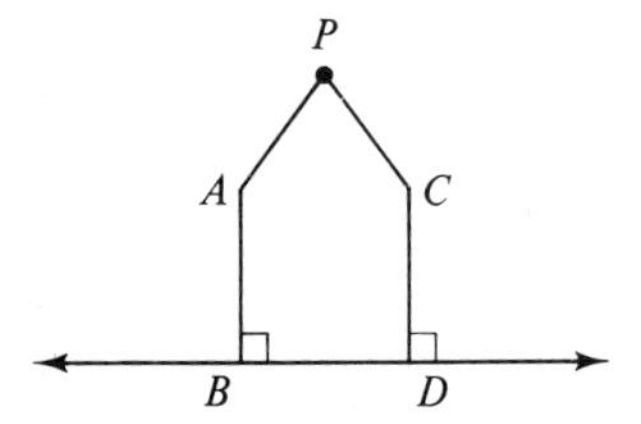

Suppose $\overleftrightarrow{AB} \perp \overleftrightarrow{BD}$ and $\overleftrightarrow{CD} \perp \overleftrightarrow{BD}$. If the two lines were in the same plane but were *not* parallel, then they would intersect in some point, P. There would then be two lines from point P which are perpendicular to line $\overleftrightarrow{BD}$. But this contradicts the assumption (axiom) that there is only one such perpendicular from a point to a line. Thus, if two lines are perpendicular to the same line and lie in the same plane, then the two lines are parallel.

Objective 9: Drawing a line parallel to a given line

(40) This statement suggests the procedure used to find a parallel to any given line: Given $\overleftrightarrow{AB}$, we first draw a perpendicular to $\overleftrightarrow{AB}$; call the perpendicular $\overleftrightarrow{CD}$.

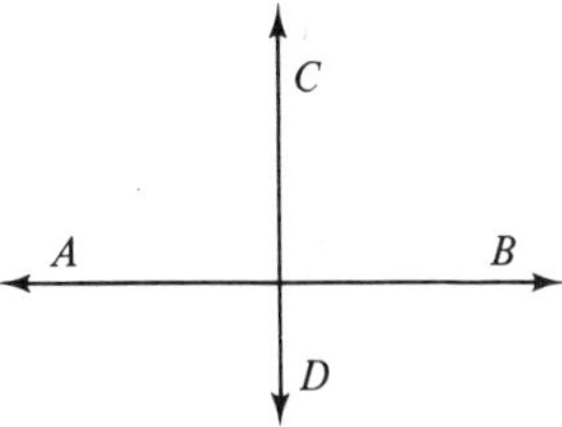

We then draw a perpendicular to $\overleftrightarrow{CD}$; call this second perpendicular $\overleftrightarrow{CE}$.

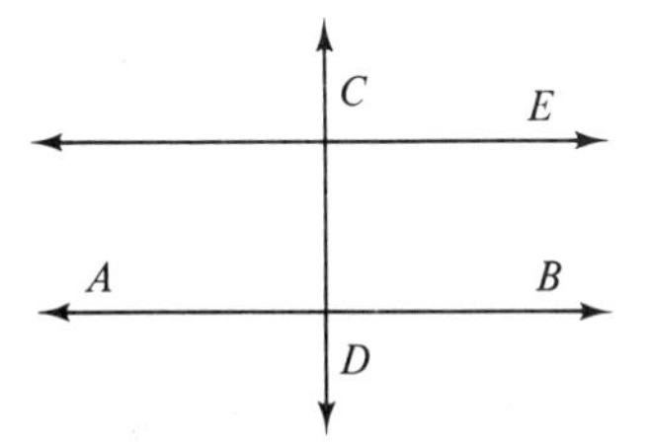

We know $\overleftrightarrow{CE} \parallel \overleftrightarrow{AB}$, because two lines perpendicular to the same line are parallel to each other.

See also Exercise 9.

Transversal

(41) Parallel lines are the subject of a number of theorems which we will prove. However, we first must define some additional terms: A *transversal* is a line which intersects two other lines in two distinct points. In each figure below, $\overleftrightarrow{l}$ is a transversal. Many angles are formed by the transversal

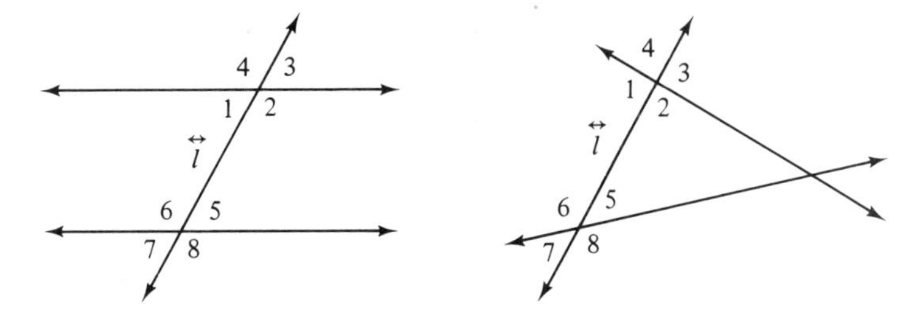

and the other two lines. Some pairs of angles have special properties when the two lines cut by the transversal are parallel. These pairs of angles are given special names, as defined below.

Alternate interior angles

(42)

Alternate interior angles are angles whose interiors lie between the parallel lines but on opposite sides of the transversal. Pairs of alternate interior angles below are $\sphericalangle 1$ and $\sphericalangle 5$, as well as $\sphericalangle 2$ and $\sphericalangle 6$. *Corre-*

Corresponding angles

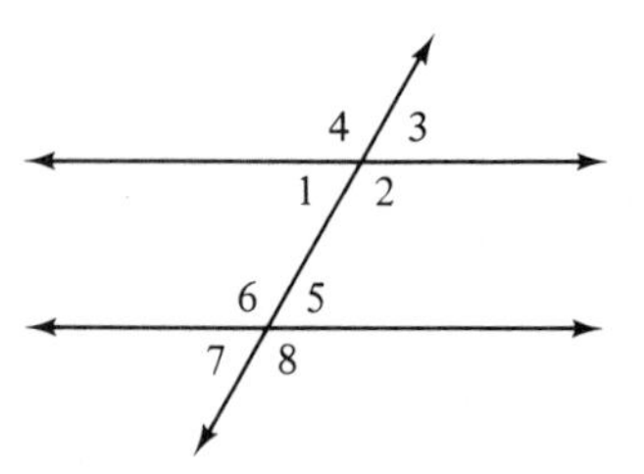

sponding angles are nonadjacent angles whose interiors lie on the same side of the transversal, but one angle lies between the parallel lines while the other is on the outside. Pairs of corresponding angles above are $\sphericalangle 1$ and $\sphericalangle 7$, $\sphericalangle 4$ and $\sphericalangle 6$, $\sphericalangle 2$ and $\sphericalangle 8$, and $\sphericalangle 3$ and $\sphericalangle 5$.

Interior angles on the same side of the transversal are two angles whose interiors lie between the parallel lines and on the same side of the transversal. In the figure above, $\sphericalangle 1$ and $\sphericalangle 6$ as well as $\sphericalangle 2$ and $\sphericalangle 5$ are pairs of interior angles on the same side of the transversal.

(43)

With this background, we are now ready to consider several theorems involving transversal and parallel lines.

Theorem *If the lines cut by a transversal are parallel lines, the alternate interior angles are equal to each other.*

The proof of this theorem is too complex for this presentation of geometry. However, once this theorem is accepted, the following theorems are easily proved.

(44)

Theorem *If the lines cut by a transversal are parallel lines, then the corresponding angles are equal.*

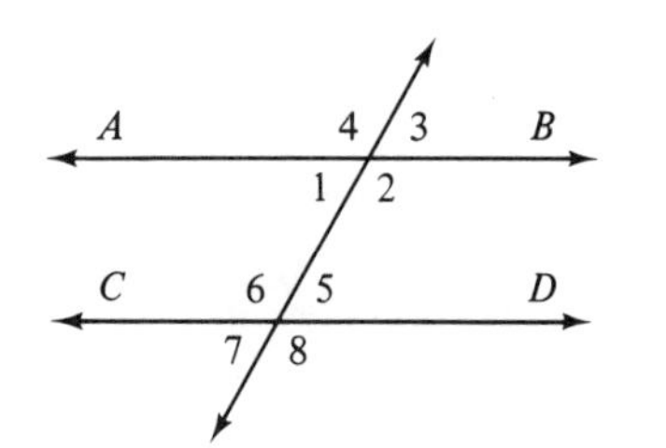

Given: $\overleftrightarrow{AB} \parallel \overleftrightarrow{CD}$

Prove: $\sphericalangle 1 = \sphericalangle 7$

Statements	*Reasons*
1. $\overleftrightarrow{AB} \parallel \overleftrightarrow{CD}$	1. Given
2. $\measuredangle 1 = \measuredangle 5$	2. If parallel lines are cut by a transversal, the alternate interior angles are equal.
3. $\measuredangle 5 = \measuredangle 7$	3. Vertical angles are equal.
4. $\measuredangle 1 = \measuredangle 7$	4. Transitive property.

In a similar manner, $\measuredangle 3 = \measuredangle 5$, $\measuredangle 4 = \measuredangle 6$, and $\measuredangle 2 = \measuredangle 8$.

Theorem *If the lines cut by a transversal are parallel lines, then the interior angles on the same side of the transversal are supplementary.* **(45)**

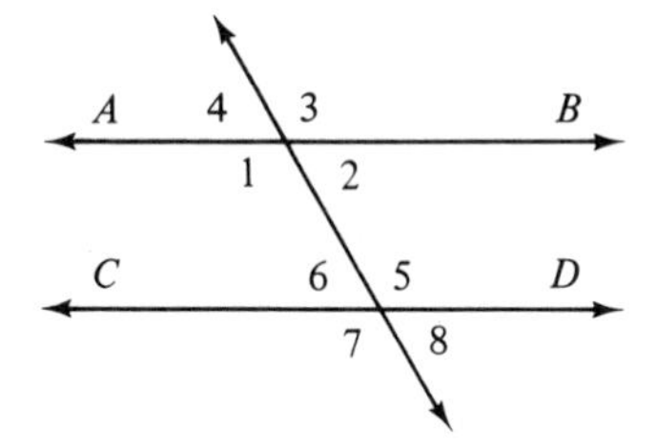

Given: $\overleftrightarrow{AB} \parallel \overleftrightarrow{CD}$

Prove: $\measuredangle 1$ and $\measuredangle 6$ are supplementary.

Statements	*Reasons*
1. $\overleftrightarrow{AB} \parallel \overleftrightarrow{CD}$	1. Given
2. $\measuredangle 1$ and $\measuredangle 2$ are supplementary	2. If the sum of two angles is a straight angle, the angles are supplementary.
3. $\measuredangle 2 = \measuredangle 6$	3. If parallel lines are cut by a transversal, then the alternate interior angles are equal.
4. $\measuredangle 1$ and $\measuredangle 6$ are supplementary	4. Substitution

In a similar manner $\measuredangle 5$ and $\measuredangle 2$ are supplementary.

Example 6 Given $\overleftrightarrow{l} \parallel \overleftrightarrow{m}$: use the three theorems above to answer the following questions: **(46)**

(a) Name all the angles supplementary to $\measuredangle 5$.

(b) Name all the angles equal to $\measuredangle 3$.

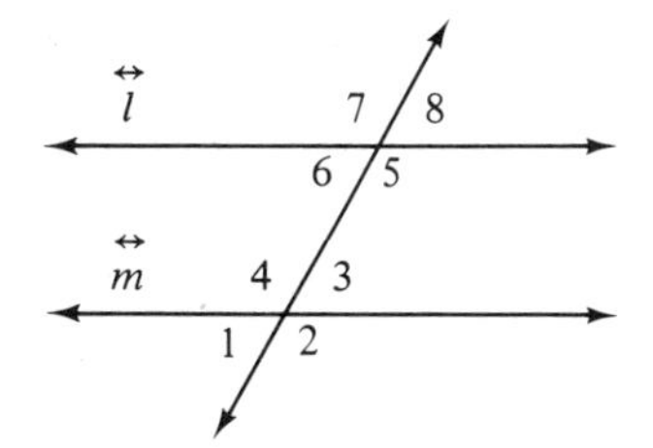

Objective 10: Parallel lines cut by a transversal

(a) and (b) Because interior angles on the same side of the transveral are supplementary, $\measuredangle 3$ is supplementary to $\measuredangle 5$. Any angle equal to $\measuredangle 3$ will be supplementary to $\measuredangle 5$.

$\measuredangle 1 = \measuredangle 3$ (Vertical angles are equal.)

$\measuredangle 6 = \measuredangle 3$ (Alternate interior angles formed by parallel lines are equal.)

$\measuredangle 8 = \measuredangle 3$ (Corresponding angles formed by parallel lines are equal.)

Thus, the angles supplementary to angle 5 are angles 3, 1, 6, and 8, which are equal angles.

● **Practice 6** Given $\overleftrightarrow{AB} \parallel \overleftrightarrow{CD}$, use the three theorems above to answer the following questions: (47)

(a) Name the angles equal to angle 6 and tell why they are equal.

(b) Name the angles supplementary to $\measuredangle 2$.

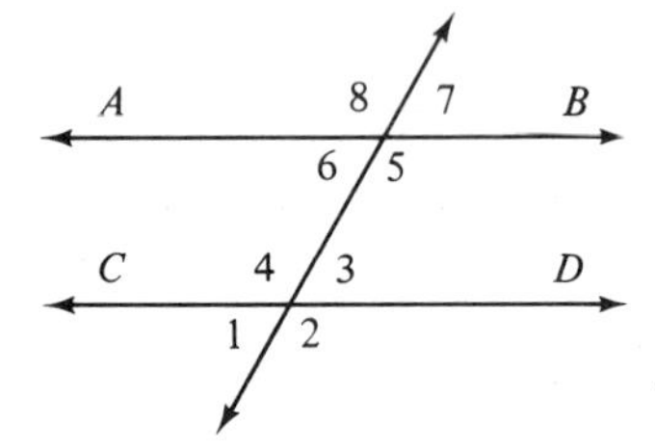

See also Exercise 10.

The converse of each of the preceding three theorems also constitutes a theorem. That is, (48)

Theorem *If two lines are cut by a transversal so that the alternate interior angles are equal, then the lines are parallel.*

Theorem *If two lines are cut by a transversal so that corresponding angles are equal, then the lines are parallel.*

Theorem *If two lines are cut by a transversal so that the interior angles on the same side of the transversal are supplementary, then the lines are parallel.*

Any of these restated theorems may now be used to prove that two lines are parallel, as illustrated below: (49)

Objective 11: Geometric proof involving parallel lines

◆ **Example 7** Supply the reasons in the following proof.

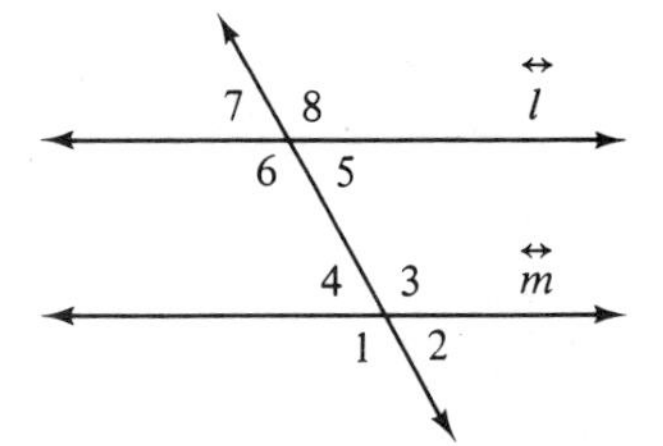

Given: $\measuredangle 1 = \measuredangle 8$

Prove: $\overleftrightarrow{l} \parallel \overleftrightarrow{m}$

Statements	Reasons
1. $\measuredangle 1 = \measuredangle 8$	1. Given
2. $\measuredangle 8 = \measuredangle 6$	2. Vertical angles are equal.
3. $\measuredangle 1 = \measuredangle 6$	3. Transitive property.
4. $\overleftrightarrow{l} \parallel \overleftrightarrow{m}$	4. If two lines are cut by a transversal so that corresponding angles are equal, the lines are parallel.

● **Practice 7** Supply the reasons in the following proof. (50)

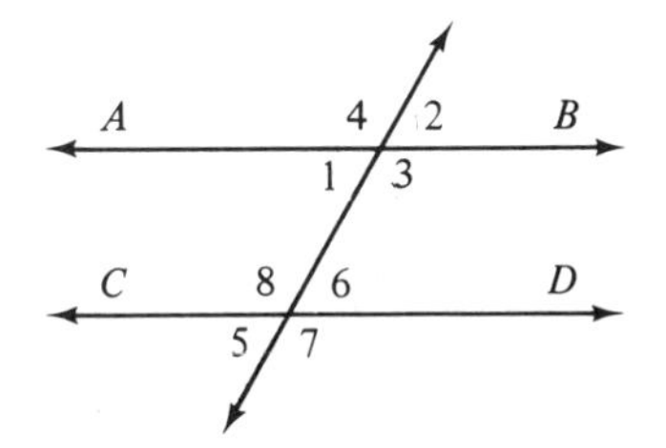

Given: $\measuredangle 7$ is supplementary to $\measuredangle 2$.

Prove: $\overleftrightarrow{AB} \parallel \overleftrightarrow{CD}$

Statements	Reasons
1. $\measuredangle 7$ is supplementary to $\measuredangle 2$	1. ______
2. $\measuredangle 7 = \measuredangle 8$, $\measuredangle 2 = \measuredangle 1$	2. ______
3. $\measuredangle 8$ is supplementary to $\measuredangle 1$	3. ______
4. $\overleftrightarrow{AB} \parallel \overleftrightarrow{CD}$	4. ______

See also Exercise 11.

F. POLYGONS

Consider the following figures. (51)

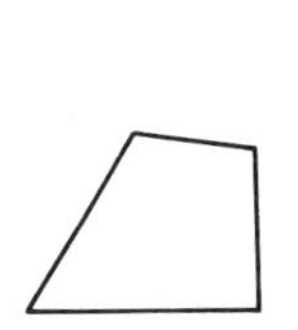
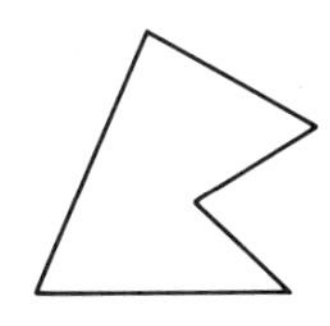
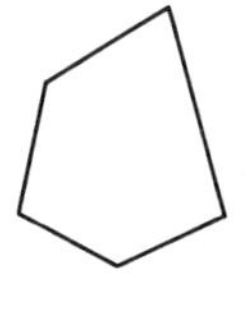

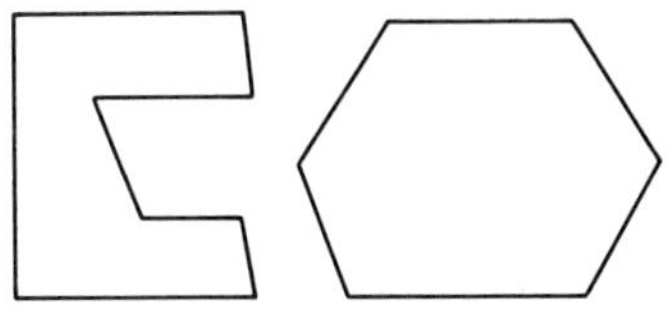

Categories of polygons

These figures are all examples of polygons. A polygon is the union of line segments that are formed by points in a plane as follows: Given a set of points $\{P_1, P_2, P_3, \ldots, P_n\}$ in a plane, the *polygon* determined by these points is the union of the line segments $\overline{P_1P_2}, \overline{P_2P_3}, \ldots, \overline{P_{n-1}P_n}, \overline{P_nP_1}$. Any one of these line segments is a *side* of the polygon.

Polygons are named according to the number of their sides: (52)

Number of sides	*Polygon name*	*Number of sides*	*Polygon name*
3	Triangle	8	Octagon
4	Quadrilateral	9	Nonagon
5	Pentagon	10	Decagon
6	Hexagon	12	Dodecagon
7	Heptagon		

An *equilateral polygon* is a polygon in which all sides are equal.

An *equiangular polygon* is a polygon in which all angles are equal.

A *reqular polygon* is a polygon which is both *equilateral and equiangular*.

Convex vs. concave polygons

Polygons may be further classified as concave or convex. A *convex* polygon is a polygon such that a line segment joining any two points inside the polygon will not intersect a side of the polygon. The following are *examples of convex polygons*: (53)

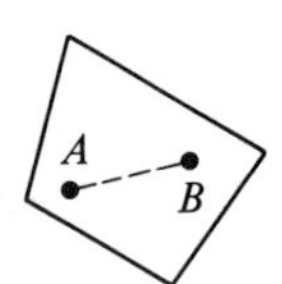

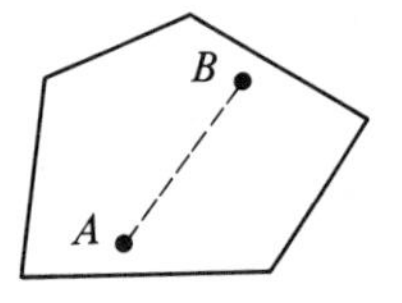

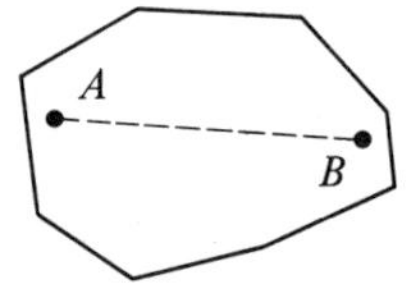

A *concave* polygon is a polygon such that a line segment joining any two given points inside the polygon may intersect sides of the polygon. The *examples below are concave polygons*:

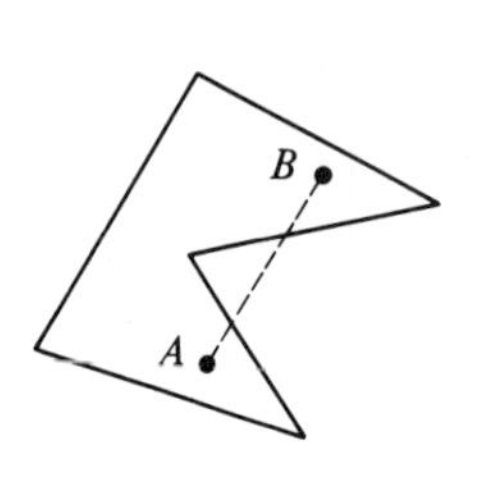

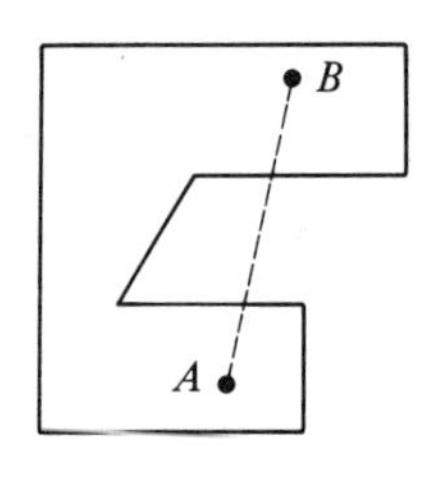

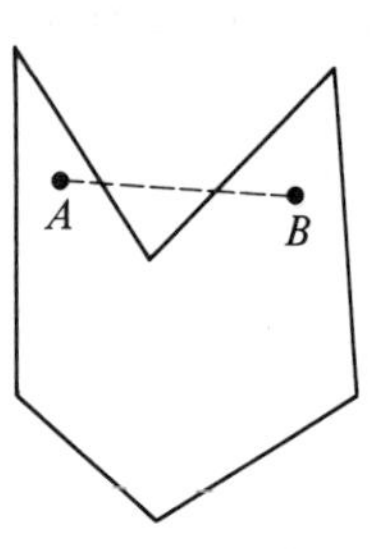

Objective 12: Classifying polygons

◆ **Example 8** For each polygon, classify the polygon by number of sides, and tell whether it is concave or convex. (54)

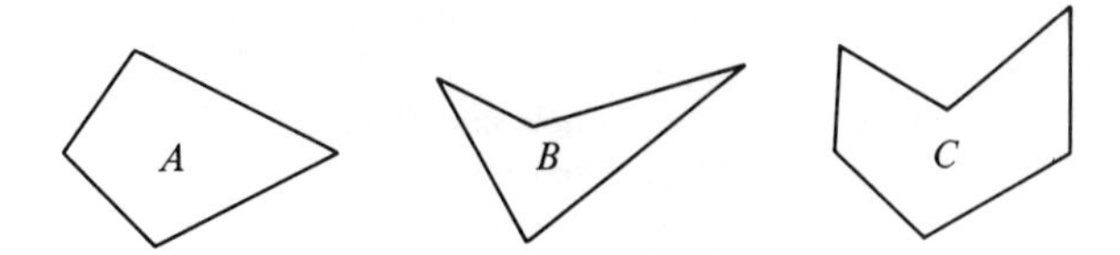

A is a convex quadrilateral; B is a concave quadrilateral; C is a concave hexagon.

● **Practice 8** Classify each polygon below according to its number of sides, and state whether it is concave or convex. (55)

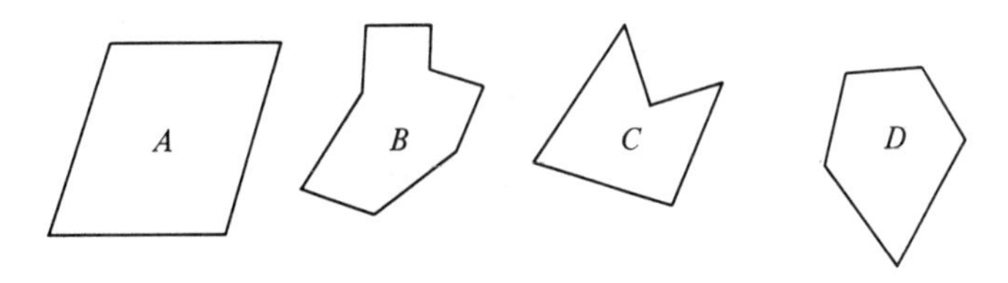

See also Exercise 12.

G. TRIANGLES

Triangle

If A, B, and C below are three points not on the same line, then *triangle* ABC (denoted $\triangle ABC$) is the union of the three line segments $\overline{AB}$, $\overline{BC}$, and $\overline{AC}$. Each point, A, B, or C, is a *vertex* of the triangle, and the line segments are *sides* of the triangle. In a triangle the side opposite an angle is often named by the lowercase letter of the angle's vertex. Thus, the side opposite $\measuredangle A$ is a; the side opposite $\measuredangle B$ is b; and the side opposite $\measuredangle C$ is c. (56)

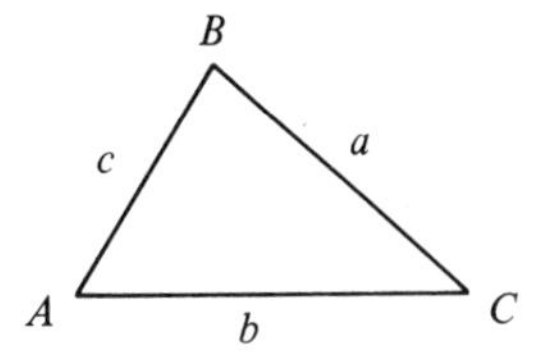

Triangles vary greatly in size and shape and can be classified in several ways as described below: (57)

Objective 13: Classifying triangles

Name	*Description*	*Example*
Acute triangle	A triangle that contains three acute angles. ($\measuredangle$'s A, B, and C are each acute $\measuredangle$'s)	
Right triangle	A triangle that contains one right angle. ($\measuredangle A$ is a right angle.)	
Obtuse triangle	A triangle that contains one obtuse angle. ($\measuredangle A$ is an obtuse angle.)	
Isosceles triangle	A triangle that contains two equal sides. ($\overline{AB} = \overline{BC}$)	
Equilateral triangle	A triangle in which all three sides are equal. ($\overline{AC} = \overline{BC} = \overline{AB}$)	

◆ **Example 9**

(a) Name two obtuse triangles in the figure below. What is the apparent relationship of the two triangles? (58)

(b) Name two acute triangles in the figure. What is the apparent relationship of the two triangles?

(c) The sides marked with a slash are equal sides. Name an equilateral triangle and an isosceles triangle.

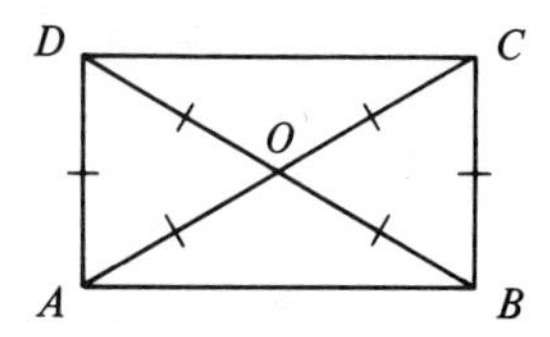

(a) $\triangle AOB$ and $\triangle DOC$ are obtuse triangles. The triangles seem to be the same size and shape.

(b) $\triangle AOD$ and $\triangle BOC$ are acute triangles. The triangles seem to be the same size and shape.

(c) $\triangle AOD$ and $\triangle BOC$ are equilateral (all the sides are equal). $\triangle COD$ and $\triangle AOB$ are isosceles (two sides are equal).

● **Practice 9**

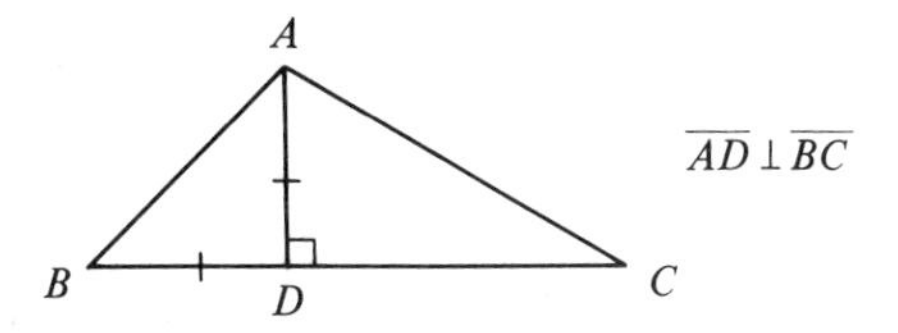

(a) Name the right triangles in the figure above. (59)

(b) Name the acute triangles in the figure.

(c) Name the obtuse triangles in the figure.

(d) If there are equilateral or isosceles triangles in the figure, name them.

See also Exercise 13.

In a triangle, there are three special kinds of line segments which can be drawn from a vertex to the opposite side—altitude, median, and angle bisector. An *altitude* of a triangle is the perpendicular line segment from a vertex to the opposite side (or the opposite side extended). (60)

Altitude

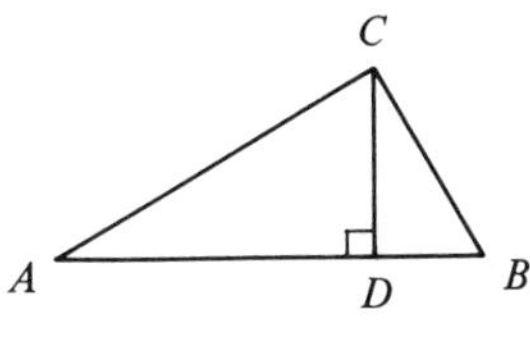

$\overline{CD}$ is the altitude from C to $\overline{AB}$.

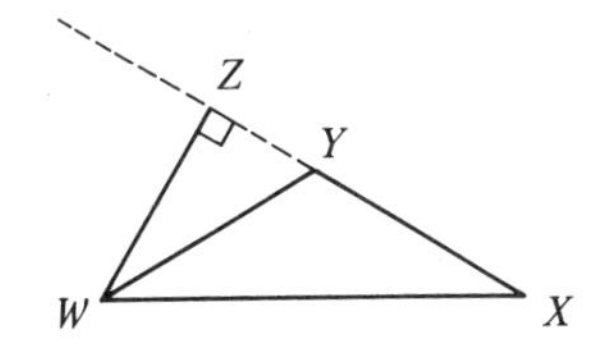

$\overline{WZ}$ is the altitude from W to $\overrightarrow{XY}$.

Median

A *median* of a triangle is a line segment from a vertex to the midpoint of the opposite side.

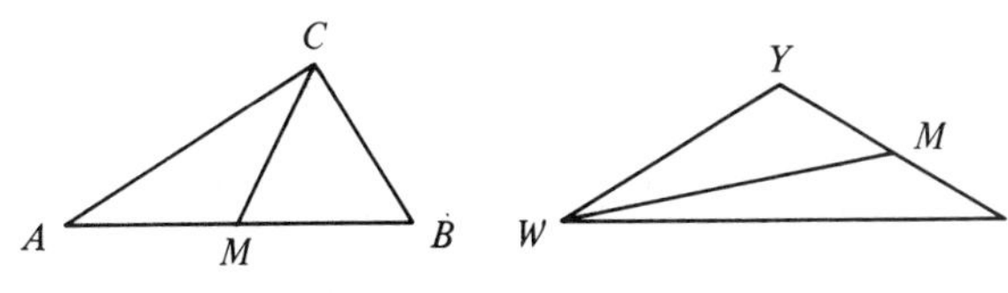

$\overline{CM}$ is the median from C to $\overline{AB}$. Thus, $\overline{AM} = \overline{MB}$.

$\overline{WM}$ is the median from W to $\overline{XY}$. Thus, $\overline{XM} = \overline{MY}$.

Angle bisector

An *angle bisector* of a triangle is a line segment from a vertex to the opposite side and which bisects the angle.

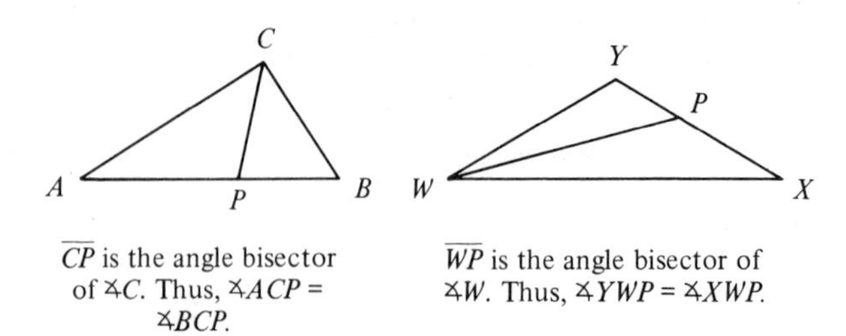

$\overline{CP}$ is the angle bisector of $\measuredangle C$. Thus, $\measuredangle ACP = \measuredangle BCP$.

$\overline{WP}$ is the angle bisector of $\measuredangle W$. Thus, $\measuredangle YWP = \measuredangle XWP$.

In a comprehensive geometry course, students learn to accurately construct the altitude, median and angle bisector using appropriate tools. In this abbreviated introduction, however, you will only be asked to approximate these line segments by sketching freehand (no tools except a pencil and, if you wish, a ruler.)

Objective 14: Median, altitude, and angle bisector

◆ **Example 10** For (a) to (c), draw the altitude, median and angle bisector from vertex A to side $\overline{BC}$. (61)

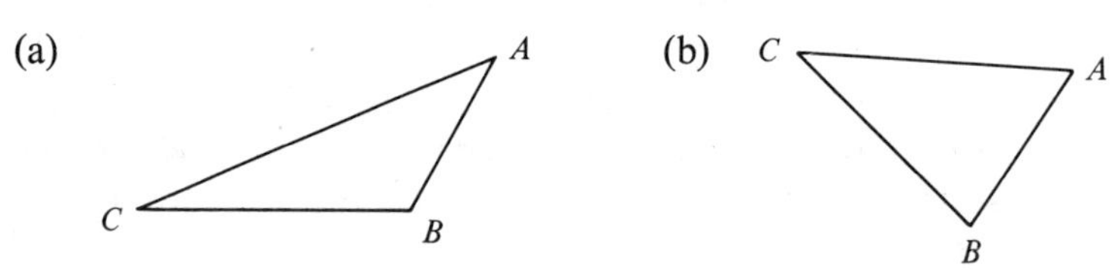

(c)

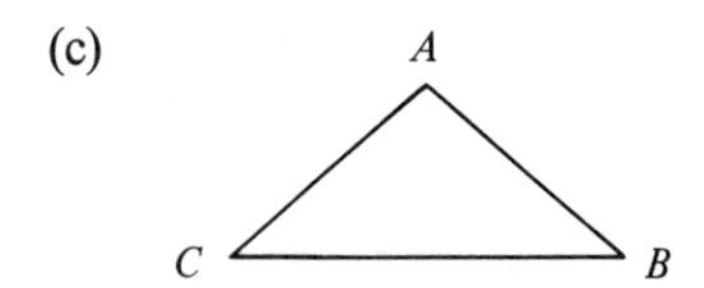

(d) What is the apparent relationship of the three line segments drawn in the isosceles triangle of part (c)?

(e) In which kind of triangle does the altitude lie in the interior of the triangle?

(f) Draw the three medians of $\triangle RST$.

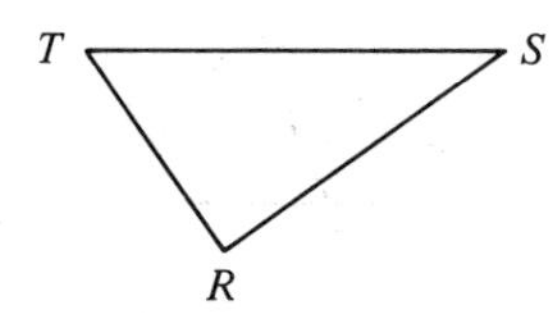

In each triangle below, $\overline{AD}$ is the altitude, $\overline{AM}$ is the median, and $\overline{AP}$ is the angle bisector:

(a)

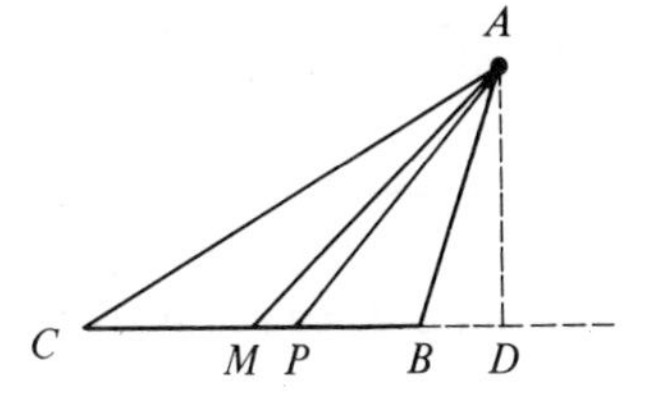

(b)

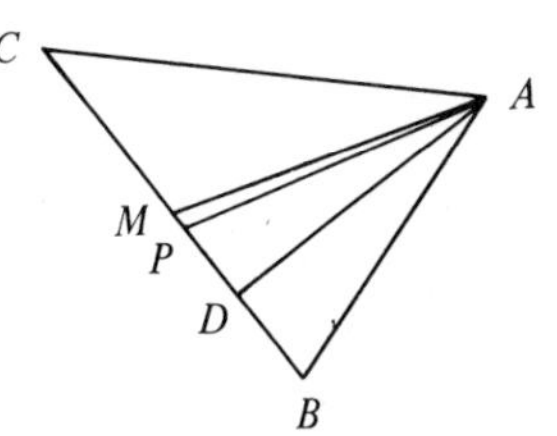

(c)

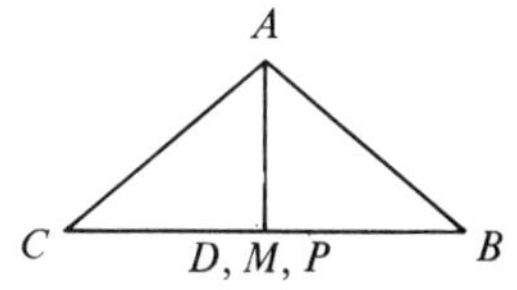

(d) When the median, altitude, and angle bisector are drawn from the vertex that lies between the equal sides of an isosceles triangle, the three line segments appear to be the same segment.

(e) In a triangle with all acute angles, the altitude lies in the interior of the triangle.

(f)

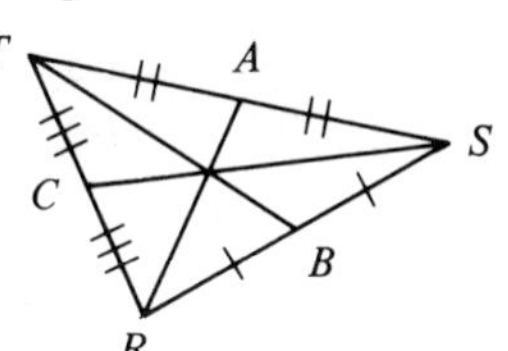

$\overline{RA}$, $\overline{TB}$, and $\overline{SC}$ are medians.

In any triangle the medians will intersect in one point.

● **Practice 10** Copy each triangle below. For (a) to (c), draw the altitude, median, and angle bisector from A to $\overline{BC}$. (62)

(a)

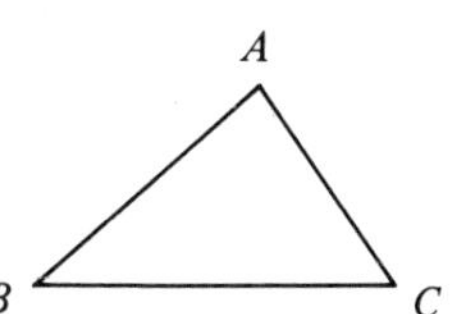

(b)

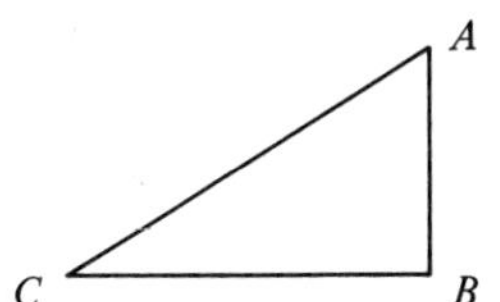

(c)

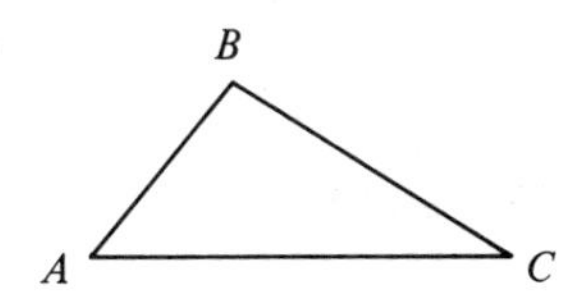

(d) Draw the three angle bisectors of $\triangle LMP$.

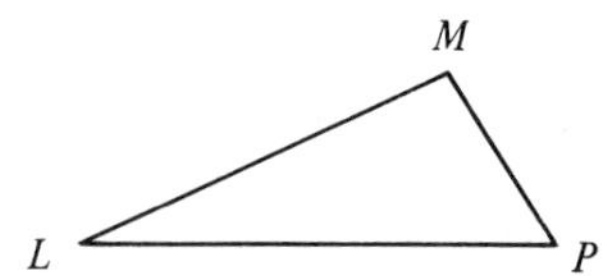

(e) Copy $\triangle LMP$ again. Now, draw the three altitudes of $\triangle LMP$.

(f) If an altitude of a triangle does not lie in the interior of the triangle, what kind of triangle is it?

See also Exercises 14 and 15.

The properties of the angles formed by parallel lines and a transversal can be used to prove an essential theorem of geometry. **(63)**

Triangle angles total 180°

Theorem *The sum of the measures of the angles of a triangle is* 180°.

To prove this theorem, we draw a line, $\overleftrightarrow{l}$, through a vertex of the triangle, parallel to an opposite side. We know such a parallel exists, by the assumption that through any given point not on a line there is one and only one parallel to the given line. **(64)**

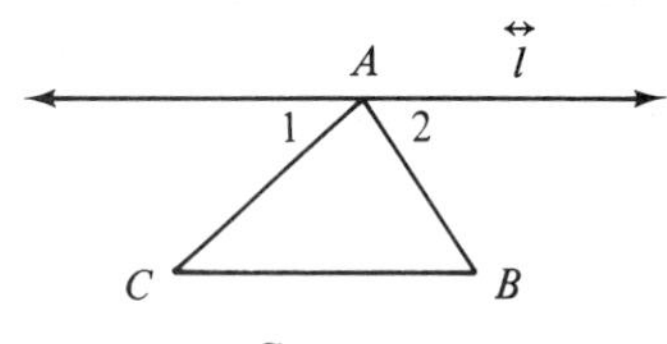

Given: $\overleftrightarrow{l} \parallel \overleftrightarrow{CB}$

Prove: $m(\measuredangle C) + m(\measuredangle B) + m(\measuredangle BAC) = 180°$

Statements	*Reasons*
1. $\overleftrightarrow{l} \parallel \overleftrightarrow{CB}$	1. Given
2. $\measuredangle 1 = \measuredangle C$, $\measuredangle 2 = \measuredangle B$	2. If two parallel lines are cut by a transversal, the alternate interior angles are equal.
3. $m(\measuredangle 1) + m(\measuredangle 2) + m(\measuredangle CAB) = 180°$	3. Sum of the angles on one side of a straight line is 180°.
4. $m(\measuredangle C) + m(\measuredangle B) + m(\measuredangle CAB) = 180°$	4. Substitution

Many useful theorems result from this theorem. Two examples of such theorems are given in Example 11(a) and Practice 11(a).

◆ **Example 11**

(a) Supply the reasons in the following proof. **(65)**

Objective 15: Geometric proof—sum of the angles of a triangle

Theorem *In a right triangle, the two acute angles are complementary.*

Given: $\triangle ABC$ is a right triangle with $\measuredangle C$ a right angle

Prove: $\measuredangle A$ is complementary to $\measuredangle B$.

Statements	*Reasons*
1. $\triangle ABC$ is a right triangle with $\measuredangle C$ a right angle.	1. Given
2. $m(\measuredangle C) = 90°$	2. Definition of right angle
3. $m(\measuredangle A) + m(\measuredangle B) + m(\measuredangle C) = 180°$	3. The sum of the angles of a triangle is 180°.
4. $m(\measuredangle A) + m(\measuredangle B) = 90°$	4. Equals subtracted from equals are equal.
5. $\measuredangle A$ and $\measuredangle B$ are complementary.	5. Definition of complementary

Objective 16: Finding measure of third angle of a triangle

(b) Find the measure of the third angle of a triangle, given the measure of two of the angles: $m(\measuredangle A) = 104°$, $m(\measuredangle B) = 16°$, $m(\measuredangle C) = ?$

$$\begin{aligned} m(\measuredangle A) + m(\measuredangle B) + m(\measuredangle C) &= 180° \\ 104° + 16° + m(\measuredangle C) &= 180° \\ 120° + m(\measuredangle C) &= 180° \\ m(\measuredangle C) &= 60° \end{aligned}$$

(c) In the figure, $\overleftrightarrow{l} \parallel \overleftrightarrow{AB}$, $\measuredangle 5 = 70°$, $\measuredangle 6 = 17°$, and $\measuredangle 7 = 50°$. Find angles 8, 9, 1, 2, 3, and 4. Since the alternate interior angles $\measuredangle 8$

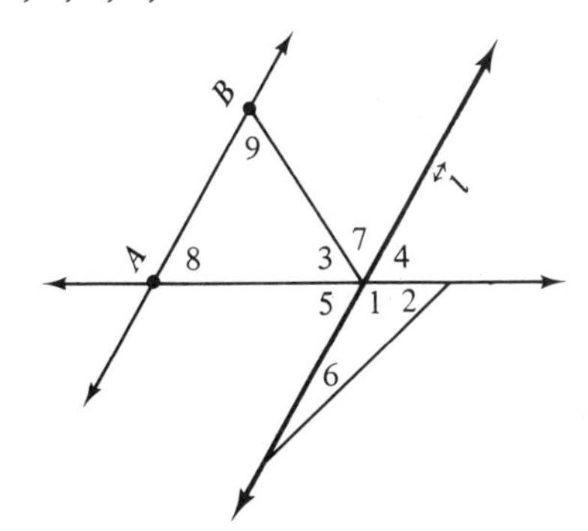

and $\measuredangle 5$ are equal, $\measuredangle 8 = 70°$. Similarly, $\measuredangle 9 = \measuredangle 7 = 50°$. Now, $\measuredangle 1$ is supplementary to $\measuredangle 5$; thus, $\measuredangle 1 = 110°$. Since the sum of the angles of a triangle is 180°,

$$m(\measuredangle 1) + m(\measuredangle 2) + m(\measuredangle 3) = 180^\circ$$
$$110^\circ + m(\measuredangle 2) + 17^\circ = 180^\circ$$
$$m(\measuredangle 2) + 127^\circ = 180^\circ$$
$$m(\measuredangle 2) = 53^\circ$$

Also, $\measuredangle 4 = \measuredangle 5$ because vertical angles are equal; thus, $m(\measuredangle 4) = 70^\circ$. Then since they form a straight line,

$$m(\measuredangle 3) + m(\measuredangle 7) + m(\measuredangle 4) = 180^\circ$$
$$m(\measuredangle 3) + 50^\circ + 70^\circ = 180^\circ$$
$$m(\measuredangle 3) + 120^\circ = 180^\circ$$
$$m(\measuredangle 3) = 60^\circ$$

● Practice 11

(a) Supply the reasons in the following proof. **(66)**

Theorem *Each of the angles of an equilateral triangle is 60° in measure.*

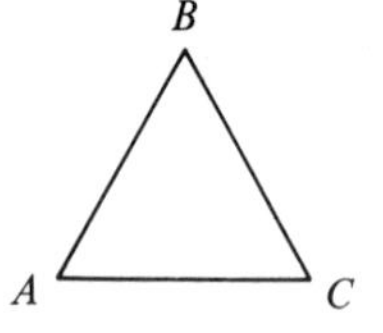

Given: $\triangle ABC$ is equilateral
$\triangle ABC$ is equiangular

Prove: $m(\measuredangle A) = m(\measuredangle B) = m(\measuredangle C) = 60^\circ$

Statements	*Reasons*
1. $\triangle ABC$ is equilateral	1. ______________
2. $\triangle ABC$ is equiangular	2. ______________
3. $\measuredangle A = \measuredangle B = \measuredangle C$	3. ______________
4. $m(\measuredangle A) + m(\measuredangle B) + m(\measuredangle C) = 180^\circ$	4. ______________
5. $m(\measuredangle C) + m(\measuredangle C) + m(\measuredangle C) = 180^\circ$	5. ______________
6. $3 \cdot m(\measuredangle C) = 180^\circ$	6. ______________
7. $m(\measuredangle C) = 60^\circ$	7. ______________

(b) Find the third angle of the triangle, given two angles
(1) 82°, 50° (2) 115°, 32°

(c) $\sphericalangle ABC$ is a right angle

$m(\sphericalangle 1) = 60°$

$\overline{BD}$ is an altitude

Find measures of angles 2, 3, and 4.

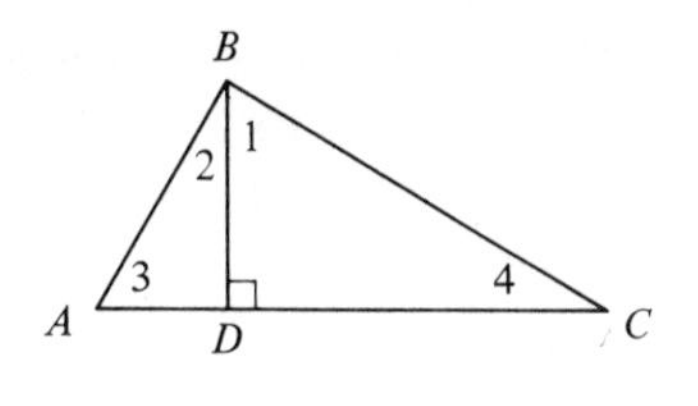

See also Exercises 16, 17, 18

Pythagorean theorem

Theorem *The square of the hypotenuse of a right triangle is equal to the sum of the squares of the legs.** **(67)**

Objective 17: Application of Pythagorean theorem

◆ **Example 12** Use the Pythagorean theorem to find the unknown side in right triangle ABC with right angle C. **(68)**

(a) $a = 6, \quad b = 8, \quad c = ?$

(b) $a = 5, \quad c = 13, \quad b = ?$

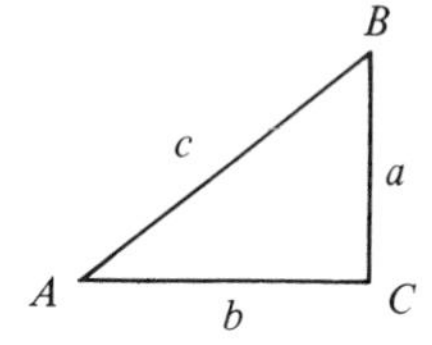

(a)
$$\begin{aligned} a^2 + b^2 &= c^2 \\ 6^2 + 8^2 &= c^2 \\ 36 + 64 &= c^2 \\ 100 &= c^2 \\ c^2 &= 100 \\ c &= \sqrt{100} \quad \text{or} \quad 10 \end{aligned}$$

(b)
$$\begin{aligned} a^2 + b^2 &= c^2 \\ 5^2 + b^2 &= 13^2 \\ 25 + b^2 &= 169 \\ b^2 &= 169 - 25 \\ b^2 &= 144 \\ b &= \sqrt{144} \quad \text{or} \quad 12 \end{aligned}$$

● **Practice 12** Use the Pythagorean theorem to find the unknown side in a right triangle ABC with right angle C. **(69)**

(a) $a = 7, \quad b = 24, \quad c = ?$

(b) $a = 8, \quad c = 10, \quad b = ?$

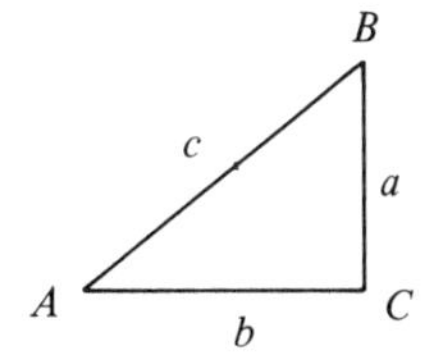

See also Exercise 19.

H. QUADRILATERALS

Objective 18: Relationships among quadrilaterals

Like triangles, quadrilaterals have also received special attention; hence, some quadrilaterals are given specific names which identify their shape and/or angle measurements as discussed in the following paragraphs. **(70)**

*The *hypotenuse* is the side opposite the right angle; the remaining sides are *legs*.

Trapezoid

A *trapezoid* is a quadrilateral which has one and *only one pair of parallel sides.*

$\overline{AB} \parallel \overline{DC}$

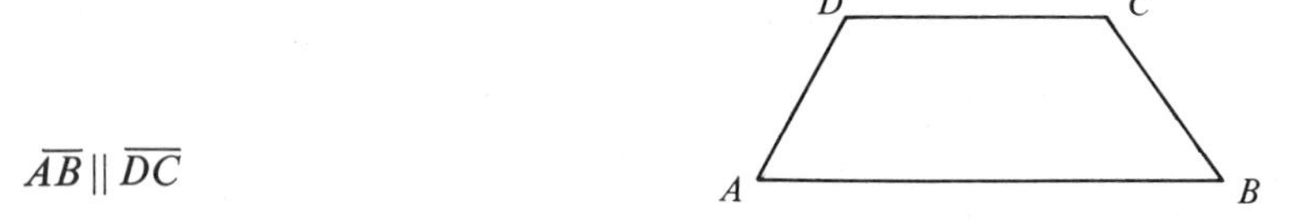

Parallelogram

A *parallelogram* is a quadrilateral in which the pairs of *opposite sides are parallel.*

$\overline{AB} \parallel \overline{DC}$ and $\overline{BC} \parallel \overline{AD}$

Rhombus

Certain special parallelograms include the rhombus, rectangle, and square. A *rhombus* is a parallelogram with four equal sides. (71)

$$\overline{AB} = \overline{BC} = \overline{CD} = \overline{DA}$$
$$\overline{AB} \parallel \overline{DC} \text{ and } \overline{AD} \parallel \overline{BC}$$

Rectangle

A *rectangle* is a parallelogram with four equal angles.

$$\measuredangle A = \measuredangle B = \measuredangle C = \measuredangle D$$
$$\overline{AB} \parallel \overline{DC} \text{ and } \overline{AD} \parallel \overline{BC}$$

The sum of the angles of any quadrilateral is 360°. Since the four angles of a rectangle are equal, each angle thus equals $\frac{360°}{4}$ or 90°.

Square

A *square* is a parallelogram with four equal sides and four equal angles. Thus the square, like the rectangle, contains four right angles.

$$\measuredangle A = \measuredangle B = \measuredangle C = \measuredangle D$$
$$\overline{AB} = \overline{BC} = \overline{CD} = \overline{DA}$$

◆ **Example 13**

(a) Is every square also a rectangle? (72)

(b) Is every rectangle a square?

(a) Every square is a rectangle because a square has four equal angles, which is the definition of a rectangle.

(b) A rectangle is not required to have equal sides; thus, many rectangles are not squares.

● **Practice 13** Given the following sets, find the indicated relationships among quadrilaterals: let P be the set of all parallelograms; R be the set of all rectangles; H be the set of all rhombuses; S be the set of all squares; and T be the set of all trapezoids. Find **(73)**

(a) $R \cap H =$

(b) $P \cap T =$

(c) $S \cap R =$

(d) $S \cup H =$

(e) Is S a subset of H?

(f) What is the subset relationship between P and R?

See also Exercise 20.

I. PERIMETER AND AREA OF POLYGONS

Two aspects of geometry that have very useful applications are perimeter and area. When a fence is built around a lot, the fence represents the perimeter. Carpet installed on a floor illustrates area. We will next consider perimeter and area measurements for several common polygons. By definition, the *perimeter* of a polygon is the sum of the lengths of its sides. **(74)**

Perimeter

Objective 19: Perimeter of a polygon

◆ **Example 14** Write an equation for the perimeter of each regular polygon and find the perimeter: **(75)**

(a)

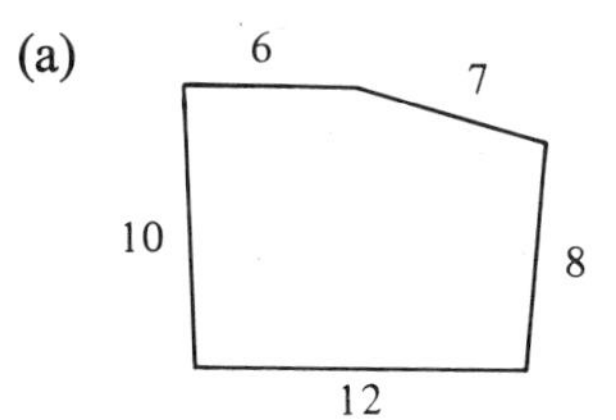

(b) *ABCD* is a parallelogram

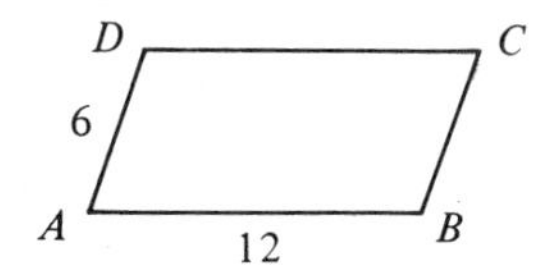

(c) A regular hexagon where one side equals 5 cm.

(d) Find the length of a fence around a rectangular field 100 feet by 40 feet.

(a) Since this is an irregular polygon, the perimeter is simply the sum of the lengths of the sides: $10 + 12 + 8 + 7 + 6 = 43$ units (feet, meters, or other measurement of length).

(b) The opposite sides of the parallelogram are equal. Thus, the sum of the two long and two short sides may be indicated by $p = 2l + 2s$. For this parallelogram, $p = 2(12) + 2(6) = 24 + 12$ or 36.

(c) A regular hexagon has six equal sides, so the perimeter, $p = 6s$. In this case, $p = 6 \cdot 5$ or 30 cm.

(d) Again, the opposite sides of a rectangle are equal; therefore, the perimeter is twice the length plus twice the width, or $p = 2l + 2w$. The length of the fence is thus $p = 2(100) + 2(40) = 200 + 80 = 280$ feet.

● **Practice 14** Write an equation to represent the perimeter of each polygon and find each perimeter: (76)

(a) *ABCD* is a rhombus

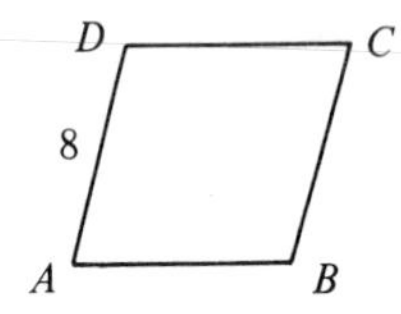

(b) *ABCDE* is a regular pentagon

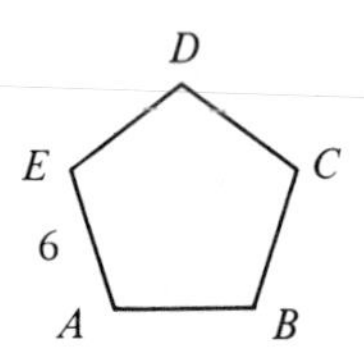

(c) A room is 8 feet by 12 feet. What is its perimeter?

See also Exercises 21 and 22.

Area

Perimeter is a measurement of length. *Area* of a polygon is a measurement of the surface enclosed by the sides of the polygon. Consider the square below whose sides measure 1 inch. The area enclosed by the square (77)

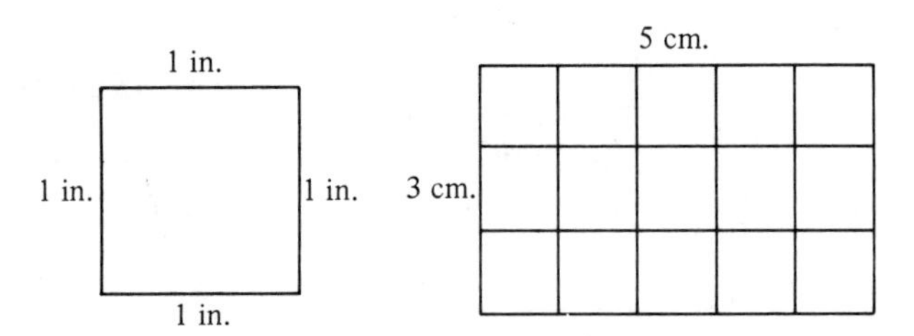

(the shaded region) is one square inch. Consider also the rectangle above which is 3 cm. by 5 cm. As illustrated, this rectangle encloses an area of $3 \cdot 5$ or 15 sq. cm.

In general, a rectangle with sides of length l and width w encloses an area of $A = lw$ square units. (This same formula is often written $A = bh$, where the sides are called "base" and "height.")

Area of rectangle

Area of a rectangle: $A = lw$

The square is a specialized rectangle which warrants a separate area formula. Since all sides of the square are equal, then $A = lw$ becomes $A = s \cdot s$ or $A = s^2$.

Area of square

Area of a square: $A = s^2$

(78)

Next consider the parallelogram $ABCD$. Suppose we draw a perpendicular from vertex D to the opposite side and also draw a perpendicular from vertex C to an extension of the opposite side. It can be shown that

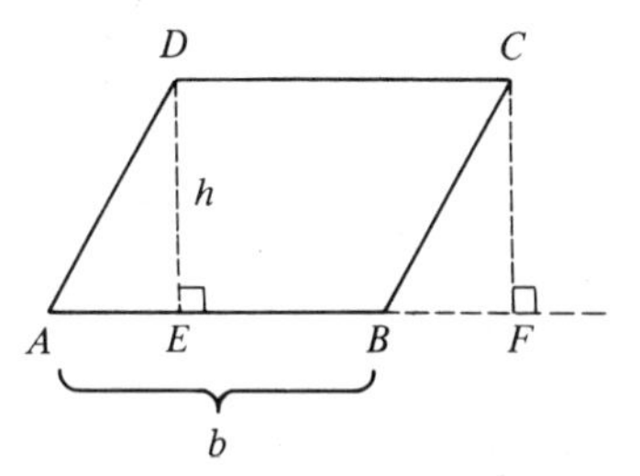

the area of $\triangle ADE$ equals the area of $\triangle BCF$ (using the concept of congruence, which we have not discussed).

Now, $EFCD$ is a rectangle, and its area is $\overline{EF} \cdot \overline{DE}$. That is, the area of the *rectangle* equals the area of the trapezoid $EBCD$ plus the area of $\triangle BCF$. However, the area of the *parallelogram* equals the area of the trapezoid $EBCD$ plus the area of $\triangle ADE$. But since the area of $\triangle ADE$ equals the area of $\triangle BCF$, the area of the parallelogram $ABCD$ is thus the same as the area of the rectangle. That is, each area is $A = bh$, where b is the base length and h is the perpendicular height from the base to the opposite side. Thus,

Area of parallelogram

Area of a parallelogram: $A = bh$

(79)

We can determine an equation for the area of any triangle in a manner similar to that used for the parallelogram. Consider any triangle, $\triangle ABC$, below. Suppose we draw a perpendicular, h, from vertex B to the base, $\overline{AC}$. Suppose we also draw $\overline{CD}$ through C parallel to $\overline{AB}$, and draw

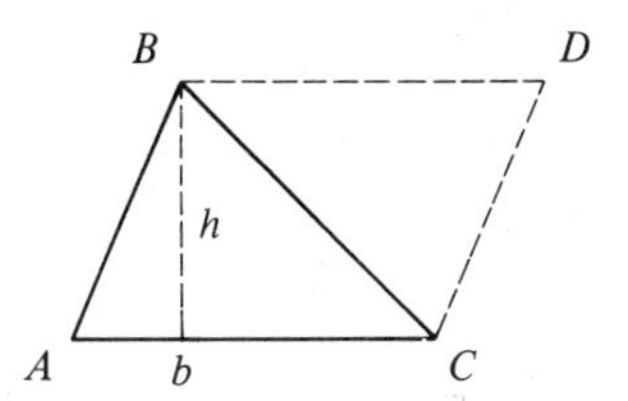

$\overline{BD}$ through B parallel to $\overline{AC}$. Now, we know that the area of parallelogram $ACDB$ is $A = bh$. The area of $\triangle ABC$ is one half of the area of the parallelogram, or $\frac{1}{2}bh$.

Area of triangle

Area of a triangle: $A = \dfrac{1}{2}bh$

(80)

An equation for the area of a trapezoid is also derived in a similar manner. Given a trapezoid $ABCD$ with parallel sides b and b', lines can be drawn to form a parallelogram $AEFD$. Now the area of the parallelogram would be $A = \text{base} \times \text{height} = (b + b')h$. It then follows that the area of either trapezoid is $A = \frac{1}{2}(b + b')h$.

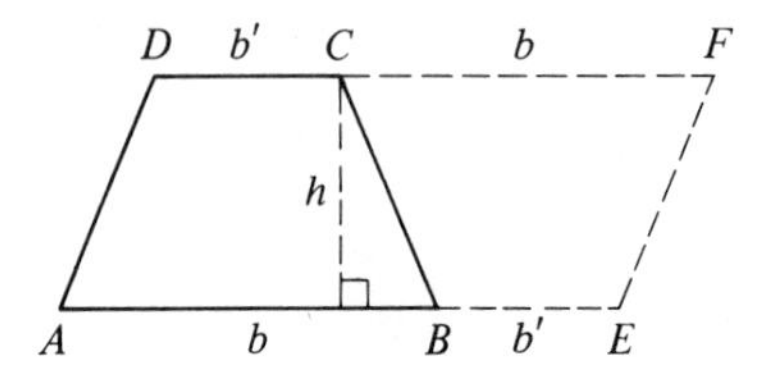

Area of trapezoid

Area of trapezoid: $A = \dfrac{1}{2}(b + b')h$

Objective 20: Area of triangle, rectangle, parallelogram

(81)

◆ **Example 15** Find the area of each figure.

(a)

(b)

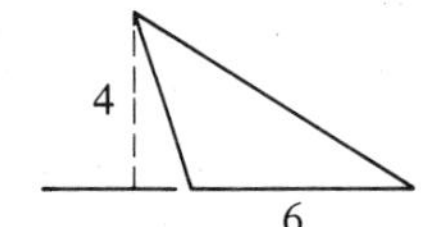

(c)

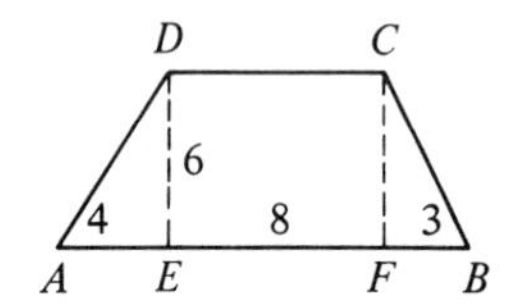

(d) Find the number of square yards of carpet needed to cover a room 5 yards by 7 yards.

(a) $A = lw = 4 \cdot 8 = 32$ square units

(b) $A = \dfrac{1}{2}bh = \dfrac{1}{2}(4 \cdot 6) = 12$ square units

(c) $A = \dfrac{1}{2}(b + b')h = \dfrac{1}{2}(15 + 8)6 = 69$ square units

(d) The number of square yards of carpet needed is $A = lw = 5 \cdot 7$ or 35 square yards.

● **Practice 15** Find the area of each figure. (82)

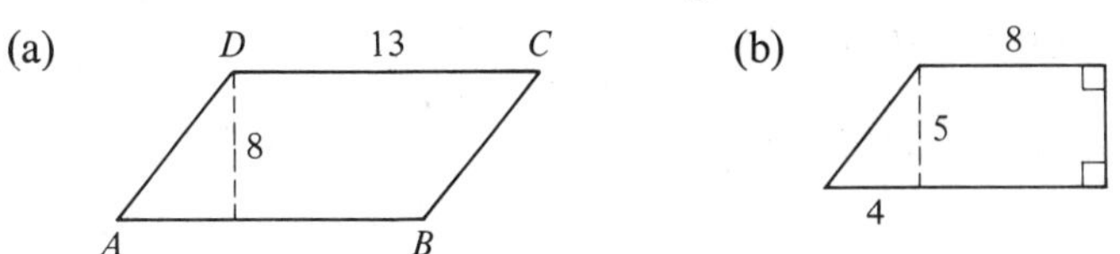

ABCD is a parallelogram.

(c) Find the number of square feet in the house whose floor plan is given below.

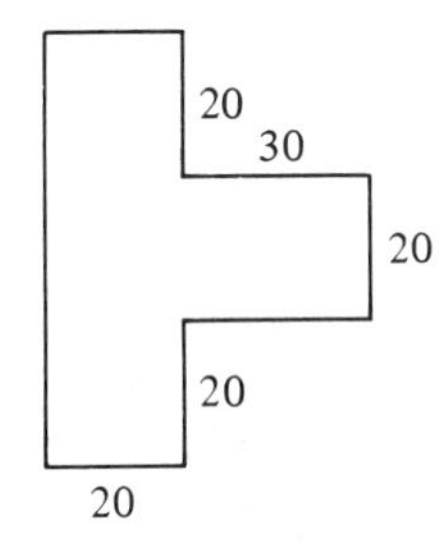

See also Exercises 23–26.

J. CIRCLE

There remains one important plane geometric figure to consider—the circle. The definition of a circle requires a specific point and a specific distance, as follows. A *circle* is the set of all points in a plane which are a given distance from a given point. This point is called the *center* of the circle, and the distance is called the *radius* of the circle. Observe from this definition that the circle consists only of the curved line of points that we visualize as a pencil-line drawing. The interior enclosed within this line is *not* considered part of the circle itself. (83)

Circle

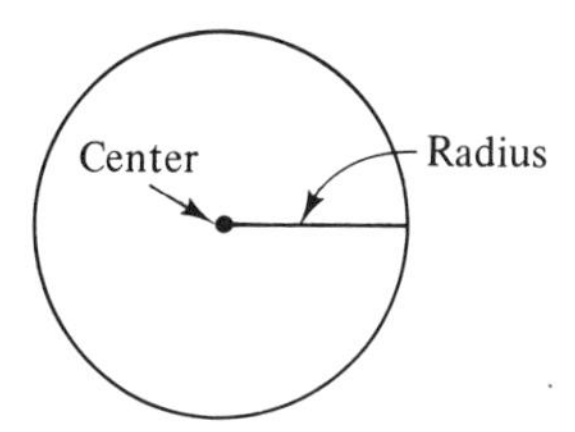

Certain line segments, lines, and subsets of the circle are given special names. Refer to the following two circles to illustrate the terms defined (84)

below. (The examples given are not necessarily every possible example from the figures.)

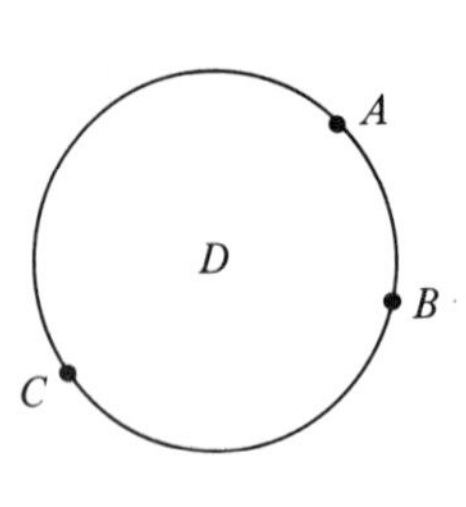

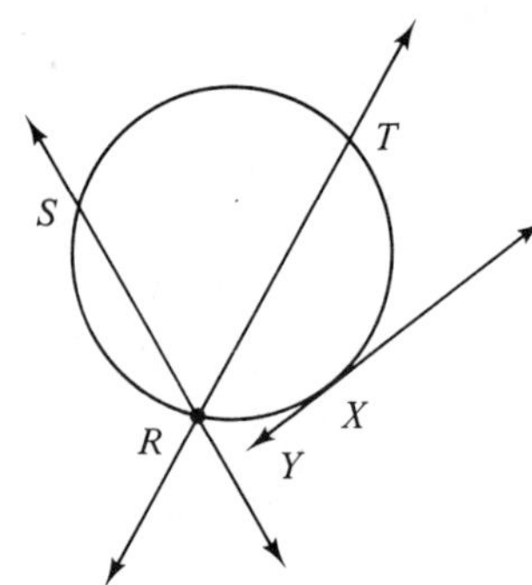

Objective 21: Identifying parts associated with circles

Term	*Definition*	*Example*
Chord	A line segment joining two points on a circle.	$\overline{AB}$ or $\overline{RS}$ or $\overline{RT}$
Radius	A line segment from the center of the circle to a point on the circle.	$\overline{DA}$ or $\overline{DB}$
Diameter	A chord of the circle which passes through the center.	$\overline{AC}$
Arc	The set of all points on a circle which lie between two given points of the circle.	$\widehat{AB}$ or $\widehat{RS}$ or $\widehat{ACB}$
Central angle (of an arc or a chord)	An angle formed by two radii; its vertex is the center, and each side of the angle passes through an endpoint of the arc or chord.	$\sphericalangle ADB$
Semicircle	The set of points on the circle between the endpoints of a diameter.	$\widehat{ABC}$
Minor arc	An arc which is a proper subset of a semicircle.	$\widehat{AB}$: A B
Major arc	An arc which has a semicircle as a proper subset.	$\widehat{ACB}$: A C B
Tangent (to a circle)	A line which intersects a circle in only one point.	$\overleftrightarrow{XY}$
Secant	A line (as opposed to a line segment) which intersects a circle in two points. (Compare to a chord, which is a line segment.)	$\overleftrightarrow{RT}$ or $\overleftrightarrow{RS}$ or $\overleftrightarrow{AC}$
Angle inscribed in a circle	An angle whose vertex lies on the circle and whose sides are rays which intersect the circle.	$\sphericalangle SRT$

◆ **Example 16** Name one of the following in the figure: (85)

(a) chord (b) diameter
(c) central angle (d) minor arc
(e) major arc (f) radius
(g) tangent (h) secant
(i) inscribed angle

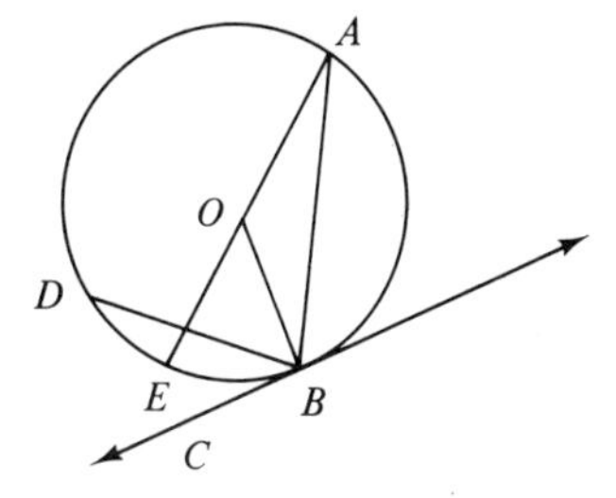

(a) Any one of the following is a chord: $\overline{AB}$, $\overline{AE}$, $\overline{BD}$.

(b) $\overline{AE}$ is a diameter.

(c) Central angles are $\measuredangle AOB$ and $\measuredangle BOE$.

(d) $\widehat{AB}$ and $\widehat{BE}$ are minor arcs of central angles, $\measuredangle AOB$ and $\measuredangle BOE$. $\widehat{BD}$ is the minor arc of chord $\overline{BD}$ (with central angle, $\measuredangle BOD$).

(e) Major arcs of the central angles are $\widehat{ADB}$ (or $\widehat{AEB}$) and $\widehat{BAE}$. The major arc of chord $\overline{BD}$ is $\widehat{BAD}$.

(f) Radii are $\overline{OA}$, $\overline{OE}$, and $\overline{OB}$.

(g) $\overleftrightarrow{BC}$ is a tangent.

(h) $\overleftrightarrow{AB}$, $\overleftrightarrow{BD}$, and $\overleftrightarrow{AE}$ are secants.

(i) Inscribed angles are $\measuredangle BAE$ and $\measuredangle ABD$.

● **Practice 16** Given the following figure, name one example of each term listed in (a)–(i) of Example 16: (86)

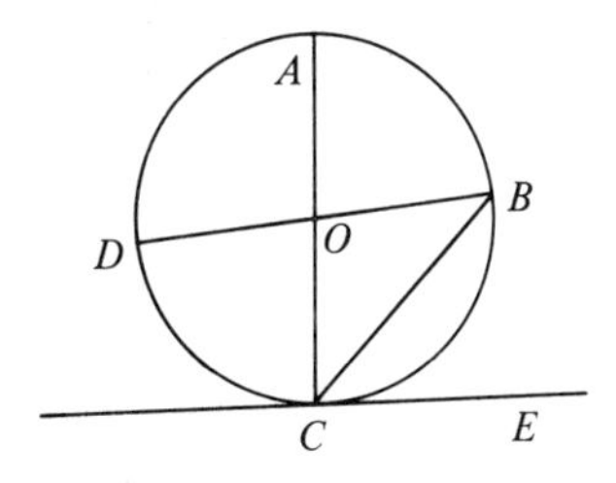

See also Exercise 27.

Arcs of circles can be measured according to length, (in inches, feet, centimeters, etc.) or in degrees. The *degree measure of a minor arc* is equal (87)

Arc degree measure

to the degree measure of its central angle. If $\measuredangle AOB$ below measures 50°, then $\widehat{AB}$ measures 50°.

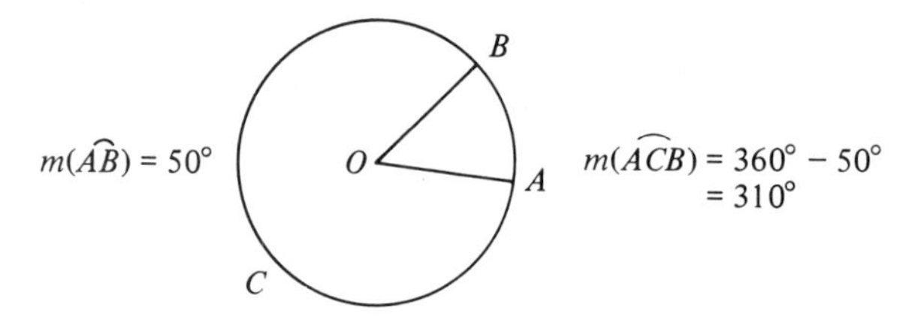

(88) The degree measure of a semicircle is 180°, since the semicircle is defined by a straight angle (diameter). The degree measure of a circle is thus $2 \cdot 180°$ or 360°. To find the *degree measure of a major arc*, subtract from 360° the degree measure of the central angle of the related minor arc. Thus, major arc ACB above measures $360° - 50° = 310°$.

(89) The measurement of an inscribed angle may also be computed from the measurement of its intercepted arc. Consider an inscribed angle which has a side that is a diameter of the circle, as shown below. We wish to find the relationship between the inscribed angle and the arc which the inscribed angle intercepts.

(90) If the degree measure of $\widehat{BC}$ is $x°$, then the central angle, $\measuredangle BOC$, also measures $x°$. Since $\measuredangle BOC$ and $\measuredangle AOB$ are supplementary, then $m(\measuredangle AOB) = (180° - x°)$. However, the sum of the measures of the angles in $\triangle AOB$ is 180°. That is,

$$m(\measuredangle A) + m(\measuredangle B) + m(\measuredangle AOB) = 180°$$
$$m(\measuredangle A) + m(\measuredangle B) + (180° - x°) = 180°$$
$$m(\measuredangle A) + m(\measuredangle B) = x°$$

(91) But triangle sides $\overline{OA}$ and $\overline{OB}$ are equal since they are both radii. That is, $\triangle AOB$ is isosceles. Thus, the angles opposite the equal sides are also equal; or $\measuredangle A = \measuredangle B$.

Thus, $$m(\measuredangle A) + m(\measuredangle B) = x°$$

becomes $$m(\measuredangle A) + m(\measuredangle A) = x°$$
$$2 \cdot m(\measuredangle A) = x°$$

or $$m(\measuredangle A) = \frac{1}{2}x°$$

This relationship can also be proved for any inscribed angle; hence, the following theorem results.

Objective 22: Measure of an inscribed angle

Theorem *The measure of an inscribed angle of a circle is equal to one-half the measure of its intercepted arc.*

◆ **Example 17**

(a) Find the measure of an inscribed angle which intercepts an arc of 78°. **(92)**

(b) How many degrees are there in a major arc $\widehat{ABC}$ if minor arc $\widehat{AB}$ measures 120°?

(c) Determine the degree measure of arcs $\widehat{AB}$, $\widehat{BC}$, and $\widehat{AC}$.

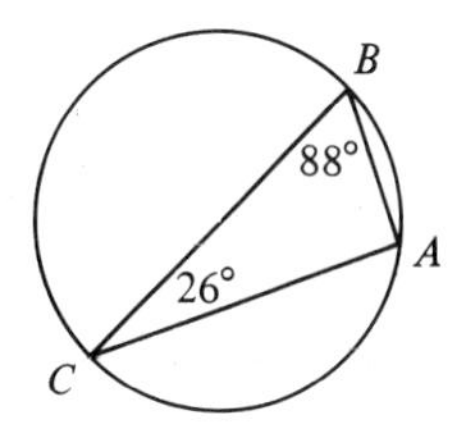

(d) If $ABCD$ is a square, how many degrees are in $\widehat{ABC}$, $\widehat{BCD}$, $\widehat{CDA}$, and $\widehat{DAB}$?

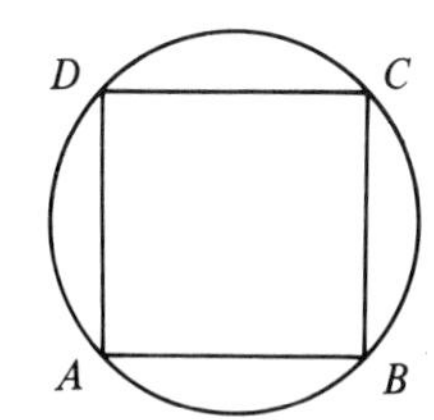

(a) The inscribed angle measures $\frac{1}{2} \cdot 78° = 39°$.

(b) The major arc measures $360° - 120° = 240°$.

(c) The other angle of $\triangle ABC$ is $180° - (88° + 26°) = 66°$. Since each inscribed angle measures one-half of its intercepted arc, then, conversely, each arc measures twice its inscribed angle. Thus,

$$\widehat{AB} = 2(26°) = 52°$$

$$\widehat{BC} = 2(66°) = 132°$$

$$\widehat{AC} = 2(88°) = 176°$$

(d) Each angle of the square is 90°. The arc which each angle intercepts must therefore be twice 90°, or 180°. Each of the given arcs is 180° in measure: $\widehat{ABC} = 180°$; $\widehat{BCD} = 180°$; $\widehat{CDA} = 180°$; $\widehat{DAB} = 180°$.

Practice 17

(a) A 40° angle is inscribed in a circle. How many degrees are there in the arc which it intercepts? (93)

(b) How many degrees are in major arc $\widehat{ACB}$, if minor arc $\widehat{AB}$ measures 75°?

(c) If triangle ABC is equilateral, how many degrees are in arcs $\widehat{AB}$, $\widehat{BC}$, and $\widehat{CA}$?

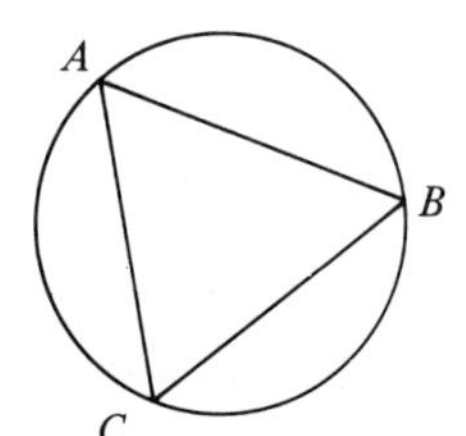

(d) Assume that $ABCD$ is a rectangle and $\widehat{AB}$ measures 24°. How many degrees are there in arcs $\widehat{BC}$, $\widehat{CD}$, and $\widehat{DA}$? (*Hint*: Use the fact that a circle equals 360°.)

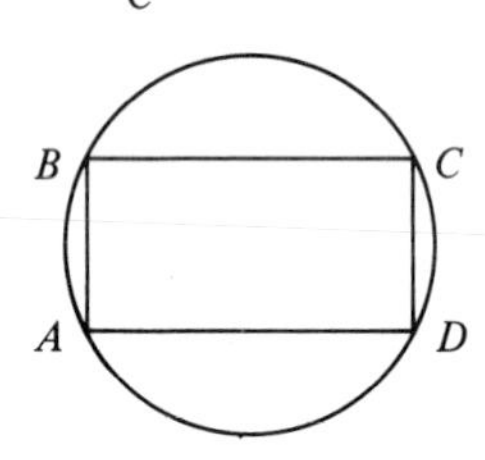

See also Exercises 28 and 29.

The equations for finding the *circumference* (length around a circle) and *area* (surface enclosed within a circle) are given here without proof. It can be shown for any circle that the ratio of the circumference to the diameter always approaches the irrational number, π, which approximately equals the decimal 3.14 or the fraction $\frac{22}{7}$. (94)

$$\pi = \frac{C}{d} \quad \text{or} \quad \frac{C}{2r}$$ where C is the circumference, d the diameter, and r the radius.

Circumference of circle

The circumference of a circle is found by the equation:

$$C = 2\pi r \quad \text{or} \quad C = \pi d$$

Area of circle

The area of a circle is found using the equation (95)

$$A = \pi r^2$$

Example 18

Objective 23: Finding area and circumference of circle

(a) Find the circumference and area of a circle with the following radius (use $\pi = \frac{22}{7}$): (96)

(1) $r = 10$ inches (2) $r = 14$ centimeters

Objective 24: Area of combination figures

(b) Find the area of the figure.

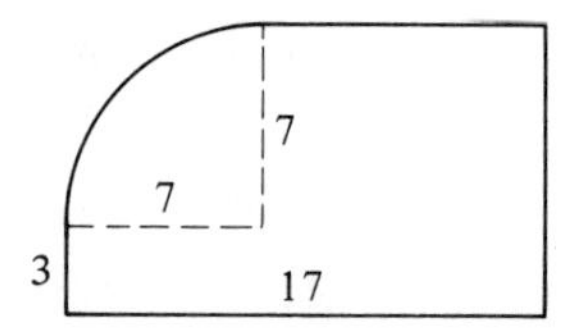

(a) (1) $C = 2\pi r = 2\left(\frac{22}{7}\right)10 = \frac{440}{7} = 62\frac{6}{7}$ inches

$A = \pi r^2 = \frac{22}{7}(10)^2 = \frac{22}{7}(100) = \frac{2200}{7}$

$= 314\frac{2}{7}$ sq. in.

(2) $C = 2\pi r = 2\left(\frac{22}{7}\right)14 = 88$ cm.

$A = \pi r^2 = \frac{22}{7}(14)^2 = \frac{22}{7}(196) = 616$ sq. cm.

(b) The curved corner is $\frac{1}{4}$ of a circle.

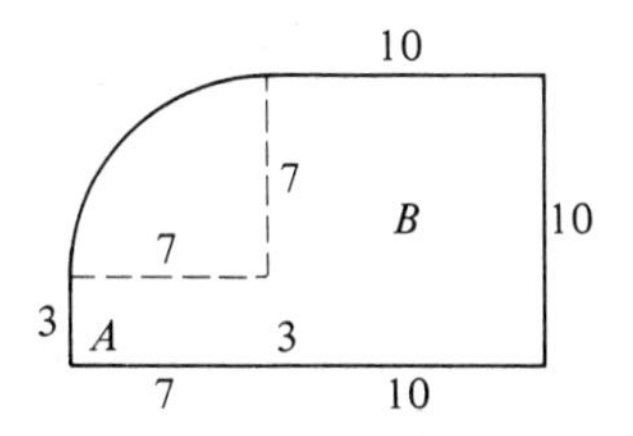

Thus,

Area of $\frac{1}{4}$ circle: $\frac{1}{4}\pi r^2 = \frac{1}{4}\left(\frac{22}{7}\right)7^2 = \frac{77}{2}$

$= 38\frac{1}{2}$ sq. units

Area of rectangle A: $lw = 7 \cdot 3 = 21$ sq. units

Area of square B: $s^2 = 10^2 = 100$ sq. units

Total area: $38\frac{1}{2} + 21 + 100 = 159\frac{1}{2}$ sq. units

- **Practice 18**

(a) Find the circumference and area of a circle with each given radius: **(97)**

(1) $r = 7$ (2) $r = 6$

(b) Find the area of the given figure.

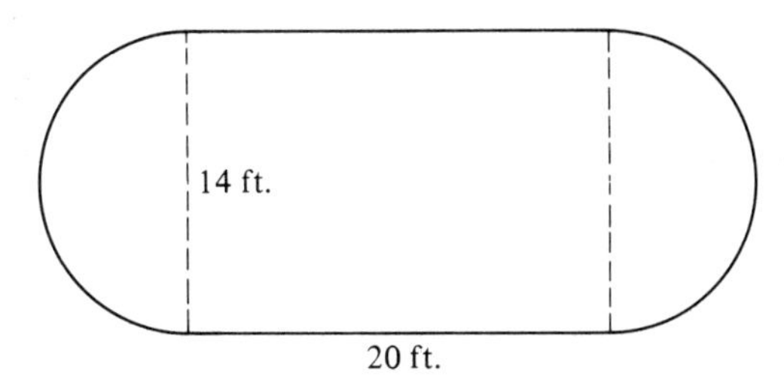

See also Exercises 30 and 31.

K. VOLUME

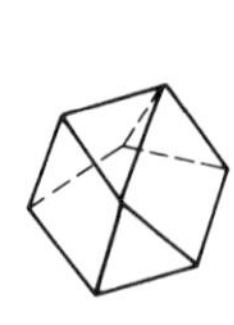
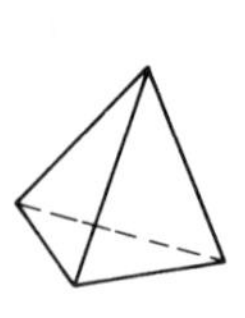
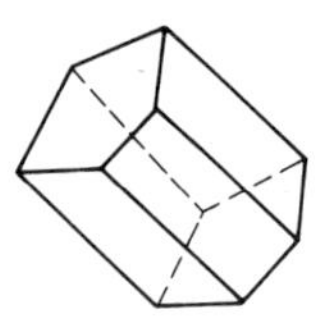
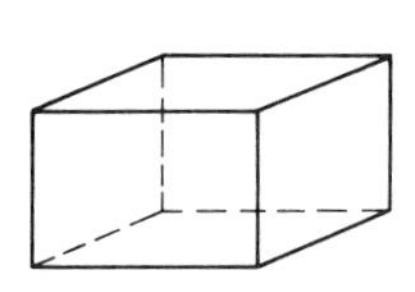

(98) Each of the above figures is a sketch of a three-dimensional figure called a polyhedron. Notice that each outside part of every figure is a polygonal region (a polygon and its interior). A *polyhedron* is thus a union of polygonal regions in different planes such that any two polygons have at most one side in common and no polygons have intersecting interiors. The polygonal regions are called *faces* of the polyhedron and any side of a polygon is called an *edge* of the polyhedron.

Polyhedron

(99) Some special kinds of polyhedra (plural of polyhedron) and volumes will be discussed. A *prism* is a polyhedron such that two of its faces are congruent identical polygonal regions which lie in parallel planes (planes which do not intersect). The parallel faces are called *bases* and the other

Prism

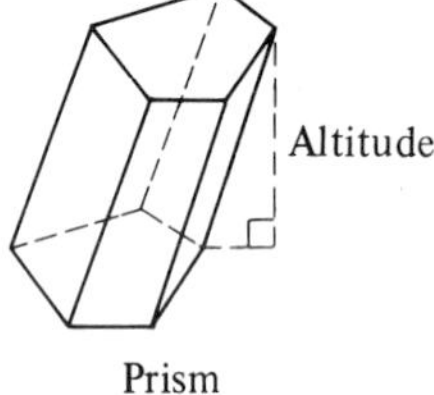

Prism

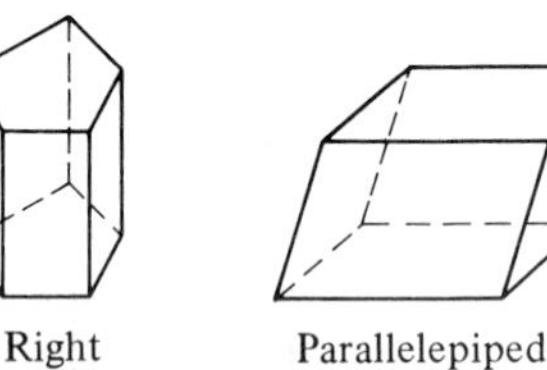
Right prism

Parallelepiped

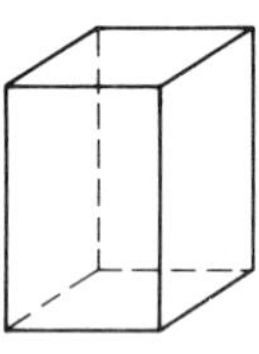
Rectangular parallelepiped

faces are called *lateral faces*. If a lateral edge is perpendicular to the base, the prism is a *right prism*. The *altitude* of a prism is the perpendicular distance between the bases. In a right prism, the altitude is the length of the equal lateral edges.

(100) When the faces of a prism are all parallelograms, the prism is a *parallelepiped*. If the parallelepiped is also a right prism, then it is a *rectangular parallelepiped*. A *cube* is a rectangular parallelepiped whose faces are congruent squares. Such figures in three-dimensional space are often called "solid" figures, and their study is the subject of "solid geometry."

Parallelepiped

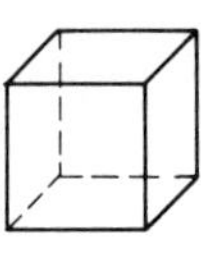
Cube

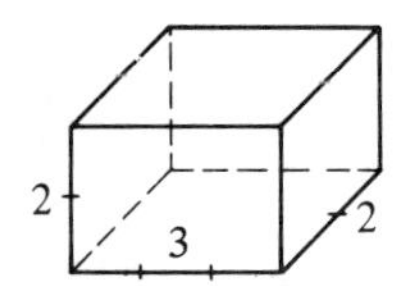

Rectangular parallelepiped

Volume

Consider a cube whose edges measure 1 inch. The *volume* (space enclosed within) of this cube is 1 cubic inch. The number of cubic inches enclosed by the rectangular parallelepiped above is the area of the base $(3 \cdot 2)$ times the altitude (2), or $(3 \cdot 2)2 = 12$ cubic inches. This procedure can be extended to any prism. In general, the volume of a prism is the area of the base times the altitude of the prism. Thus, $V = Ba$, where B is the area of the base and a is the altitude. (101)

◆ **Example 19** Find the volume of each prism, using the dimensions given below. (102)

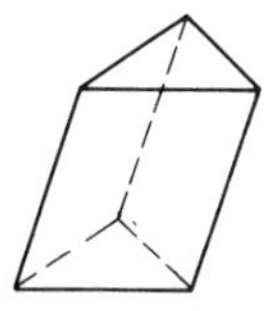

(b)

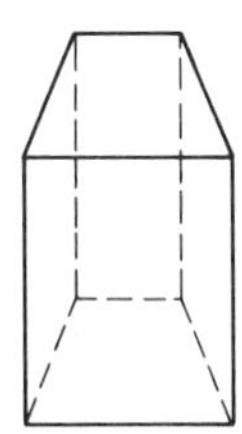

(c)

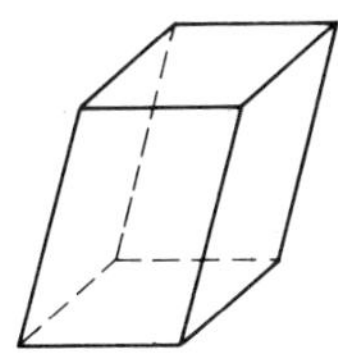

(d)

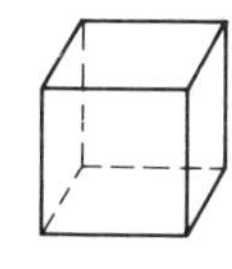

(a) The base is a triangle with base 6″ and height 3″. The altitude is 8″.

(b) The base is a trapezoid with bases 3 cm. and 5 cm. and height 4 cm. The altitude is 10 cm.

(c) The base of the prism is a rhombus with sides 2 ft. and height of 1 ft. The altitude is 8 ft.

(d) The prism is a cube with edges of 3 meters each.

(a) The formula for the area of a triangle is $\frac{1}{2}bh$. Thus, the area of the base is $A = \dfrac{1}{2}bh = \dfrac{1}{2} \cdot 6 \cdot 3 = 9$ sq. in. Then $V = Ba = 9 \cdot 8 = 72$ cubic inches.

(b) The area of the trapezoid base is $\frac{1}{2}(b + b')h = \frac{1}{2}(3 + 5)4 = 16$ sq. cm. The volume of the prism is thus $V = Ba = 16 \cdot 10 = 160$ cu. cm.

(c) The area of the rhombus base is $bh = 2 \cdot 1 = 2$ sq. ft. The volume of the prism is $Ba = 2 \cdot 8 = 16$ cu. ft.

(d) The area of the base is $bh = 3 \cdot 3 = 9$ sq. m. The volume of the cube is then $Ba = 9 \cdot 3 = 27$ cu. m.

● **Practice 19** Find the volume of each prism below. **(103)**

(a) (b)

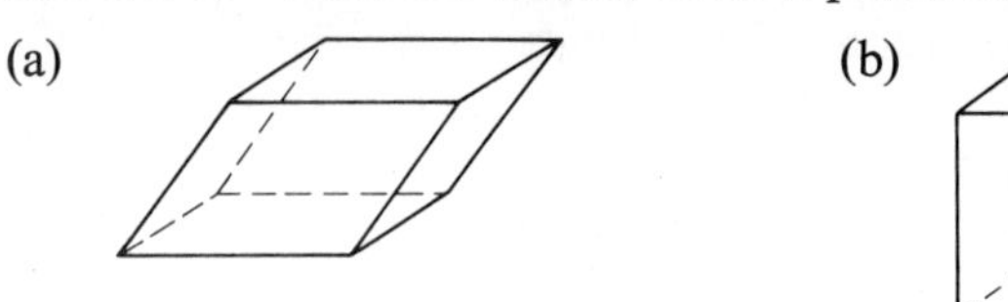

(c) (d)

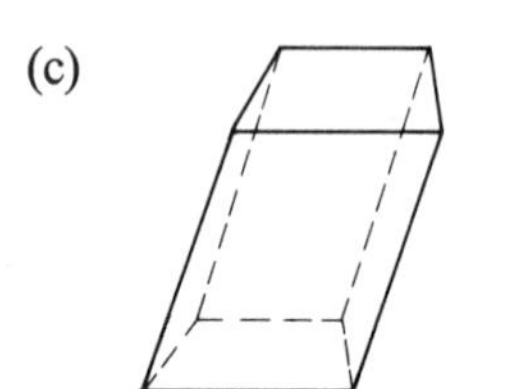

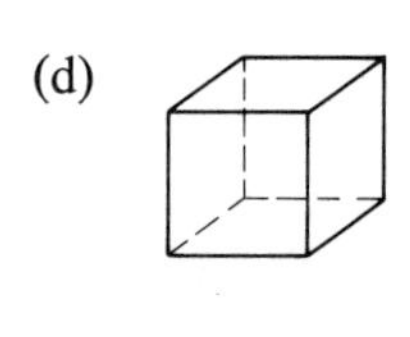

Objective 25: Volume of a prism

(a) The base is a rectangle with sides of 4″ and 9″. The altitude is 6″.

(b) The base is a triangle with base of 8 cm. and height of 5 cm. The altitude is 10 cm.

(c) The base is a trapezoid with bases of 9″ and 5″ and height of 7″. The altitude of the prism is 15″.

(d) The cube has edges of 12 mm.

See also Exercise 32.

(104)

Pyramid

Given a polygon and a point not in the plane of the polygon, line segments drawn from the given point to each of the vertices of the polygon, together with the polygon itself, form another polyhedron called a *pyramid.* Three pyramids are illustrated below. The base of each pyramid is the given polygonal region. Notice that, regardless of what type of polygon forms the base, the lateral faces of a pyramid are all triangles. The perpendicular distance from the given point to the plane of the polygon is the altitude of the pyramid. A *right pyramid* is one such that the angles formed by each face with the plane of the base each have the same degree measure. The volume of a pyramid is given by the formula $V = \frac{1}{3}Ba$, where B is the area of the base and a is the altitude.

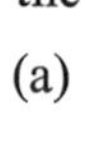

(a)

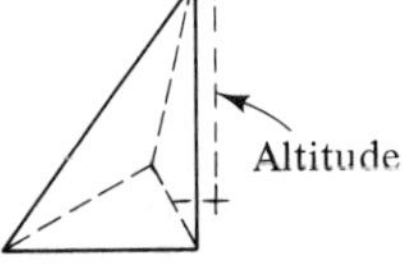

Pyramid

(b)

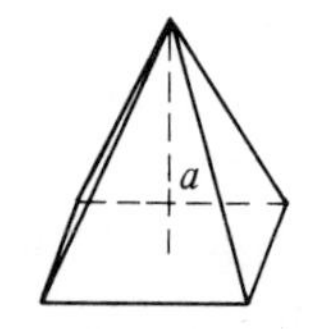

Right pyramid

(c)

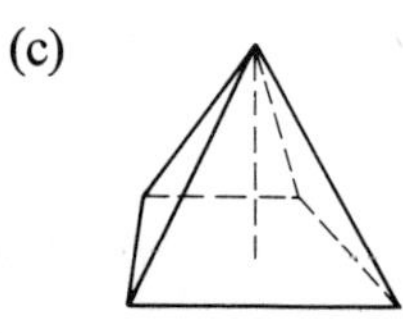

Right pyramid

Objective 25: Volume of a pyramid

◆ **Example 20** Find the volume of each pyramid above, given the following information:

(a) The base is a triangle with base of 9 cm. and height of 12 cm. The altitude is 20 cm.

(b) The base is a parallelogram with base of 18″ and height of 16″. The pyramid altitude is 9″.

(c) The base is a trapezoid with bases of 18 cm. and 10 cm. and height of 6 cm. The altitude is 15 cm.

(a) The area of the base is $\frac{1}{2}bh = \frac{1}{2}\cdot 9\cdot 12 = 54$ sq. cm. The volume is $\frac{1}{3}Ba = \frac{1}{3}\cdot 54\cdot 20 = 360$ cu. cm.

(b) The area of the base is $bh = 18\cdot 16 = 288$ sq. in. The volume is $\frac{1}{3}Ba = \frac{1}{3}\cdot 288\cdot 9 = 864$ cu. in.

(c) The area of the base is $\frac{1}{2}(b + b')h = \frac{1}{2}(18 + 10)6 = 84$ sq. cm. The volume is $\frac{1}{3}Ba = \frac{1}{3}\cdot 84\cdot 15 = 420$ cu. cm.

● **Practice 20** Find the volume of the following pyramids. **(105)**

(a)

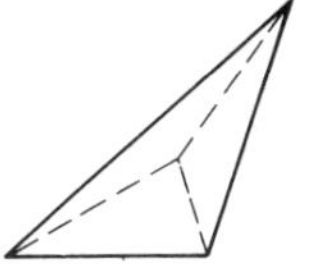

(b)

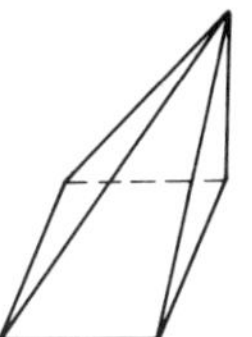

(c)

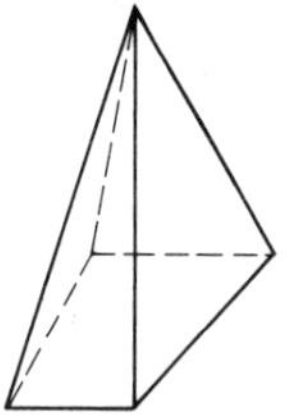

(a) The base is a triangle with height of 17 ft. and base of 12 ft. The altitude is 10 ft.

(b) The base is a rhombus with sides of 8″ and height of 6″. The altitude of the pyramid is 12″.

(c) The base of the pyramid is a trapezoid with bases of 6 m. and 5 m. and height 6 m. The altitude is 10 m.

See also Exercise 33.

(106)

Some three-dimensional figures which include circles are shown below.

(a) (b)

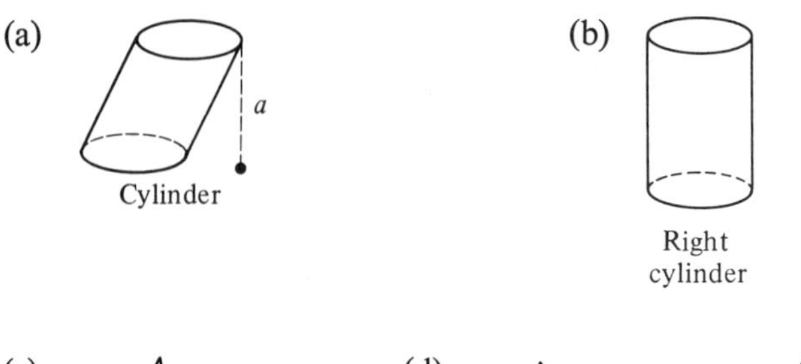

(c) (d) (e)

a

Cone

Right cone

Sphere

Cylinder

Given two equal circles (circles with equal radii) which lie in parallel planes, the line segment which joins the centers, P and P', is denoted $\overline{PP'}$. For any point A on one circle there is a point A' on the other circle such that $\overline{AA'}$ is parallel to $\overline{PP'}$. A *cylinder* is the union of the two circles which lie in parallel planes, the interiors of the circles, and all of the line segments $\overline{AA'}$ (which are parallel to $\overline{PP'}$). If $\overline{PP'}$ is perpendicular to the planes of the circles, the cylinder is a *right cylinder*.

The volume of a cylinder is the area of the base circle times the altitude of the cylinder (the perpendicular distance between the parallel planes). That is, $V = Ba$. But the area of a circle is πr^2, so thus the volume of a cylinder is $V = \pi r^2 a$.

(107)

Cone

Given a circle and a point not in the plane of the circle, line segments drawn from the given point to each point on the circle, together with the circle itself and its interior, form a *cone*. The *altitude* of a cone is the perpendicular distance from the given point to the plane of the circle. A *right cone* is one such that the line segment joining the given point with the center of the circle is also the altitude of the cone. The volume of a cone is $V = \frac{1}{3}Ba$ or $\frac{1}{3}\pi r^2 a$, where r is the radius of the circle and a is the altitude.

(108)

Sphere

Recall that a circle is the set of all points *in a plane* which are a given distance from a given point. A *sphere* is the set of all points in space which lie a given distance from a given point. The volume of a sphere is $V = \frac{4}{3}\pi r^3$.

Objectives 26 and 27:

◆ **Example 21** Find the volume of the figures above, given the following dimensions: (109)

(a) $r = 7''$, $a = 10''$ (b) $r = 5$ m., $a = 14$ m.
(c) $r = 5$ cm., $a = 12$ cm. (d) $r = 4$ ft, $a = 7$ ft.
(e) $r = 12$ cm.

(a) $V = \pi r^2 a = \frac{22}{7} \cdot 7^2 \cdot 10 = 1{,}540$ cu. in.

(b) $V = \pi r^2 a = \frac{22}{7} \cdot 5^2 \cdot 14 = 1{,}100$ cu. m.

(c) $V = \frac{1}{3}\pi r^2 a = \frac{1}{3} \cdot \frac{22}{7} \cdot 5^2 \cdot 12 = 314\frac{2}{7}$ cu. cm.

(d) $V = \frac{1}{3}\pi r^2 a = \frac{1}{3} \cdot \frac{22}{7} \cdot 4^2 \cdot 7 = 117\frac{1}{3}$ cu. ft.

(e) $V = \frac{4}{3}\pi r^3 = \frac{4}{3} \cdot \frac{22}{7} \cdot 12^3 = 7{,}241\frac{1}{7}$ cu. cm.

● **Practice 21** Find the volume of the same five figures above, if their dimensions are as follows: (110)

(a) $r = 8$ cm., $a = 7$ cm. (b) $r = 6''$, $a = 8''$
(c) $r = 14$ cm., $a = 10$ cm. (d) $r = 2$ ft., $a = 3$ ft.
(e) $r = 6''$

See also Exercise 34.

EXERCISES

Objective 1; Example 1

1. Use the six basic axioms of points, lines, and planes to reach the following conclusions:
   (a) How many lines are determined by five points, no three of which lie on the same line?
   (b) How many planes are determined by two intersecting lines?
2. Use axioms to show why there must be at least 6 lines.
3. Answer each of the following questions. Explain each "no" response:

Objectives 2, 3 Example 2

   (a) Can one ray be the subset of another ray?
   (b) Is the intersection of two rays always a point?
   (c) How many angles are formed by four rays with a common endpoint in a plane, no two of the rays being part of the same line? (Draw a sketch.)
   (d) Can a ray be a subset of a line segment?
   (e) Can a ray be a subset of an angle?

4. (a) Draw a sketch of two adjacent acute angles.

Objective 4
Example 3

(b)

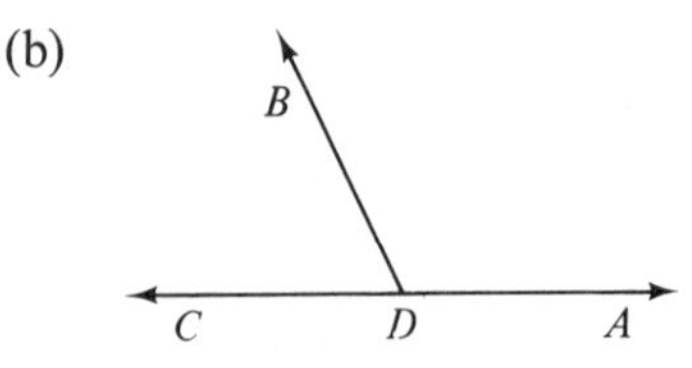

(i) Name the acute angle in the above figure.
(ii) Name the obtuse angle.

5. Answer the following questions. Explain any "no" responses:

Objective 5;
Example 4

(a) Are two complementary angles necessarily adjacent?
(b) Find the supplement of an angle whose measure is 80°.
(c) Find the complement of an angle whose measure is 64°.
(d) Does an obtuse angle have a complement?
(e) If two angles are supplementary, must one angle be an acute angle and the other an obtuse angle?

6. Suppose $\measuredangle A$ is supplementary to $\measuredangle B$, and $\measuredangle A = \measuredangle C$. What can you conclude about $\measuredangle C$ and $\measuredangle B$? (Draw a sketch if necessary.)

7. Use the sketch below to answer the following questions:

Objectives 6, 7;
Example 5

(a) Name the perpendicular lines.
(b) Name each pair of vertical angles.
(c) Name a pair of complementary angles which are adjacent.
(d) Name a pair of angles which are complementary but not adjacent.
(e) Name a pair of supplementary angles.

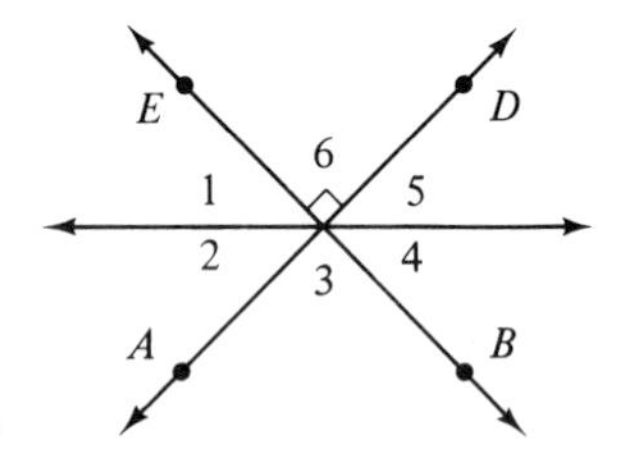

Objective 8:
Items (36) and (37)

8. State the difference between parallel lines and skew lines.

Objective 9;
Item (40)

9. Explain how to draw a line parallel to a given line.

10. Given $\overleftrightarrow{l} \parallel \overleftrightarrow{m}$.

Objective 10;
Example 6

(a) Name all the angles equal to $\measuredangle 5$.
(b) Name all the angles supplementary to $\measuredangle 4$.

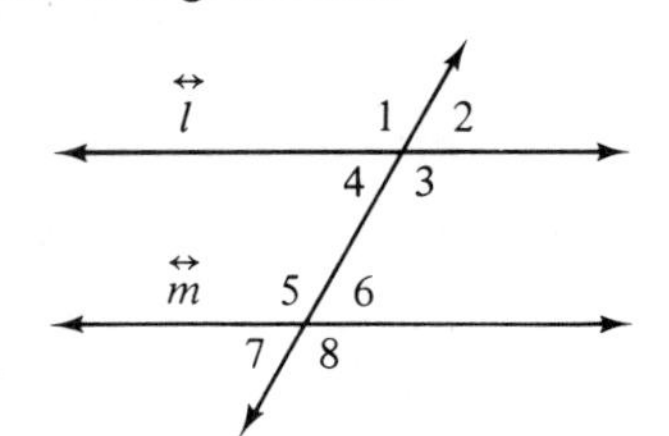

11. Supply the reasons in the proof.

Objective 11;
Example 7

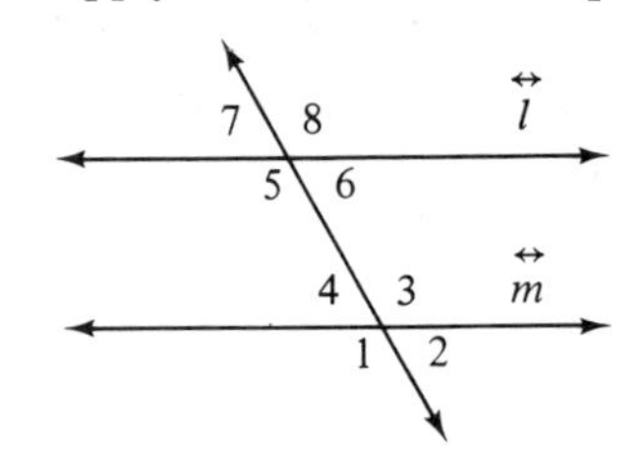

Given: $\measuredangle 1 = \measuredangle 8$
Prove: $\overleftrightarrow{l} \parallel \overleftrightarrow{m}$

Statements	*Reasons*
1. $\measuredangle 1 = \measuredangle 8$	1. ________
2. $\measuredangle 1 = \measuredangle 3$	2. ________
3. $\measuredangle 3 = \measuredangle 8$	3. ________
4. $\overleftrightarrow{l} \parallel \overleftrightarrow{m}$	4. ________

Objective 12; Example 8

12. (a) For each polygon, classify the polygon by the number of sides and tell whether it is concave or convex.

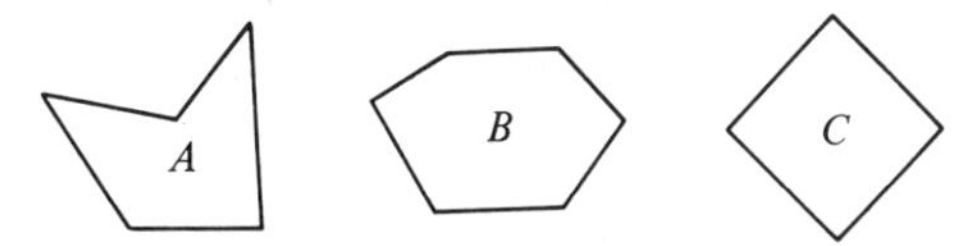

(b) Draw a sketch of a convex pentagon.

Objective 13; Example 9

13. In the figure, name each of the following:
(a) right triangle,
(b) acute triangle,
(c) obtuse triangle,
(d) equilateral triangle,
(e) isosceles triangle.

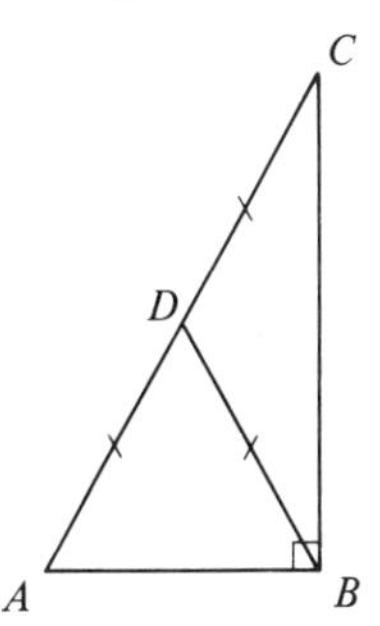

Objective 14; Example 10

14. Sketch the altitude, median and angle bisector from A to $\overline{BC}$.

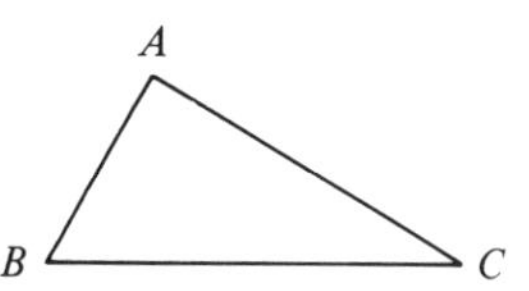

Objective 14; Example 10

15. In what kind of triangle will an altitude also be a side of the triangle?

Objectives 15, 16; Example 11

16. Find the measure of the third angle of a triangle if two of the angles measure 70° and 32°.

17. Find the two acute angles of a right isosceles triangle.

18. In the given figure, $AB \parallel DE$. Find angles A, B, and D.

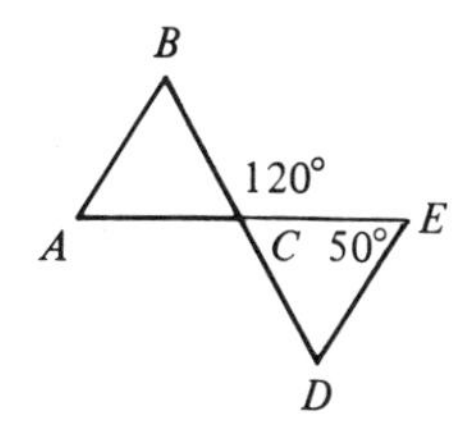

Objective 17; Example 12

19. (a) The two legs of a right triangle are 5 and 12. Find the hypotenuse.
(b) A leg of a right triangle is 12; the hypotenuse is 20. Find the other leg of the triangle.

Objective 18; Example 13

20. (a) What is another name for a regular (equiangular and equilateral) parallelogram?
(b) What is another name for an equiangular parallelogram?
(c) What is another name for an equilateral parallelogram?

Objective 19; Example 14

21. Write a perimeter equation and find the perimeter of the following polygons:
(a) a regular octagon, if one side is 12 inches;
(b) a regular nonagon if one side is 8 cm.

Objective 19; Example 14

22. Write an equation and find the following lengths:
(a) a fence which encloses a field 60 meters by 80 meters;
(b) a rope which encloses a square boxing/wrestling "ring" with a side of 15 feet.

Objective 20 Example 15

23. Find the area of a triangle with base 10 cm. and height 6 cm.

24. Find the area of a square if one side is 11 in.

25. Find the area of the rhombus with side 8 cm. and height 3 cm.

26. Find the area of each figure.

(a)

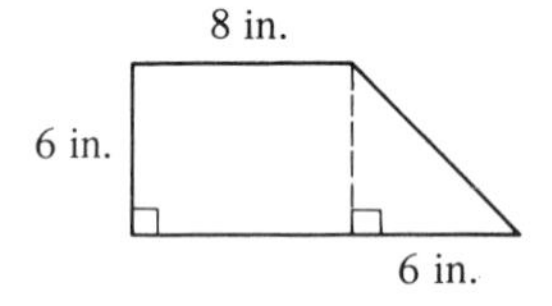

(b)

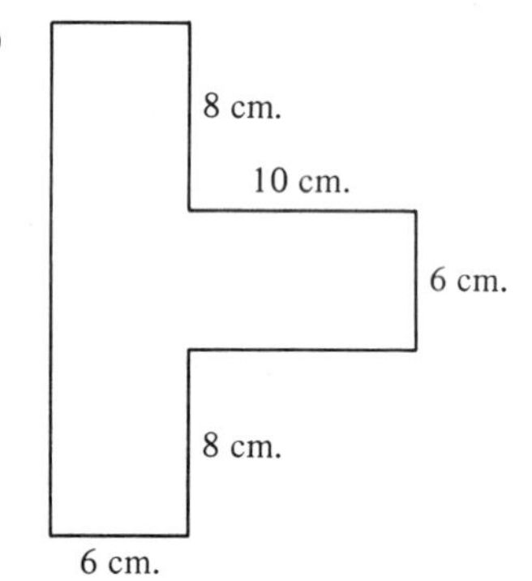

Objective 21; Example 16

27. Given the adjoining figure, name one example of each of the following terms:

(a) chord (b) central angle
(c) inscribed angle (d) secant
(e) minor arc (f) major arc

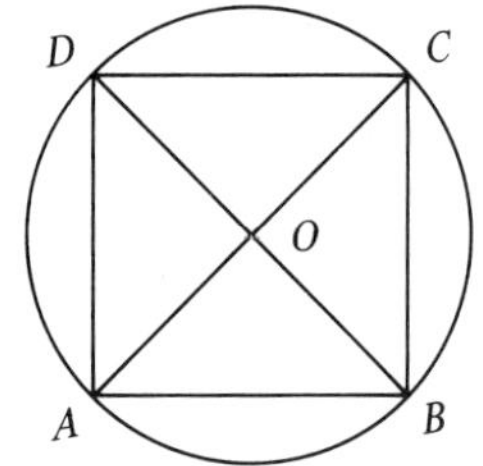

Objective 22; Example 17

28. Assume in the figure in Exercise 27 that $\measuredangle AOB$ is 90°. What are the measures of $\widehat{AB}$, $\widehat{ADB}$, and $\measuredangle ACB$?

29. An equilateral triangle is inscribed in a circle (that is, its vertices lie on the circle). What is the measure of each minor arc associated with each side?

Example 23; Example 18

30. Find the circumference and area of a circle with
(a) a radius of 12 cm. (b) a diameter of 8 inches

Objective 24; Example 19

31. Find the area of the figure below.

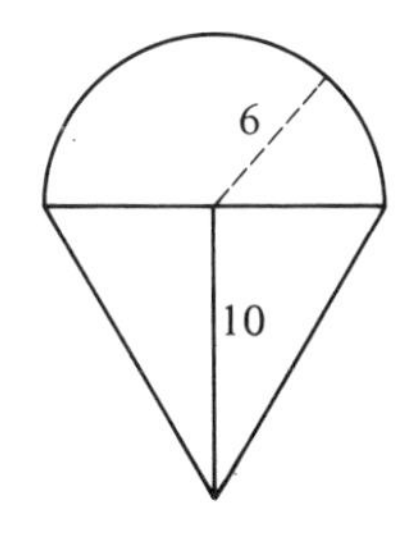

Objective 25; Example 19

32. Find the volume of each prism, given the following dimensions:
(a) A rectangular parallelopiped with base sides of 6″ and 4″ and altitude of 5″.
(b) A prism with a triangular base which has height of 4 cm. and base of 3 cm. and with prism altitude of 6 cm.
(c) A prism with a trapezoidal base which has height of 12 m. and bases of 7 m. and 5 m., and with prism altitude of 6 m.
(d) A cube with cdgcs of 9 ft.

Objective 25; Example 20

33. Find the volume of the following pyramids:
(a) A pyramid with triangular base where the base is 10 cm., height is 6 cm., and pyramid altitude is 4 cm.
(b) A pyramid has a square base with sides of 3″, and the pyramid altitude is 7″.
(c) A pyramid has a trapezoidal base, with bases of 3 m. and 7 m., and height of 4 m. The altitude of the pyramid is 6 m.

Objectives 26, 27; Example 21

34. Find the volume of the following figures:
(a) A cylinder with base radius of 3″ and altitude of 14″.
(b) A right cylinder with base radius of 10 m. and altitude of 8 m.
(c) A cone with base radius of 3 cm. and altitude of 14 cm.
(d) A right cone with base radius of 7″ and altitude of 6″.
(e) A sphere with a radius of 21 ft.

OBJECTIVES

After completing this chapter, you should be able to

A. Points, lines, and planes

1 Apply the six basic axioms of points, lines, and planes (given a list of the axioms). **(2–11)**

B. Rays, line segments, and angles

2 Name an angle using its vertex and two points on each side. **(12–16)**

3 Given a figure with intersecting lines, identify adjacent angles, straight angles, and interior of an angle. **(17–21)**

C. Measurement of angles

4 Given a sketch of an angle, tell whether it is acute, right, or obtuse. **(22–25)**

5 Given the measure of an angle, find the measure of its complement or supplement. **(26–29)**

D. Vertical angles and perpendicular lines

6 Use the fact that vertical angles are equal to find unknown angles in a figure. **(30–32)**

7 Determine properties of perpendicular lines using the definition of perpendicular lines. **(33–35)**

E. Parallel lines

8 State the difference between parallel lines and skew lines. **(36–37)**

9 State the procedure for sketching a line parallel to a given line, and draw the sketch. **(38–40)**

10 Given two parallel lines cut by a transversal, find the angles which are equal to a given angle or which are supplementary to a given angle and give the reason for the answer. **(41–47)**

11 Supply the reasons in a geometric proof involving parallel lines. **(48–50)**

F. Polygons

12 Given a polygon, classify the polygon according to its number of sides, and state whether the polygon is concave or convex. **(51–55)**

G. Triangles

13 Classify a given triangle as acute, obtuse, right, isosceles, or equilateral. **(56–59)**

14 Given a triangle, sketch the median, altitude, and angle bisector from an indicated vertex. **(60–62)**

15 Supply the reasons in a geometric proof involving the sum of the angles of a triangle. **(63–66)**

16 Given the measure of two angles of a triangle, find the measure of the third angle. **(63–66)**

17 Given two sides of a right triangle, use the Pythagorean theorem to find the third side. **(67–69)**

H. Quadrilaterals

18 Use the definitions of trapezoid, parallelogram, rectangle, rhombus, and square to show relationships among quadrilaterals. **(70–73)**

I. Perimeter and area of polygons

19 Given a polygon and information about the lengths of its sides, write an equation to represent the perimeter of the polygon, and compute the perimeter. **(74–76)**

20 Given base and height, use the proper equation to find the area of a triangle $(A = \frac{1}{2}bh)$, a rectangle $(A = bh)$, a square $(A = s^2)$, a parallelogram $(A = bh)$, or a trapezoid $(A = \frac{1}{2}(b + b')h)$. **(77–82)**

J. Circles

21 Given a sketch of a circle and lines intersecting the circle, identify each of the following terms: chord, minor arc, major arc, radius, diameter, central angle, inscribed angle, tangent, and secant. **(83–86)**

22 Given the measure of an arc, find the measure of its central angle or inscribed angle. Conversely, given the measure of a central angle or inscribed angle, find the measure of its intercepted arc. **(87–93)**

23 Given the radius of a circle, find its area or circumference using the equations $A = \pi r^2$ and $C = 2\pi r$. **(93–97)**

24 Find the area of given geometric figures composed of polygons and circles. **(97)**

K. Volume

25 Find the volume of a prism or pyramid, given the dimensions of its base and its altitude. **(98–105)**

26 Find the volume of a cylinder or cone, given the radius of its base and its altitude. **(106–110)**

27 Find the volume of a sphere, given its radius. **(108)**

Appendix: Percent

The fundamental concept of percent is one with which everyone is familiar: that is, numbers which may be quite different in size are compared by using a common denominator of 100 and comparing the number of parts out of 100 that each is equivalent to. (1)

Percent is one of the most frequently used mathematical concepts in almost every career field and is therefore a topic that everyone needs a certain facility with. Since it is an elusive topic that many students have never mastered adequately, a brief review may be in order. (2)

A. CONVERSIONS: PERCENTS, DECIMALS, AND FRACTIONS

Percent means hundredths

The key aspect to keep in mind when working with percents is that *percent* means *hundredths*. That is, the expression "36%" may just as correctly be stated "36 hundredths." The term "hundredths" denotes either a common fraction with 100 as the denominator or else a decimal fraction of two decimal places. Thus, (3)

$$36\% \quad = \quad 36 \text{ hundredths} \quad = \quad \frac{36}{100} \quad \text{or} \quad 0.36$$

Objective 1: Converting percent to a decimal

When the subject of percent is mentioned, most people visualize the most typical percents—those between 1% and 99%. These are the percents that occupy the first two places in the decimal representation of a percent. (4)

For example, 99% = 99 hundredths = 0.99, and 1% = 1 hundredth = 0.01. When converting a percent to a decimal, be sure the whole percents between 1% and 99% are written in the first two decimal places; then, the remaining digits will naturally be aligned in their correct places. Keep in mind that "percent" means "hundredths".

◆ **Example 1** Express each given percent as a decimal: (5)

(a) 8% = 8 hundredths = 0.08

(b) 71% = 71 hundredths = 0.71

(c) 267.25% = 2.6725 (since 67% = 0.67)

(d) 5.09% = 0.0509 (since 5% = 0.05)

(e) 0.6% = 0.006 (since 0% = 0.00)

Fractional percents to decimals

Some percents are written using common fractions (which are rational numbers.) Recall that one definition of a rational number is "any number that can be expressed as the quotient of two integers." This definition also describes the process for converting a *fractional percent* to a decimal: divide the numerator by the denominator to obtain a decimal percent; and then convert the decimal percent to an ordinary percent. Recall also that, for example, $\frac{1}{8}\%$ means "$\frac{1}{8}$ of 1%" (which is less than 1%). (6)

(f) $\frac{1}{8}\%$ Now, $\frac{1}{8}$ means $8\overline{)1.000}^{\,.125} = 0.125$ (7)

And since $\frac{1}{8} = 0.125$

then $\frac{1}{8}\% = 0.125\% = 0.00125$

(g) $\frac{4}{7}\%$ First, $\frac{4}{7}$ means $7\overline{)4.00}^{\,.57\frac{1}{7}} = 0.57\frac{1}{7}$

Then $\frac{4}{7} = 0.57\frac{1}{7}$

implies $\frac{4}{7}\% = 0.57\frac{1}{7}\% = 0.0057\frac{1}{7}$

Parts (f) and (g) illustrate the customary procedure used for expressing the decimal equivalent of a fraction: (1) continue the division process until it terminates provided it will terminate in three (or four) places; or (2) use two decimal places and a fractional remainder otherwise.

If a given fraction is one for which the decimal equivalent is already known, the answer can be written immediately, remembering that a percent less than 1% is indicated by zeros in the first two decimal places: (8)

(h) $\frac{1}{4}\% \quad = \quad \frac{1}{4}$ of $1\% \quad = \quad 0.0025$

● **Practice 1** Convert each percent to its decimal equivalent: (9)

(a) 42%	(b) 3%	(c) 15.2%	(d) 287%
(e) 109.75%	(f) 6.5%	(g) 0.7%	(h) 0.02%
(i) $\frac{3}{8}\%$	(j) $\frac{2}{7}\%$	(k) $\frac{3}{4}\%$	(l) $\frac{2}{3}\%$

See also Exercise 4.

Ordinary percents can be converted to a fraction by again applying the fact that "percent" means "hundredths," as follows: (10)

Objective 2: Converting percent to a fraction

◆ **Example 2** Convert each percent to its fractional equivalent:

(a) $48\% \quad = \quad 48 \text{ hundredths} \quad = \quad \frac{48}{100} \quad = \quad \frac{12}{25}$

(b) $125\% \quad = \quad 125 \text{ hundredths} \quad = \quad \frac{125}{100} \quad = \quad \frac{5}{4}$

Notice that fractions are reduced to lowest terms. (An improper* fraction is considered to be in lowest terms provided it cannot be further reduced.)

Percents containing fractions

Recall that $\frac{x}{100}$ is the same as $x \cdot \frac{1}{100}$. Thus, the fact that "percent" means "hundredths" may be shown as (11)

$$\% \quad = \quad \text{Hundredth} \quad = \quad \frac{1}{100}$$

This fact is applied to convert *percents containing fractions* to their ordinary common fraction equivalents:

(c) $44\frac{4}{9}\% \quad = \quad 44\frac{4}{9} \times \frac{1}{100} \quad = \quad \frac{400}{9} \times \frac{1}{100} \quad = \quad \frac{400}{900} \quad = \quad \frac{4}{9}$

(d) $\frac{1}{3}\% \quad = \quad \frac{1}{3} \times \frac{1}{100} \quad = \quad \frac{1}{300}$

Percents containing decimals

Decimal percents may most easily be converted to fractions by applying the following procedure: (12)

*Recall an "improper" fraction, $\frac{a}{b}$, is a fraction where $a > b$.

1. Change the percent to its decimal equivalent.
2. Pronounce the value of the decimal; then write this number as a fraction and reduce it.

(e) 17.5%
  1. As a decimal, 17.5% = 0.175
  2. Then, 0.175 = 175 thousandths = $\frac{175}{1000}$ = $\frac{7}{40}$

● **Practice 2** Convert the following percents to their fractional equivalents: (13)

(a) 8% (b) 55% (c) 240% (d) 172%

(e) $37\frac{1}{2}\%$ (f) $83\frac{1}{3}\%$ (g) $14\frac{2}{7}\%$ (h) $\frac{3}{4}\%$

(i) 24.8% (j) 37.5% (k) 1.25% (l) 0.5%

See also Exercise 4.

Objective 3: Converting a decimal to percent

Just as "percent" means "hundredths," so does the reverse also hold: *hundredths = percent.* When a decimal value is to be expressed as a percent, first isolate the whole percents between 1% and 99%, and the remaining digits can easily be aligned correctly. Consider the following example, remembering that "hundredths" means "percent." (14)

◆ **Example 3** Express each decimal value as its equivalent percent: (15)

(a) 0.07 = 7 hundredths = 7%
(b) $0.45\frac{1}{3}$ = $45\frac{1}{3}$ hundredths = $45\frac{1}{3}\%$

It may be helpful to circle the first two decimal places, as an aid in identifying the whole percents between 1% and 99%:

(c) 1.523 Given 1.(52)3, the .52 represents 52%.
  Thus 1.523 = 152.3%

(d) 0.4 Given 0.(4), the .4 or .40 means 40%.
  Hence, 0.4 = 40%

(e) 0.0017 Given 0.(00)17, the 0.00 denotes 0%.
  Therefore, 0.0017 = 0.17%

● **Practice 3** Convert each of the following to a percent: (16)

(a) 0.69 (b) 0.08 (c) $0.13\frac{1}{2}$ (d) $0.82\frac{2}{5}$

(e) 0.231 (f) 0.7604 (g) 1.919 (h) 2.057

(i) 0.5 (j) 0.0034 (k) 0.008 (l) 1.002

See also Exercise 5.

Objective 4: Converting a fraction to percent

To express a fraction as its equivalent percent, apply the following procedure: (17)

1. First convert the fraction to its decimal equivalent (by dividing the numerator by the denominator.)
2. Then, convert this decimal to a percent, as illustrated in the preceding example.

◆ **Example 4** Convert each fraction to its equivalent percent: (18)

(a) $\frac{5}{11} = 11\overline{)5.00}^{\;.45\frac{5}{11}} = 0.45\frac{5}{11} = 45\frac{5}{11}\%$

(b) $2\frac{1}{8}$ This may be done in either of two ways:

Since $\frac{1}{8} = 8\overline{)1.000}^{\;.125} = 0.125$, then $2\frac{1}{8} = 2.125$ $= 212.5\%$

Or, $2\frac{1}{8} = \frac{17}{8} = 8\overline{)17.000}^{\;2.125} = 2.125 = 212.5\%$

● **Practice 4** Express each of the following as a percent: (19)

(a) $\frac{5}{8}$ (b) $\frac{1}{6}$ (c) $\frac{9}{13}$

(d) $\frac{7}{4}$ (e) $\frac{15}{8}$ (f) $\frac{17}{11}$

(g) $1\frac{2}{3}$ (h) $1\frac{5}{9}$ (i) $2\frac{3}{7}$

See also Exercise 5.

B. PERCENT EQUATION FORMS

All problems that involve the use of percent are some variation of the basic percent form: "Some percent of one number equals another number." Observe that this basic percent form, which can be expressed in equation form as $x\%$ of $y = z$, involves exactly three elements. (20)

Objective 5: Basic percent form

The easiest and most successful method for solving percentage problems is to consider them as first-degree equations in one variable, such as those introduced in the algebra unit. Since there can be only one (21) unknown in these equations, there are only three possible variations of the basic equation form: the unknown can be either (1) the percent (or *rate*), (2) the first number (or *base*), or (3) the second number (often called the *percentage*).

Recall that before a percent can be used in computation, it must first be converted to either a fractional or decimal equivalent. Conversely, if the variable in the equation represents a percent, then the solution (decimal or fraction) must be converted to its equivalent percent. (22)

◆ **Example 5** Find each of the following: (23)

(a) 70% of 36 is what number?

$$0.7 \times 36 = n$$
$$25.2 = n$$

(b) What percent of 16 is 6?

$$r \times 16 = 6$$

or

$$16r = 6$$
$$\frac{16r}{16} = \frac{6}{16}$$
$$r = \frac{3}{8}$$
$$r = 37\frac{1}{2}\%$$

(c) 25% of what number is 7?

$$\frac{1}{4} \times n = 7$$

or

$$\frac{n}{4} = 7$$
$$4 \times \frac{n}{4} = 7 \times 4$$
$$n = 28$$

● **Practice 5** Determine the unknown amount in each of the following: (24)

(a) 9% of 120 is what number?

(b) $4\frac{1}{2}\%$ of $80 is what amount?

(c) How much is $\frac{1}{4}\%$ of 360?

(d) 0.35% of 600 is what number?
(e) What percent of 45 is 36?
(f) What percent of 36 is 45?
(g) 12 is what percent of 42?
(h) 6.2 is what percent of 155?
(i) 45% of what number is 108?
(j) 3% of what number is 4.2?
(k) $66\frac{2}{3}\%$ of what number is 54?
(l) 4.5 is 0.5% of what amount?

See also Exercise 6.

Objective 6: Percent of change

(25) Frequent references are made to the percentage increase in the cost of living, increase in unemployment, decrease in rural population, decrease in school-age children, etc. Such "percents of change" (that is, percents of increase or decrease) are found using a variation of the basic percent equation. This word formula, "What percent of the original number is the change?" may be expressed

What % of Original = Change?

(26) ◆ **Example 6** Determine each percent of increase or decrease:

(a) What percent more than 30 is 42?

Original number To have "more than 30" implies that we originally had 30; thus, 30 is the "original number." (In general, the original number will be the number which follows the words "more than" or "less than.")

Change The change is the numerical difference between the given amounts. (Whether this change is an increase or decrease—positive or negative—is not significant to the formula.) Thus, the change in this example is $42 - 30 = 12$.

What % of Original = Change

$$?\% \text{ of } 30 = 12$$

$$30r = 12$$

$$\frac{30r}{30} = \frac{12}{30}$$

$$r = \frac{2}{5}$$

$$r = 40\%$$

Thus, 42 is a 40% increase over 30.

(b) 36 is what percent less than 48? (27)

Original = 48

Change = 48 − 36
= 12

What % of Original = Change

$$?\% \text{ of } 48 = 12$$
$$48r = 12$$
$$\frac{48r}{48} = \frac{12}{48}$$
$$r = \frac{1}{4}$$
$$r = 25\%$$

Hence, 48 decreased by 25% leaves 36.

● **Practice 6** Find each percent of increase or decrease: (28)

(a) What percent more than 25 is 35?
(b) What percent less than 35 is 25?
(c) 42 is what percent less than 60?
(d) 88 is what percent more than 64?
(e) 80 increased by what percent gives 92?
(f) $390 is what percent decrease from $400?

See also Exercise 7.

C. ENGLISH SENTENCES TO MATH SENTENCES

Objective 7: Word equations with %'s

The algebra unit included a section of problems for which there are no prescribed formulas but which are best solved by restating the English sentences as equations. These problems, which are typical of situations in many different professions, frequently include some reference to percentages. The following example provides practice with such problems. (29)

◆ **Example 7** Express a word equation for each of the following, and use it to write a math equation which solves the problem. (30)

(a) A food company distributed 1,200 questionnaires to sample consumer satisfaction with a new product; 720 of the questionnaires were returned. What percent of the questionnaires were returned?

What percent ↓ of ↓ questionnaires ↓ were ↓ returns? ↓

$$?\% \times 1{,}200 = 720$$
$$1{,}200r = 720$$
$$\frac{1{,}200r}{1{,}200} = \frac{720}{1{,}200}$$
$$r = \frac{3}{5}$$
$$r = 60\%$$

(b) Thirty-five percent of the cost of a mental health project was salaries. If $1,400 was spent for salaries, what was the total amount of the project? **(31)**

$$\underline{35\%}\ \underline{\text{of}}\ \underline{\text{the cost}}\ \underline{\text{was}}\ \underline{\text{salaries.}}$$
$$\downarrow \quad \downarrow \quad \downarrow \quad \downarrow \quad \downarrow$$
$$35\% \times c = \$1{,}400$$
$$0.35c = 1{,}400$$
$$\frac{0.35c}{0.35} = \frac{1{,}400}{0.35}$$
$$c = \$4{,}000$$

(c) Last year the historical society enrolled 2,500 members; this year's membership is 2,650. What was the percent of increase in membership? **(32)**

$$\text{Original} = 2{,}500$$
$$\text{Change} = 2{,}650 - 2{,}500$$
$$= 150$$

$$?\% \text{ of Original} = \text{Change}$$
$$?\% \text{ of } 2{,}500 = 150$$
$$2{,}500r = 150$$
$$\frac{2{,}500r}{2{,}500} = \frac{150}{2{,}500}$$
$$r = 0.06$$
$$r = 6\%$$

(d) Home construction has decreased by 15%, with only 68 new homes started this year. How many homes were constructed last year? **(33)**

$$\underline{\text{Past construction}}\ \underline{\text{decreased by}}\ \underline{15\%\text{ (of itself)}}\ \underline{\text{gives}}\ \underline{\text{present construction.}}$$
$$\downarrow \quad \downarrow \quad \downarrow \quad \downarrow \quad \downarrow$$
$$c - (0.15 \times c) = 68$$
$$c^* - 0.15c = 68$$
$$0.85c = 68$$
$$\frac{0.85c}{0.85} = \frac{68}{0.85}$$
$$c = 80$$

"% of itself"

Note: Percents are never complete in themselves. In an equation, any percent must be a percent *of* ("times") some other number or variable. In the example above, the statement "decreased by 15%" implied "15% of itself." As illustrated above, however, in math sentences the variable must be *stated*, rather than just implied.

*Recall that $c = 1c$. Thus, $c - 0.15c = 1c - 0.15c$
$= 1.00c - 0.15c$
$= 0.85c$

● **Practice 7** State each of the following in a concise English equation. Then convert this to a math equation and solve: (34)

(a) Some $33\frac{1}{3}\%$ of those surveyed indicated they normally have only coffee for breakfast. How many persons were surveyed, if 280 reported they have only coffee for breakfast?

(b) A total of $1,200 was spent to stage a concert by a community orchestra. Of this amount, the sheet music itself totaled $180. What percent of the cost was the sheet music?

(c) Fifty-five percent of families with young children indicated a preference for station wagons. On this basis, if 132 station wagons are sold to young families, how many such families have purchased new cars?

(d) A homeowner has paid a total of $2,000 in mortgage payments. A statement from the savings and loan association informs him that this amount included $1,250 in interest. What percent of his payments was interest?

(e) The homeowner above purchased his home for $32,000. When he was transferred to a new city four years later, he resold the home for $36,000. What was the percent increase in selling price?

(f) By utilizing several energy-saving measures, a family reduced its electrical consumption from 1,200 kilowatt-hours per month to 900 kilowatt-hours. What percent decrease in consumption does this represent?

(g) What number decreased by 18% of itself yields 123?

(h) What number increased by 24% gives 62?

(i) The cost of public utilities increased by 20% during one year. If the current cost is $96, how much would the same consumption have cost previously?

(j) In order to conserve gasoline, a driver reduced her mileage by 30%. If she now averages 420 miles per month, what would her mileage have been during a normal month earlier?

See also Exercise 8.

D. SUMMARY OF CONVERSIONS

(35)

Given:	To convert to:	
1. *Percents*	*Decimal*	*Fraction*
(a) Basic % (1% to 99%)	Basic percents become the first two (hundredths) decimal places. (Other digits align accordingly.) Example: 76% = 76 hundredths = 0.76.	Percent sign is dropped and given number is placed over a denominator of 100; reduce. Example: 76% = 76 hundredths = $\frac{76}{100} = \frac{19}{25}$.
(b) Decimal % (125.73%; 0.45%)	Digits representing percents greater than 99% or less than 1% are aligned around the first two (hundredths) decimal places (see also "basic percents"). Example: 0.45% = 0.0045.	Convert the percent to a decimal (as indicated at left). Pronounce the decimal value; then write this as a fraction and reduce. Example: 0.45% = 0.0045 = 45 ten-thousandths = $\frac{45}{10{,}000} = \frac{9}{2{,}000}$.
(c) Fractional % $\left(\frac{1}{4}\%;\ \frac{7}{5}\%;\ 1\frac{2}{3}\%\right)$	Convert the fractional percent to a decimal percent (by dividing the numerator by the denominator). Then convert this decimal percent to an ordinary decimal (as shown above). Example: $\frac{7}{5}\% = 1.4\% = 0.014$.	Percent sign is replaced by $\frac{1}{100}$; multiply and reduce. Example: $\frac{7}{5}\% = \frac{7}{5} \times \frac{1}{100} = \frac{7}{500}$.
2. *Decimals* (0.473; 1.2; 0.0015)	*Percent* Isolate the first two (hundredths) decimal places to identify the basic percents (1% to 99%). Other percents will then be aligned correctly. Example: 0.473 = 0.(47)3 = 47.3%.	
3. *Fractions* $\left(\frac{2}{9};\ \frac{7}{3};\ 1\frac{1}{2}\right)$	*Percent* Convert the fraction to a decimal (by dividing the numerator by the denominator). Then follow the above procedure for converting a "decimal to percent." Example: $\frac{2}{9} = 0.22\frac{2}{9} = 0.(22)\frac{2}{9} = 22\frac{2}{9}\%$.	

EXERCISES

Percents to decimals and fractions

Objective 8a; Items 3–13

1. (a) "Percent" means ____.
   (b) Given a percent such as $ab\%$, "hundredths" may be substituted for "%" by using either a decimal with ____ decimal places or a fraction with ____.
   (c) The first two decimal places are used to represent the value of the common percents between ____% and ____%.
   (d) Given a decimal percent (such as $0.ab\%$), the first two decimal places of its decimal equivalent would be ____ because the percent is less than ____%.
   (e) Express the decimal equivalent of $0.ab\%$.
   (f) A decimal percent (such as $ab.c\%$ or $0.ab\%$) is converted to an equivalent fraction by ____.
   (g) Express the fractional equivalent of $ab.c\%$; of $0.ab\%$.
   (h) Express 1.25% as a decimal; as a fraction.
   (i) Given a fractional percent $\left(\text{such as } \frac{1}{a}\%\right)$, this means "$\frac{1}{a}$ of ____%."
   (j) When a given percent contains a fraction $\left(\text{such as } \frac{a}{b}\%\right)$, then the decimal equivalent is found by ____.
   (k) When given a percent as described in (j), then the equivalent fraction is found by replacing "%" with ____.
   (l) If $\frac{a}{b} = 0.xyz$, then express the decimal value of $\frac{a}{b}\%$.
   (m) If $\frac{a}{b} = 0.xyz$, then express $\frac{a}{b}\%$ as its equivalent fraction.
   (n) Express $\frac{3}{4}\%$ as a decimal; as a fraction.

Decimals and fractions to percents

Objective 8b; Items 14–19

2. (a) Given a decimal number such as $0.abc$, the first two decimal places represent ____.
   (b) The number $0.0a$ represents ____%.
   (c) The number $0.ab$ represents ____%.
   (d) The number 1 represents ____%.
   (e) Given $0.00a$, we know this is less than ____%.
   (f) Given $a.bc$, we know this is more than ____%.
   (g) Given any fraction $\frac{a}{b}$, this can be converted to a percent by ____.
   (h) Given $\frac{a}{b}$ where $a < b$ (and $b \neq 0$), then the equivalent percent is less than ____%.

(i) Given $\frac{a}{b}$ where $a > b$ (and $b \neq 0$), then the equivalent percent is greater than ____%.

(j) If $\frac{a}{b} = 0.xyz$, express $\frac{a}{b}$ as a percent.

(k) If $\frac{a}{b} = x.yz$, express the equivalent percent for $\frac{a}{b}$.

(l) If $a = b$ (and $\neq 0$), express $\frac{a}{b}$ as a percent.

(m) Express $\frac{7}{8}$ as a percent.

(n) Express $\frac{8}{7}$ as a percent.

Calculations with percents

Objective 8c; Items 20–34

3. (a) Before a percent can be used in a calculation, it first must be ____.

(b) All calculations involving percent are some variation of the basic percent equation ____.

(c) When the percent equation is used to find an unknown percent, the initial solution obtained represents ____.

(d) A variation of the basic percent equation that is used to find percent of decrease is the formula ____.

(e) The "original number" of an equation is the smaller number when the "change" has been a(n) ____.

(f) A student begins an equation as follows:

$$x - 28\% = 36$$
$$x - 0.28 = 36$$

Explain what is wrong with this equation.

(g) What % of 60 is 45?

(h) What % less than 60 is 45?

(i) Suppose that a is 80% of b. Then, b decreased by what percent (of itself) would give a? (Or, what percent less than b gives a?)

(j) What % of 45 is 60?

(k) What % increase over 45 gives 60?

(l) Suppose that 125% of a would give b. Then, b is what percent more than a?

(m) Suppose that a is 80% of b. Then, b is what percent of a? (Hint: express the given information in an equation, substituting a fraction for the percent. Then solve for b.)

(n) Suppose that a is $66\frac{2}{3}\%$ of b. Then b is what percent of a?

(o) Substitute your solution to (g) in the equation "a is ?% of b," and solve for b. Does your (j) solution verify this result?

(p) When you know "$a = x\%$ of b" and you find "$b = y\%$ of a," how are the fractional equivalents of $x\%$ and $y\%$ related to each other? Thus, if "a is $62\frac{1}{2}\%$ of b," then find "b is what percent of a?" without solving an equation.

Objectives 1 and 2; Examples 1 and 2

4. Express each percent as (1) a decimal and (2) a fraction in its lowest terms:

(a) 24% (b) 59% (c) 86% (d) 5%
(e) 6% (f) 145% (g) 14.5% (h) 1.45%
(i) 37.2% (j) 7.08% (k) 6.4% (l) 41.75%
(m) 0.4% (n) 0.28% (o) 162.5% (p) $12\frac{1}{2}\%$
(q) $42\frac{6}{7}\%$ (r) $16\frac{2}{3}\%$ (s) $77\frac{7}{9}\%$ (t) $8\frac{1}{3}\%$
(u) $7\frac{1}{2}\%$ (v) $\frac{1}{5}\%$ (w) $\frac{5}{6}\%$ (x) $\frac{7}{4}\%$
(y) $\frac{8}{3}\%$

Objectives 3 and 4; Examples 3 and 4

5. Express each of the following as its equivalent percent:

(a) 0.65 (b) 0.43 (c) 0.09 (d) 0.739
(e) 0.108 (f) 0.001 (g) 0.082 (h) 0.0034
(i) 0.9502 (j) 0.6 (k) 2.7 (l) 1.57
(m) 1.042 (n) $\frac{1}{15}$ (o) $\frac{4}{5}$ (p) $\frac{5}{9}$
(q) $\frac{5}{4}$ (r) $\frac{13}{8}$ (s) $1\frac{5}{12}$ (t) $2\frac{4}{7}$

Objective 5; Example 5

6. Find the missing element in each of the following:

(a) 53% of 80 is what amount?
(b) 7.5% of 140 is how much?
(c) What is $\frac{1}{2}\%$ of $830?
(d) 0.42% of 500 is what number?
(e) What percent of 54 is 36?
(f) What percent of 36 is 54?
(g) $12 is what percent of $42?
(h) 3.6 is what percent of 72?
(i) 84% of what number is 189?
(j) 9% of what amount is 6.48?
(k) 18 is $37\frac{1}{2}\%$ of what number?
(l) 8.4% of what number is 21?

Objective 6; Example 6

7. Determine each indicated percent of change:

(a) What percent more than 48 is 72?
(b) What percent less than 72 is 48?
(c) 69 is what percent more than 50?
(d) 45 is what percent less than 54?
(e) What percent increase over 120 is 198?
(f) 66 is what percent decrease from 72?

Objective 7; Example 7

8. Express each of the following in a word equation; then translate that into a math equation and solve:

(a) City tax records indicate that 65% of all local families live in apartments. If 13,000 apartments are occupied, how many family units does the city have altogether?

(b) A cost-of-living report indicates that an "average" family of four spends 22% of its spendable income for housing. If the typical housing expense is $2,310 annually, what is the income of this "average" family?

(c) A travel brochure advertising a trip to Hawaii states that the $720 "package price" includes $600 for air transportation. What percent of the total price is air fare?

(d) Fund-raising projects by the school band collected $4,500 toward the purchase of new uniforms. If the uniforms will cost $6,000, what percent of the total cost has been raised?

(e) Following relocation of a mobile library station, the monthly circulation increased from 2,100 books to 2,700 books. By what percent did the circulation increase?

(f) Following a compaign to encourage measles vaccinations for all children, the number of reported cases dropped from 450 cases to 315 cases annually. What was the percent of decrease in cases?

(g) What number increased by 12.5% of itself gives 36?

(h) What amount decreased by 5% leaves $114?

(i) This year, the number of pledges to the United Fund was increased by 8%. If 27,000 pledges were made this year, how many people had pledged previously?

(j) After the installation of new safety equipment, the number of work-related accidents at a large industrial plant were reduced by 25%. If 21 accidents occurred during the current reporting period, how many accidents had occurred during the previous reporting period?

OBJECTIVES

After completing this appendix, you should be able to

A to D. Percent conversions

1	Express any given percent in decimal form.	**(3–8)**
2	Express any given percent as a fraction.	**(9–13)**
3	Express any given decimal as a percent.	**(14–16)**
4	Express any given fraction as a percent.	**(17–19)**

E. Percent equation forms

5 Use an equation to find the unknown element in a percentage relationship. (20–24)

6 Use the equation form "What % of Original = Change?" to find a percent of change (increase or decrease). (25–28)

F. Word problems containing percents

7 Given a word problem containing percent, express the problem in a word equation which translates into a math equation that solves the problem. (29–34)

8 Explain the procedures required to
(a) convert percents to decimals and fractions; (Exercise 1)
(b) convert decimals and fractions to percents; and (Exercise 2)
(c) perform calculations with percents. (Exercise 3)

Tables

Table 1 Six Basic Axioms of Points, Lines, and Planes

Axiom 1:	Every line is an infinite set of points that includes at least two distinct points.
Axiom 2:	If A and B are two distinct points, then there is one and only one line through A and B.
Axiom 3:	A plane is a set of points which contains at least three points that are not all on the same line.
Axiom 4:	If A, B, and C are three distinct points which do not lie on the same line, there is one and only one plane through A, B, and C.
Axiom 5:	If two points of a line lie in a plane, then all the points of the line lie in the plane.
Axiom 6:	There are at least four points, not all of which are in the same plane.

Table 2 **Compound Amount of $1 (5% Annual Interest, Compounded Monthly)**

No. of mos.	$(1 + i)^n$	*No. of mos.*	$(1 + i)^n$	*No. of mos.*	$(1 + i)^n$	*No. of mos.*	$(1 + i)^n$
1	1.004 166 6667	25	1.109 545 2578	49	1.225 982 4190	73	1.354 638 6514
2	1.008 350 6944	26	1.114 168 3630	50	1.231 090 6791	74	1.360 282 9791
3	1.012 552 1557	27	1.118 810 7312	51	1.236 220 2236	75	1.365 950 8249
4	1.016 771 1230	28	1.123 472 4426	52	1.241 371 1412	76	1.371 642 2867
5	1.021 007 6693	29	1.128 153 5778	53	1.246 543 5209	77	1.377 357 4629
6	1.025 261 8680	30	1.132 854 2177	54	1.251 737 4523	78	1.383 096 4523
7	1.029 533 7924	31	1.137 574 4436	55	1.256 953 0250	79	1.388 859 3542
8	1.033 823 5165	32	1.142 314 3371	56	1.262 190 3293	80	1.394 646 2681
9	1.038 131 1145	33	1.147 073 9802	57	1.267 449 4556	81	1.400 457 2943
10	1.042 456 6608	34	1.151 853 4551	58	1.272 730 4950	82	1.406 292 5330
11	1.046 800 2303	35	1.156 652 8445	59	1.278 033 5388	83	1.412 152 0852
12	1.051 161 8979	36	1.161 472 2313	60	1.283 358 6785	84	1.418 036 0522
13	1.055 541 7391	37	1.166 311 6990	61	1.288 706 0063	85	1.423 944 5358
14	1.059 939 8297	38	1.171 171 3310	62	1.294 075 6147	86	1.429 877 6380
15	1.064 356 2457	39	1.176 051 2116	63	1.299 467 5964	87	1.435 835 4615
16	1.068 791 0633	40	1.180 951 4250	64	1.304 882 0447	88	1.441 818 1093
17	1.073 244 3594	41	1.185 872 0559	65	1.310 319 0533	89	1.447 825 6847
18	1.077 716 2109	42	1.190 813 1895	66	1.315 778 7160	90	1.453 858 2917
19	1.082 206 6952	43	1.195 774 9111	67	1.321 261 1273	91	1.459 916 0346
20	1.086 715 8897	44	1.200 757 3066	68	1.326 766 3820	92	1.465 999 0181
21	1.091 243 8726	45	1.205 760 4620	69	1.332 294 5753	93	1.472 107 3473
22	1.095 790 7221	46	1.210 784 4639	70	1.337 845 8026	94	1.478 241 1279
23	1.100 356 5167	47	1.215 829 3992	71	1.343 420 1602	95	1.484 400 4660
24	1.104 941 3356	48	1.220 895 3550	72	1.349 017 7442	96	1.490 585 4679

Table 3 Compound Amount of $1 (6% Annual Interest, Compounded Monthly)

No. of mos.	$(1+i)^n$	*No. of mos.*	$(1+i)^n$	*No. of mos.*	$(1+i)^n$	*No. of mos.*	$(1+i)^n$
1	1.005 000 0000	25	1.132 795 5751	49	1.276 841 6069	73	1.439 204 4999
2	1.010 025 0000	26	1.138 459 5530	50	1.283 225 8149	74	1.446 400 5224
3	1.015 075 1250	27	1.144 151 8507	51	1.289 641 9440	75	1.453 632 5250
4	1.020 150 5006	28	1.149 872 6100	52	1.296 090 1537	76	1.460 900 6876
5	1.025 251 2531	29	1.155 621 9730	53	1.302 570 6045	77	1.468 205 1911
6	1.030 377 5094	30	1.161 400 0829	54	1.309 083 4575	78	1.475 546 2170
7	1.035 529 3969	31	1.167 207 0833	55	1.315 628 8748	79	1.482 923 9481
8	1.040 707 0439	32	1.173 043 1187	56	1.322 207 0192	80	1.490 338 5678
9	1.045 910 5791	33	1.178 908 3343	57	1.328 818 0543	81	1.497 790 2607
10	1.051 140 1320	34	1.184 802 8760	58	1.335 462 1446	82	1.505 279 2120
11	1.056 395 8327	35	1.190 726 8904	59	1.342 139 4553	83	1.512 805 6080
12	1.061 677 8119	36	1.196 680 5248	60	1.348 850 1525	84	1.520 369 6361
13	1.066 986 2009	37	1.202 663 9274	61	1.355 594 4033	85	1.527 971 4843
14	1.072 321 1319	38	1.208 677 2471	62	1.362 372 3753	86	1.535 611 3417
15	1.077 682 7376	39	1.214 720 6333	63	1.369 184 2372	87	1.543 289 3984
16	1.083 071 1513	40	1.220 794 2365	64	1.376 030 1584	88	1.551 005 8454
17	1.088 486 5070	41	1.226 898 2077	65	1.382 910 3092	89	1.558 760 8746
18	1.093 928 9396	42	1.233 032 6987	66	1.389 824 8607	90	1.566 554 6790
19	1.099 398 5843	43	1.239 197 8622	67	1.396 773 9850	91	1.574 387 4524
20	1.104 895 5772	44	1.245 393 8515	68	1.403 757 8550	92	1.582 259 3896
21	1.110 420 0551	45	1.251 620 8208	69	1.410 776 6442	93	1.590 170 6866
22	1.115 972 1553	46	1.257 878 9249	70	1.417 830 5275	94	1.598 121 5400
23	1.121 552 0161	47	1.264 168 3195	71	1.424 919 6801	95	1.606 112 1477
24	1.127 159 7762	48	1.270 489 1611	72	1.432 044 2785	96	1.614 142 7085

Table 4 Number System Requirements (Including Properties)

Require-ments	CHARACTERISTICS OF SPECIFIC NUMBER SYSTEMS	*Reference paragraphs*, by chapter* *Ch.* 3, N	*Ch.* 3, W	*Ch.* 4, I	*Ch.* 5, F	*Ch.* 7, R
1. Set of elements	Set of natural numbers: $N = \{1, 2, 3, 4, \ldots\}$	3				
	Set of whole numbers: $W = \{0, 1, 2, 3, 4, \ldots\}$		50			
	Set of integers: $I = \{\ldots, -3, -2, -1, 0, 1, 2, 3, \ldots\}$			17–22		
	Set of rational numbers: $F = \left\{\frac{a}{b} \text{ where } a \text{ and } b \text{ are integers and } b \neq 0\right\}$				4–16	
	Set of real numbers: $R = \{a \text{ where } a \text{ is either rational or irrational}\}$					32; 61
2. Operations	Addition	4–10	52–55		45–52	
	Multiplication	11–15	56–58		23–35	
3. Relations	Equals relation	2			18–21	
	Order relation			6–11	59	
4. Properties	(See following list of 11 *field properties.* Notice certain properties do not hold in early systems.)	(8 properties hold)	(9 properties hold)	(10 properties hold)	(11 properties hold) 66	(12 properties hold) 67
	PROPERTIES OF SYSTEM OF REAL† NUMBERS					
	For any numbers† a, b, and c, the following properties hold:					
	Closure, $+$: There is one and only one number,† $a + b$.	4–5		36–38	65	
	Closure, $\times$: There is one and only one number,† $a \times b$.	4–5		43–47	65	
	Commutative, $+$: $a + b = b + a$	17–31				
	Commutative, $\times$: $a \times b = b \times a$	32–40				
	Associative, $+$: $(a + b) + c = a + (b + c)$	18–31				
	Associative, $\times$: $(a \cdot b)c = a(b \cdot c)$	33–40				
	Distributive, $\times$ over $+$: $a(b + c) = a \cdot b + a \cdot c$	41–43				
	Identity, $+$: $a + 0 = 0 + a = a$	‡	60			
	Identity, $\times$: $a \times 1 = 1 \times a = a$	45–47				
	Inverse, $+$: $a + (-a) = (-a) + a = 0$	‡	‡	23–33		
	Inverse, $\times$: $a \times \frac{1}{a} = \frac{1}{a} \times a = 1$ (where $a \neq 0$)	‡	‡	‡	4–8; 33–35; 39	
	Two other operations are not considered fundamental because they are defined in terms of the operations above:					
	Subtraction		61–64	52–60	53–57	
	Division		65–71	61–63	36–44	

*Once established, a characteristic or property applies in all subsequent systems, even if no paragraphs elaborate further.

†The terms "natural number," "whole number," "integer," "rational number," or "real number" apply for their corresponding systems.

‡This property does *not* hold in the indicated system(s).

Table 5 Development of the Real Number System

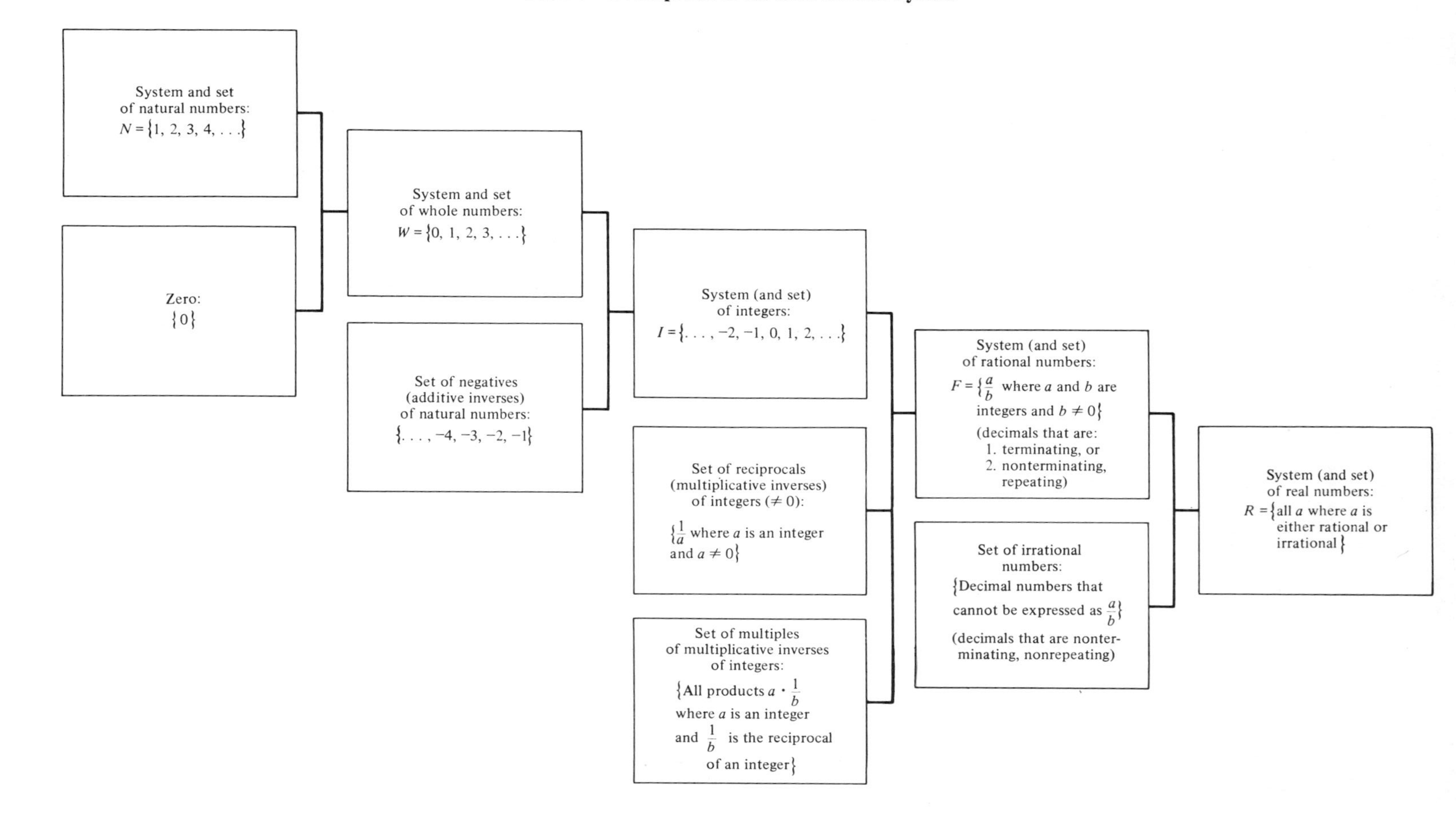

Table 6 **Annual Percentage Rate for Monthly Payment Plans**

Annual percent-age rate	*Number of monthly payments* 6	12	18	24	30	36	*Annual percent-age rate*
	Finance charge per $100 of amount financed						
14.00	4.12	7.74	11.45	15.23	19.10	23.04	14.00
14.25	4.20	7.89	11.66	15.51	19.45	23.48	14.25
14.50	4.27	8.03	11.87	15.80	19.81	23.92	14.50
14.75	4.35	8.17	12.08	16.08	20.17	24.35	14.75
15.00	4.42	8.31	12.29	16.37	20.54	24.80	15.00
15.25	4.49	8.45	12.50	16.65	20.90	25.24	15.25
15.50	4.57	8.59	12.72	16.94	21.26	25.68	15.50
15.75	4.64	8.74	12.93	17.22	21.62	26.12	15.75
16.00	4.72	8.88	13.14	17.51	21.99	26.57	16.00
16.25	4.79	9.02	13.35	17.80	22.35	27.01	16.25
16.50	4.87	9.16	13.57	18.09	22.72	27.46	16.50
16.75	4.94	9.30	13.78	18.37	23.08	27.90	16.75
17.00	5.02	9.45	13.99	18.66	23.45	28.35	17.00
17.25	5.09	9.59	14.21	18.95	23.81	28.80	17.25
17.50	5.17	9.73	14.42	19.24	24.18	29.25	17.50
17.75	5.24	9.87	14.64	19.53	24.55	29.70	17.75
18.00	5.32	10.02	14.85	19.82	24.92	30.15	18.00
18.25	5.39	10.16	15.07	20.11	25.29	30.60	18.25
18.50	5.46	10.30	15.28	20.40	25.66	31.05	18.50
18.75	5.54	10.44	15.49	20.69	26.03	31.51	18.75
19.00	5.61	10.59	15.71	20.98	26.40	31.96	19.00
19.25	5.69	10.73	15.93	21.27	26.77	32.42	19.25
19.50	5.76	10.87	16.14	21.56	27.14	32.87	19.50
19.75	5.84	11.02	16.36	21.86	27.52	33.33	19.75
20.00	5.91	11.16	16.57	22.15	27.89	33.79	20.00
20.25	5.99	11.31	16.79	22.44	28.26	34.25	20.25
20.50	6.06	11.45	17.01	22.74	28.64	34.71	20.50
20.75	6.14	11.59	17.22	23.03	29.01	35.17	20.75
21.00	6.21	11.74	17.44	23.33	29.39	35.63	21.00
21.25	6.29	11.88	17.66	23.62	29.77	36.09	21.25
21.50	6.36	12.02	17.88	23.92	30.14	36.56	21.50
21.75	6.44	12.17	18.09	24.21	30.52	37.02	21.75
22.00	6.51	12.31	18.31	24.51	30.90	37.49	22.00
22.25	6.59	12.46	18.53	24.80	31.28	37.95	22.25
22.50	6.66	12.60	18.75	25.10	31.66	38.42	22.50
22.75	6.74	12.75	18.97	25.40	32.04	38.89	22.75

Table 6 (cont.) Annual Percentage Rate for Monthly Payment Plans

Annual percent-age rate	*Number of monthly payments* 6	12	18	24	30	36	*Annual percent-age rate*
	Finance charge per $100 of amount financed						
23.00	6.81	12.89	19.19	25.70	32.42	39.35	23.00
23.25	6.89	13.04	19.41	25.99	32.80	39.82	23.25
23.50	6.96	13.18	19.62	26.29	33.18	40.29	23.50
23.75	7.04	13.33	19.84	26.59	33.57	40.77	23.75
24.00	7.12	13.47	20.06	26.89	33.95	41.24	24.00
24.25	7.19	13.62	20.28	27.19	34.33	41.71	24.25
24.50	7.27	13.76	20.50	27.49	34.72	42.19	24.50
24.75	7.34	13.91	20.72	27.79	35.10	42.66	24.75
25.00	7.42	14.05	20.95	28.09	35.49	43.14	25.00
25.25	7.49	14.20	21.17	28.39	35.88	43.61	25.25
25.50	7.57	14.34	21.39	28.69	36.26	44.09	25.50
25.75	7.64	14.49	21.61	29.00	36.65	44.57	25.75
26.00	7.72	14.64	21.83	29.30	37.04	45.05	26.00
26.25	7.79	14.78	22.05	29.60	37.43	45.53	26.25
26.50	7.87	14.93	22.27	29.90	37.82	46.01	26.50
26.75	7.95	15.07	22.50	30.21	38.21	46.49	26.75
27.00	8.02	15.22	22.72	30.51	38.60	46.97	27.00
27.25	8.10	15.37	22.94	30.82	38.99	47.45	27.25
27.50	8.17	15.51	23.16	31.12	39.38	47.94	27.50
27.75	8.25	15.66	23.39	31.43	39.77	48.42	27.75
28.00	8.32	15.81	23.61	31.73	40.17	48.91	28.00
28.25	8.40	15.95	23.83	32.04	40.56	49.40	28.25
28.50	8.48	16.10	24.06	32.34	40.95	49.88	28.50
28.75	8.55	16.25	24.28	32.65	41.35	50.37	28.75
29.00	8.63	16.40	24.51	32.96	41.75	50.86	29.00
29.25	8.70	16.54	24.73	33.27	42.14	51.35	29.25
29.50	8.78	16.69	24.96	33.57	42.54	51.84	29.50
29.75	8.85	16.84	25.18	33.88	42.94	52.33	29.75

Table 7 Metric System Equivalents of English System Measurements

Length:	1 inch	= 2.54 centimeters
		= 0.0254 meter
		= 25.4 millimeters
	1 foot	= 30.48 centimeters
		= 0.3048 meter
	1 yard	= 0.914 meter
	1 mile	= 1.61 kilometers
Area:	1 square inch	= 6.45 square centimeters
	1 square foot	= 929.03 square centimeters
		= 0.09 square meter
Volume (space):	1 cubic inch	= 16.39 cubic centimeters
	1 cubic foot	= 0.028 cubic meter
Weight:	1 ounce	= 28.35 grams
	1 pound	= 454. grams
		= 4.54 hectograms
		= 45.4 decagrams
		= 0.454 kilogram
Volume (liquid):	1 liquid pint	= 0.47 liter
	1 liquid quart	= 0.95 liter
	1 liquid gallon	= 3.8 liters

Table 8 English System Equivalents of Metric System Measurements

Length:	1 centimeter	= 0.39 inch
	1 meter	= 39.37 inches
		= 3.28 feet
		= 1.09 yards
	1 kilometer	= 0.62 mile
Area:	1 square centimeter	= 0.16 square inch
	1 square meter	= 10.76 square feet
		= 1.2 square yards
Volume (space):	1 cubic centimeter	= 0.061 cubic inch
	1 cubic meter	= 35.31 cubic feet
Weight:	1 gram	= 0.035 ounce
	1 kilogram	= 2.2 pounds
		= 35.27 ounces
Volume (liquid):	1 liter	= 1.06 liquid quarts

Answers to Practices and Odd-numbered Exercises

CHAPTER 1: PRACTICE SETS

1. (a) Well defined: $\{a, b, c, d, e, f, g, h, i, j, k, l\}$.
   (b) Well defined: {Tuesday, Thursday}.
   (c) Not well defined.
   (d) Not well defined.
   (e) Well defined: $\{6, 7, 8, 9, 10\}$.

2. (a) Roster: $D = \{8, 9, 10, \ldots\}$
   Set builder: $D = (x \mid x > 7$, x is a whole number$)$
   (b) $E = \{1, 2, 3, 4, 5, 6\}$
   (c) $A = \varnothing$ or $A = \{\ \}$
   (d) Roster: $B = \{1, 2\}$
   Set builder: $B = \{x \mid x < 3$, x is a natural number$\}$
   (e) $B = \{3, 4, 5, 6, 7, 8, 9\}$
   (f) Roster: $F = \{1\}$
   Set builder: $F = \{x \mid x < 2$, x is a natural number$\}$

3.

	Proper subset relationships	*Subset relationships*	
(a)	$A \subset U$	$A \subseteq U$	$U \subseteq U$
	$B \subset U$	$B \subseteq U$	$A \subseteq A$
	$C \subset U$	$C \subseteq U$	$B \subseteq B$
	$D \subset U$	$D \subseteq U$	$C \subseteq C$
	$A \subset B$	$A \subseteq B$	$D \subseteq D$
	$D \subset C$	$D \subseteq C$	

(b) $B \subset U$ $\quad$ $B \subseteq U$ $\quad$ $U \subseteq U$
$D \subset U$ $\quad$ $D \subseteq U$ $\quad$ $A \subseteq A$
$D \subset A$ $\quad$ $A \subseteq U$ $\quad$ $B \subseteq B$
$B \subset A$ $\quad$ $D \subseteq A$ $\quad$ $D \subseteq D$
$B \subseteq A$

4. (These are examples. You could write others that are just as appropriate. Check your answers with your instructor.)

(a) The set of all people in Sante Fe.
The set of all people in New Mexico.
The set of all people in the USA.

(b) The set of the numbers 3, 4, 5.
The set of the natural numbers.
The set of the whole numbers.

(c) The set of people in the building where you are now.
The set of people in the city where you are now.
The set of people in the country where you are now.

5. (a) $B \subseteq A$ but $A \nsubseteq B$ $\quad$ (b) $C \nsubseteq D$ and $D \nsubseteq C$
(c) $E \subseteq F$ and $F \subseteq E$ $\quad$ (d) $G \subseteq H$ but $H \nsubseteq G$

6. (a) $\{1\}$ $\quad$ $\{1, 2\}$ $\quad$ $2^2 = 4$ subsets
$\{2\}$ $\quad$ $\varnothing$

(b) {○} $\quad$ {○, □} $\quad$ {○, □, △} $\quad$ $2^3 = 8$ subsets
{□} $\quad$ {○, △} $\quad$ $\varnothing$
{△} $\quad$ {□, △}

(c) $\{a\}$ $\quad$ Only the set itself $\quad$ $2^1 = 2$ subsets
$\varnothing$ $\quad$ and the null set are subsets.

(d) $\{1\}$ $\quad$ $\{1, 2\}$ $\quad$ $\{1, 2, 3\}$ $\quad$ $\{1, 2, 3, 4\}$ $\quad$ $2^4 = 16$ subsets
$\{2\}$ $\quad$ $\{1, 3\}$ $\quad$ $\{1, 2, 4\}$ $\quad$ $\varnothing$
$\{3\}$ $\quad$ $\{1, 4\}$ $\quad$ $\{1, 3, 4\}$
$\{4\}$ $\quad$ $\{2, 3\}$ $\quad$ $\{2, 3, 4\}$
$\{2, 4\}$
$\{3, 4\}$

(e) {apple} $\quad$ {apple, pear} $\quad$ {apple, pear, peach} $\quad$ $2^3 = 8$ subsets
{pear} $\quad$ {apple, peach} $\quad$ $\varnothing$
{peach} $\quad$ {pear, peach}

(f) {Jane} $\quad$ {Jane, Sarah} $\quad$ $2^2 = 4$ subsets
{Sarah} $\quad$ $\varnothing$

7. (a) $A' = \{3, 7\}$ $\quad$ (b) $A' = \{d\}$ $\quad$ (c) $A' = \varnothing$

8. (a) $A \cup B = \{a, x, y\}$ $\quad$ (b) $D \cup E = \{1, 2, 3, 4, 6, 7\}$
(c) $F \cup G = \{3, 6, 9, 12\}$

(d) $(A \cup B) \cup C = \{1, 2, 3, 8, 9\} \cup \{1, 2, 3, 5\}$
$= \{1, 2, 3, 5, 8, 9\}$

(e) $A \cup (B \cup C) = \{3, 8, 9\} \cup \{1, 2, 3, 5\}$
$= \{1, 2, 3, 5, 8, 9\}$

9. (a) $A \cap B = \{12, 18\}$ (b) $K \cap L = \{\square\}$ (c) $H \cap I = \varnothing$

(d) $M \cap N = \{\text{Ted}\}$ (e) $P \cap Q = \{4, 6, 8\}$

(f) $(A \cap B) \cap C = \{6, 12\} \cap \{1, 10, 12\} = \{12\}$

(g) $A \cap (B \cap C) = \{6, 8, 10, 12\} \cap \{1, 12\} = \{12\}$

10. (a) $A \times B = \{(a, 3), (a, 7), (b, 3), (b, 7)\}$

(b) $D \times C = \{(1, 3), (1, 7), (1, 10), (6, 3), (6, 7), (6, 10)\}$

(c) $A \times B = \{(7, 2)\}$
$B \times A = \{(2, 7)\}$

(d) $E \times F = \{(a, 1), (a, 2), (a, 3), (a, 4)\}$
$F \times E = \{(1, a), (2, a), (3, a), (4, a)\}$

11. (a) Finding $A \times B$:

Elements in A	*Elements in B*	*Ordered Pairs in A × B*
7 →	a →	(7, a)
10 →	a →	(10, a)

$A \times B = \{(7, a), (10, a)\}$

Finding $B \times A$:

Elements in B	*Elements in A*	*Ordered pairs in B × A*
a →	7 →	(a, 7)
a →	10 →	(a, 10)

$B \times A = \{(a, 7), (a, 10)\}$

(b) Finding $D \times C$:

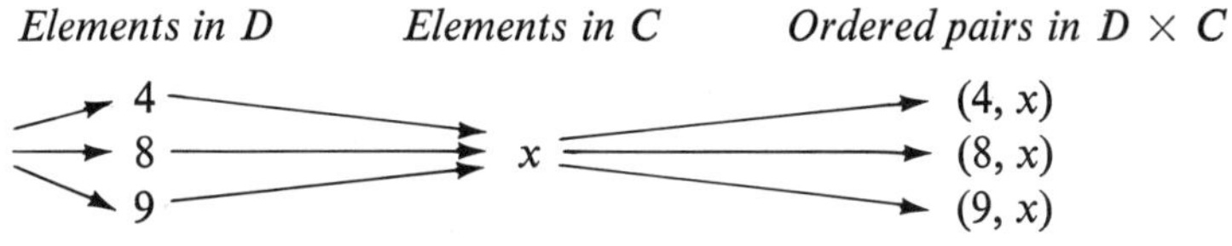

$D \times C = \{(4, x), (8, x), (9, x)\}$

(c) Finding $U \times U$:

Elements in U	*Elements in U*	*Ordered pairs in U × U*
1 →	1 →	(1, 1)
	2 →	(1, 2)
	3 →	(1, 3)
2 →	1 →	(2, 1)
	2 →	(2, 2)
	3 →	(2, 3)
3 →	1 →	(3, 1)
	2 →	(3, 2)
	3 →	(3, 3)

$U \times U = \{(1, 1), (1, 2), (1, 3), (2, 1), (2, 2), (2, 3), (3, 1), (3, 2), (3, 3)\}$

(d) Finding $E \times F$:

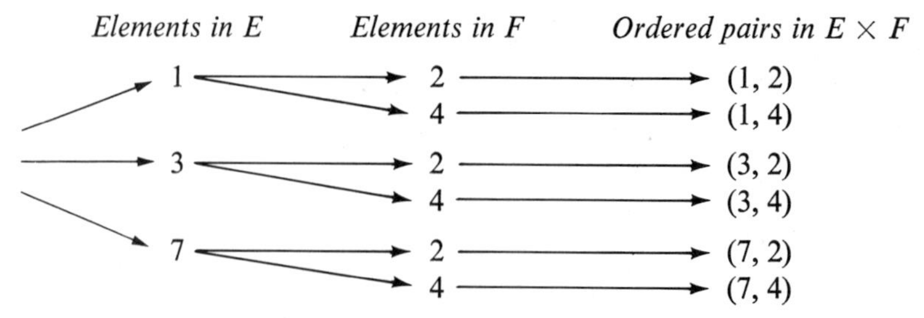

$E \times F = \{(1, 2), (1, 4), (3, 2), (3, 4), (7, 2), (7, 4)\}$

12. (a) $A \times B = \{(1, a), (1, b), (3, a), (3, b), (5, a), (5, b)\}$

Elements of set B →				
	b	$(1, b)$	$(3, b)$	$(5, b)$
	a	$(1, a)$	$(3, a)$	$(5, a)$
		1	3	5 ← Elements of set A

$B \times A = \{(a, 1), (a, 3), (a, 5), (b, 1), (b, 3), (b, 5)\}$

Elements of set A →			
	5	$(a, 5)$	$(b, 5)$
	3	$(a, 3)$	$(b, 3)$
	1	$(a, 1)$	$(b, 1)$
		a	b ← Elements of set B

(b) $A \times B = \{(a, 1), (a, 8)\}$

Elements of set B →		
	8	$(a, 8)$
	1	$(a, 1)$
		a ← Elements of set A

13. (a)

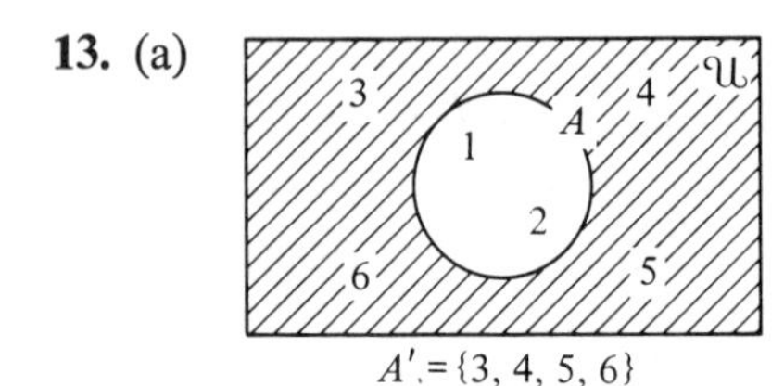

$A' = \{3, 4, 5, 6\}$

(b)

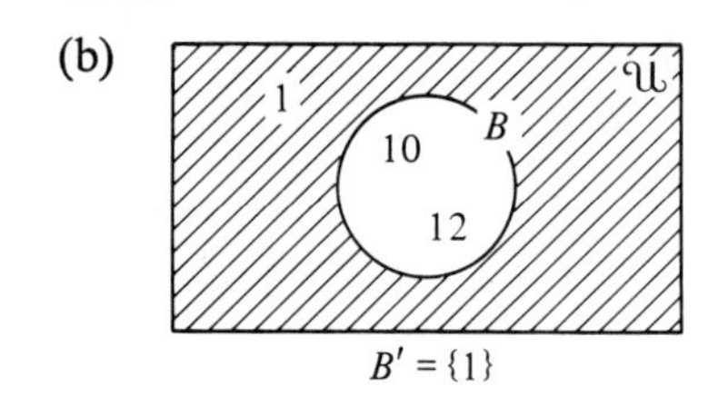

$B' = \{1\}$

(c)

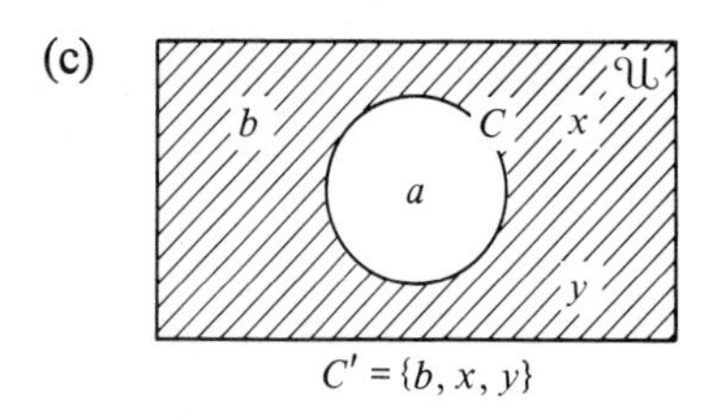

$C' = \{b, x, y\}$

14. (a)

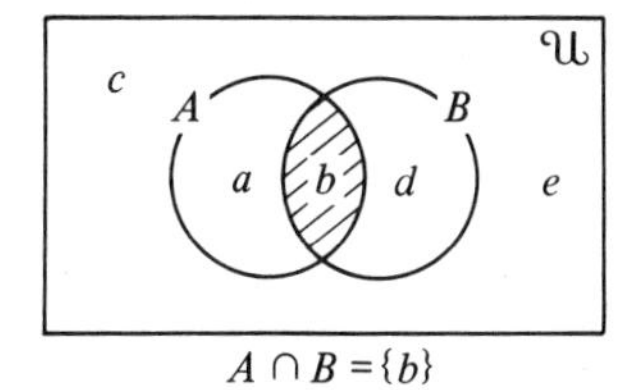

$A \cap B = \{b\}$

(b)

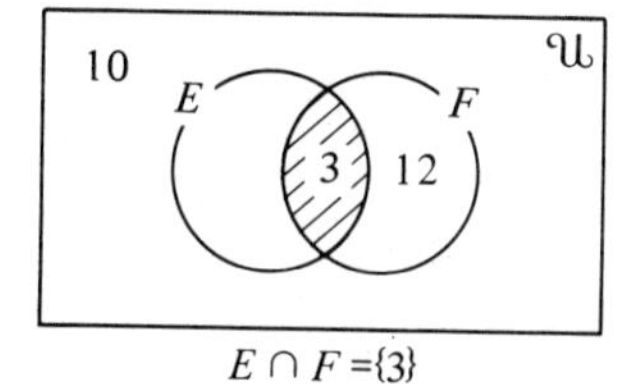

$E \cap F = \{3\}$

(c)

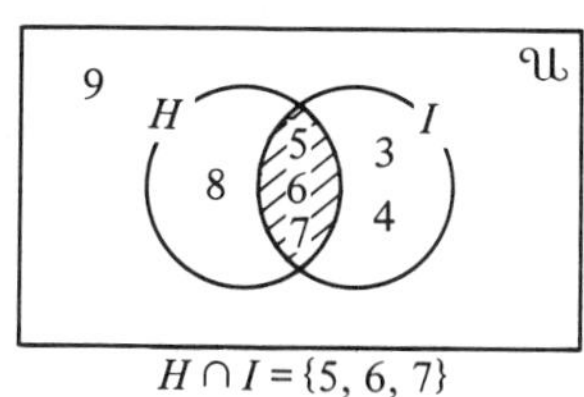

$H \cap I = \{5, 6, 7\}$

15. (a) $U = \{2, 4, 6, 8, 10, 12\}$, $A = \{4\}$, $B = \{4, 8, 12\}$

(1)

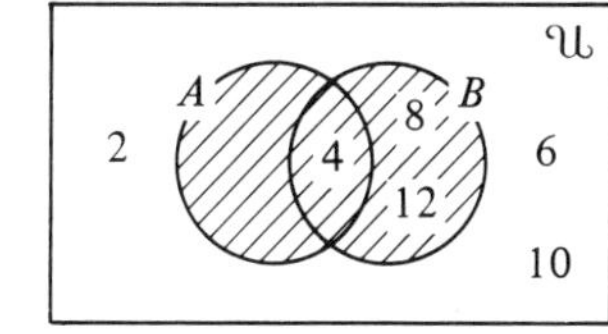

$A \cup B = \{4, 8, 12\}$

(2)

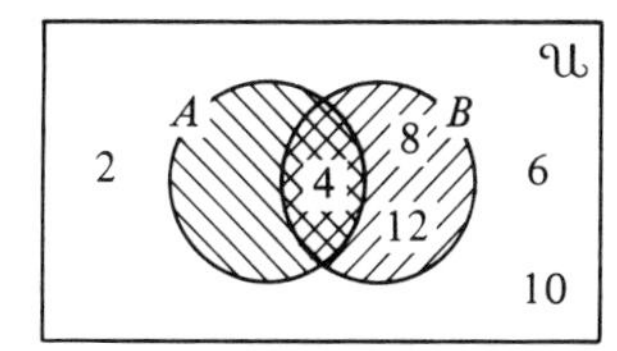

$A \cap B = \{4\}$

(3)

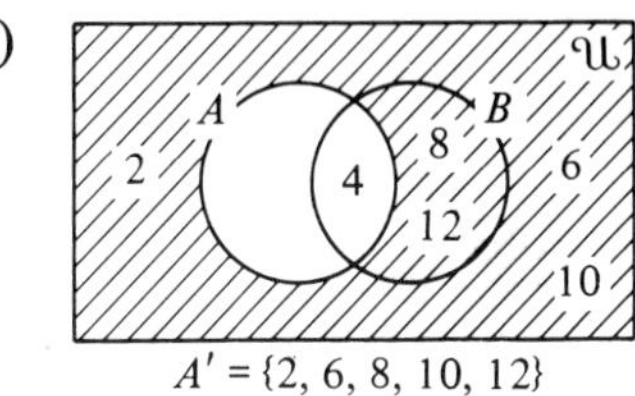

$A' = \{2, 6, 8, 10, 12\}$

(4)

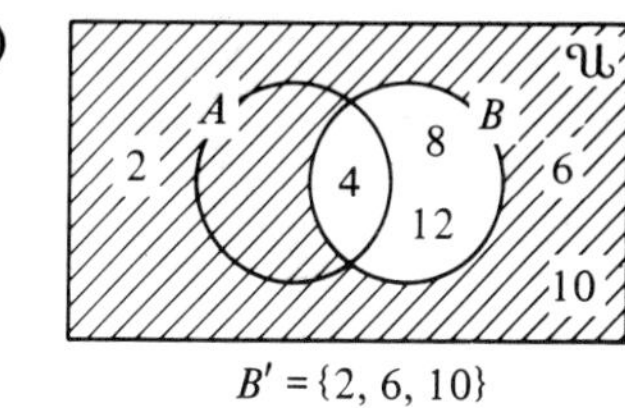

$B' = \{2, 6, 10\}$

(b) $U = \{a, b, c, d, e, f\}$, $E = \{e, f\}$, $F = \{a, b, e, f\}$

(1)

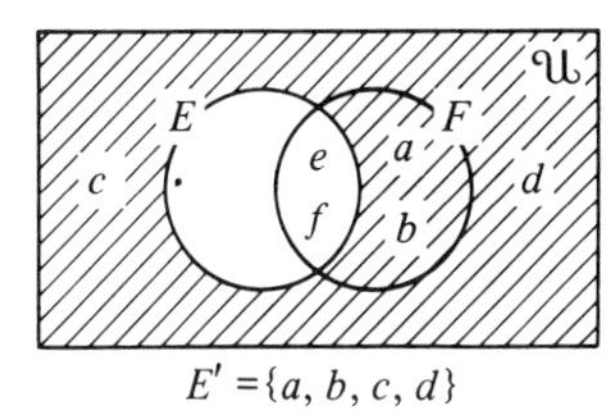

$E' = \{a, b, c, d\}$

(2)

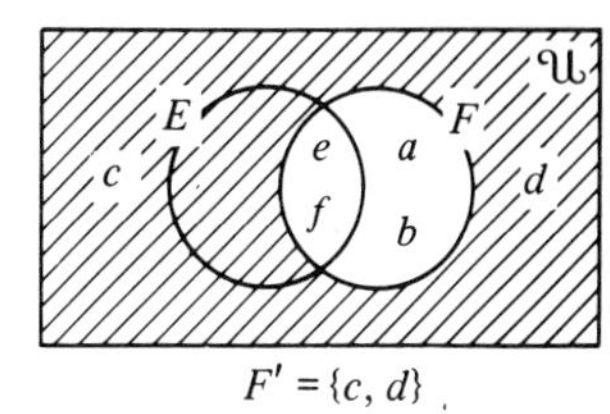

$F' = \{c, d\}$

16. (a)

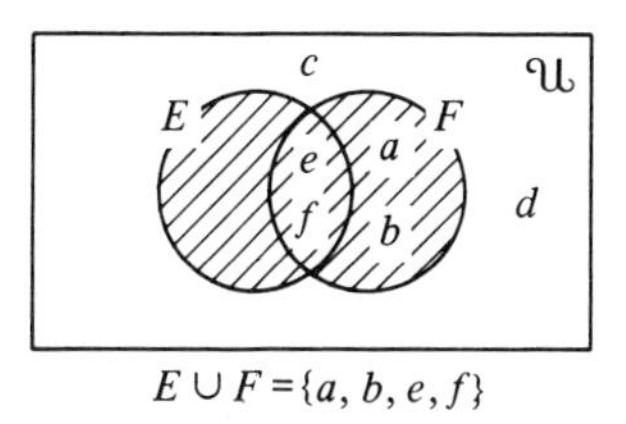

$E \cup F = \{a, b, e, f\}$

(b)

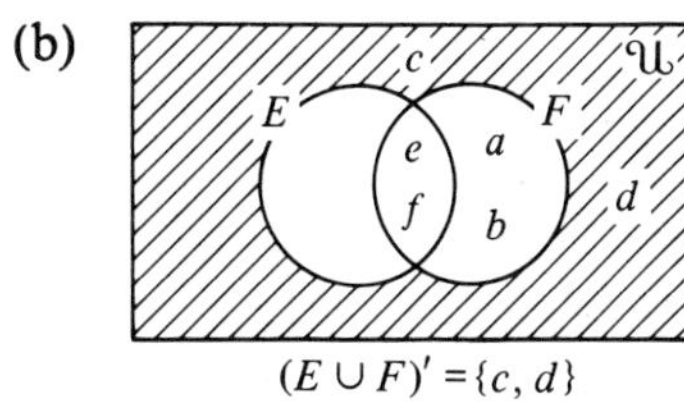

$(E \cup F)' = \{c, d\}$

(c)

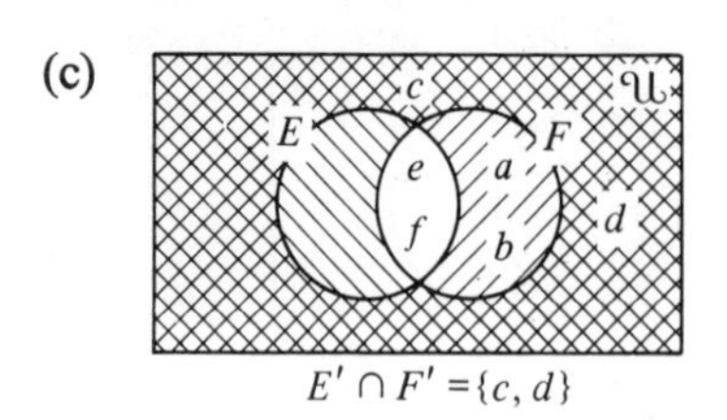

$E' \cap F' = \{c, d\}$

(d)

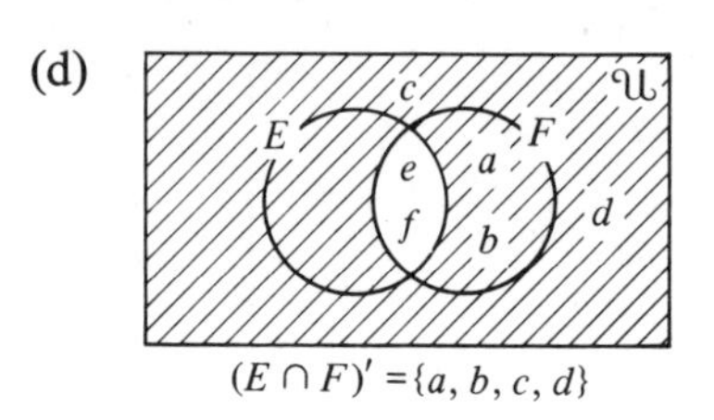

$(E \cap F)' = \{a, b, c, d\}$

17. (a) $\{6, 8, 10, 12\}$ ↔ $\{6, 8, 10, 14\}$

Each element in both sets is paired with only one element in the other set. The sets are in one-to-one correspondence.

(b) $\{2, 3, 8\}$ ↔ $\{a, x, \square\}$

Each element in both sets is paired with only one element in the other set. Thus, the sets are in one-to-one correspondence.

(c) $\{3, 10, 15\}$ ↔ $\{3, f\}$

Not a one-to-one correspondence. There is not an element in the second set that can be paired with the element 15 in the first set.

(d) $\{x, y\}$ ↔ $\{a, b, c\}$

Not a one-to-one correspondence. There is not an element in the first set that can be paired with the element c in the second set.

18. (a)

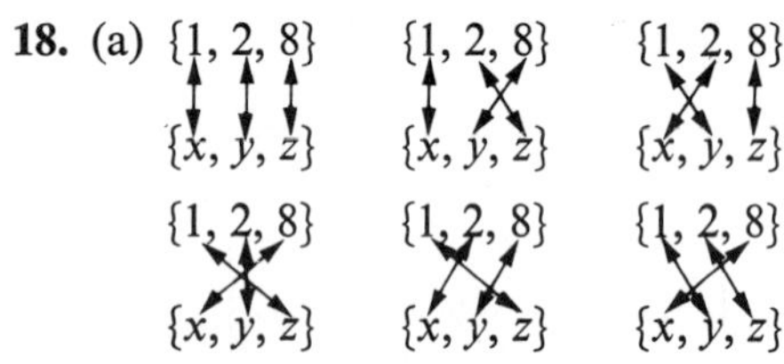

(b) {apple, pear} ↔ $\{b, c\}$ {apple, pear} ↔ $\{b, c\}$

(c) $\{x\}$ ↔ $\{a\}$

(d) $\{1, 7\}$ ↔ $\{7, 10\}$ $\{1, 7\}$ ↔ $\{7, 10\}$

19. (a) $A = \{a, b, c, d, e, f\}$ ↔ $\{1, 2, 3, 4, 5, 6, \ldots\}$

$n(A) = 6$

(b) $B = \{3, 10, 14, 16\}$ ↔ $\{1, 2, 3, 4, 5, 6, \ldots\}$

$n(B) = 4$

(c) $C = \{$glass, tire, aluminum$\}$ ↔ $\{1, 2, 3, 4, 5, \ldots\}$

$n(C) = 3$

(d) $D = \{$David$\}$ ↔ $\{1, 2, 3, \ldots\}$

$n(D) = 1$

(e) $E = \{3, \square, x, 7\}$ ↔ $\{1, 2, 3, 4, \ldots\}$

$n(E) = 4$

(f) $F = \{1, 2, 3, 4, 5\}$ ↔ $\{1, 2, 3, 4, 5, 6, \ldots\}$

$n(F) = 5$

20.

	One-to-one correspondence?	*Same cardinal number?*	*Equivalent?*
(a)	yes	yes	yes
(b)	no	no	no
(c)	yes	yes	yes
(d)	yes	yes	yes
(e)	no	no	no

21. (a) Equivalent and equal
(b) Equivalent and not equal
(c) Equivalent and not equal
(d) Not equivalent and not equal
(e) Equivalent and not equal
(f) Equivalent and equal

CHAPTER 1: EXERCISES

1. (a) Well defined; {9, 10, 11}
(b) Well defined; {Washington, Wisconsin, Wyoming}
(c) Not well defined
(d) Well defined; {0, 1, 2, 3, 4, 5}
(e) Well defined; $\{f, r, i, e, n, d\}$
(f) Well defined; {3, 4, 5, 6, . . .}
(g) Well defined; {13}

3. *Proper subset relationships* / *Subset relationships*

(a) Proper: $A \subset U$, $B \subset U$, $C \subset U$, $D \subset U$, $A \subset B$, $A \subset C$, $A \subset D$, $B \subset D$, $C \subset D$
Subset: $A \subseteq U$, $B \subseteq U$, $C \subseteq U$, $D \subseteq U$, $A \subseteq B$, $A \subseteq C$, $A \subseteq D$, $B \subseteq D$, $C \subseteq D$, $U \subseteq U$, $A \subseteq A$, $B \subseteq B$, $C \subseteq C$, $D \subseteq D$

(b) Proper: $A \subset U$, $B \subset U$, $A \subset C$, $B \subset C$
Subset: $A \subseteq U$, $B \subseteq U$, $C \subseteq U$, $A \subseteq C$, $B \subseteq C$, $U \subseteq U$, $A \subseteq A$, $B \subseteq B$, $C \subseteq C$, $U \subseteq C$

(c) Proper: $A \subset U$, $B \subset U$, $C \subset U$, $A \subset C$, $B \subset C$
Subset: $A \subseteq U$, $B \subseteq U$, $C \subseteq U$, $A \subseteq C$, $B \subseteq C$, $U \subseteq U$, $A \subseteq A$, $B \subseteq B$, $C \subseteq C$

(d) Proper: $A \subset U$, $B \subset U$, $A \subset C$, $B \subset C$, $A \subset B$
Subset: $A \subseteq U$, $B \subseteq U$, $C \subseteq U$, $A \subseteq B$, $A \subseteq C$, $B \subseteq C$, $U \subseteq U$, $A \subseteq A$, $B \subseteq B$, $C \subseteq C$, $U \subseteq C$

5. (a) $A \subseteq B$ but $B \nsubseteq A$
(b) $C \subseteq D$ and $D \subseteq C$
(c) $E \nsubseteq F$ but $F \subseteq E$
(d) $G \nsubseteq H$ but $H \subseteq G$
(e) $I \nsubseteq J$ but $J \subseteq I$
(f) $K \subseteq L$ and $L \subseteq K$

7. (a) $A' = \{8, 10\}$
(b) $C' = \varnothing$
(c) $B' = U$ or $B' = \{1, 2, 3\}$
(d) $D' = \{e, g, h\}$
(e) $E' = \{1, 2, 3, 4\}$

9. (a) $A \cap B = \{7\}$
(b) $(C \cap D) \cap E = \{10\} \cap \{9, 10\} = \{10\}$
(c) $A \cap B = \varnothing$
(d) $A \cap (B \cap C) = \{1, 6\} \cap \{2\} = \varnothing$
(e) $B \cap C = \varnothing$
(f) $A \cap B = \{1, 10, 15\}$, or set B, or set A
(g) $(E \cap F) \cap G = \{1, 3, 5\} \cap \{2, 4\} = \varnothing$

11. (a) $A \times B = \{(a, c), (a, d), (a, e), (b, c), (b, d), (b, e)\}$
$B \times A = \{(c, a), (c, b), (d, a), (d, b), (e, a), (e, b)\}$

(b) $E \times F = \{(1, 1), (1, 2), (2, 1), (2, 2)\}$
$F \times E = \{(1, 1), (1, 2), (2, 1), (2, 2)\}$

(c) $B \times C = \{(3, 5)\}$
$C \times B = \{(5, 3)\}$

(d) $A \times B = \{(1, b), (1, 7), (1, 10), (a, b), (a, 7), (a, 10)\}$
$B \times A = \{(b, 1), (b, a), (7, 1), (7, a), (10, 1), (10, a)\}$

13. (a)

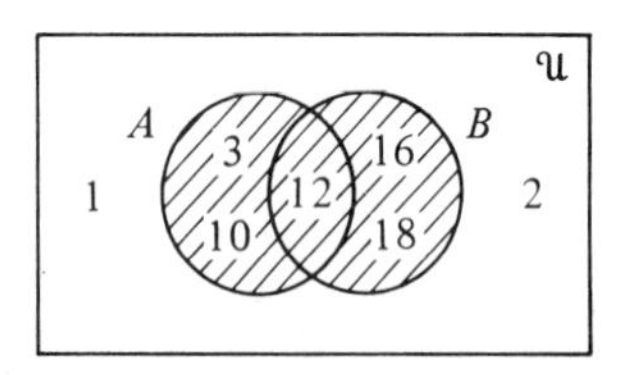

$A \cup B = \{3, 10, 12, 16, 18\}$

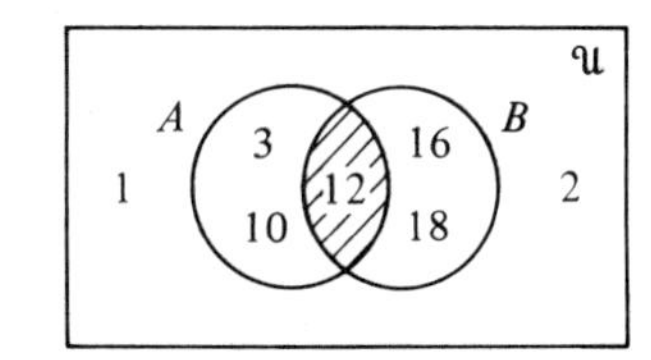

$A \cap B = \{12\}$

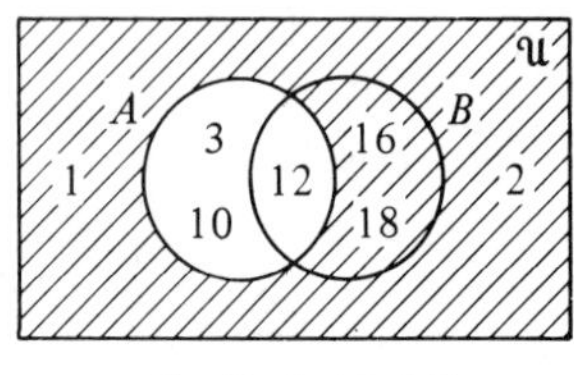

$A' = \{1, 2, 16, 18\}$

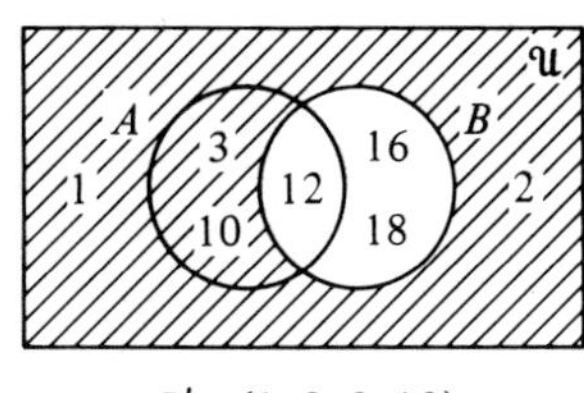

$B' = \{1, 2, 3, 10\}$

(b)

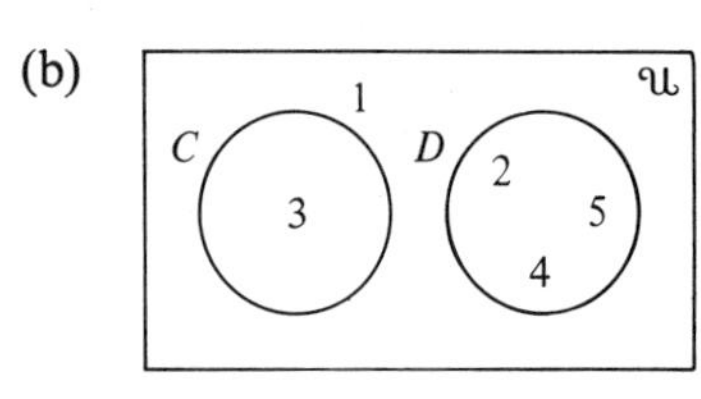

$C \cap D = \phi$

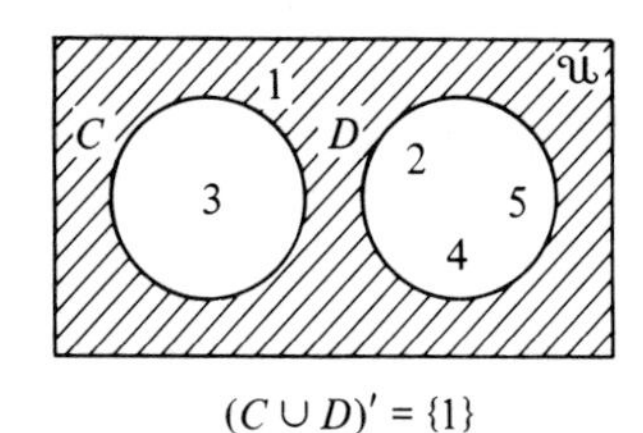

$(C \cup D)' = \{1\}$

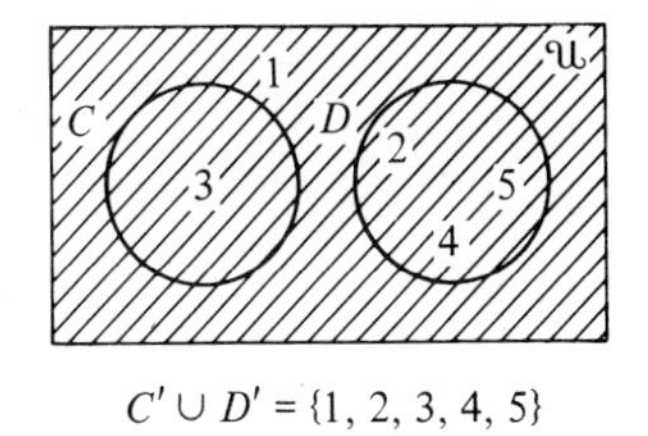

$C' \cup D' = \{1, 2, 3, 4, 5\}$

(c)

$(A \cap C)' = \phi' = \mathcal{U} = \{a, b, c\}$

$A \cap B = \phi$

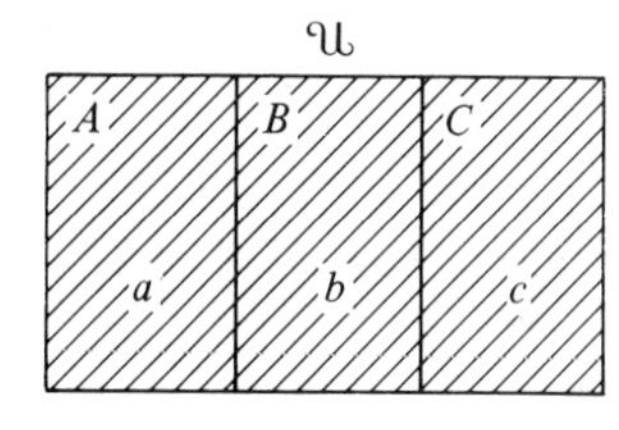

$(A \cup B) \cup C = \{a, b\} \cup \{c\} = \{a, b, c\} = \mathcal{U}$

(d)

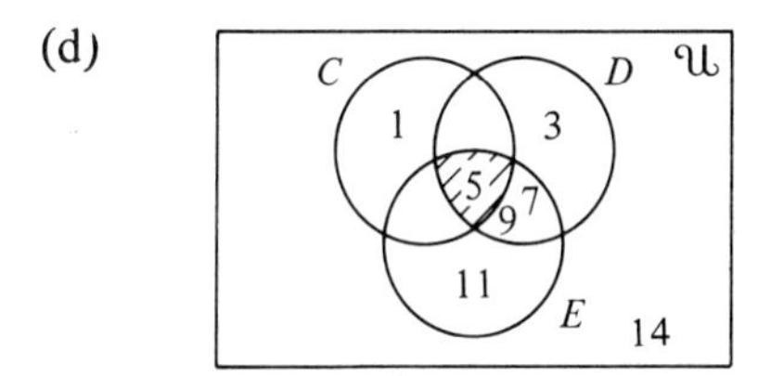

$C \cap (D \cap E) = \{1, 5\} \cap \{5, 7, 9\} = \{5\}$

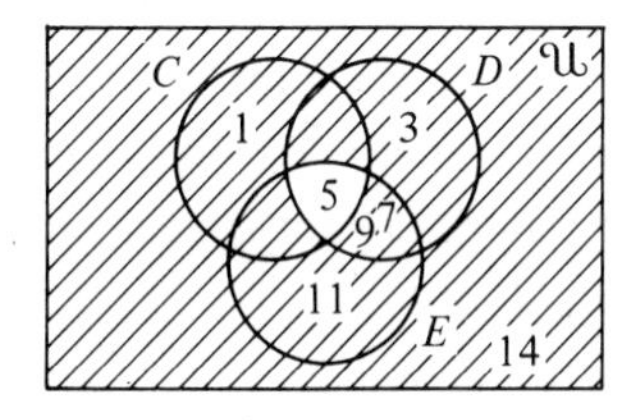

$(C \cap D \cap E)' = \{1, 3, 7, 9, 11, 14\}$

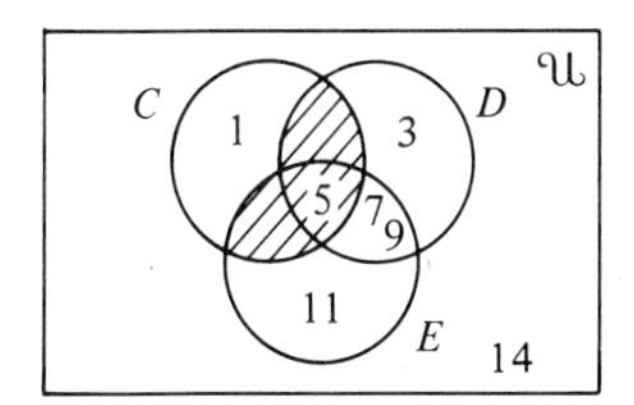

$C \cap (D \cup E) = \{1, 5\} \cap \{3, 5, 7, 9, 11\}$
$= \{5\}$

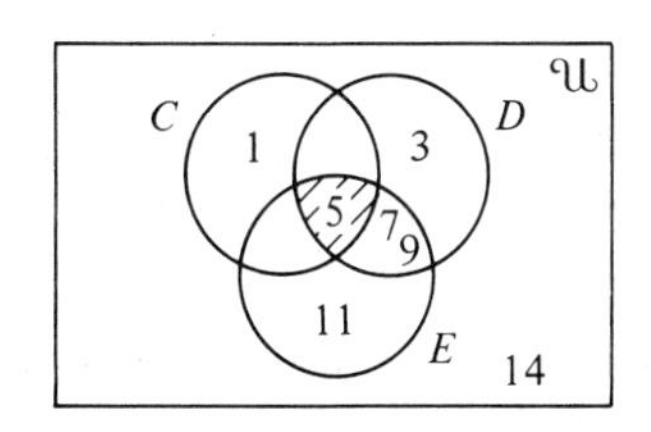

$(C \cap D) \cup (C \cap E) = \{5\} \cup \{5\} = \{5\}$

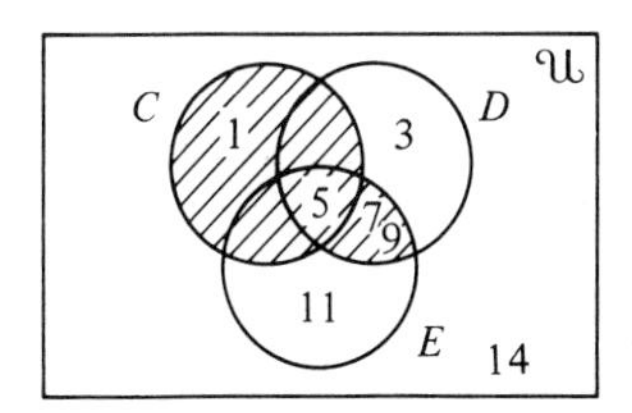

$C \cup (D \cap E) = \{1, 5\} \cup \{5, 7, 9\}$
$= \{1, 5, 7, 9\}$

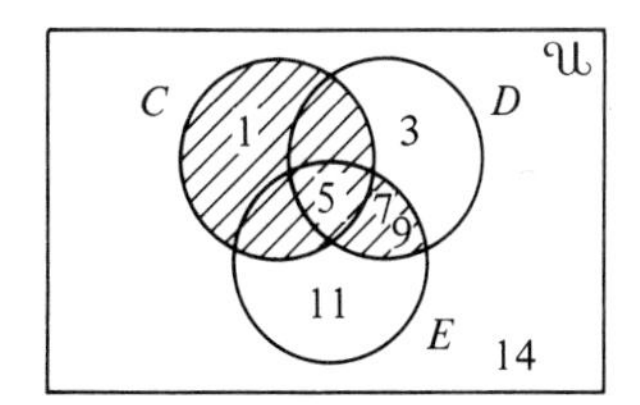

$(C \cup D) \cap (C \cup E) =$
$\{1, 3, 5, 7, 9\} \cap \{1, 5, 7, 9, 11\} =$
$\{1, 5, 7, 9\}$

15.
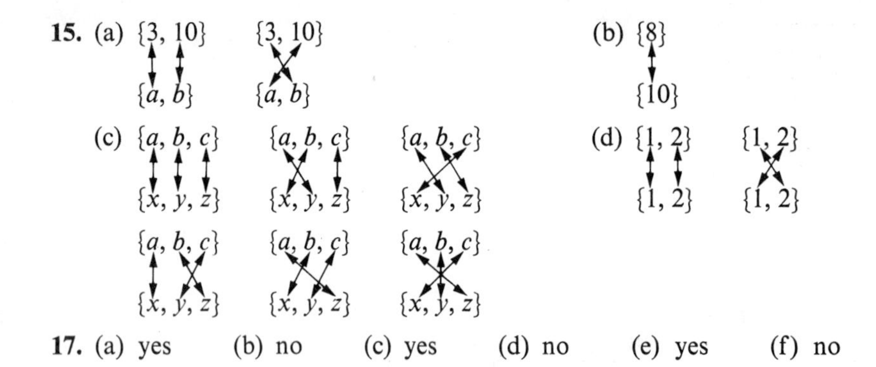

17. (a) yes (b) no (c) yes (d) no (e) yes (f) no

CHAPTER 2: PRACTICE SETS

1. (a) Yes; yes (domain is set of all real numbers); no
   (b) No ; no; yes (true) (c) No ; no; yes (false)
   (d) No ; no; yes (true)

2. (a) It is not true that I will go to the ballgame Friday.
   I will not go to the ballgame Friday.

   (b) It is not true that Benjamin Franklin was President of the United States.
   Benjamin Franklin was not President of the United States.

   (c) It is not true that James is not sixteen years old.
   James is sixteen years old.

3. (a) $A \cap B \subseteq C$ (statement)
   $A \cap B \nsubseteq C$ (negation)
   (b) $2 \cdot 5 - 4 > 8$ (statement)
   $2 \cdot 5 - 4 \ngtr 8$ (negation)

   (c) $3 \in X$ (statement)
   $3 \notin X$ (negation)

4. (a) Conjunction: Bill is older than Jack and Bill will inherit the estate.
   Disjunction: Bill is older than Jack or Bill will inherit the estate.
   Conditional: If Bill is older than Jack, then Bill will inherit the estate.

   (b) (1) If a rectangle has 4 right angles, then a square is a rectangle.
   (2) A rectangle does not have 4 right angles.
   (3) A rectangle does not have 4 right angles or a square is not a rectangle.
   (4) A rectangle does not have 4 right angles and a square is a rectangle.

   (c) (1) $p \rightarrow q$ (2) $\sim q$ (3) $\sim p \wedge q$

5. (a)

p	q	$\sim q$	$p \wedge \sim q$
T	T	F	F
T	F	T	T
F	T	F	F
F	F	T	F

(b)

p	q	$\sim p$	$\sim p \vee q$	$(\sim p \vee q) \rightarrow q$
T	T	F	T	T
T	F	F	F	T
F	T	T	T	T
F	F	T	T	F

(c)

p	q	$\sim p$	$(\sim p \wedge q)$	$(\sim p \wedge q) \longrightarrow p$
T	T	F	F	T
T	F	F	F	T
F	T	T	T	F
F	F	T	F	T

(d)

p	q	$p \longrightarrow q$	$\sim p$	$\sim p \wedge q$	$(p \longrightarrow q) \vee (\sim p \wedge q)$
T	T	T	F	F	T
T	F	F	F	F	F
F	T	T	T	T	T
F	F	T	T	F	T

6. (a)

p	q	$\sim p$	$(p \longrightarrow q)$	$(q \vee \sim p)$	$(p \longrightarrow q) \longleftrightarrow (q \vee \sim p)$
T	T	F	T	T	T
T	F	F	F	F	T
F	T	T	T	T	T
F	F	T	T	T	T

The statements are logically equivalent.

(b)

p	q	$\sim p$	$\sim q$	$p \vee q$	$\sim(p \vee q)$	$(\sim p \wedge \sim q)$	$\sim(p \vee q) \longleftrightarrow (\sim p \wedge \sim q)$
T	T	F	F	T	F	F	T
T	F	F	T	T	F	F	T
F	T	T	F	T	F	F	T
F	F	T	T	F	T	T	T

The statements are logically equivalent.

(c)

p	q	$\sim p$	$\sim q$	$p \wedge q$	$\sim(p \wedge q)$	$\sim p \wedge \sim q$	$\sim(p \wedge q) \longleftrightarrow (\sim p \wedge \sim q)$
T	T	F	F	T	F	F	T
T	F	F	T	F	T	F	F
F	T	T	F	F	T	F	F
F	F	T	T	F	T	T	T

The statements are not logically equivalent.

(d)

p	q	$\sim p$	$\sim q$	$p \longrightarrow q$	$\sim q \longrightarrow \sim p$	$(p \longrightarrow q) \longleftrightarrow (\sim q \longrightarrow \sim p)$
T	T	F	F	T	T	T
T	F	F	T	F	F	T
F	T	T	F	T	T	T
F	F	T	T	T	T	T

The statements are logically equivalent.

7. (a) It is not true that $x = 6$ or that $y = 4$.
(b) $x \ngtr 6$ or $x \nless -6$
(c) It is not true that $y \nless 6$ and $y = 2$.
(d) Jack is not older than Frank and Jack is not 25.

8. (a)

p	$\sim p$	$(p \wedge \sim p)$	$\sim(p \wedge \sim p)$
T	F	F	T
F	T	F	T

The statement is a tautology.

(b)

p	$\sim p$	$\sim(\sim p)$
T	F	T
F	T	F

The statement is not a tautology.

(c)

p	q	$\sim p$	$\sim q$	$(p \rightarrow q)$	$[(p \rightarrow q) \vee \sim p]$	$[(p \rightarrow q) \vee \sim p] \rightarrow \sim q$
T	T	F	F	T	T	F
T	F	F	T	F	F	T
F	T	T	F	T	T	F
F	F	T	T	T	T	T

The statement is not a tautology.

(d)

p	q	$p \rightarrow q$	$[(p \rightarrow q) \wedge p]$	$[(p \rightarrow q) \wedge p] \rightarrow q$
T	T	T	T	T
T	F	F	F	T
F	T	T	F	T
F	F	T	F	T

The statement is a tautology.

9. (a) Converse: If a number is greater than 0, then a number is positive.
Inverse: If a number is not positive, then the number is not greater than 0.
Contrapositive: If a number is not greater than 0, then a number is not positive.

(b) Converse: If we do not go to the ballgame, then it is raining.
Inverse: If it is not raining, then we do go to the ballgame.
Contrapositive: If we do go to the ballgame, then it is not raining.

10. See top of page 451.

11. (a) (1) Analog (2) Analog (3) Digital
(b) (1) iii (2) i (3) vi (4) v (5) ii (6) iv

12. (a) (1) $X = 3(5A - B^2)$ (2) $Y = \frac{A}{2} + \frac{B}{2} + \frac{C}{2}$ (3) $C = 4(A^2 - B^2)$

(4) $A = \frac{4X + 2Y}{6}$ (5) $D = \frac{A^2}{B^2 - C^2}$

(b) (1) $X = (2 * B) \uparrow 3$ (2) $A = 3 * B + C/2$ (3) $X = (Y - 4) \uparrow 2$

13. (a) (1) i (2) ii (3) iii

(b) (1) Legal
(2) Not legal because the THEN must refer the computer to another statement
(3) Not legal because 3 times M must be written 3 * M

(c) (1) The computer would compute 2 times A, compare this result with B, and let X equal the larger of the two numerical values.

(2) The computer would print the numerical value of A, then the words "cubed is," and then the numerical value of A^3.

(d) 45 PRINT X = (2 * B MIN D)

10.

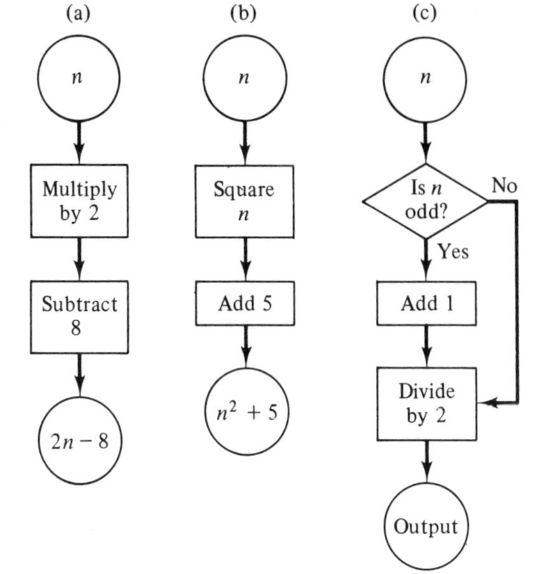

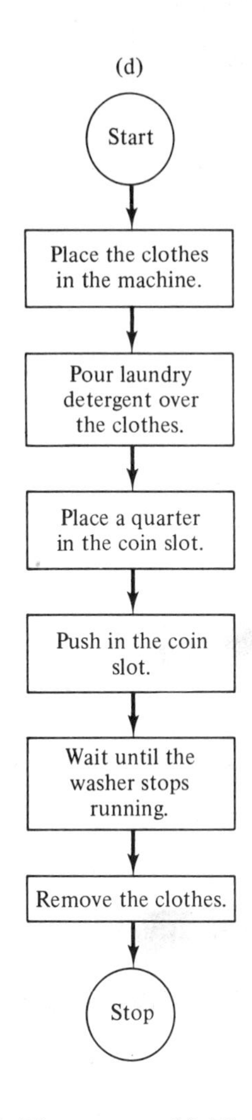

CHAPTER 2: EXERCISES

1. (a) Yes (b) No (c) No (d) Yes (e) Yes

3. (a) Not p or q

(b) p implies not q
(or If p, then not q.)

(c) p and q

(d) Not q implies not p
(or If not q, then not p.)

(e) p if and only if q

(f) p implies q and not q
(or If p then q and not q.)

5. (a) $q \longrightarrow \sim p$ (b) $\sim p \wedge \sim q$ (c) $q \vee p$

7. (a) It is not true that I am older than my sister.
I am not older than my sister.

(b) It is not true that George is not taking history this semester.
George is taking history this semester.

(c) It is not true that the number 3 is a whole number.
The number three is not a whole number.

(d) It is not true that seven times six is equal to fifty.
Seven times six is not equal to fifty.

9. (a)

p	q	$\sim p$	$\sim q$	$\sim p \longrightarrow q$	$\sim q \longrightarrow p$	$(p \longrightarrow q) \longleftrightarrow (\sim q \longrightarrow \sim p)$
T	T	F	F	T	T	T
T	F	F	T	T	T	T
F	T	T	F	T	T	T
F	F	T	T	F	F	T

The statements are equivalent.

(b)

p	q	$\sim q$	$p \longrightarrow q$	$\sim(p \longrightarrow q)$	$p \wedge \sim q$	$\sim(p \longrightarrow q) \longleftrightarrow (p \wedge \sim q)$
T	T	F	T	F	F	T
T	F	T	F	T	T	T
F	T	F	T	F	F	T
F	F	T	T	F	F	T

The statements are equivalent.

11. (a) $A \cup D = \{4, 7, 10, 12, 14\}$
$A \cup D \neq \{4, 7, 10, 12, 14\}$

(b) $6 \cdot 3 < 22 \qquad 6 \cdot 3 \nless 22$

(c) $D \subseteq E \qquad D \nsubseteq E$

(d) $11 \notin F \qquad 11 \in F$

13. (a) Converse: If two lines are not perpendicular, they are parallel.
Inverse: If two lines are not parallel, they are perpendicular.
Contrapositive: If two lines are perpendicular, they are not parallel.

(b) Converse: If my travel expenses are lower, then I joined a car pool.
Inverse: If I do not join a car pool, then my travel expenses will not be lower.
Contrapositive: If my travel expenses are not lower, then I did not join a car pool.

15. Equation: $4(x + 2) = 32$ Solution set: $\{6\}$

17. (a)

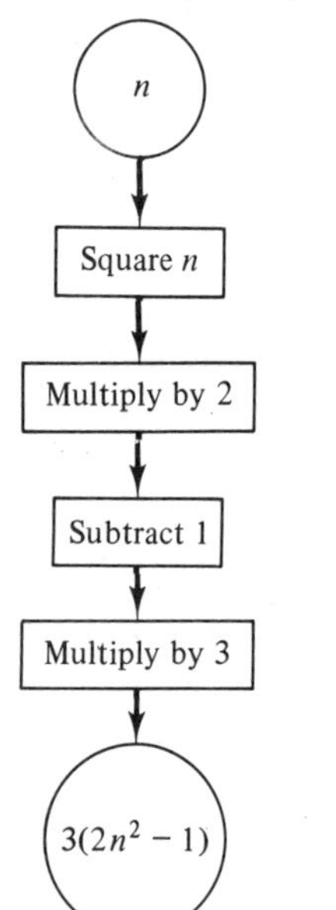

(b)

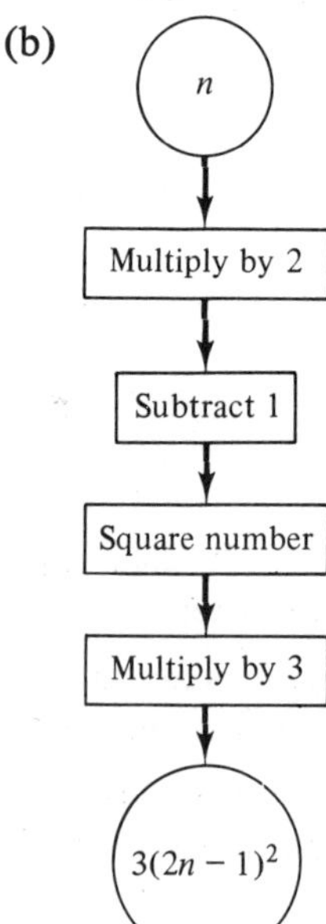

19. (a) An hour glass is an analog computer because it measures the sand to determine time.

(b) A mileage gauge is a digital computer because it counts the miles traveled.

21. Arithmetic unit: performs the computations.
Memory unit: stores the data and the results.
Control unit: acts as supervisor transferring information from one unit to another.

23. (a) $Y = 2A + B^3$ (b) $X = \frac{4A - C}{3}$ (c) $X \leq 2B^3$

(d) $Y = 3B + \frac{7D}{4}$ (e) $X = 3(A^3 + B^2)$ (f) $Y = \frac{B}{2} + 3C$

25. (a) Input/output statement (b) Control statement (c) Arithmetic statement

27. (a) The computer would cube A and cube B and let X equal the sum of these two results.

(b) The computer would compare 20 with C and would print the larger numerical value.

(c) The computer would compare B and 7. If B is greater than or equal to 7, the computer would go to statement 65. If B is less than 7, the computer would go to the next statement (63).

(d) The computer would square X, multiply that result by 3, and let that final product equal Y.

29.
```
80  PRINT "THE AREA IS"; A ↑ 2
```

CHAPTER 3: PRACTICE SETS—THE NATURAL NUMBERS

1. (a) $n(D) = 3$; $n(E) = 5$; $n(D \cup E) = 8$; So, $n(D) + n(E) = n(D \cup E)$

$$3 + 5 = 8$$

(b) If $n(D) = d$ and $n(E) = e$, then the binary operation of addition, $d + e$, produces one and only one element, f, that is the $n(D \cup E)$:

$$n(D) + n(E) = n(D \cup E)$$
$$d + e = f$$

2. (a) $T \times V = \{(g, r), (g, s), (h, r), (h, s), (k, r), (k, s)\}$

Now, $n(T) = 3$, $n(V) = 2$, and $n(T \times V) = 6$

Then $n(T) \times n(V) = n(T \times V)$

$$3 \times 2 = 6$$

(b) If $n(T) = t$ and $n(V) = v$, then the binary operation of multiplication, $t \times v$, produces one and only one element, u, which is the $n(T \times V)$:

$$n(T) \times n(V) = n(T \times V)$$
$$t \times v = u$$

3. (a) C, + (b) A, + (c) C, + (d) C, + (e) C, + (f) A, +
(g) A, +
(h) C, + (See Example 3c.)
(i) C, + (See Example 3b.)
(j) C, + (See Example 3d.)
(k) A, + (See Example 3d.)
(l) C, +
(m) A, + (See Example 3d.)
(n) A, + (See Example 4d.)

4. (a) C, +
A, +

(b) C, +
A, +

(c) A, +
C, +
A, +

(d) C, +
A, +
C, +

(e) A, +
A, +
C, +
A, +
A, +

(f)
$$\begin{aligned}(1+7)+9 &= 1+(7+9) && \text{A}, + \\ &= 1+(9+7) && \text{C}, +\end{aligned}$$

(g)
$$\begin{aligned}(3+5)+2 &= 3+(5+2) && \text{A}, + \\ &= 3+(2+5) && \text{C}, +\end{aligned}$$

(h)
$$\begin{aligned}(7+9)+3 &= 7+(9+3) && \text{A}, + \\ &= 7+(3+9) && \text{C}, + \\ &= (7+3)+9 && \text{A}, +\end{aligned}$$

OR

$$\begin{aligned}(7+9)+3 &= 3+(7+9) && \text{C}, + \\ &= (3+7)+9 && \text{A}, + \\ &= (7+3)+9 && \text{C}, +\end{aligned}$$

(i)
$$\begin{aligned}4+(2+6) &= (2+6)+4 && \text{C}, + \\ &= 2+(6+4) && \text{A}, + \\ &= 2+(4+6) && \text{C}, +\end{aligned}$$

OR

$$\begin{aligned}4+(2+6) &= (4+2)+6 && \text{A}, + \\ &= (2+4)+6 && \text{C}, + \\ &= 2+(4+6) && \text{A}, +\end{aligned}$$

(j)
$$\begin{aligned}(8+1) + (2+9) &= [(8+1)+2]+9 && \text{A}, + \\ &= [8+(1+2)]+9 && \text{A}, + \\ &= [8+(2+1)]+9 && \text{C}, + \\ &= [(8+2)+1]+9 && \text{A}, + \\ &= (8+2) + (1+9) && \text{A}, +\end{aligned}$$

OR

$$\begin{aligned}(8+1) + (2+9) &= 8+[1+(2+9)] && \text{A}, + \\ &= 8+[(1+2)+9] && \text{A}, + \\ &= 8+[(2+1)+9] && \text{C}, + \\ &= 8+[2+(1+9)] && \text{A}, + \\ &= (8+2) + (1+9) && \text{A}, +\end{aligned}$$

5. (a) C, $\times$ (b) A, $\times$
(c) C, $\times$ (d) C, $\times$
(e) C, $\times$ (See Example 5b.) (f) C, $\times$ (See Example 5c.)
(g) A, $\times$ (h) C, $\times$
(i) A, $\times$

6. (a) $(5\cdot7)2 = (7\cdot5)2$ C, $\times$
$= 7(5\cdot2)$ A, $\times$

(b) $3(4\cdot8) = 3(8\cdot4)$ C, $\times$
$= (3\cdot8)4$ A, $\times$

(c) $(6\cdot9)5 = (9\cdot6)5$ C, $\times$
$= 9(6\cdot5)$ A, $\times$

(d) $(8\cdot7)5 = 8(7\cdot5)$ A, $\times$
$= 8(5\cdot7)$ C, $\times$
$= (8\cdot5)7$ A, $\times$
or $(8\cdot7)5 = 5(8\cdot7)$ C, $\times$
$= (5\cdot8)7$ A, $\times$
$= (8\cdot5)7$ C, $\times$

(e) $6(2\cdot7) = (2\cdot7)6$ C, $\times$
$= 2(7\cdot6)$ A, $\times$
$= 2(6\cdot7)$ C, $\times$
or $6(2\cdot7) = (6\cdot2)7$ A, $\times$
$= (2\cdot6)7$ C, $\times$
$= 2(6\cdot7)$ A, $\times$

(f) $25(3\cdot4) = (3\cdot4)25$ C, $\times$
$= 3(4\cdot25)$ A, $\times$
$= 3(25\cdot4)$ C, $\times$
or $25(3\cdot4) = (25\cdot3)4$ A, $\times$
$= (3\cdot25)4$ C, $\times$
$= 3(25\cdot4)$ A, $\times$

7. (a) D, $\times$ over $+$ (b) C, $+$ (c) C, $\times$ (d) C, $\times$ (e) D, $\times$ over $+$
(f) C, $+$

8. (a) $5(8+1) = 5\cdot8 + 5\cdot1$ D, $\times$ over $+$
$= 5\cdot8 + 5$ I, $\times$

(b) $6(5+1) = 6\cdot5 + 6\cdot1$ D, $\times$ over $+$
$= 6\cdot5 + 6$ I, $\times$

CHAPTER 3: EXERCISES—THE NATURAL NUMBERS

1. (a) Elements (b) Set of natural numbers
Operations — Addition and subtraction
Relation — Equality (or equals) relation
Properties — Closure, $+$ and $\times$; commutative, $+$ and $\times$; associative, $+$ and $\times$; distributive, $\times$ over $+$; identity, $\times$

(c) An operation that combines (or associates) two elements

3. (a) $n(M) + n(R) = n(M \cup R)$
$6 + 2 = 8$

(b) If $n(M) = m$ and $n(R) = r$, then the binary operation of addition, $m + r$, produces one and only one element, z, that is the $n(M \cup R)$:

$n(M) + n(R) = n(M \cup R)$
$m + r = z$

5. (a) $U \times T = \{(b, 1), (b, r), (b, 3), (b, z), (2, 1), (2, r), (2, 3), (2, z), (y, 1), (y, r), (y, 3), (y, z)\}$

Now $n(U) = 3$, $n(T) = 4$, and $n(U \times T) = 12$

Then $n(U) \times n(T) = n(U \times T)$
$3 \times 4 = 12$

(b) If $n(U) = u$ and $n(T) = t$, then the binary operation of multiplication, $u \times t$, produces one and only one element, v, which is the $n(U \times T)$:

$n(U) \times n(T) = n(U \times T)$
$u \times t = v$

7. (a) $B \times A = \{(w, g), (w, h), (x, g), (x, h), (y, g), (y, h), (z, g), (z, h)\}$

Now $n(B) = 4$, $n(A) = 2$, and $n(B \times A) = 8$

Then $n(B) \times n(A) = n(B \times A)$

Or $4 \times 2 = 8$

(b) No.

9. Examples are: (a) $7 - 4 \neq 4 - 7$ (b) $6 \div 3 = 3 \div 6$

11. $42 \neq 70$; no property exists

13. (a) No; no identity element for multiplication
(b) No; not closed under addition
(c) No; no identity element for multiplication

15. Other sequences of steps may also be possible:

(a) C, +
A, +

(b) C, ×
A, ×

(c) A, +
C, +
A, +

(d) A, ×
C, ×
A, ×

(e) D, × over +
I, ×

(f) A, +
A, +
C, +
A, +
A, +

(g) C, +
A, +
C, +

(h) D, × over +
I, ×

(i) C, ×
A, ×

(j) A, +
A, +
C, +
A, +
A, +

(k) C, +
A, +

(l) A, ×
C, ×
A, ×

CHAPTER 3: PRACTICE SETS—THE WHOLE NUMBERS

1. (a) Here $n(P) = 3$, $n(Z) = 0$, and $n(P \cup Z) = 3$.

Then, $n(P) + n(Z) = n(P \cup Z)$
$3 + 0 = 3$

(b) If $n(P) = p$ and $n(Z) = 0$, then the binary operation of addition, $p + 0$, produces one and only one element, p, which is the $n(P \cup Z)$:

$$\begin{aligned} n(P) + n(Z) &= n(P \cup Z) \\ p \ + \ 0 &= \ p \end{aligned}$$

2. (a) Now, $n(P) = 3$, $n(Z) = 0$, and $n(P \times Z) = 0$.

Then, $$\begin{aligned} n(P) \times n(Z) &= n(P \times Z) \\ 3 \ \times \ 0 &= \ 0 \end{aligned}$$

(b) If $n(P) = p$ and $n(Z) = 0$, then the binary operation of multiplication, $p \times 0$, produces one and only one element, 0, which is the $n(P \times Z)$. That is,

$$\begin{aligned} n(P) \times n(Z) &= n(P \times Z) \\ p \ \times \ 0 &= \ 0 \end{aligned}$$

CHAPTER 3: EXERCISES—THE WHOLE NUMBERS

1. (a) Now $n(D) = 4$, $n(Z) = 0$, and $n(D \cup Z) = 4$

So, $n(A) + n(Z) = n(D \cup Z)$

or $4 \ + \ 0 \ = \ 4$

(b) If $n(D) = d$ and $n(Z) = 0$, then the binary operation of addition, $d + 0$, produces one and only one element, d, which is the $n(D \cup Z)$. That is,

$$\begin{aligned} n(D) + n(Z) &= n(D \cup Z) \\ d \ + \ 0 &= \ d \end{aligned}$$

3. (a) Since $D \times Z = \varnothing$, then $n(D) = 4$, $n(Z) = 0$, and $n(D \times Z) = 0$

So, $$\begin{aligned} n(D) \times n(Z) &= n(D \times Z) \\ 4 \ \times \ 0 &= \ 0 \end{aligned}$$

(b) If $n(D) = d$ and $n(Z) = 0$, then the binary operation of multiplication, $d \times 0$, produces one and only one element, 0, which is the $n(D \times Z)$. That is,

$$\begin{aligned} n(D) \times n(Z) &= n(D \times Z) \\ d \ \times \ 0 &= \ 0 \end{aligned}$$

5. The set of whole numbers includes the element zero; it is the union of zero with the set of natural numbers: $W = N \cup Zero$. The set of natural numbers is a proper subset of the set of whole numbers: $N \subset W$

7. (a) and (b) If r and s are whole numbers, then $(r + s)$ and $(r \cdot s)$ are whole numbers.

6	⟵ is a whole number ⟶	0
+0	⟵ is a whole number ⟶	×4
6	⟵ is a whole number ⟶	0

(c) $r + s = s + r$
$13 = 13$

(d) $(r + s) + t = r + (s + t)$
$14 = 14$

(e) $r \cdot s = s \cdot r$
$21 = 21$

(f) $r(s \cdot t) = (r \cdot s)t$
$64 = 64$

(g) $r(s + t) = r \cdot s + r \cdot t$
$48 = 48$

(h) $r + 0 = 0 + r = r$
$9 = 9 = 9$

(i) $r \cdot 1 = 1 \cdot r = r$
$4 = 4 = 4$

9. (a) Yes (b) No

11. (a) D, × over + (b) A, + (c) C, × (d) C, ×; A, ×

(e) I, × (f) C, +; A, +; C, + (g) C, + (h) A, ×

(i) D, × over +; I, ×; I, + (j) A, ×; C, ×; A, × (k) C, × (l) C, +; A, +

13. (a) Because it is defined in terms of addition.

(b) No. Since $N \subset W$, subtraction is not closed for the set of natural numbers because it is not closed for the set of whole numbers.

15. (a) $r(s - t) = r \cdot s - r \cdot t$
(b) $36 = 36$

17. (a) Because it is defined in terms of multiplication.
(b) Yes, because every natural number is a whole number.
(c) The restriction against division by zero (since $0 \notin$ N).
(d) Because a whole number q such that $a = b \cdot q$ often does *not exist.*

19. Examples may vary: (a) No (b) No (c) No (d) No (e) No

21. (a) $(r + s) \div t = r \div t + s \div t$
$5 = 5$

(b) $(r - s) \div t = r \div t - s \div t$
$1 = 1$

CHAPTER 4: PRACTICE SETS

1. (a)

(b)
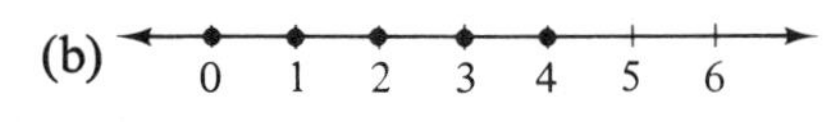

(c)
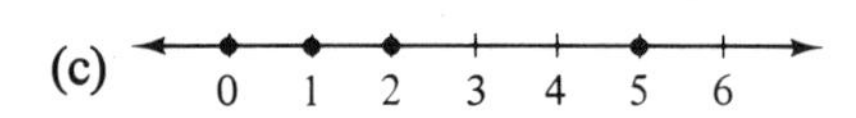

(d)
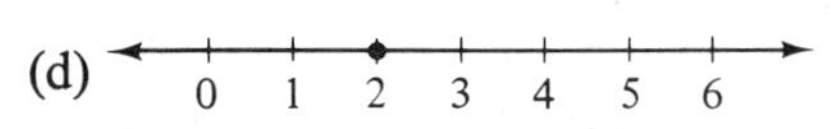

2. For some $k \neq 0$, either $8 + k = 3$ or $3 + k = 8$.
When $k = 5$, we have $3 + 5 = 8$
Thus, if $3 + 5 = 8$, then $3 < 8$

3. (a) $<$ (b) $=$ (c) $>$ (d) $>$
(e) $=$ (f) $<$ (g) $7 \geq 6$; T (h) $5 \leq 5$; T
(i) $9 \leq 8$; F (j) $6 \geq 6$; T (k) $4 \geq 5$; F (l) $5 \leq 8$; T

4. (a) $0 < 5$ (b) $0 > -3$ (c) $-2 < 2$ (d) $3 > -4$
(e) $-7 < -1$ (f) $-3 > -6$

(g) −6 −5 −4 −3 −2 −1 0 1 2 3 4 5 6

(h) −6 −5 −4 −3 −2 −1 0 1 2 3 4 5 6

(i) −6 −5 −4 −3 −2 −1 0 1 2 3 4 5 6

(j) −6 −5 −4 −3 −2 −1 0 1 2 3 4 5 6

(k) −6 −5 −4 −3 −2 −1 0 1 2 3 4 5 6

5. (a) −2 (b) 3 (c) 0
(d) −7 (e) 4 (f) 0

6. (a) Yes (b) Yes (c) No
(d) Yes (e) No (f) Yes

7. (a) 9 (b) −8 (c) 3
(d) −3 (e) 6 (f) −2

8. (a) −40 (b) −36 (c) 42 (d) 0
(e) 0 (f) 24 (g) −14 (h) −6
(i) −32 (j) 12 (k) 9 (l) −28

9. (a) −2 (b) 6 (c) −4 (d) −13
(e) −8 (f) −3 (g) 10 (h) 6
(i) −2 (j) −11 (k) 9 (l) 7
(m) −5 (n) 13 (o) −13 (p) 5

10. (a) −4 (b) −2 (c) 6 (d) 0
(e) −1 (f) −2 (g) 3 (h) −2

CHAPTER 4: EXERCISES

1. (a) Yes; 1. No. (b) Yes; 0. No. (c) No. No.

3. (a) Yes. (b) Yes. (c) Yes.
(d) No; negative integers are not whole numbers.
(e) No; integer zero is neither positive nor negative.
(f) No; whole number zero is not positive.
(g) Yes.
(h) Yes; the two sets are exactly the same.

5. (a) Point
(b) No; there are an infinite number of points on the number line besides those points which match with integers.
(c) larger; smaller.

7. (a) $a < b$ (b) $b > a$
(c) One; $a = b$, $a < b$, or $a > b$ (d) $a < b$ or $a > b$
(e) $a > b$ (f) $b < c$

9. (a) $-1 < 4$ (b) $0 > -2$ (c) $-8 < -3$
(d) $-2 > -6$ (e) $4 > -8$ (f) $>$, since $6 > 5$
(g) $=$, since $-3 = -3$ (h) $<$, since $-5 < 0$
(i) $>$, since $5 > -3$ (j) $=$, since $2 = 2$

11. (a) -4 (b) 0 (c) 1 (d) -5
(e) -2 (f) 4 (g) 0 (h) 7
(i) 5 (j) -4 (k) -3 (l) -4

13. (a) 6 (b) -10 (c) 4 (d) -4
(e) -2 (f) 5

15. (a) Yes (b) Yes (c) Yes (d) No

17. (a) Yes; a negative integer plus a negative integer always equals a negative integer.
(b) No; a negative integer times a negative integer equals a positive integer.

19. Examples will vary.
(a) Yes (b) No (c) No (d) Yes

21. (a) No (b) No (c) Either $a = b$ or else a and b are additive inverses.

***23.** Sequence of steps: commutative, $+$; associative, $+$; associative, $+$; inverse, $+$; identity, $+$; inverse, $+$; renaming $7 + 4$; thus $-7 + (-4) = -11$, by inverse, $+$.

***25.** Sequence of steps: distributive, $\times$ over $+$; inverse, $+$; renaming (4×0); renaming (4×3); thus $(4 \times {}^{-}3) = -12$, by inverse, $+$.

CHAPTER 5: PRACTICE SETS

1. (a) Yes; $16 = 16$ (b) No; $75 \neq 80$
(c) Yes; $816 = 816$ (d) No; $792 \neq 825$

2. (a) $\frac{4}{5}$ (b) $\frac{-3}{5}$ (c) $\frac{2}{3}$
(d) $\frac{1}{10}$ (e) $\frac{1}{-32}$ (f) $\frac{1}{12}$
(g) $\frac{8}{45}$ (h) $\frac{9}{20}$ (i) $\frac{-6}{35}$

3. (a) $\frac{5}{1} \cdot \frac{1}{5} = 1$ (b) $\frac{-8}{1} \cdot \frac{1}{-8} = 1$ (c) $\frac{1}{2} \cdot \frac{2}{1} = 1$

(d) $\frac{1}{-4} \cdot \frac{-4}{1} = 1$ (e) $\frac{3}{5} \cdot \frac{5}{3} = 1$ (f) $\frac{8}{5} \cdot \frac{5}{8} = 1$

(g) $\frac{-3}{7} \cdot \frac{7}{-3} = 1$ (h) $\frac{4}{-3} \cdot \frac{-3}{4} = 1$

4. (a) $\frac{3}{4}$ (b) -10 (c) $\frac{1}{-6}$

(d) $\frac{8}{9}$ (e) $\frac{3}{-10}$ (f) $\frac{4}{3}$

(g) Not permitted: $\frac{c}{d} = 0$ (h) 0 (i) $\frac{14}{15}$

5. (a) $\frac{5}{7}$ (b) $\frac{1}{5}$ (c) $\frac{-5}{9}$

(d) $\frac{11}{15}$ (e) $\frac{-11}{35}$ (f) $\frac{-20}{21}$

(g) $\frac{13}{-15}$ (h) $\frac{31}{-35}$ (i) $\frac{-11}{15}$

6. (a) $\frac{2}{5}$ (b) $\frac{-3}{7}$ (c) $\frac{-5}{9}$

(d) $\frac{1}{14}$ (e) $\frac{9}{20}$ (f) $\frac{24}{35}$

(g) $\frac{-19}{28}$ (h) $\frac{8}{45}$ (i) $\frac{-17}{12}$

7. (a) $\frac{3}{26}$ (b) $\frac{5}{52}$ (c) $\frac{15}{34}$

(d) $\frac{29}{68}$ (e) $\frac{7}{12}$ (f) $\frac{53}{112}$

(g) $\frac{31}{40}$

CHAPTER 5: EXERCISES

1. (a) A rational number (b) Yes (c) No
(d) No (e) Yes (f) No; no
(g) No. Despite how small was the positive rational number named, by the density property another rational could be found between 0 and that number.
(h) Yes; 0 (i) An infinite number
(j) By the density property, between any two rational numbers there is always another rational number.
(k) Yes

(l) No; the number line contains an infinite number of other points which do not match with any rational number.

3. (a) $1 = \frac{1}{1}$

(b) $0 = \frac{0}{1}$ or $\frac{0}{a}$ for any integer $a \neq 0$.

(c) Yes.

(d) No; by definition, if $\frac{a}{b}$ is rational then $b \neq 0$.

(e) Yes.

(f) No; any integer except zero may be a value for b.

(g) No; if $\frac{b}{a} = 0$, then $b = 0$. But b cannot be zero or else $\frac{a}{b}$ would not have been a rational number.

(h) No; $\frac{a}{a} = 1$ only when $a \neq 0$.

(i) If $\frac{a}{b}$ represents a nonzero integer then $b = 1$.

(j) Yes.

(k) No; division by zero is not permitted.

(l) No; if $c = 0$ then $\frac{c}{d} = 0$ and this would be division by zero, which is not allowed.

(m) No; if $d = 0$, then $\frac{c}{d}$ would not be a rational number.

5. (a) Definition of addition
(b) Renaming $y \times 0$
(c) Identity, +
(d) Definition of multiplication
(e) Renaming $\frac{z}{z}$
(f) Identity, x

7. Sequence of steps: definition of multiplication; commutative, $\times$; definition of multiplication; renaming $\frac{x}{x}$ and $\frac{y}{y}$; identity, $\times$.

9. Sequence of steps: definition of multiplication, $a \times \frac{1}{b}$; associative, $\times$; inverse, $\times$; identity $\times$.

11. (a) Definition of addition
(b) Commutative, $\times$
(c) Distributive, $\times$ over +
(d) Definition of multiplication, $\frac{a}{b} \cdot \frac{c}{d}$
(e) Renaming $\frac{c}{c}$
(f) Identity, $\times$

13. (a) Definition of subtraction, $\frac{a}{c} - \frac{b}{c}$
(b) Definition of addition, $\frac{a}{c} + \frac{b}{c}$
(c) Associative, +
(d) Inverse, +
(e) Identity, +

15.

Number	Naturals	Wholes	Integers	Rationals
(a) -3			✓	✓
(b) $\frac{-6}{7}$				✓
(c) 0		✓	✓	✓
(d) -2			✓	✓
(e) 12	✓	✓	✓	✓
(f) $\frac{7}{10}$				✓
(g) 6	✓	✓	✓	✓
(h) $\frac{3}{4}$				✓
(i) $\frac{8}{5}$				✓
(j) -1			✓	✓
(k) 2	✓	✓	✓	✓

17.

Property	Naturals	Wholes	Integers	Rationals
Closure, +	✓	✓	✓	✓
Closure, ×	✓	✓	✓	✓
Commutative, +	✓	✓	✓	✓
Commutative, ×	✓	✓	✓	✓
Associative, +	✓	✓	✓	✓
Associative, ×	✓	✓	✓	✓
Distributive, × over +	✓	✓	✓	✓
Identity, +		✓	✓	✓
Identity, ×	✓	✓	✓	✓
Inverse, +			✓	✓
Inverse, ×				✓

19. (a) Yes; (+) plus (+) equals (+)
(b) No; (+) minus (+) may equal (−)
(c) Yes; (+) times (+) equals (+)
(d) Yes; (+) divided by (+) equals (+)
(e) Yes; (−) plus (−) equals (−)
(f) No; (−) minus (−) may equal (+)
(g) No; (−) times (−) equals (+)
(h) No; (−) divided by (−) equals (+)

21. (a) $\frac{-4}{7}$ (b) $\frac{1}{-8}$ (c) $\frac{1}{12}$ (d) $\frac{8}{-21}$

(e) $\frac{12}{35}$ (f) $\frac{18}{-49}$ (g) $\frac{0}{14}$ or 0 (h) $\frac{gj}{hk}$

23. (a) $\frac{4}{9}$ (b) $\frac{1}{-10}$ (c) $\frac{15}{-16}$ (d) $\frac{0}{12}$ or 0

(e) $\frac{-20}{-21} = \frac{20}{21}$ (f) $\frac{-18}{55}$ (g) $\frac{-35}{36}$ (h) $\frac{ps}{qr}$

25. (a) $\frac{2}{9}$ (b) $\frac{4}{5}$ (c) $\frac{-17}{20}$ (d) $\frac{-7}{24}$

(e) $\frac{1}{-21}$ (f) $\frac{9}{40}$ (g) $\frac{8}{35}$

(h) $\frac{wz + x(-y)}{xz}$ (i) $\frac{w(z + x)}{xz}$

27. (a) $\frac{7}{24}$ (b) $\frac{37}{48}$ (c) $\frac{23}{40}$ (d) $\frac{103}{180}$

***29.** Sequence of steps: state $\frac{w}{x} \div \frac{y}{z}$ using definition of division for integers; multiply $\frac{z}{y}$ times both sides, by closure, x; associative, $\times$; inverse, $\times$; identity, $\times$; commutative, $\times$; thus q equal to both operations implies they are equal to each other (the desired conclusion).

***31.** Sequence of steps: definition of subtraction (integers); definition of multiplication, $a \times \frac{1}{b}$; commutative, $\times$; distributive, $\times$ over $+$; commutative, $\times$; definition of multiplication, $a \times \frac{1}{b}$.

CHAPTER 6: PRACTICE SETS

1. (a) $D_{20} = \{1, 2, 4, 5, 10, 20\}$ and $D_{30} = \{1, 2, 3, 5, 6, 10, 15, 30\}$
(b) 1; 10
(c) $D_{20} \cap D_{30} = \{1, 2, 5, 10\} = D_{10}$

2. (a) $D_8 = \{1, 2, 4, 8\}$
$D_{12} = \{1, 2, 3, 4, 6, 12\}$
GCD(8, 12) = 4

(b) $D_{16} = \{1, 2, 4, 8, 16\}$
$D_{24} = \{1, 2, 3, 4, 6, 8, 12, 24\}$
GCD(16, 24) = 8

(c) $D_{18} = \{1, 2, 3, 6, 9, 18\}$
$D_{27} = \{1, 3, 9, 27\}$
GCD(18, 27) = 9

(d) $D_{26} = \{1, 2, 13, 26\}$
$D_{35} = \{1, 5, 7, 35\}$
GCD(26, 35) = 1
Thus, 26 and 35 are relatively prime.

3. (a) $M_5 = \{5, 10, 15, 20, 25, 30, \ldots\}$; 10.
(b) $M_2 \cap M_5 = \{10, 20, 30, 40, \ldots\} = M_{10}$.
(c) Any number that is a multiple of both 2 and 5 is also a multiple of 10.

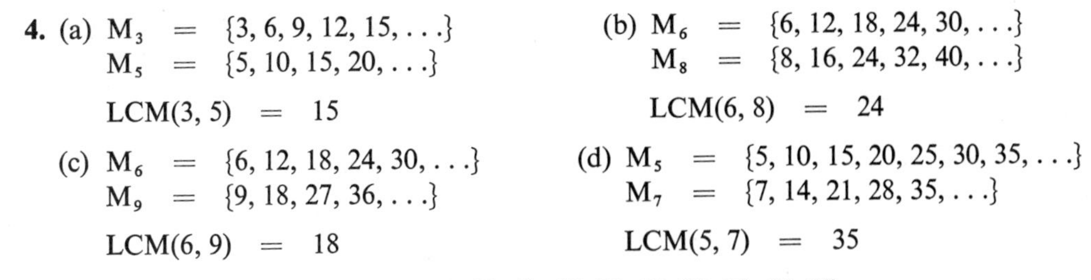

4. (a) $M_3 = \{3, 6, 9, 12, 15, \ldots\}$
$M_5 = \{5, 10, 15, 20, \ldots\}$

LCM(3, 5) = 15

(b) $M_6 = \{6, 12, 18, 24, 30, \ldots\}$
$M_8 = \{8, 16, 24, 32, 40, \ldots\}$

LCM(6, 8) = 24

(c) $M_6 = \{6, 12, 18, 24, 30, \ldots\}$
$M_9 = \{9, 18, 27, 36, \ldots\}$

LCM(6, 9) = 18

(d) $M_5 = \{5, 10, 15, 20, 25, 30, 35, \ldots\}$
$M_7 = \{7, 14, 21, 28, 35, \ldots\}$

LCM(5, 7) = 35

5b. (a) 10. (b) Primes > 50: 53, 59, 61, 67, 71, 73, 79, 83, 89, 97
(c) 3 and 5; 5 and 7; 11 and 13; 17 and 19; 29 and 31; 41 and 43; 59 and 61; 71 and 73.
(d) 13 and 31; 17 and 71; 37 and 73; 79 and 97.

6. (a) $8 = 2^3$ (b) $10 = 2 \times 5$ (c) $21 = 3 \times 7$
(d) $18 = 2 \times 3^2$ (e) $30 = 2 \times 3 \times 5$ (f) $36 = 2^2 \times 3^2$

7. (a) GCD(15, 35) = 5 (b) GCD(12, 20) = 2^2 = 4
(c) GCD(30, 70) = $2 \cdot 5$ = 10 (d) GCD(36, 90) = $2 \cdot 3^2$ = 18
(e) GCD(300, 360) = $2^2 \cdot 3 \cdot 5$ = 60 (f) Relatively prime

8. (a) $\frac{3}{5}$ (b) $\frac{2}{5}$ (c) $\frac{2}{3}$ (d) $\frac{3}{4}$ (e) $\frac{7}{9}$ (f) $\frac{9}{14}$

9. (a) LCM(10, 14) = $2 \cdot 5 \cdot 7$ = 70 (b) LCM(8, 12) = $2^3 \cdot 3$ = 24
(c) LCM(18, 24) = $2^3 \cdot 3^2$ = 72 (d) LCM(30, 45) = $2 \cdot 3^2 \cdot 5$ = 90
(e) LCM(56, 60) = $2^3 \cdot 3 \cdot 5 \cdot 7$ = 840
(f) LCM(9, 16) = $2^4 \cdot 3^2$ = 144

10. (a) $\frac{7}{12}$ (b) $\frac{4}{15}$ (c) $\frac{13}{24}$ (d) $\frac{5}{8}$ (e) $\frac{2}{3}$ (f) $\frac{34}{45}$ (g) $\frac{4}{15}$

(h) $\frac{31}{35}$ (i) $\frac{23}{30}$

11. (a) 12 (b) 24 (c) 45 (d) 30

12. Add using even $2k$ plus odd $(2p + 1)$. Sequence of steps: A, +; D, × over +; closure, +; substitution; conclusion.

CHAPTER 6: EXERCISES

1. (a) Finite (b) Infinite (c) Infinite (d) Finite
(e) Infinite (f) Infinite (g) Infinite

3. (a) No; natural number 1 is neither prime nor composite.
(b) One—itself
(c) Since no prime has any prime divisor except itself, primes cannot be expressed as a "product" of primes.

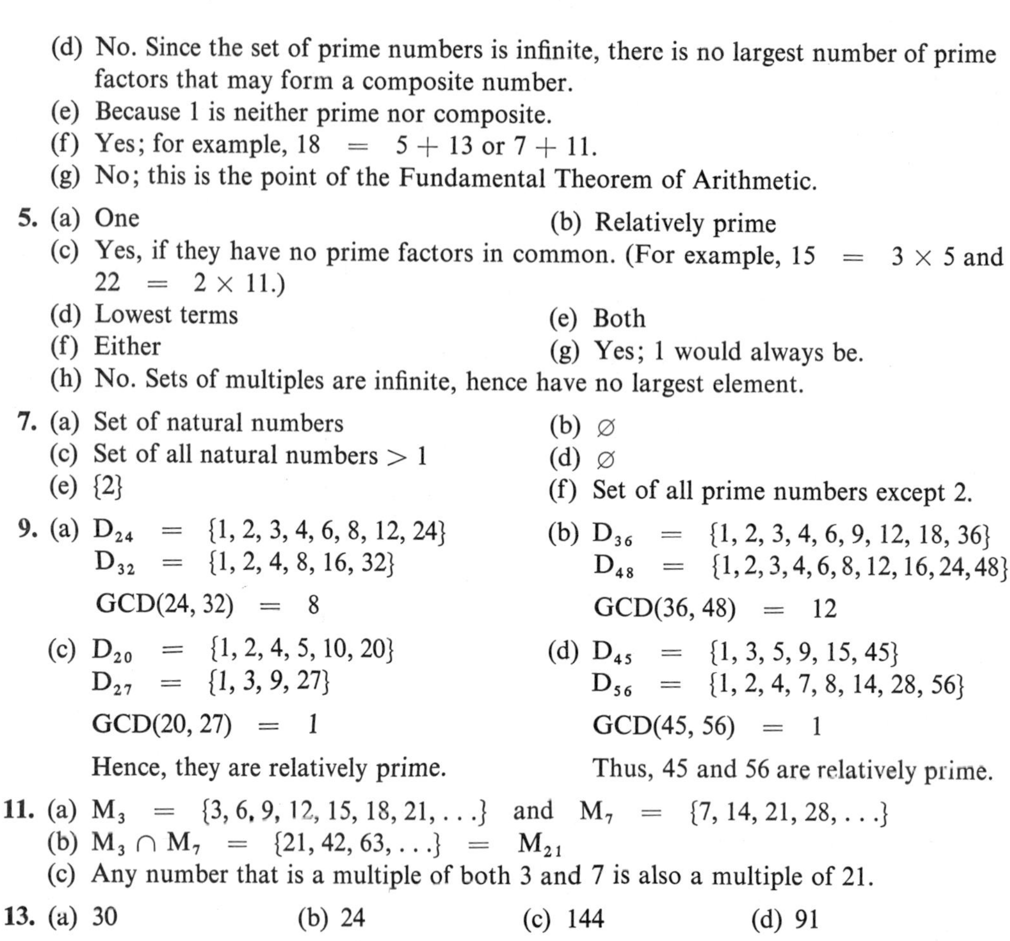

(d) No. Since the set of prime numbers is infinite, there is no largest number of prime factors that may form a composite number.
(e) Because 1 is neither prime nor composite.
(f) Yes; for example, $18 = 5 + 13$ or $7 + 11$.
(g) No; this is the point of the Fundamental Theorem of Arithmetic.

5. (a) One
(b) Relatively prime
(c) Yes, if they have no prime factors in common. (For example, $15 = 3 \times 5$ and $22 = 2 \times 11$.)
(d) Lowest terms
(e) Both
(f) Either
(g) Yes; 1 would always be.
(h) No. Sets of multiples are infinite, hence have no largest element.

7. (a) Set of natural numbers
(b) $\varnothing$
(c) Set of all natural numbers > 1
(d) $\varnothing$
(e) $\{2\}$
(f) Set of all prime numbers except 2.

9. (a) $D_{24} = \{1, 2, 3, 4, 6, 8, 12, 24\}$
$D_{32} = \{1, 2, 4, 8, 16, 32\}$
$GCD(24, 32) = 8$

(b) $D_{36} = \{1, 2, 3, 4, 6, 9, 12, 18, 36\}$
$D_{48} = \{1, 2, 3, 4, 6, 8, 12, 16, 24, 48\}$
$GCD(36, 48) = 12$

(c) $D_{20} = \{1, 2, 4, 5, 10, 20\}$
$D_{27} = \{1, 3, 9, 27\}$
$GCD(20, 27) = 1$
Hence, they are relatively prime.

(d) $D_{45} = \{1, 3, 5, 9, 15, 45\}$
$D_{56} = \{1, 2, 4, 7, 8, 14, 28, 56\}$
$GCD(45, 56) = 1$
Thus, 45 and 56 are relatively prime.

11. (a) $M_3 = \{3, 6, 9, 12, 15, 18, 21, \ldots\}$ and $M_7 = \{7, 14, 21, 28, \ldots\}$
(b) $M_3 \cap M_7 = \{21, 42, 63, \ldots\} = M_{21}$
(c) Any number that is a multiple of both 3 and 7 is also a multiple of 21.

13. (a) 30 (b) 24 (c) 144 (d) 91

15. (a) Stop at 15. Primes > 100: 101, 103, 107, 109, 113, 127, 131, 137, 139, 149, 151, 157, 163, 167, 173, 179, 181, 191, 193, 197, 199
(b) 15; 10; 10; 11
(c) 101 and 103; 107 and 109; 137 and 139; 149 and 151; 179 and 181; 191 and 193; 197 and 199
(d) 101, 131, 181, 191
(e) Any number ending in a multiple of two is composite. Any number beginning with these digits would be a multiple of two when reversed.
(f) $\{3, 5, 7\}$. Given any other three consecutive odd numbers, one of them will be a multiple of 3 and hence not prime. (Three consecutive even numbers would be multiples of 2.)

17. (a) $GCD(42, 70) = 2 \cdot 7 = 14$
(b) $GCD(30, 75) = 3 \cdot 5 = 15$
(c) $GCD(54, 81) = 3^3 = 27$
(d) Relatively prime
(e) $GCD(90, 108) = 2 \cdot 3^2 = 18$
(f) $GCD(120, 252) = 2^2 \cdot 3 = 12$
(g) $GCD(360, 450) = 2 \cdot 3^2 \cdot 5 = 90$
(h) Relatively prime

19. (a) $\text{LCM}(14, 21) = 2 \cdot 3 \cdot 7 = 42$
(b) $\text{LCM}(15, 25) = 3 \cdot 5^2 = 75$
(c) $\text{LCM}(16, 21) = 2^4 \cdot 3 \cdot 7 = 336$
(d) $\text{LCM}(54, 72) = 2^3 \cdot 3^3 = 216$
(e) $\text{LCM}(24, 42) = 2^3 \cdot 3 \cdot 7 = 168$
(f) $\text{LCM}(75, 90) = 2 \cdot 3^2 \cdot 5^2 = 450$
(g) $\text{LCM}(48, 60) = 2^4 \cdot 3 \cdot 5 = 240$
(h) $\text{LCM}(50, 63) = 2 \cdot 3^2 \cdot 5^2 \cdot 7 = 3{,}150$

21. (a) 24 (b) 36 (c) 80 (d) 150

23. (a) Add using even numbers $2k$ plus $2p$. Sequence of steps: D, $\times$ over $+$; closure, $+$; substitution; conclusion.
(b) No; the sum of two odd numbers is even.

25. Multiply even number $2n$ times itself. Sequence of steps: A, $\times$; closure, $\times$; substitution; conclusion.

CHAPTER 7: PRACTICE SETS

1. (a) 0.8 (b) 0.375 (c) 0.65 (d) 0.3125
(e) $0.7\bar{7}$ (f) $0.416\bar{6}$ (g) 0.5454... (h) $0.\overline{285714}$

2. (a) $\frac{4}{5}$ (b) $\frac{7}{20}$ (c) $\frac{7}{8}$ (d) $\frac{18}{125}$ (e) $\frac{11}{40}$ (f) $\frac{7}{16}$

3. (a) $\frac{7}{9}$ (b) $\frac{61}{99}$ (c) $\frac{41}{333}$ (d) $\frac{13}{18}$ (e) $\frac{5}{12}$ (f) $\frac{8}{33}$

4. Possible solutions:
(a) 0.64644644464444...
(b) 2.543543354333...
(c) 0.141441444...
(d) $1.357357\underline{1}357\underline{2}357\underline{3}357\underline{4}\ldots$

5. (a) ≈ 2.23 (b) ≈ 5.38

6. (a) $c = 10$ (b) $c = \sqrt{13}$ (c) $a = 12$ (d) $b = \sqrt{7}$

7. (a) $6;\ 6^2 = 36$
(b) Irrational
(c) $3;\ 3^4 = 81$
(d) $9;\ 9^2 = 81$
(e) Not defined
(f) $3;\ 3^3 = 27$
(g) Irrational
(h) Irrational
(i) $-5;\ (-5)^3 = -125$
(j) Not defined
(k) Irrational
(l) $12;\ 12^2 = 144$

8. (a) $8 = 8$ implies $11 = 11$
(b) $8 = 8$ implies $24 = 24$
(c) $-4 < 2$ implies $-1 < 5$
(d) $-4 < 2$ implies $-12 < 6$
(e) $-4 < 2$ implies $12 - 6$

9.

	R	*S*	*T*	*Eqv.*
(a)	✓	✓	✓	Yes
(b)			✓	No
(c)	✓	✓		No

Examples should also be written out for each property.

10.

	R	*S*	*T*	*Eqv.*
(a)			✓	No
(b)	✓		✓	No
(c)	✓	✓	✓	Yes
(d)		✓		No

Examples should be written.

11. (a) 3 (b) $3i$ (c) $8i$ (d) $7i$
(e) $i\sqrt{3}$ (f) $i\sqrt{2}$ (g) $i\sqrt{11}$ (h) $i\sqrt{23}$
(i) -1 (j) -49 (k) -16 (l) -3
(m) ± 2 (n) $\pm 2i$ (o) $\pm 9i$ (p) $\pm i\sqrt{6}$

12. (a) $3 + 4i$ (b) $-8 + 1\sqrt{2}$ (c) $\frac{1}{2} - 9i$
(d) $\pi - i\sqrt{14}$ (e) $4 + 0i$ (f) $-2 + 0i$
(g) $0 + 0i$ (h) $\frac{-3}{8} + 0i$ (i) $0 + 6i$
(j) $\sqrt{15} + 0i$ (k) $0 + i\sqrt{22}$ (l) $0 + i\sqrt{7}$

CHAPTER 7: EXERCISES

1. (a) Ten (b) Rational numbers
(c) Power of ten (d) Quotient
(e) Dividing the numerator by the denominator.
(f) Terminating; nonterminating, repeating
(g) Terminating (h) Nonterminating, repeating
(i) Rational number (j) Appending repeating zeros
(k) $0.35 = 0.35\bar{0}$

3. (a) Yes; no; yes. Every square root is either an integer or an irrational number.
(b) Yes. The square root of a prime cannot be an integer because the only factors of a prime are itself and one.
(c) No. Some composites have square roots that are integers (ex: $\sqrt{36} = \pm 6$). But other composites have square roots that are irrational (ex: $\sqrt{24}$ is irrational).
(d) Yes.
(e) No. We cannot measure accurately enough to determine whether a length is irrational. We are always forced to substitute a rational number that approximately equals the value of any irrational length.
(f) By using the (irrational) hypotenuse of a right triangle as the radius of a circle and letting the circle intersect the number line at the point that corresponds to the irrational number.
(g) Rational number
(h) Yes.
(i) No. We know $3q$ is irrational, but we cannot compute it because we cannot finish writing q in order to multiply it by 3.
(j) If we measure C and d, then we obtain rational numbers that are only approximate. Thus, $\frac{C}{d}$ will give some rational value that approximately equals π, but will not exactly equal π.

5. (a) Closure, $+$ and $\times$
Commutative, $+$ and $\times$
Associative, $+$ and $\times$
Distributive, $\times$ over $+$
Identity, $+$ and $\times$
Inverse, $+$
Inverse, $\times$ (except 0)

(b) Yes
(c) No; only when $c \neq 0$
(d) Yes (if *first* $ac = bc$).
(e) No. If $a < b$, then $ac > bc$ when $c < 0$.
(f) Equivalence
(g) Yes
(h) No; $a \nless a$
(i) Yes

7. (a) 0.6 (b) 0.175 (c) 0.625 (d) 0.34375
(e) 0.7333... (f) $0.\overline{384615}$ (g) $0.\overline{783}$ (h) 0.121212...

9. (a) $\frac{8}{9}$ (b) $\frac{17}{30}$ (c) $\frac{8}{11}$ (d) $\frac{12}{37}$ (e) $\frac{3}{11}$ (f) $\frac{7}{12}$

11. Possible solutions: (a) 0.868668666... (b) 3.952952295222...
(c) 2.474774777... (d) $0.248248\underline{1}248\underline{2}...$

13. (a) $c = 15$ (b) $c = \sqrt{34}$ (c) $b = 7$ (d) $a = \sqrt{5}$

15. (a) 0.6666... is rational. (b) Yes; $6 = 6$ and $12 = 12$.
(c) $\sqrt{2} \cdot \sqrt{8} = \sqrt{16} = 4$, which is rational.
(d) No, because the irrational numbers are not closed under addition or multiplication.
(e) No; $7 \neq 5$; $14 \neq 10$; $5 \neq \sqrt{13}$

17.

Properties	*Natural numbers*	*Whole numbers*	*Integers*	*Rational numbers*	*Real numbers*
Closure, $+$	✓	✓	✓	✓	✓
Closure, $\times$	✓	✓	✓	✓	✓
Commutative, $+$	✓	✓	✓	✓	✓
Commutative, $\times$	✓	✓	✓	✓	✓
Associative, $+$	✓	✓	✓	✓	✓
Associative, $\times$	✓	✓	✓	✓	✓
Distributive, $\times$ over $+$	✓	✓	✓	✓	✓
Identity, $+$		✓	✓	✓	✓
Identity, $\times$	✓	✓	✓	✓	✓
Inverse, $+$			✓	✓	✓
Inverse, $\times$				✓	✓
Density				✓	✓
Completeness					✓

19.

	R	*S*	*T*	*Eqv.*
(a)	✓		✓	No
(b)			✓	No
(c)	✓	✓	✓	Yes
(d)		✓		No
(e)	✓		✓	No

21.

Number	*Natural numbers*	*Whole numbers*	*Integers*	*Rational numbers*	*Irrational numbers*	*Real numbers*	*Imaginary numbers*	*Complex numbers*
(a) -7			✓	✓		✓		✓
(b) 0		✓	✓	✓		✓		✓
(c) $\frac{3}{8}$				✓		✓		✓
(d) $\sqrt{5}$					✓	✓		✓
(e) $\sqrt{-9}$							✓	✓
(f) $3 + 2i$								✓
(g) $\sqrt[3]{7}$					✓	✓		✓
(h) 12	✓	✓	✓	✓		✓		✓
(i) $\frac{-7}{8}$				✓		✓		✓
(j) $0.\overline{12}$				✓		✓		✓
(k) π					✓	✓		✓
(l) 0.8				✓		✓		✓
(m) 0.020020002...					✓	✓		✓
(n) $3i$							✓	✓
(o) $\sqrt[3]{-27}$			✓	✓		✓		✓

CHAPTER 8: PRACTICE SETS

1. (a) $8{,}346 = (8 \times 10^3) + (3 \times 10^2) + (4 \times 10^1) + (6 \times 10^0)$
(b) $1{,}327 = (1 \times 10^3) + (3 \times 10^2) + (2 \times 10^1) + (7 \times 10^0)$
(c) $98{,}234 = (9 \times 10^4) + (8 \times 10^3) + (2 \times 10^2) + (3 \times 10^1) + (4 \times 10^0)$
(d) $470 = (4 \times 10^2) + (7 \times 10^1) + (0 \times 10^0)$

2. (a) $3405_{\text{five}} = (3 \times 5^3) + (4 \times 5^2) + (0 \times 5^1) + (5 \times 5^0)$
(b) $10023_{\text{five}} = (1 \times 5^4) + (0 \times 5^3) + (0 \times 5^2) + (2 \times 5^1) + (3 \times 5^0)$
(c) $10204_{\text{five}} = (1 \times 5^4) + (0 \times 5^3) + (2 \times 5^2) + (0 \times 5^1) + (4 \times 5^0)$

3. (a) $11101_{\text{two}} = (1 \times 2^4) + (1 \times 2^3) + (1 \times 2^2) + (0 \times 2^1) + (1 \times 2^0)$
(b) $11001_{\text{two}} = (1 \times 2^4) + (1 \times 2^3) + (0 \times 2^2) + (0 \times 2^1) + (1 \times 2^0)$
(c) $1001_{\text{two}} = (1 \times 2^3) + (0 \times 2^2) + (0 \times 2^1) + (1 \times 2^0)$

4. (a) $1102_{\text{three}} = (1 \times 3^3) + (1 \times 3^2) + (0 \times 3^1) + (2 \times 3^0)$
(b) $20112_{\text{three}} = (2 \times 3^4) + (0 \times 3^3) + (1 \times 3^2) + (1 \times 3^1) + (2 \times 3^0)$
(c) $2001_{\text{three}} = (2 \times 3^3) + (0 \times 3^2) + (0 \times 3^1) + (1 \times 3^0)$

5. (a) 86 (b) 447 (c) 43 (d) 23 (e) 15

6. (a) 0.1 (b) centi (c) milli (d) 100
(e) 10 (f) .001 (g) deci (h) 1000
(i) hecto (j) deca

7. (a) 95° (b) 104° (c) 71.6° (d) 117.8°
(e) 15° (f) 35° (g) 70° (h) 105°

8. (a) 5.08 (b) 2724 (c) 1.90 (d) 510.3
(e) 3.048 (f) 426.72 (g) 76.2 (h) 635.6
(i) 13.3 (j) 3742.20

9. (a) 6.2 (b) 505.05 (c) 110 (d) 61.48
(e) 67.84 (f) 3.51 (g) 1.925 (h) 2.964
(i) 122.96 (j) 4.4

10. (a) 4 (b) 1.5 (c) .2 (d) 30; 3
(e) 3.8; .38 (f) 5; .5 (g) 68; 6.8; .68 (h) 92; 9.2; .92

CHAPTER 8: EXERCISES

1. (a) $(1 \times 2^3) + (0 \times 2^2) + (0 \times 2^1) + (1 \times 2^0)$
(b) $(1 \times 3^2) + (2 \times 3^1) + (0 \times 3^0)$
(c) $(4 \times 5^1) + (1 \times 5^0)$
(d) $(1 \times 10^3) + (9 \times 10^2) + (8 \times 10^1) + (3 \times 10^0)$
(e) $(1 \times 2^3) + (0 \times 2^2) + (1 \times 2^1) + (1 \times 2^0)$
(f) $(3 \times 10^2) + (8 \times 10^1) + (9 \times 10^0)$
(g) $(1 \times 5^3) + (0 \times 5^2) + (0 \times 5^1) + (2 \times 5^0)$
(h) $(1 \times 3^2) + (1 \times 3^1) + (2 \times 3^0)$
(i) $(2 \times 3^1) + (2 \times 3^0)$
(j) $(5 \times 10^4) + (1 \times 10^3) + (0 \times 10^2) + (3 \times 10^1) + (7 \times 10^0)$
(k) $(1 \times 2^2) + (1 \times 2^1) + (1 \times 2^0)$
(l) $(3 \times 5^2) + (2 \times 5^1) + (1 \times 5^0)$

3. (a) .001 (b) hecto (c) 10 (d) centi
(e) .1 (f) 10 (g) 1000 (h) 1000
(i) 1 (j) 100 (k) deca (l) kilo

5. (a) 169.164 (b) 152.4 (c) 25

7. (a) 85.05 (b) 2270 (c) 22,700

9. (a) Approx. 5 pounds (b) Approx. 108 square feet (c) Approx. 20 gallons

CHAPTER 9: PRACTICE SETS

1. (a) $y = 22$ (b) $k = -3$ (c) $d = 36$ (d) $s = \frac{2}{3}$

2. (a) $d = 42$ (b) $y = 3$ (c) $k = 2$ (d) $28 = t$
(e) $n = 4$ (f) $c = 5$ (g) $p = 1$ (h) $s = 20$
(i) $\frac{2}{3} = g$ (j) $b = \frac{5}{3}$ (k) $r = -4$ (l) $z = 3$
(m) $q = -2$ (n) $18 = x$ (o) $c = 6$ (p) $t = -2$

3. (a) $t > 2$ (b) $x \leq 3$ (c) $y \leq -2$ (d) $c > 4$

4. (a) $y > -3$ and $y < 0$ (b) $k \geq -2$ and $k < 5$
(c) $p > 1$ and $p \leq 4$ (d) $d \geq -4$ and $d \leq 2$

5. (a) $n = 43$ (b) $n = 97$ (c) $s = 61$
(d) $i = 178$ (e) $L = 320$ (f) $s = 840$
(g) $215 = b$ (h) $\$12,500 = a$ (i) $\$100 = u$
(j) $9 = s$ (k) $a = \$360$; $f = \$1,440$ (l) $T = 36$; $S = 108$
(m) $f < \$80$; $t = \$240$ (approx.); $L = \$160$ (approx.)
(n) $x > \$60$; $r = \$120$ (approx.); $d = \$240$ (approx.)
(o) $E = 12$; $H = 15$ (p) $a = 200$; $c = 500$

CHAPTER 9: EXERCISES

1. (a) Algebra (b) Yes; yes (c) Yes; no—open sentences contain variables.
(d) Yes; yes (e) Yes; yes; yes (f) Yes; yes
(g) Equation (h) Inequality (i) Yes.
(j) No. It's neither true nor false until the variable is replaced by some number.

3. (a) $\{n \mid n > 2\}$ (b) $\{n \mid -3 < n < 5\}$ (c) $\{n \mid 0 \leq n < 4\}$
(d) Yes; no—because -2 is not a whole number.
(e) Sense (or direction)
(f) The same number must be obtained on each side of the inequality.
(g) A second point verifies that the direction of the inequality is correct.

4. (a) $y = 60$ (c) $r = 4$ (e) $k = 24$ (g) $h = -4$
(i) $14 = c$ (k) $a = 6$ (m) $z = -2$ (o) $8 = p$
(q) $q = 7$ (s) $24 = t$

5. (a) $d \leq 3$ (b) $y < 4$ (c) $g \leq -5$
(d) $r > 2$ (e) $w \leq -3$ (f) $x < -6$

7. (a) $n = 46$ (b) $n = 82$ (c) $a = 39$
(d) $i = 1200$ (e) $51 = c$ (f) $L = 96$
(g) $g = \$89{,}800$ (h) $g < \$80{,}000$ (approx.) (i) $t = 15$
(j) $a = 900$; $c = 3{,}600$ (k) $g = 1{,}000$; $b = 2{,}000$
(l) $p > \$200$; $s = \$400$ (approx.); $c = \$500$ (approx.)
(m) $w = 32$ (n) $p = 180$ (o) $g = 12$; $n = 20$
(p) $t = 12$; $r = 16$

CHAPTER 10: PRACTICE SETS

1. (a) (1) Domain $= \{r, s\}$
Range $= \{u, v\}$

(2) Domain $= \{r, s, t\}$
Range $= \{x, y, z\}$

(3) Domain $= \{p, q, r, s\}$
Range $= \{t, u, v\}$

(4) Domain $= \{x, y, z\}$
Range $= \{m, n, o, p\}$

(5) Domain $= \{t, u, v\}$
Range $= \{p, q, r\}$

(b) Functions are (1), (3), and (5).

2. Functions are (a), (c), (d), (f), (g), and (h).

3.

(a) (x, y)	(b) (x, y)	(c) (x, y)	(d) (x, y)
$(-2, 1)$	$(-2, -8)$	$(-2, 7)$	$(-2, 4)$
$(0, 3)$	$(0, -2)$	$(0, 3)$	$(0, 5)$
$(2, 5)$	$(2, 4)$	$(2, -1)$	$(2, 6)$
$(4, 7)$	$(4, 10)$	$(4, -5)$	$(4, 7)$

4. (e) Yes (f) No

5.

(a) (x, y)	(b) (x, y)	(c) (x, y)	(d) (x, y)
$(-3, -5)$	$(-3, 10)$	$(-3, -3)$	$(-3, 6)$
$(0, -2)$	$(0, 4)$	$(0, 0)$	$(0, 7)$
$(3, 1)$	$(3, -2)$	$(3, 3)$	$(3, 8)$
$(6, 4)$	$(6, -8)$	$(6, 6)$	$(6, 9)$

6.

	y-int.	x-int.
(a)	$(0, -6)$	$(2, 0)$
(c)	$(0, 5)$	$(-5, 0)$

	y-int.	x-int.
(b)	$(0, 4)$	$(2, 0)$
(d)	$(0, 3)$	$(-6, 0)$

CHAPTER 10: EXERCISES

1. (a) $A \times B = \{(1, 3), (1, 4), (2, 3), (2, 4)\}$; yes; no—because same first component forms more than one ordered pair.
(b) $A \times B = \{(2, 3), (2, 4)\}$; yes; no [See part (a)].
(c) $A \times B = \{(1, 3), (2, 3)\}$; yes; yes. (d) Functions: (1), (2), (4), (5).

(e) Yes; yes (f) 7; (3, 7) (g) Yes
(h) No—many functions are either curved lines or even distinct points.
(i) No—straight vertical lines are not functions.
(j) Yes (k) Yes
(l) No—inequalities cannot pass the vertical-line test.

3. (a) No; the root of a first-degree equation in one variable is a single point on the number line.
(b) Yes; the root of a first-degree inequality in one variable starts at some point on the number line and includes every point beyond, in one direction or the other. That is, ←—o—→ or ←—o—→
(c) Yes. The solution to a first-degree equation in two variables graphs as a straight line, which has an infinite number of points (each of which is a solution to the equation).
(d) Yes. The solution to each first-degree inequality in two variables includes every point above (or below) a line; thus, the solution set is infinite.
(e) Number lines (f) The abscissa is 3.
(g) (0, 0) (h) $\geq$

5. Functions: (a), (b), (f), and (h).

7.

	y-int.	*x*-int.
(a)	(0, 2)	$\left(-\frac{1}{2}, 0\right)$
(b)	(0, 2)	(−2, 0)
(c)	(0, 2)	(−4, 0)
(d)	(0, 2)	(1, 0)

	y-int.	*x*-int.
(e)	(0, −8)	(4, 0)
(f)	(0, −2)	(1, 0)
(g)	(0, 4)	(−2, 0)
(h)	(0, 6)	(−3, 0)

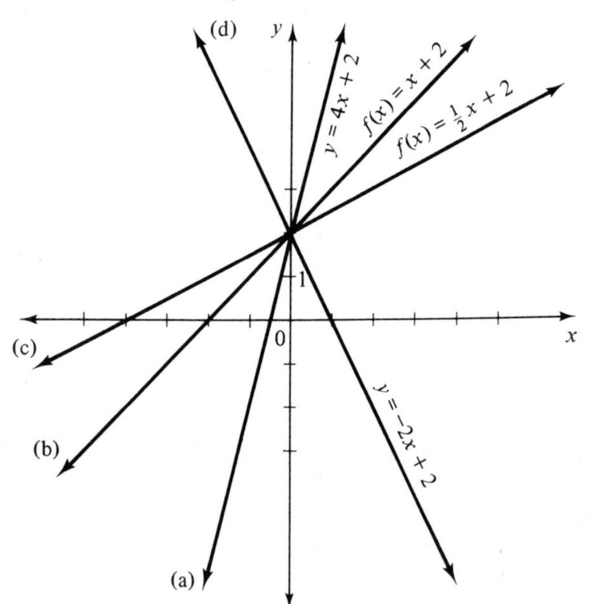

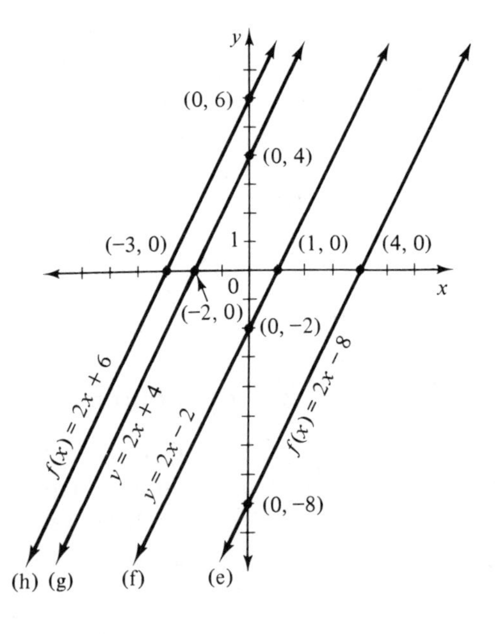

CHAPTER 11: PRACTICE SETS

1. (a) (2) Parameter
(b) (3) Sample
(c) (1) Parameter
(d) (3) Population

2. (a)

Score	*Frequency of score*
85	2
90	2
100	3
105	3
120	4

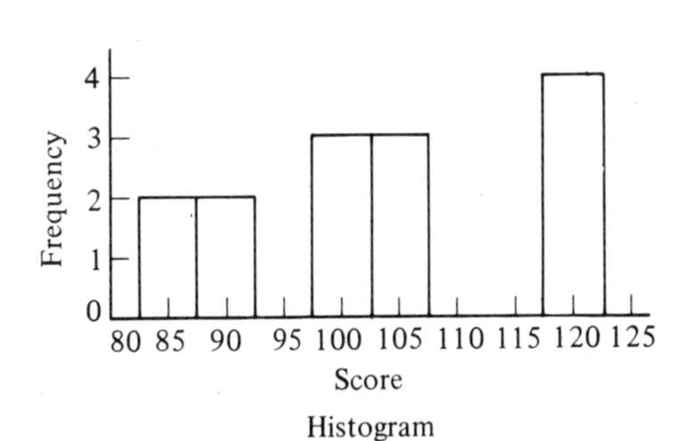

Histogram

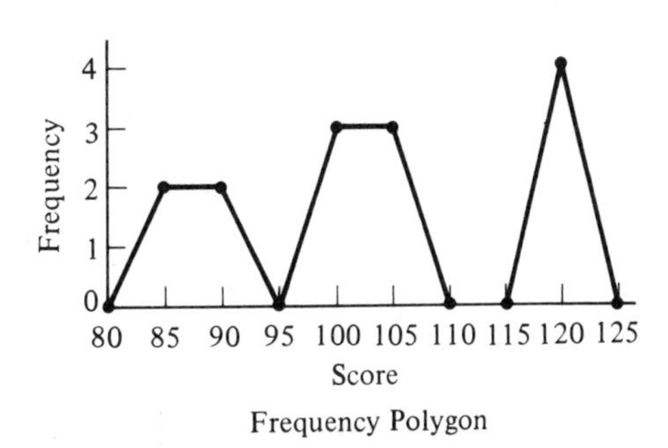

Frequency Polygon

(b)

Score	*Frequency of score*
15	1
17	3
21	2
25	3
30	4
32	2
35	3

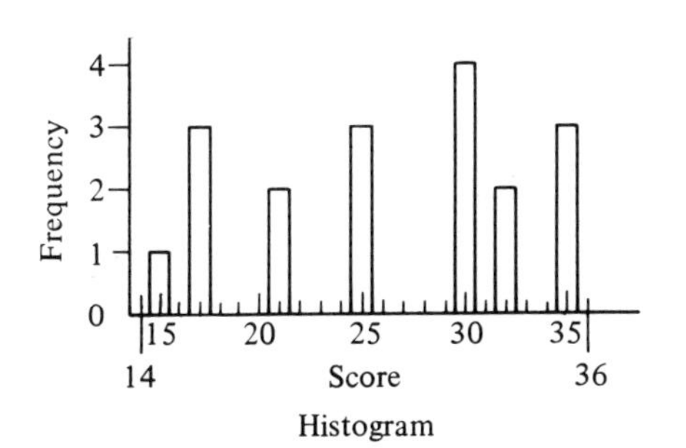

Histogram

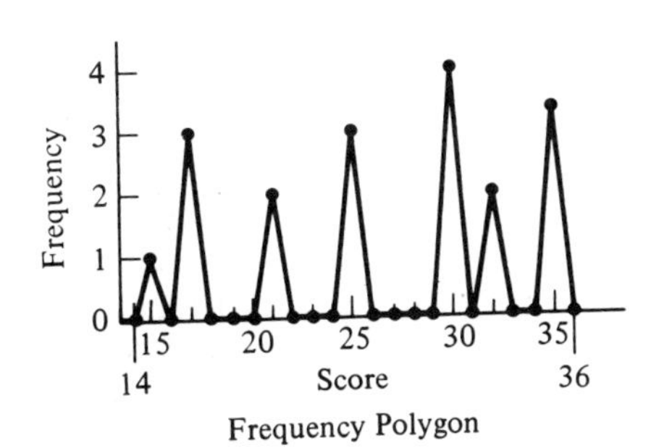

Frequency Polygon

3. (a) 32 (b) 97

4. (a) 5 (b) 84 (c) 6 (d) 7.5

5. (a) 51 (b) 60 (c) 37 (d) 170

6. (a) (2)

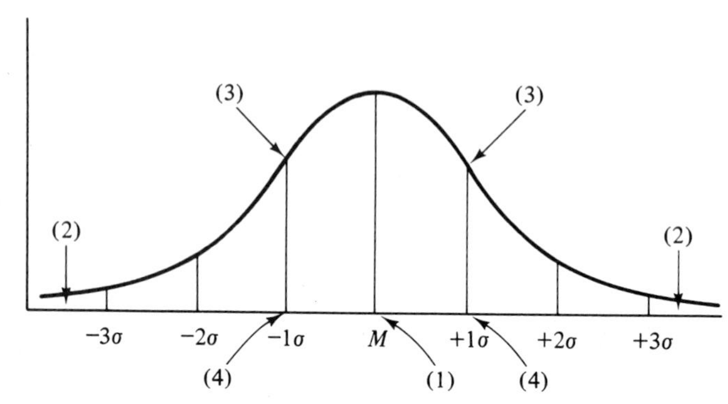

7. (a)

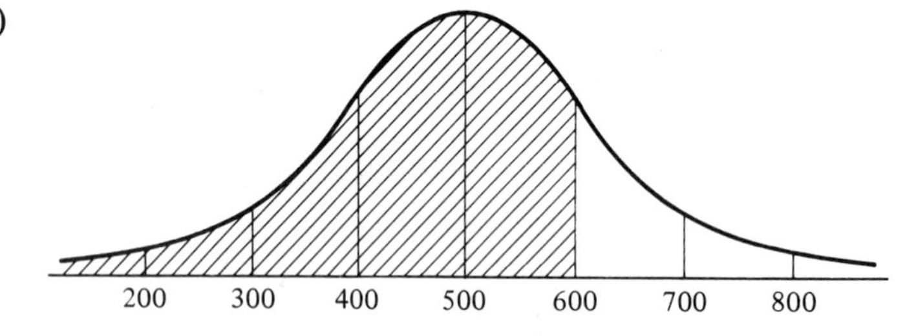

A person with an SAT score of 600 equals or exceeds 84% (or 50% + 34%) of the test population.

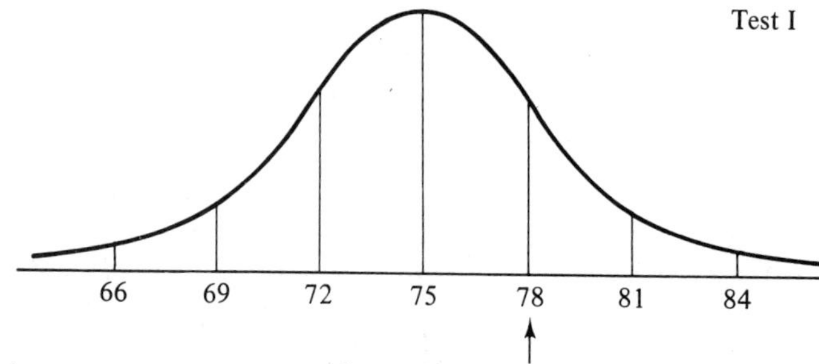

Student's score, 78, is one standard deviation above the mean.

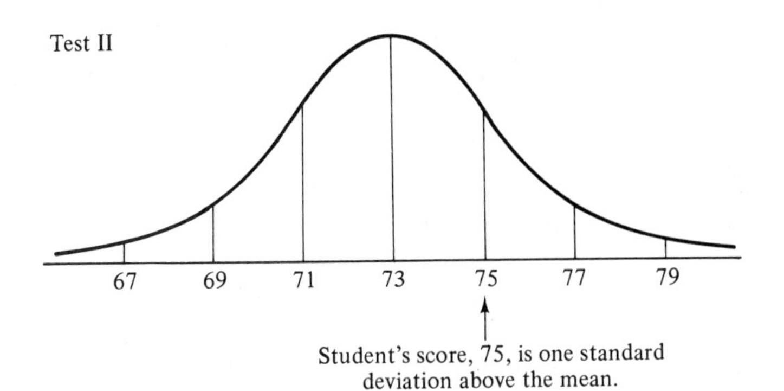

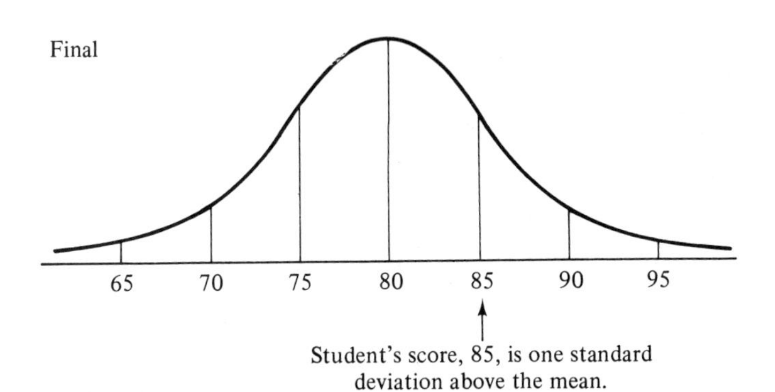

Conclusion: No, the student did equally well on all three tests.

8. (a) 1′0″ (b) 251 miles

9. (a) $\overline{X} = 7$
AD = 1.5

Score	*Score minus mean*	*Deviation of score from mean*
7	7 − 7	0
9	9 − 7	+2
6	6 − 7	−1
10	10 − 7	+3
8	8 − 7	+1
7	7 − 7	0
5	5 − 7	−2
4	4 − 7	−3

(b) $\overline{X} = 46$
AD = 20.8

Score	*Score minus mean*	*Deviation of score from mean*
10	10 − 46	−36
70	70 − 46	+24
60	60 − 46	+14
60	60 − 46	+14
30	30 − 46	−16

(c) $\overline{X} = 20$
AD = 7.7

Score	Score minus mean	Deviation of score from mean
27	27 − 20	+7
20	20 − 20	0
13	13 − 20	−7
5	5 − 20	−15
35	35 − 20	+15
15	15 − 20	−5
25	25 − 20	+5

(d) $\overline{X} = 10$
AD = 4.4

Score	Score minus mean	Deviation of score from mean
10	10 − 10	0
2	2 − 10	−8
18	18 − 10	+8
13	13 − 10	+3
7	7 − 10	−3

10. (a)

Score	Score minus mean	Deviation of score from mean	Square of deviation of score from mean
10	10 − 15	−5	25
20	20 − 15	+5	25
15	15 − 15	0	0
25	25 − 15	+10	100
5	5 − 15	−10	100

$\sigma^2 = 50; \quad \sigma = 7.07$

(b)

Score	Score minus mean	Deviation of score from mean	Square of deviation of score from mean
6	6 − 6	0	0
6	6 − 6	0	0
6	6 − 6	0	0
6	6 − 6	0	0

$\sigma^2 = 0; \quad \sigma = 0$

(c)

Score	Score minus mean	Deviation of score from mean	Square of deviation of score from mean
4	4 − 4	0	0
3	3 − 4	−1	1
3	3 − 4	−1	1
6	6 − 4	+2	4
5	5 − 4	+1	1
5	5 − 4	+1	1
2	2 − 4	−2	4

$\sigma^2 = 1.71; \quad \sigma = 1.3$

11. (a) yes (b) yes (c) no

12.

Salesperson	Test score	Sales
Anderson (A)	30	$60
Brown (B)	50	80
Cook (C)	60	75
Davis (D)	70	90
Eastman (E)	90	100

Relationship tends to vary directly. That is, when the test score is high the sales are high; when the test score is low, the sales are low.

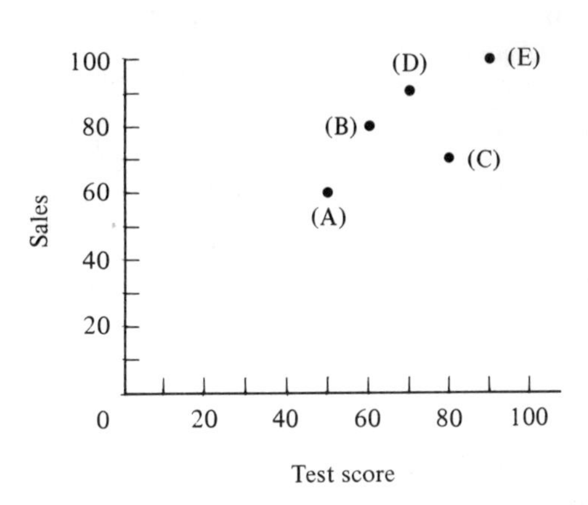

Amount of relationship: high
Direction of relationship: positive

13. 2.25

14. (a) The findings would have occurred by chance only 1 time out of 100.
(b) The findings would have occurred 2 times out of 100 by chance alone.
(c) The findings were not significant. (Results failed to reach the .05 level or some lower level of significance.)

CHAPTER 11: EXERCISES

1. Example: The average income of 45-year-old physicians in the U.S. is to be estimated from measures made on 100 of them.
Population: *All* 45-year-old physicians in the U.S.
Parameter: Average income if *all* 45-year-old physicians were polled.
Sample: Randomly chosen 100 physicians who are 45-years-old.
Statistic: Average income calculated from the sample of 100 45-year-old physicians.
In general: The statistic is to the sample as the parameter is to the population.

3. (a)

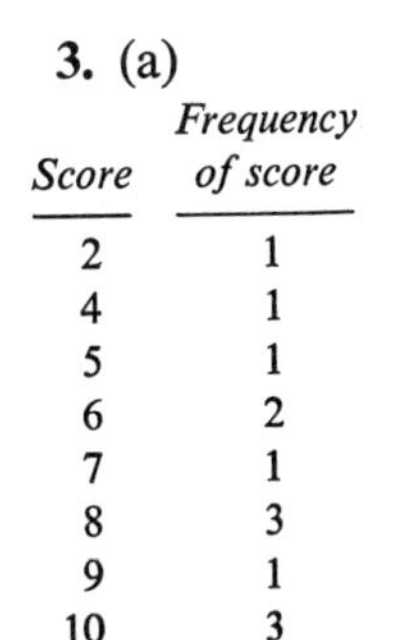

Score	*Frequency of score*
2	1
4	1
5	1
6	2
7	1
8	3
9	1
10	3

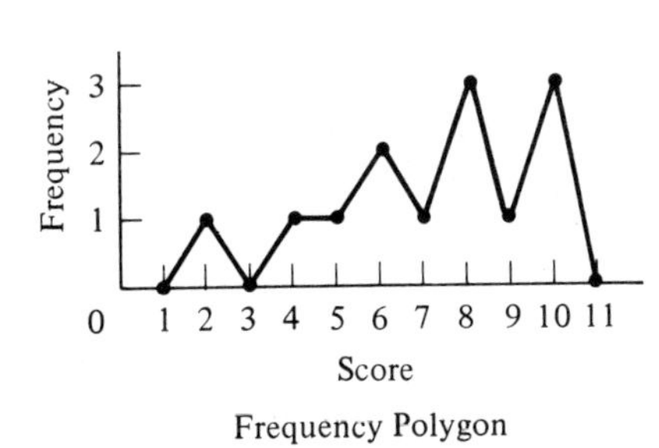

Frequency Polygon

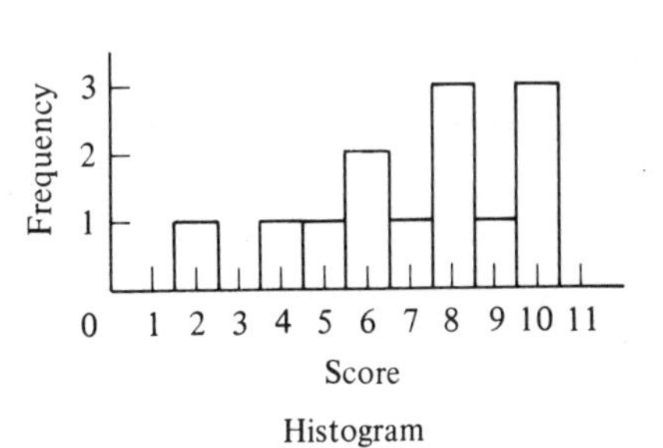

Histogram

(b)

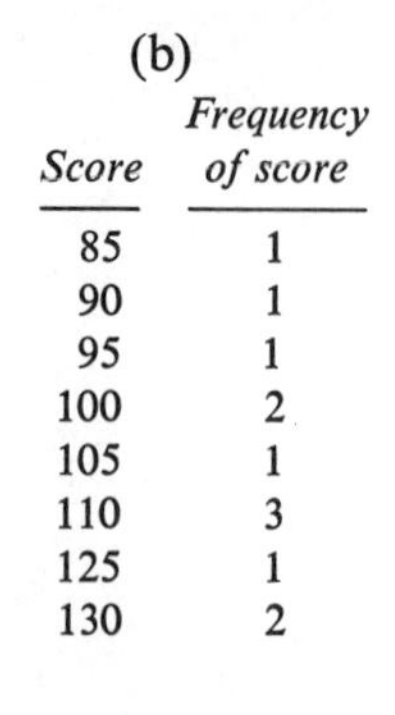

Score	Frequency of score
85	1
90	1
95	1
100	2
105	1
110	3
125	1
130	2

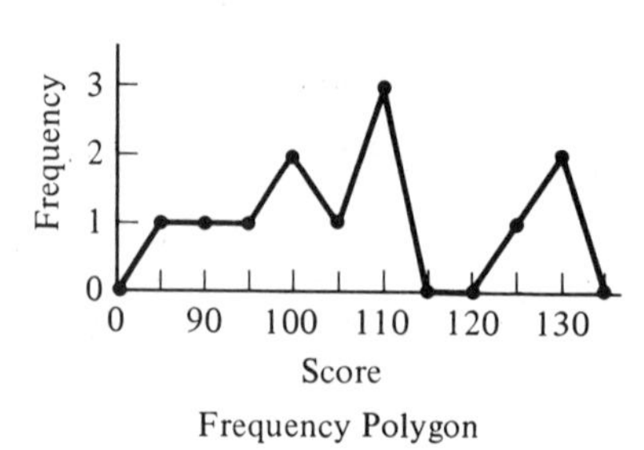

Frequency Polygon

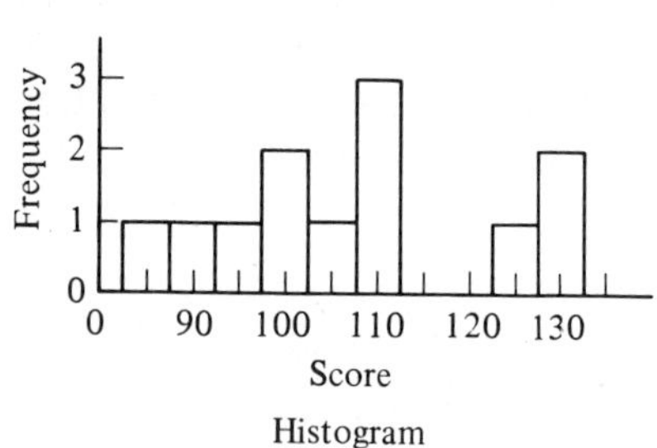

Histogram

(c)

Score	Frequency of score
1	1
2	1
3	2
4	2
5	3
6	2
7	1
8	1
9	1

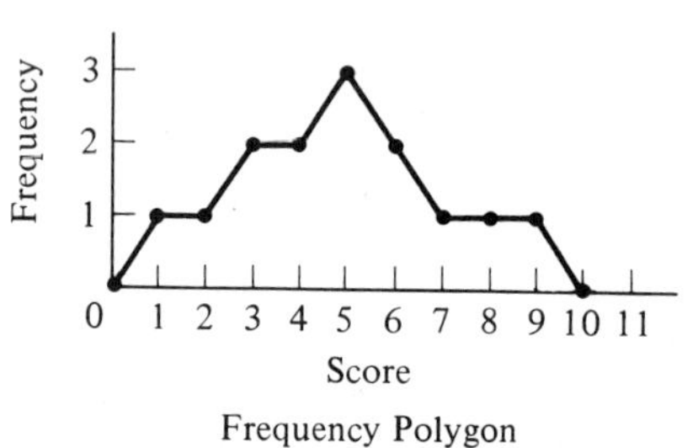

Frequency Polygon

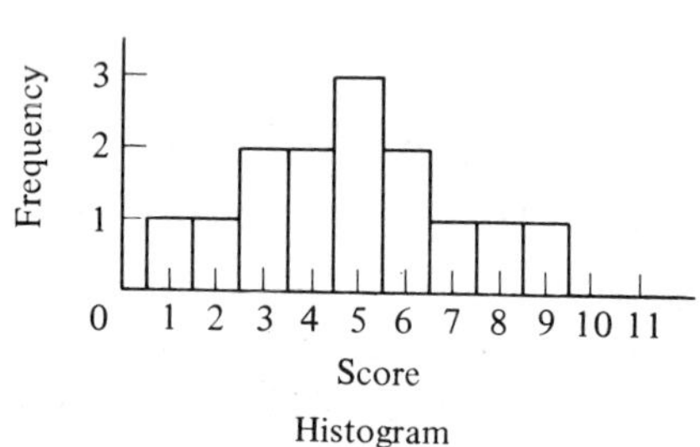

Histogram

5. (a) 7 (b) 31 (c) 210 (d) 4

7. *Four characteristics of a normal curve*

1. The normal curve is symmetrical about the mean.
2. The tails of a normal curve are asymptatic to the baseline.
3. There are two points of inflection on the normal curve.
4. A perpendicular line from the points of inflection to the baseline mark off a standard distance which is used in dividing the baseline into equal segments.

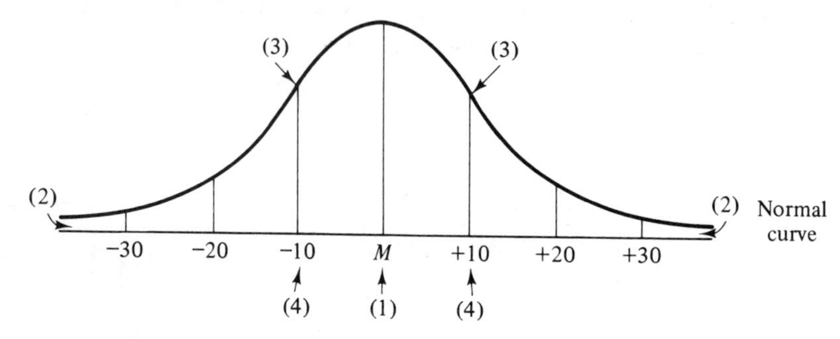

9.

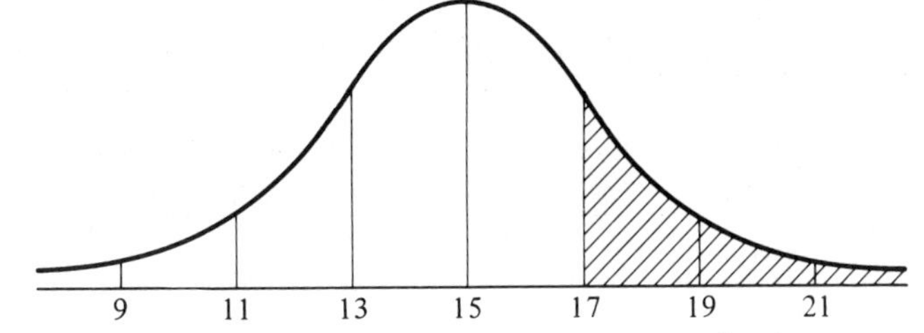

The portion of normal children who learn to walk when they are 17 months old or older is 16% (or 50% − 34%).

11. (a) \$675 (b) 24 years (c) 11 (d) 75 pounds

13. (a) $\bar{x} = \frac{\sum_{i=1}^{4} x_i}{4} = \frac{45 + 50 + 60 + 85}{4} = \frac{240}{4} = 60$

Score	*Score minus mean*	*Deviation of score from mean*	*Square of deviation of score from mean*
45	45 − 60	−15	225
50	50 − 60	−10	100
60	60 − 60	0	0
85	85 − 60	+25	625

$\sigma^2 = 237.5$ $\sigma = 15.4$

(b) $\bar{x} = \frac{\sum_{i=1}^{7} x_i}{7} = \frac{10 + 10 + 15 + 20 + 60 + 70 + 95}{7} = \frac{280}{7} = 40$

Score	*Score minus mean*	*Deviation of score from mean*	*Square of deviation of score from mean*
10	10 − 40	−30	900
10	10 − 40	−30	900
15	15 − 40	−25	625
20	20 − 40	−20	400
60	60 − 40	+20	400
70	70 − 40	+30	900
95	95 − 40	+55	3025

$\sigma^2 = 1021$ $\sigma = 31.9$

(c) $\bar{x} = \dfrac{\sum_{i=1}^{8} x_i}{8} = \dfrac{2+2+3+3+6+7+8+9}{8} = \dfrac{40}{8} = 5$

Score	*Score minus mean*	*Deviation of score from mean*	*Square of deviation of score from mean*
2	2 − 5	−3	9
2	2 − 5	−3	9
3	3 − 5	−2	4
3	3 − 5	−2	4
6	6 − 5	+1	1
7	7 − 5	+2	4
8	8 − 5	+3	9
9	9 − 5	+4	16

$\sigma^2 = 7$ $\sigma = 2.6$

15. (a)

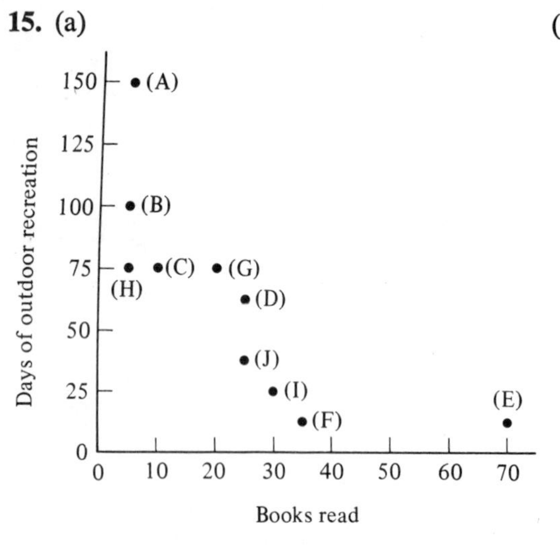

Amount of relationship: high
Direction of relationship: negative

(b)

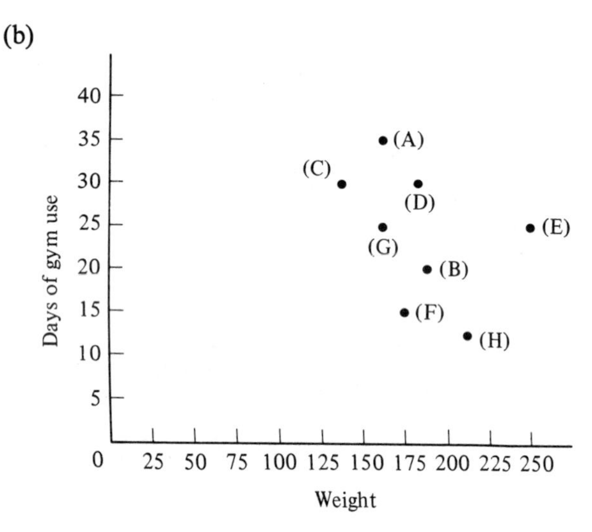

Amount of relationship: low
Direction of relationship: negative

17. +.70

19. 1. Through observation, logic, previous knowledge, etc., an individual develops a certain view or theory concerning a phenomenon which leads to certain questions which are stated as hypotheses.
2. These logical structures are then operationalized which means ultimately stating them in numerical form. This then produces a mathematical model.
3. Data are gathered and an attempt is made to see how clearly they fit the model.
4. As a result of 3, the hypotheses are rejected or they are not, depending upon how well the model and the data fit together.

CHAPTER 12: PRACTICE SETS

1. (a) General principle of counting: $m = 3, n = 3, mn = 3 \cdot 3 = 9$ two-digit numbers (repetitions allowed).
Tree diagram: There are 9 possible two-digit numbers that can be formed if repetition of digits is allowed: 33, 34, 35, 43, 44, 45, 53, 54, 55.
(b) General principle of counting: $m = 3, n = 2, r = 1, mnr = 3 \cdot 2 \cdot 1 = 6$ three-digit numbers (repetitions not allowed).
Tree diagram: There are 6 possible three-digit numbers that can be formed if repetition of digits is not allowed: 345, 354, 435, 453, 534, 543.
(c) General principle of counting: $m = 2, n = 2, mn = 2 \cdot 2 = 4$ two-person teams of one man and one woman.
Tree diagram: There are 4 possible two-person teams of one man and one woman: Jack, Ann; Jack, Sarah; George, Ann; George, Sarah.

2. (a) 24; 1; 2; 720 (b) 12 (c) 21 (d) 120

3. (a) 120 (b) 56 (c) 462

4. (a) 720 (b) 3024 (c) 360

5. (a) 116,280 (b) 120 (c) 165 (d) 120 (e) 24

6. (a) $\frac{1}{3}$ or (1, 3) or 1:3 (b) $\frac{6}{25}$ or (6, 25) or 6:25
$\frac{17}{88}$ or (17, 88) or 17:88
$\frac{9}{62}$ or (9, 62) or 9:62

7. (a) $\frac{1}{4}$ (b) $\frac{1}{2}$ (c) $\frac{5}{6}$ (d) $\frac{2}{3}$

8. (a) $\frac{1}{6}$ (b) $\frac{5}{12}$

9. (a) $\frac{4}{7}$ (b) .55 (c) $\frac{1}{5}$

10. (a) .58 (b) $\frac{5}{14}$ (c) 3; 7; 20 (d) $\frac{1}{2}$ (e) $\frac{4}{13}$

11. (a) .45 (b) $\frac{19}{30}$ (c) .2 (d) $\frac{4}{13}$ (e) $\frac{6}{7}$

12. (a) Mutually exclusive but not exhaustive
(b) Mutually exclusive and exhaustive
(c) Mutually exclusive but not exhaustive

13. (a) $\frac{3}{5}$ (b) .18 (c) .1; 30

14. (a) $\frac{1}{4}$ (b) $\frac{1}{8}$ (c) $\frac{1}{8}$ (d) $\frac{1}{4}$

15. (a) $\frac{10}{17}$; $\frac{1}{2}$; $\frac{17}{20}$ (b) $\frac{7}{10}$; $\frac{14}{25}$; $\frac{4}{5}$

CHAPTER 12: EXERCISES

1. (a) General principle of counting: $m = 3, n = 2, r = 2, mnr = 3 \cdot 2 \cdot 2 = 12$ different outfits.

Tree diagram: There are 12 possible outfits that can be formed of one jacket, one pair of pants, one pair of shoes: J1, P1, S1; J1, P1, S2; J1, P2, S1; J1, P2, S2; J2, P1, S1; J2, P1, S2; J2, P2, S1; J2, P2, S2; J3, P1, S1; J3, P1, S2; J3, P2, S1; J3, P2, S2.

(b) General principle of counting: $m = 4, n = 4, mn = 4 \cdot 4 = 16$ two-digit numbers if repetitions allowed.

Tree diagram: There are 16 two-digit numbers that can be formed from the set of natural numbers $N = \{1, 2, 3, 4\}$ if repetition of digits is allowed: 11, 12, 13, 14, 21, 22, 23, 24, 31, 32, 33, 34, 41, 42, 43, 44.

General principle of counting: $m = 4, n = 3, mn = 4 \cdot 3 = 12$ two-digit numbers if repetitions not allowed.

Tree diagram: There are 12 two-digit numbers that can be formed from the set of natural numbers $N = \{1, 2, 3, 4\}$ if repetition of digits is not allowed: 12, 13, 14, 21, 23, 24, 31, 32, 34, 41, 42, 43.

(c) General principle of counting: $m = 3, n = 3, r = 3, mnr = 3 \cdot 3 \cdot 3 = 27$ three-digit numbers if repetitions allowed.

Tree diagram: There are 27 three-digit numbers that can be formed from the set of natural numbers $N = \{2, 3, 4\}$ if repetition of digits is allowed: 222, 223, 224, 232, 233, 234, 242, 243, 244, 322, 332, 324, 332, 333, 334, 342, 343, 344, 422, 423, 424, 432, 433, 434, 442, 443, 444.

General principle of counting: $m = 3, n = 2, r = 1, mnr = 3 \cdot 2 \cdot 1 = 6$ three-digit numbers if any number can only appear once.

Tree diagram: There are 6 three-digit numbers that can be formed from the set $N = \{2, 3, 4\}$ if a digit can only appear once in the three-digit number: 234, 243, 324, 342, 423, 432.

(d) General principle of counting: $m = 4, n = 4, mn = 4 \cdot 4 = 16$ two-letter "words" may be formed if repetitions allowed.

Tree diagram: There are 16 two-letter "words" that can be formed from the set $L = \{t, e, p, m\}$ if repetitions are allowed: *tt, te, tp, tm, et, ee, ep, em, pt, pe, pp, pm, mt, me, mp, mm.*

General principle of counting: $m = 4, n = 3, mn = 4 \cdot 3 = 12$ two-letter "words" may be formed if repetitions are not allowed.

Tree diagram: There are 12 two-letter "words" that can be formed from the set $L = \{t, e, p, m\}$ if repetitions are not allowed: *te, tp, tm, et, ep, em, pt, pe, pm, mt, me, mp.*

3. Probability: the probability of success for a given event is defined as the ratio of the number of possible successes to the total number of all possible outcomes of that event. If the event is designated A, then the formula for probability of success of an event A is: $P(A) = \frac{n_A}{n}$. This formula is read: the probability of success (occurrence) of event A is the ratio of the number of possible successes of event A, n_A, to the total number of possible outcomes of that event, n.

5. (a) Coins on table: HHH; HHT; HTH; HTT; THH; THT; TTH; TTT

n_A = number of cases of one or two heads = 6; $n = 8$; $P(A) = \frac{n_A}{n} = \frac{6}{8} = \frac{3}{4}$

(b) Die on table:

1, 1	2, 1	3, 1	4, 1	5, 1	6, 1
1, 2	2, 2	3, 2	4, 2	5, 2	6, 2
1, 3	2, 3	3, 3	4, 3	5, 3	6, 3
1, 4	2, 4	3, 4	4, 4	5, 4	6, 4
1, 5	2, 5	3, 5	4, 5	5, 5	6, 5
1, 6	2, 6	3, 6	4, 6	5, 6	6, 6

n_A = number of cases of one or both die being a six = 11; $n = 36$; $P(A) = \frac{n_A}{n} = \frac{11}{36}$

(c) Balls drawn:

red, green	green, red	white, red	orange, red
red, white	green, white	white, green	orange, green
red, orange	green, orange	white, orange	orange, white

n_A = number of cases where one ball is green = 6; $n = 12$; $P(A) = \frac{n_A}{n} = \frac{6}{12} = \frac{1}{2}$

7. (a) 3; 4; 10 (b) $\frac{6}{35}$; $\frac{12}{35}$; $\frac{18}{35}$ (c) .72 (d) 35; $\frac{9}{35}$; $\frac{4}{35}$

(e) $\frac{21}{32}$ (f) $\frac{1}{2}$ (g) $\frac{7}{15}$

9. Exhaustive categories: categories of outcomes where every possible relevant outcome is included in one of the categories.

Example: (a) Drawing a red or a black card from a standard deck of 52 cards

(b)

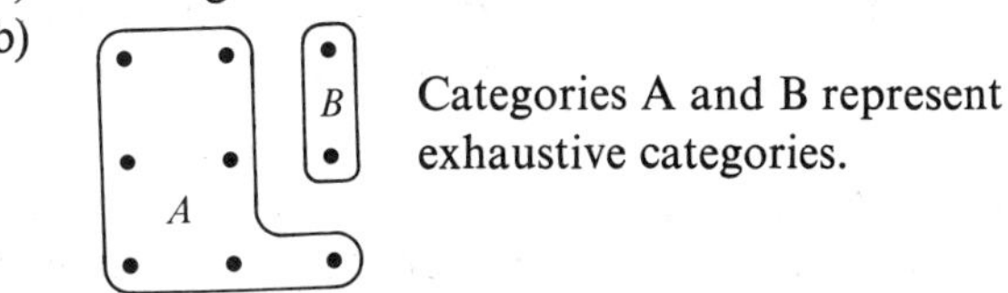

Categories A and B represent exhaustive categories.

11. (a) $\frac{1}{9}$ (b) .1 (c) $\frac{1}{3}$; $\frac{1}{3}$ (d) .2; .2
(e) $\frac{1}{5}$ (f) $\frac{9}{13}$ (g) $\frac{1}{2000}$; $\frac{6}{2000}$; $\frac{1993}{2000}$

13. (a) $\frac{1}{216}$ (b) $\frac{1}{8}$ (c) $\frac{1}{169}$ (d) $\frac{1}{64}$

15. (a) 2,598,960 (b) 45 (c) 495 (d) 5005

17. (a) 12,376 (b) 210 (c) 40,320 (d) 435 (e) 56

CHAPTER 13: PRACTICE SETS

1. (a) \$.47; \$9.79 (b) \$1.91; \$33.70

2. (a) \$52 (b) \$29 (c) \$68

3. (a) $T = \$178.50$ (b) $T = \$164.50$ (c) $V = \$19{,}000$
(d) $V = \$14{,}000$ (e) $R = 1.6\%$ (f) $R = \$2.65$ per C
(g) $T = \$295$ (h) $V = \$3{,}500$ (i) $V = \$32{,}500$
(j) $R = 1.5\%$ (k) $R = \$1.88$ per C

4. *Key figures:*

Net sales	\$400,000	100%	Total expense	\$120,000	30%
Cost of goods	260,000	65	Net profit	20,000	5
Gross profit	140,000	35	Net income	16,000	4

5. *Key figures:*

Current assets	\$45,000	50%	Current liabilities	\$15,300	17%
Fixed assets	45,000	50	Total liabilities	37,800	42
Total assets	90,000	100	Capital	52,200	58

6. (a) \$20; $\frac{5}{4}$ or 1.25 (b) \$42; $\frac{6}{5}$ or 1.2
(c) \$58; 1.45 (d) \$69; 1.38

7. (a) \$40; $\frac{5}{4}$ (b) \$60; $\frac{4}{3}$ (c) \$175; $\frac{1}{.52}$
(d) \$90; $\frac{5}{3}$ or $\frac{1}{.6}$ (e) \$125; $\frac{1}{.72}$

8. (a) 50%; $33\frac{1}{3}\%$ (b) $33\frac{1}{3}\%$; 25% (c) 40%; $28\frac{4}{7}\%$ (d) 25%; 20%

9. (a) \$20; \$620 (b) \$73.50; \$1,473.50 (c) \$12.50; \$812.50
(d) \$18.40; \$738.40 (e) 11% (f) $\frac{1}{3}$ yr. or 4 mo.
(g) \$450 (h) \$720 (i) 9% (j) 91 days

10. (a) \$15 (b) \$12.40 (c) 18.25% (d) 27.0%
(e) \$162; \$59; 21.75% (f) \$150; \$25; 18.0%

11. (a) \$1,283.36; \$283.36 (b) \$1,196.68; \$196.68
(c) \$22.09; \$44.18; \$88.36. Doubling the principal also doubles the interest.
(d) \$127.16; \$270.49; \$614.14. Doubling the time more than doubles the interest.
(e) \$60; \$61.68 (f) \$300; \$348.85 (g) \$420; \$520.37

CHAPTER 13: EXERCISES

1. (a) Percent calculation (b) Yes (c) Yes
(d) No; 5% of \$16.00 = \$.80; but 5% of \$16.80 = \$.84.
(e) Real and personal
(f) Assessed

3. (a) Gross profit
(b) No; selling price must more than cover operating expenses as well.
(c) 1.2; 1.25 (d) 1.4; 1.5 (higher) (e) yes

5. (a) \$.31; \$8.06
\$1.24; \$26.02
(b) \$7.40; \$56; \$25; \$17

7. *Key figures:*

Net sales	\$500,000	100%	Total expenses	\$140,000	28%
Cost of goods	300,000	60	Net profit, oper.	60,000	12
Gross profit	200,000	40	Net income	25,000	5

9. (a) \$54; $\frac{6}{5}$ or 1.2 (b) \$35; $\frac{5}{4}$ or 1.25
(c) \$69; 1.15 (d) \$32; $\frac{4}{3}$ or $1.33\frac{1}{3}$ or $\frac{1}{.75}$

11. (a) \$42; \$842 (b) \$24.20; \$624.20 (c) $\frac{1}{4}$ yr. or 3 mos.
(d) 10% (e) \$1,200 (f) \$2,000

13. (a) \$1,270.49; \$270.49 (b) \$2,766.19; \$766.19 (c) \$976.64; \$176.64
(d) \$100; \$104.94 (e) \$250; \$283.36 (f) \$400; \$490.59

CHAPTER 14: PRACTICE SETS

1. (a)

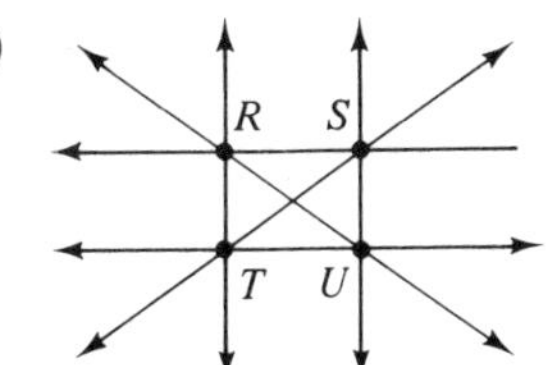

There are six lines determined by the four points.

(b) There are an infinite number of planes which pass through one line.
(c) Axiom: If two points of a line lie in a plane, the whole line is in the plane.

2. (a), (d)

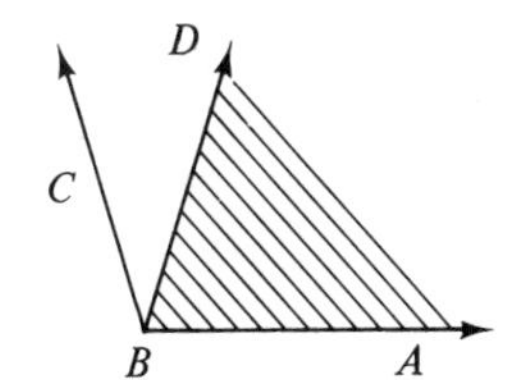

(b) Three angles are formed: $\angle ABC$, $\angle ABD$, $\angle DBC$.
(c) One pair of adjacent angles are formed: $\angle ABD$ and $\angle CBD$.

3. (a) $\angle 1$ and $\angle 3$ (b) $\angle 2$ and $\angle 4$ (c) $\angle ABC$
(d) The acute angles seem to be equal. (e) The obtuse angles seem to be equal.

4. (a) 20° (b) 70° (c) Yes (d) $m(\measuredangle A) = m(\measuredangle B)$ because
$m(\measuredangle A) + m(\measuredangle D) = 180°$
$m(\measuredangle B) + m(\measuredangle D) = 180°$
so $m(\measuredangle A) = m(\measuredangle D)$

5. (a) The angles equal to $\measuredangle 3$ are: angles 1, 5, and 7.
(b) The angles supplementary to $\measuredangle 3$ are: 2, 4, 8, and 6.
(c) The perpendicular lines form 4 right angles. Each pair of angles will be equal, and they will be supplementary.

6. (a) $\measuredangle 7 = \measuredangle 6$ (Vertical angles are equal.)
$\measuredangle 1 = \measuredangle 6$ (Corresponding angles are equal.)
$\measuredangle 3 = \measuredangle 6$ (Alternate interior angles are equal.)

(b) $\measuredangle 3$ is supplementary to $\measuredangle 2$ because their sum is a straight angle. The angles equal to $\measuredangle 3$ are also supplementary to $\measuredangle 2$. These are the angles 1, 7, and 6.

7. 1. Given. 2. Vertical angles are equal. 3. Substitution.
4. If the interior angles on the same side of a transversal are supplementary, then the lines cut by the transversal are parallel.

8. *A*—convex quadrilateral *B*—concave octagon
C—concave pentagon *D*—convex pentagon

9. (a) The right triangles are $\triangle ABD$ and $\triangle ADC$.
(b) There are no acute triangles in the figure.
(c) $\triangle ABC$ is obtuse.
(d) $\overline{BD} = \overline{AD}$, thus $\triangle ADB$ is an isosceles triangle. There are no equilateral triangles in the figure.

10. (a)

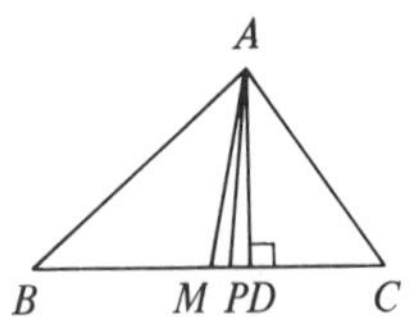

$\overline{AD}$ is altitude
$\overline{AM}$ is median
$\overline{AP}$ is angle bisector

(b)

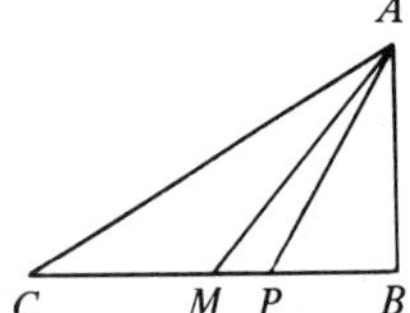

$\overline{AB}$ is altitude
$\overline{AM}$ is median
$\overline{AP}$ is angle bisector

(c)

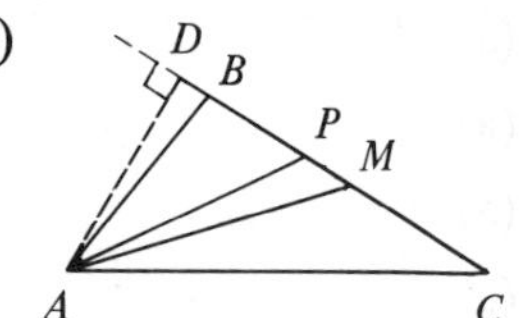

$\overline{AD}$ is altitude
$\overline{AM}$ is median
$\overline{AP}$ is angle bisector

(d)

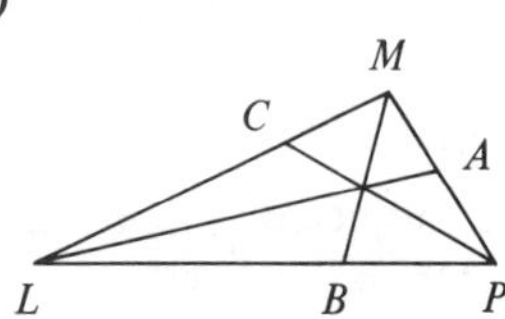

$\overline{LA}$ is angle bisector
$\overline{MB}$ is angle bisector
$\overline{PC}$ is angle bisector

(e)

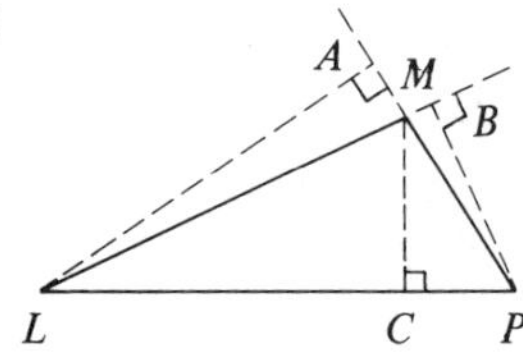

$\overline{LA}$ is altitude
$\overline{PB}$ is altitude
$\overline{MC}$ is altitude

(f) The triangle must be an obtuse triangle.

11. (a) 1. Given.
2. Given.
3. Definition of equiangular.
4. The sum of the angles of a triangle is 180°.
5. Substitution.
6. Renaming.
7. Multiplication property of equality.

(b) (1) 48°
(2) 33°

(c) $\measuredangle 2 = 30°$
$\measuredangle 3 = 60°$
$\measuredangle 4 = 30°$

12. (a) 25 (b) 6

13. (a) S (b) $\varnothing$ (c) S (d) H (e) Yes (f) $R \subset P$

14. (a) $p = 4s$; 32 (b) $p = 5s$; 30 (c) $p = 2l + 2w$; 40 feet

15. (a) 104 (b) 50 (c) 1800 square feet

16. (a) $\overline{BC}$, $\overline{AC}$ or $\overline{BD}$ (b) $\overline{AC}$ or $\overline{BD}$
(c) $\measuredangle AOB$, $\measuredangle BOC$, $\measuredangle COD$, or $\measuredangle DOA$ (d) $\widehat{AB}$, $\widehat{BC}$, $\widehat{CD}$, or $\widehat{AD}$
(e) Of central angles: $\widehat{ADB}$ (or $\widehat{ACB}$), $\widehat{BAC}$ (or $\widehat{BDC}$), $\widehat{CBD}$ (or $\widehat{CAD}$), or $\widehat{DCA}$ (or $\widehat{DBA}$).
(f) $\overline{OA}$, $\overline{OB}$, $\overline{OC}$, or $\overline{OD}$ (g) $\overleftrightarrow{CE}$
(h) $\overleftrightarrow{AC}$ or $\overleftrightarrow{BD}$ or $\overleftrightarrow{BC}$ (i) $\measuredangle ACB$ or $\measuredangle CBD$

17. (a) 80° (b) 285° (c) 120°; 120°; 120° (d) 156°; 24°; 156°

18. (a) (1) 44; 154 (2) $37\frac{5}{7}$; $113\frac{1}{7}$ (b) 434 square feet

19. (a) 216 cu. in. (b) 200 cu. cm. (c) 735 cu. in. (d) 1728 cu. mm.

20. (a) 340 cu. ft. (b) 192 cu. in. (c) 110 cu. m.

21. (a) 1408 cu. cm. (b) $905\frac{1}{7}$ cu. in. (c) $2053\frac{1}{3}$ cu. cm.
(d) $12\frac{4}{7}$ cu. ft. (e) $905\frac{1}{7}$ cu. in.

CHAPTER 14: EXERCISES

1. (a) Five points determine 10 lines.

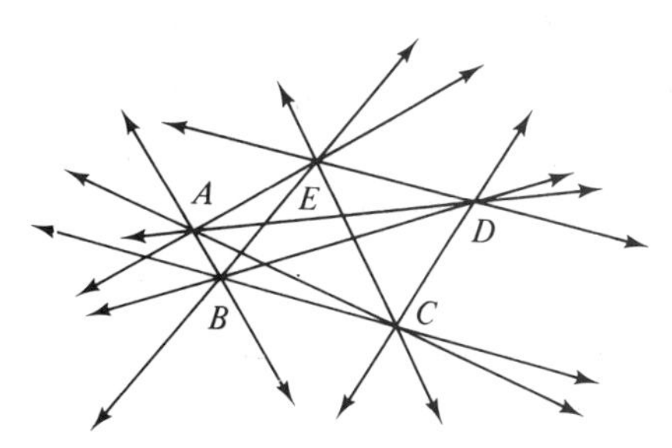

(b) 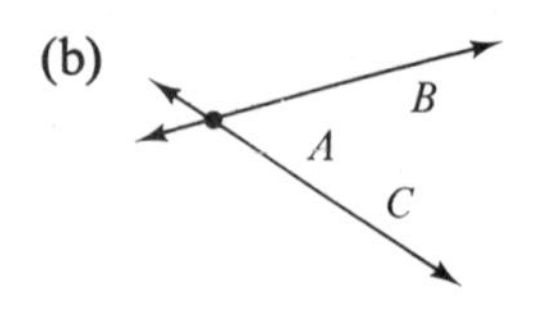

Consider the point, A, of intersection of the lines and one other point on each of the lines, B and C. There is only one plane through these three points. This plane will contain all of the two lines because if a plane contains two points of a line, it contains the whole line.

3. (a) Yes.

(b) No, the intersection of two rays which are part of the same line might be a line segment. For example, $\overrightarrow{AB} \cap \overrightarrow{CD} = \overline{CB}$.

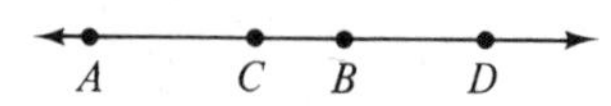

(c) Six angles are formed. $\measuredangle 1$, $\measuredangle 2$, $\measuredangle 3$, $\measuredangle AEC$, $\measuredangle BED$, $\measuredangle AED$

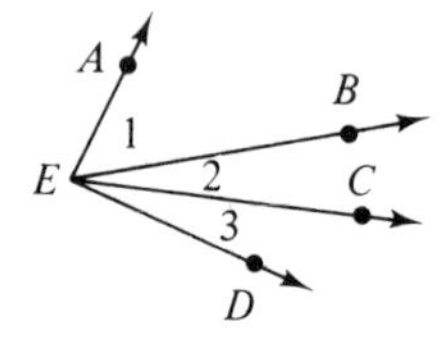

(d) No, a ray must extend infinitely in one direction. A line segment has endpoints.

(e) Yes.

5. (a) No, complementary angles must only total 90°. They do not have to be next to each other.

(b) The sum of the measures of two supplementary angles is 180°. Thus, if the measure of one angle is 80°, the measure of the other angle must be 180° − 80° or 100°.

(c) The sum of the measures of two complementary angles is 90°. Thus, if the measure of one angle is 64°, the measure of the other angle is 90° − 64° or 26°.

(d) No, an obtuse angle is more than 90°.

(e) No, two right angles are also supplementary.

7. (a) $\overleftrightarrow{AD} \perp \overleftrightarrow{BE}$

(b) $\measuredangle 1$ and $\measuredangle 4$, $\measuredangle 2$ and $\measuredangle 5$, $\measuredangle 3$ and $\measuredangle 6$, $\measuredangle ADB$ and $\measuredangle EOD$, $\measuredangle AOE$ and $\measuredangle BOD$

(c) $\measuredangle 1$ and $\measuredangle 2$ or $\measuredangle 4$ and $\measuredangle 5$

(d) $\measuredangle 1$ and $\measuredangle 5$ or $\measuredangle 2$ and $\measuredangle 4$

(e) $\measuredangle AOE$ and $\measuredangle EOD$, $\measuredangle AOB$ and $\measuredangle DOB$, $\measuredangle AOE$ and $\measuredangle AOB$, $\measuredangle EOD$ and $\measuredangle DOB$, $\measuredangle AOE$ and $\measuredangle BOD$, or $\measuredangle DOE$ and $\measuredangle AOB$

9. First, draw a perpendicular, $\overleftrightarrow{CD}$, to the given line, $\overleftrightarrow{AB}$.

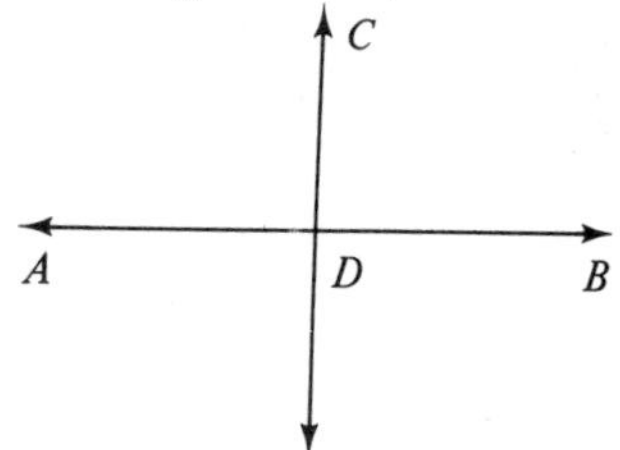

Then, draw a perpendicular, $\overleftrightarrow{CE}$, to $\overleftrightarrow{CD}$. $\overleftrightarrow{CE}$ is then parallel to $\overleftrightarrow{AB}$.

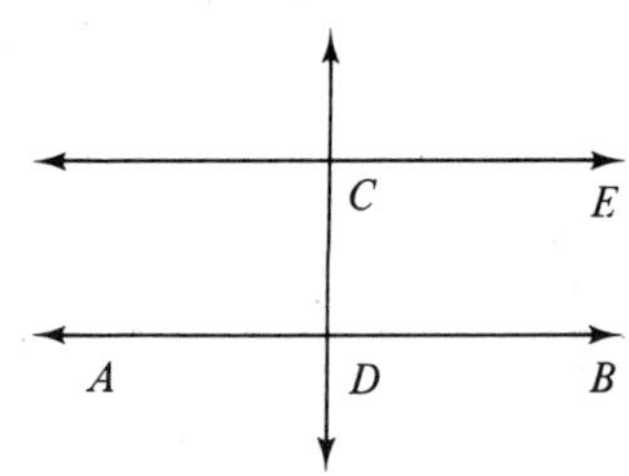

11. 1. Given. 2. Vertical angles are equal. 3. Substitution.
4. If two lines are cut by a transversal so that corresponding angles are equal, then the lines are parallel.

13. (a) $\triangle ABC$ is a right triangle because it has one right angle.
(b) $\triangle ABD$ is an acute triangle because it has three acute angles.
(c) $\triangle BCD$ is an obtuse triangle because it has an obtuse angle.
(d) $\triangle ABD$ is an equilateral triangle because it has three equal sides.
(e) $\triangle BCD$ is an isosceles triangle because it has two equal sides.

15. In a right triangle, an altitude drawn from the vertex of either acute angle will also be a leg of the triangle.

17. 45°, 45°

19. (a)
$$\begin{aligned} a^2 + b^2 &= c^2 \\ 5^2 + 12^2 &= c^2 \\ 25 + 144 &= c^2 \\ 169 &= c^2 \end{aligned}$$
$$c = \sqrt{169} = 13$$

(b)
$$\begin{aligned} a^2 + b^2 &= c^2 \\ 12^2 + b^2 &= 20^2 \\ 144 + b^2 &= 400 \\ b^2 &= 400 - 144 \\ b^2 &= 256 \end{aligned}$$
$$b = \sqrt{256} = 16$$

21. 13; 8; 80°; 100°; 100°

23. 30 sq. cm.

25. 24 sq. cm.

27. (a) $\overline{AD}$, $\overline{DC}$, $\overline{CB}$, $\overline{BA}$, $\overline{AC}$, or $\overline{BD}$
(b) $\angle DOA$, $\angle DOC$, $\angle COB$, $\angle AOB$
(c) $\angle DAC$, $\angle CAB$, $\angle ABD$, $\angle DBC$, $\angle DCA$, $\angle ACB$, $\angle ADB$, $\angle CDB$, $\angle DAB$, $\angle ABC$, $\angle BCD$, $\angle CDA$
(d) $\overleftrightarrow{AB}$, $\overleftrightarrow{BC}$, $\overleftrightarrow{CD}$, $\overleftrightarrow{AD}$, $\overleftrightarrow{AC}$, $\overleftrightarrow{BD}$
(e) $\overset{\frown}{AB}$, $\overset{\frown}{BC}$, $\overset{\frown}{CD}$, $\overset{\frown}{AD}$
(f) $\overset{\frown}{ACB}$ (or $\overset{\frown}{ADB}$), $\overset{\frown}{BAC}$ (or $\overset{\frown}{BDC}$), $\overset{\frown}{CBD}$ (or $\overset{\frown}{CAD}$), $\overset{\frown}{DCA}$ (or $\overset{\frown}{DBA}$)

29. 120°

31. $116\frac{4}{7}$

33. (a) 40 cu. cm. (b) 21 cu. in. (c) 40 cu. in.

APPENDIX: PRACTICE SETS

1. (a) 0.42 (b) 0.03 (c) 0.152 (d) 2.87
(e) 1.0975 (f) 0.065 (g) 0.007 (h) 0.0002
(i) 0.00375 (j) $0.0028\frac{4}{7}$ (0.00286) (k) 0.0075 (l) $0.0066\frac{2}{3}$ (0.00667)

2. (a) $\frac{2}{25}$ (b) $\frac{11}{20}$ (c) $\frac{12}{5}$ (d) $\frac{43}{25}$
(e) $\frac{3}{8}$ (f) $\frac{5}{6}$ (g) $\frac{1}{7}$ (h) $\frac{3}{400}$
(i) $\frac{31}{125}$ (j) $\frac{3}{8}$ (k) $\frac{1}{80}$ (l) $\frac{1}{200}$

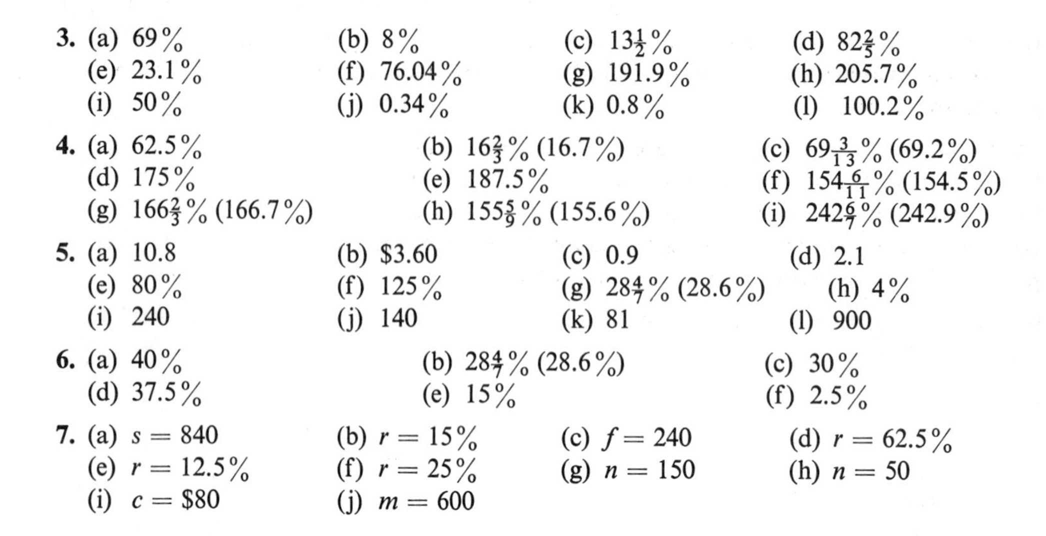

3. (a) 69% (b) 8% (c) $13\frac{1}{2}\%$ (d) $82\frac{2}{5}\%$
(e) 23.1% (f) 76.04% (g) 191.9% (h) 205.7%
(i) 50% (j) 0.34% (k) 0.8% (l) 100.2%

4. (a) 62.5% (b) $16\frac{2}{3}\%$ (16.7%) (c) $69\frac{3}{13}\%$ (69.2%)
(d) 175% (e) 187.5% (f) $154\frac{6}{11}\%$ (154.5%)
(g) $166\frac{2}{3}\%$ (166.7%) (h) $155\frac{5}{9}\%$ (155.6%) (i) $242\frac{6}{7}\%$ (242.9%)

5. (a) 10.8 (b) \$3.60 (c) 0.9 (d) 2.1
(e) 80% (f) 125% (g) $28\frac{4}{7}\%$ (28.6%) (h) 4%
(i) 240 (j) 140 (k) 81 (l) 900

6. (a) 40% (b) $28\frac{4}{7}\%$ (28.6%) (c) 30%
(d) 37.5% (e) 15% (f) 2.5%

7. (a) $s = 840$ (b) $r = 15\%$ (c) $f = 240$ (d) $r = 62.5\%$
(e) $r = 12.5\%$ (f) $r = 25\%$ (g) $n = 150$ (h) $n = 50$
(i) $c = \$80$ (j) $m = 600$

APPENDIX: EXERCISES

1. (a) Hundredths. (b) Two; a denominator of 100.
(c) 1% and 99%. (d) Zeros; 1%. (e) $0.00ab$.
(f) Writing the decimal equivalent of the percent; then writing the decimal fraction as a common fraction and reducing.
(g) $ab.c\% = 0.abc = \dfrac{abc}{1000}$; $0.ab\% = 0.00ab = \dfrac{ab}{10{,}000}$.
(h) 0.0125; $\frac{1}{80}$. (i) 1%.
(j) Dividing b into a to obtain the equivalent decimal percent; then converting the decimal percent to an ordinary decimal.
(k) $\frac{1}{100}$ and multiplying. (l) $\dfrac{a}{b}\% = 0.xyz\% = 0.00xyz$.
(m) $\dfrac{a}{b}\% = 0.xyz\% = 0.00xyz = \dfrac{xyz}{100{,}000}$.
(n) $\frac{3}{4}\% = 0.75\% = 0.0075$; $\frac{3}{4}\% = \frac{3}{4} \cdot \frac{1}{100} = \frac{3}{400}$.

3. (a) Converted to an equivalent decimal or fraction.
(b) $x\%$ of $y = z$.
(c) The decimal or fractional equivalent of the percent.
(d) What % of original is change?
(e) Increase.
(f) Percents are not used independently in an equation. The 28% must be 28% *of* (times) some variable or number.
(g) 75% (h) 25% (i) 20% (j) $133\frac{1}{3}\%$
(k) $33\frac{1}{3}\%$ (l) 25% (m) 125% (n) 150%
(o) Yes, b is $133\frac{1}{3}\%$ of a.

(p) The fractional equivalents of $x\%$ and $y\%$ are reciprocals. Now $62\frac{1}{2}\% = \frac{5}{8}$; the reciprocal is $\frac{8}{5}$, which equals 1.6 or 160%. Thus, b is 160% of a.

5. (a) 65% (b) 43% (c) 9% (d) 73.9%
(e) 10.8% (f) 0.1% (g) 8.2% (h) 0.34%
(i) 95.02% (j) 60% (k) 270% (l) 157%
(m) 104.2% (n) $6\frac{2}{3}\%$ (6.7%) (o) 80% (p) $55\frac{5}{9}\%$ (55.6%)
(q) 125% (r) 162.5% (s) $141\frac{2}{3}\%$ (141.7%) (t) $257\frac{1}{7}\%$ (257.1%)

7. (a) 50% (b) $33\frac{1}{3}\%$ (c) 38%
(d) $16\frac{2}{3}\%$ (16.7%) (e) 65% (f) $8\frac{1}{3}\%$ (8.3%)

Index